GW00703127

THE RESTORATION WARSHIP

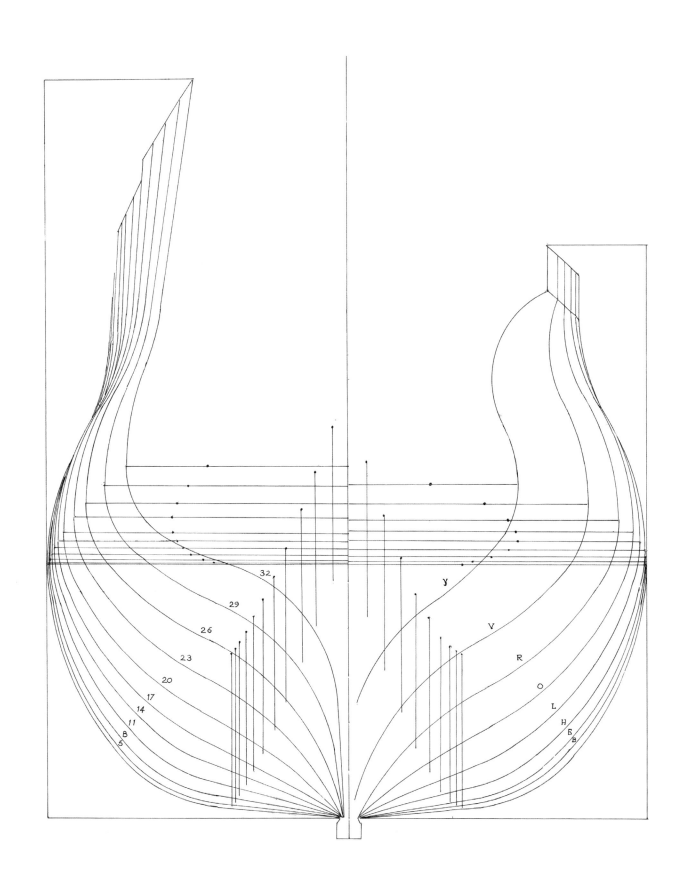

THE RESTORATION WARSHIP

THE DESIGN, CONSTRUCTION AND CAREER
OF A THIRD RATE OF CHARLES II'S NAVY

RICHARD ENDSOR

CONWAY

ACKNOWLEDGEMENTS

About a quarter of a century ago I decided to make a model of a particularly attractive seventeenth-century ship illustrated in a recently published book, *Great Ships* by Frank Fox (Conway, 1980). After the model was complete I wrote an article for a well known ship modeller's magazine about the research necessary to create the draught. Upon submission the editor of the time persuaded me, with generous flattery, to expand the article into a book with an estimated completion date of two years. This target may have been achieved had I not had the fortune, or misfortune, to meet and befriend the few historians and enthusiasts who take a strange interest in the seventeenth-century navy.

Foremost in thinking of obscure archives in which I could spend my time was Dr Peter LeFevre. My task was often onerous but Peter's advice and outstanding knowledge was always available and greatly appreciated. He was ably assisted in the difficult task of educating me in the wonders of naval history by Dr David Davies and Dr Ann Coats. Seventeenth century shipbuilding technology is a subject studied by few but I was fortunate in having the experts Frank Fox, David Roberts and the late John Franklin as colleagues with whom I could spend endless hours researching and solving problems. Another specialist subject was artillery and again I was blessed with the immense contribution made by the retired Army Officers Charles Trollope and the late Adrian Caruanna, while Dewit Bailey and Sim Comfort told me all I needed to know about small arms. I was helped in the study of contemporary ship models by Simon Stephens of the National Maritime Museum and spent many enjoyable hours visiting models with him. I also extend my gratitude to Arnold Kriegstein for the hospitality and access to his collection of models. My ignorance extended into the field of archaeology and diving but again I received the help and understanding of exponents of these arts from Robert Peacock, Dr Doug McElvogue, Dr Peter Marsden and Kroum Batchvarov.

All these people are exceptional in their fields and most are among my closest friends. This work would not have been possible without their contribution although I accept full responsibility for any errors that must inevitably have been made.

My indebtedness and thanks extends to the staff of the many institutions visited over the years, most memorable being Dr Richard Luckett of the Pepys Library, Magdalene College, Cambridge, whose madcap idea for me to build a replica Pepys bookcase was actually carried out. Another is Julie Ash of the National Archives, a place I must have visited a couple of hundred times. Her good humoured help and patience over many years was greatly appreciated.

I also thank other friends, the Hon. Don Rosenberg, Randy Mafit and Herbert Thomesen for their support. I also thank my family and my late father who accompanied me on many expeditions, with the kind permission of my mother. The tolerance of my children was also appreciated. Finally my wife, Ilona, although no more interested in seventeenth century ships than most women, her support, help and encouragement were fundamental to the completion of this work. I could not have produced it without her.

I would also like to thank John Lee and Matthew Jones of Conway Maritime Press for their encouragement and help in this work.

A Conway Maritime Book

© Richard Endsor, 2009
Volume © Conway, 2009

First published in Great Britain in 2009 by Conway,
An imprint of Anova Books Company Ltd.
10 Southcombe Street
London,
W14 0RA

www.anovabooks.com
www.conwaymaritime.com

All rights reserved. No part of this publication may be reproduced, stored in a retrieval system, or transmitted in any form or by any means electronic, mechanical, photocopying, recording or otherwise, without the prior written permission of the publishers and copyright owner.

Richard Endsor has asserted his moral right to be identified as the author of this work.

British Library Cataloguing in Publication Data:
A record of this title is available on request from the British Library.

10 9 8 7 6 5 4 3 2 1

ISBN 978 1 844860 88 3

Designed by John Heritage
Printed and bound by 1010 Printing International Ltd, China

To receive regular email updates on forthcoming Conway titles, email conway@anovabooks.com with Conway Update in the subject field.

A NOTE ON PICTURE CREDITS

All drawings and plans are the work of the Author unless indicated otherwise. Other illustrations have been credited appropriately in their respective captions. Engravings reproduced in the appendices of this book are taken from Thomas Blanckley's *A Naval Expositor, Shewing and Explaining the Words and Terms of Art Belonging to the Parts, Qualities, and Proportions of Building, Rigging, Furnishing & Fitting a Ship for Sea.*, 1750 (Author's collection).

Frontispiece: (half-title) *Lenox* general arrangement, body lines (1:72). **(title page)** *Lenox at Sea in a Gale*, painting by the Author (Author's collection).

CONTENTS

FOREWORD

It was with considerable interest that I was informed of the proposed publication of a book about a ship named *Lenox*, in honour of my ancestor Charles Lenox, the 1st Duke of Richmond and Lennox. After being acquainted more fully with the book, I am very impressed by the high level of research that has gone into it.

Charles II's decision to bestow his own son's name upon the first of his "Thirty New Ships" shows how interested he was in the navy, and in maritime affairs. It also demonstrates the great affection he felt for his young son Charles Lenox, and for the boy's mother, his own beloved mistress Louise de Keroualle, a beautiful and sophisticated Frenchwoman. The pomp and circumstance that attended the ship's launching was no doubt deserved – it was an admirable feat of shipbuilding expertise and competent naval administration. I am intrigued that the famous Samuel Pepys features so strongly in the book.

Important elements of our national history have been admirably brought to the fore by the author. He has provided a complete career history of *Lenox* herself, as well as a comprehensive analysis of naval and social affairs of the seventeenth century. The detailed illustrations and illuminating insights contained within these pages undoubtedly constitute a major contribution to maritime scholarship. The chapters devoted to ship design, construction, armament, fitting out and sea service are particularly fascinating.

As I am the 10th Duke, directly descended from the 1st Duke, and my family name is still Lennox (actually Gordon Lennox), history really comes very close. At Goodwood we live not far from King Charles's beloved maritime city of Portsmouth, and we can even see the sea from the House. I welcome this very impressive new publication on an historical link about which we previously knew little.

THE DUKE OF RICHMOND,
LENNOX AND GORDON

INTRODUCTION

The 70-gun third-rate *Lenox* represents the pinnacle of Restoration ship-building practice. She was the first ship to be completed of King Charles II's thirty ship building programme of 1677. During her lifetime England progressed from a position of inferiority to the French and Dutch to become the world's leading maritime power. By examining this one rather typical ship in detail all third-rate ships of the period can be understood. She was not an exceptional ship nor had any great claim to fame but she did fight in battles and survive harrowing storms. *Lenox* was built by John Shish, the Master Shipwright at Deptford dockyard and was launched on 12 April 1678. She survived until 1756 when she was sunk as a break-water at Sheerness. During this period she was rebuilt twice, firstly in 1701 and again in 1723.

This work is concerned with her life from conception until the period of her first rebuild in 1701. She was chosen as a subject for study because a survey of surviving evidence showed there is probably more extant material concerning her than any other third-rate ship of the period.

Lenox and her sisters represent a period when ships were not only expected to sail and perform well but also reflect the glory and majesty of the king. The resulting vessels were powerful two-deckers with spectacular decoration that gave the seventeenth century navy perhaps the most beautiful and graceful warships ever built.

During the seventeenth century only the most technologically advanced countries were able to produce and maintain fleets of warships. They had to be financed by an economy large and rich enough to support them. The effort in terms of manpower and organisation was enormous. Vast numbers of highly skilled shipwrights and mariners, with years of experience, were required for their building and manning. Perhaps the nearest equivalent today is the aerospace industry. Indeed, much of the terminology used in old shipbuilding rumbles on in the most advanced aircraft designs. The structures that were created using a balk of oak and an adze are now fashioned from a billet of aluminium by a computer-controlled milling machine. As different as these processes may seem, the man who programmes the milling machine shares the same type of mental skills as his predecessor who wielded the adze.

The design of 1677 was so successful that its form remained virtually unchanged until it was replaced by the 74-gun ship in 1755[1], when larger ships became necessary to match the size of French and Spanish ships and to sail ever greater distances as British interests expanded worldwide. Indeed, the battlegrounds of the Dutch Wars of the time were so close that the sound of naval gunfire could be heard in London.

THE EVOLUTION OF THE THIRD-RATE SHIP
The lessons of the Dutch Wars, which ended in 1674, were incorporated into *Lenox*'s design. The third-rate, two-decker warship had already established itself as the most cost-effective vessel to fight in the recently introduced line of battle. It sailed better than the three decker first- and second-rate ships, yet it was strong enough to carry a heavy battery of guns on its lower gun deck. *Lenox*'s immediate ancestry began during the Commonwealth in 1650, when Peter Pett and his brother Christopher built the first two of the *Speaker*-class frigates. They were the prototype of the third-rate ships that were to remain the backbone of the fleet until the end of the days of sailing warships.

Christopher Pett built *Speaker*, renamed *Mary* ten years later, at the restoration of Charles II at Woolwich dockyard, while Peter built the second ship, the *Fairfax*, at Deptford dockyard. Peter's assistant at Deptford was Jonas Shish, the father of John Shish who would build

Below: *A broadside view taken from off the starboard bow of a Commonwealth third-rate ship of 50-56 guns, dating from c. 1650.(Courtesy United States Naval Academy Museum. Photographer: Dr. Richard Bond)*

Lenox. The *Speaker* class broke away from the heavy construction of the 'great ship' that had become fashionable during the first half of the seventeenth century, and included some of the principles of Elizabethan light, fast sailing ships and so-called "Dunkirk frigates". An unidentified model of the type exists in the National Maritime Museum, Greenwich, which shows the fine lines and light construction.[2] Firepower, however, was not sacrificed and the Speaker class had thirteen gun ports on the gun deck and a full tier on the upper deck. The design was undoubtedly a great success. Being lightly built, they were fast sailers of about 730 tons yet still strong enough to withstand the firing of their own guns. Within four years of the completion of *Speaker* and *Fairfax* eleven similar ships were constructed.[3] During this period the English Royal Navy was not alone in developing warships. Across the Channel the French had also been at work. Samuel Pepys, Secretary to the Admiralty, wrote:

> "In the years 63 and 64 the Dutch and French built another sort of ships with two decks which carried from 60 to 70 guns, and were so contrived that they carried their lower guns 4 foot from the water, and to stow 4 months provision; whereas our frigates from the Dunkirk-built, which were narrower and sharper, carried their guns but little more than 3 foot from the water, and but 10 weeks provision, which was to be avoided. Observing of this, Anthony Deane [Pepys' friend and a Master Shipwright] built the *Rupert* and *Resolution*, Mr Shish the *Cambridge*, Mr Johnson the *Warspite* and Mr Castle the *Defiance*."

Below: *Detail of the figurehead and beakhead bulkhead of a Commonwealth third-rate. Note the carved brackets on the head rails and the pierced trailboard. (Courtesy of the United States Naval Academy Museum. Photographer: Dr. Richard Bond)*

The king normally built ships in the royal yards, but the *Warspite* and *Defiance* were built by contract agreed with the Commissioners of the Navy by Mr Johnson and Mr Castle, both private shipbuilders. The ships were built to carry six months' provisions and carry their guns 4ft 6in from the water.[4] These ships were constructed between 1665 and 1667 during the Second Dutch War and were about 100 tons heavier than *Speaker*, allowing some of them to carry as many as seventy guns and considerably more stores. The ability to carry stores should not be underestimated; for example, each man aboard was entitled to eight pints of beer a day, which amounted to 143 tons for three months' supply. Shortly after the war, Francis Baylie at Bristol built the next third-rate, *Edgar*. She continued the trend for growth in size, being longer, beamier and heavier, of 994 tons.

King Charles II was secretly sympathetic toward Catholicism and, wishing to remain independent of Parliament, entered into an extraordinary secret treaty with King Louis XIV of France. In return for a grant of £200,000 a year he issued a "Declaration of Indulgence toward Catholics and Dissenters" and entered into war against the Dutch. He wanted revenge for the raid on the Medway in 1667, in which a Dutch fleet had bombarded and captured Sheerness, sailed up the River Thames to Gravesend, then up the River Medway to Chatham, where they burned three capital ships and ten lesser naval vessels and towed away the *Unity* and the *Royal Charles*. The King also wanted to reduce Dutch political and economic power. Thus began the Third Dutch War in 1672, with England fighting as an ally of France.

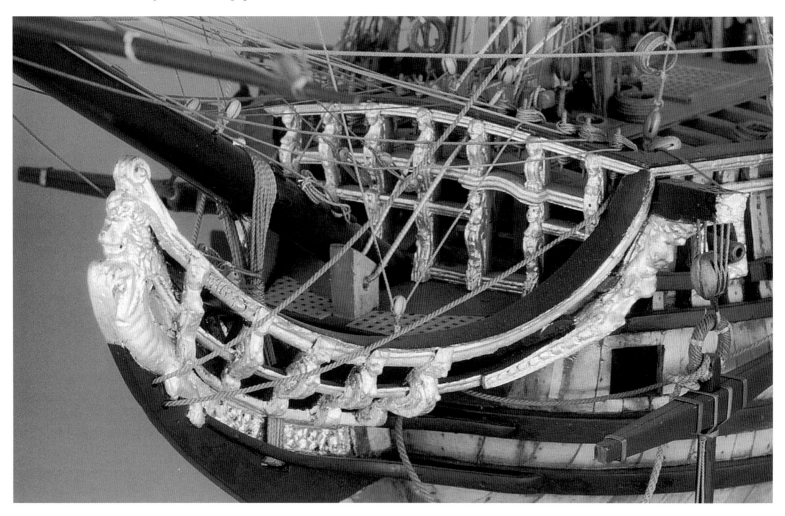

PARLIAMENTARY APPROVAL

As a French ally the opportunity was taken by the English to observe some of the French fleet. "In 1672 and 1673 the French brought a squadron of about 35 ships to Spithead, which King Charles II visited. He went aboard the *Superbe*, a two-deck ship measuring 40ft broad and carrying seventy-four guns and six months' provisions. It was noticed that our ships being narrower, could not stow so much provision nor carry their guns so far from the water".[1]

To increase the strength of the English fleet four new third-rate ships were ordered. Sir Anthony Deane at Harwich built the *Swiftsure*, launched in 1673, and the *Harwich* the following year. At Chatham, Phineas Pett II built the *Defiance*, launched in 1675, while Jonas Shish built the *Royal Oak* at Deptford. Pepys gave credit to Sir Anthony Deane for his response to the *Superbe*; Deane had the opportunity to measure her while she was being repaired in England. Pepys tells us that when Deane gave King Charles her dimensions the king commanded him to build his ships as near as possible to *Superbe*'s design. In fact Deane's ships and that of Pett were only about the same size as *Edgar* built some five years earlier.

The real response came from Jonas Shish at Deptford whose *Royal Oak*, launched in 1674, matched the description given by Pepys. She was the first English 74-gun ship and had fourteen gun ports per side on the lower gun deck. At 1107 tons she was more than 100 tons heavier than any other ship of her type. *Royal Oak* was considered very special, for she was armed entirely with 'Rupertinoe' patent guns. These strong, high-quality guns, nealed and turned, cost three times the amount of normal iron guns of the same calibre. Among the few other ships to be armed with them were the prestigious first-rates *Royal James* and *Royal Charles*.[2]

POLITICAL TURMOIL

The Third Dutch War and the Declaration of Indulgence aroused considerable animosity towards King Charles and his court and government. When Parliament met in February 1673 they forced the withdrawal of the Declaration of Indulgence and reversed its principles by imposing the Test Act, which required all Crown office holders to acknowledge the Anglican Church's primacy. The most famous casualty of this Act turned out to be the king's brother, the Duke of York and Lord High Admiral. Charles was also disappointed by the conduct of the war, as the Dutch, in a series of brilliant defensive battles, avoided defeat. Faced by such circumstances Charles had no option but to withdraw from the war after which he prorogued Parliament. The vacancy left by the Duke of York and Privy Council Committee of the Navy was replaced by an Admiralty Commission, with Charles effectively taking over from his brother as Lord High Admiral and with Samuel Pepys acting as Secretary. The political situation was serious for Charles but he set about creating a court party led by the able and loyal Sir Thomas Osborne, who rose to become the Earl of Danby by 1674.

Above: *Charles II (1630–85), painted c. 1675 by Sir Peter Lely, a Dutch-born artist who specialised in portraiture. Charles II initiated the ambitious thirty ships building programme. (The Trustees of the Goodwood Collection)*

THE KING'S NEED FOR NEW SHIPS

At the end of the Third Dutch War, Samuel Pepys, the Secretary to the Admiralty Commission, received a copy of a letter dated 12 October 1674 written by Sir Anthony Deane to the Lord Treasurer Danby. In it, Deane, in his usual gloomy and sceptical way, reflected on the times and described the poor condition of many warships and the need for their repair. He went on to compare the size of the English fleet with that of the Dutch and French and with some exaggeration concluded the English were little more than one-third the strength of the Dutch and only three-fifths of the French. For England to regain its position as a maritime force, Deane suggested that the royal dockyards repair the existing fleet of fifty ships and build five new large third-rates by contract each year for the next four years.[3] Pepys agreed with Deane's advice and forwarded his friend's paper to the Admiralty Commission. For the king to repair and build new ships on such a scale he needed Parliament to pass an Act to raise money by tax.

During this period the Admiralty Commissioners usually met in the Robes Chamber at Whitehall with the king attending more meetings than anyone else. Among his pleasures was a keen interest in sailing, which he practised during his exile in Jersey and probably in Holland. After the Restoration the Dutch gave him the yacht *Mary* as a gift, with which he was so pleased he ordered others at regular intervals. He enjoyed sailing them and in 1661 John Evelyn, Pepys' friend and fellow diarist, accompanied him when he raced his brother James the Duke of York's yacht from Greenwich to Gravesend and back. Charles lost the outward leg against the wind, but won the return.[4] John Sheffield, Earl of Mulgrave and a warship commander during the Third Dutch War, wrote of him: "Besides the great and almost only pleasure of mind he appeared addicted to, was shipping and sea affairs; which seemed to be so much his talent both for knowledge, as well as inclination, that a war of that kind was rather an entertainment, than any disturbance to his thoughts".[5] John Evelyn agreed and said the king was "a lover of the sea and skilful in shipping",[6] while William Sutherland remarked that he was "able to discourse and examine most of the principal shipbuilders".[7] He had sufficient technical understanding to design with his own hand the original draught of a yacht, probably the *Fubbs*, built by "his great favourite" Phineas Pett II at Greenwich in 1682.[8] Pett earned his right to Charles's favour, for his contemporaries regarded him as "the greatest scholar of a shipwright".[9] Pepys later wrote, concerning Charles and the Duke of York, that they "encouraged men to bring their draughts to them, and themselves vouchsafed to administer occasion of discoursing and debating the same and the reasons appertaining thereto. Not only to the great and universal encouragement of the men, but improvement of their art to the benefit of the state: themselves taking delight to visit the merchant yards as well as their own, and doth honour and assist with their presence no less the merchant builders at his launchings of a new ship of any tolerable consideration, and enquiring after the proofs of them at their return from sea, than his own master builders".[10]

Deane's paper about the state of the navy was discussed on 23 December 1674 and it was directed it be written up fair for a sitting after Christmas.[11] This duly took place on 16 January 1675 when it was viewed and considered by some of the Commissioners before being agreed and signed. It was to be tendered for signing by the Lord Treasurer and the rest of the Commissioners at the next meeting before being presented to His Majesty.[12]

PARLIAMENTARY APPROVAL

With the acceptance of Deane's paper by King Charles and the Admiralty Commissioners the project gained momentum. However, although it was well enough to have both the Admiralty and king's approval, finance could only be provided by Parliament. This was potentially problematic. Yet Parliament was fully aware of the importance of the navy and only a week later, on 22 January, passed an order: "That Mr Pepys do on Saturday morning next at ten of the clock, bring into the House a true state of the present condition of the navy, and of the stores and provisions thereof".[13] Samuel Pepys, who was a Member of Parliament for Castle Rising as well as Secretary to the Admiralty, presented his papers two days later as ordered. The case presented was much the same as that proposed by Deane except that the comparison with the Dutch and French navies was altered to appear more credible. It now stated that the English fleet contained 92

warships, the French 96 and the Dutch 136. Pepys argued the government's case to Parliament but the opposition, expressing their supposed concern over the navy's ability to spend money without apparent result, persuaded Parliament not to vote new money. Instead, part of the revenue from customs was to be used, a move that effectively stopped the bill as the customs money was already allocated for other purposes. This was a potentially dangerous conflict between the court party and the opposition, who now wanted war against France in support of the Dutch. Soon after, a dispute between the Houses of Lords and Commons resulted in Parliament business grinding to a halt and being prorogued until the autumn.

With the new sitting of Parliament on 13 October 1675, King Charles again asked the House to approve money for new ships. Pepys had spent the recess preparing for the debate. Among the papers he had drawn up was an estimate of the charge for building ships of each rate in His Majesty's yards compared with building in a private yard by contract. Pepys estimated that a third-rate would cost £10 per ton in royal yards and £9 per ton by contract. He also estimated the number of workmen required to build a ship in a year: seventy-five shipwrights, four caulkers, five sawyers and six labourers, a total of ninety men. A list of principal scantling of timber (the size of the ship's timbers) was made, dated November 1675.[14] Pepys made such a convincing case that a resolution was passed to build twenty ships.

However, the opposition was not yet finished. During discussions of Ways and Means they proposed that the money voted be lodged with the City of London rather than with the Crown, suggesting that the king could not be trusted to spend the money for its true purpose. The debate raged for days on end without resolution, until eventually Charles relented and he prorogued Parliament on 16 November.

The expansionist war that Louis XIV was still waging against the Dutch was very unpopular in England. Louis was holding off Charles with a new secret subsidy of £100,000 a year, a sizeable sum, but not enough for Charles to continue indefinitely without additional money from Parliament. The lack of money resulted in nothing being done in England to repair the old ships of the fleet or to build new ones. In stark contrast, Louis XIV's minister of finance, Jean Baptiste Colbert, had built a modern navy, which proved to be the equal of the Dutch at the battle of Augusta in 1676.

Charles survived without calling Parliament until 15 February 1677. Pepys once again took up the cause but this time, because of the size of the French fleet and the success it was having against the Dutch, sought funding for even more new ships. The implacable opposition led by Shaftesbury tried to force a general election on the grounds that Parliament was illegal after such a long recession. The Lords, outraged at the opposition's presumption, strongly disagreed and ordered Shaftesbury and Buckingham thrown in the Tower. With the main opposition removed Pepys, proceeded to pour forth to Parliament all the arguments for new ships that had been building up within him over the preceding years.

Pepys evidently took no chances and covered every detail concerning shipbuilding as far as he could. It was even suggested by some that he bored any opposition into submission. He proposed building one first-rate of 1460 tons, nine second-rates of 1300 tons each and twenty third-rates of 978 tons each. The third-rates were to be the same design as the *Harwich* and *Swiftsure*, the last two ships built by Sir Anthony Deane. Pepys carried the day, and on 23 February 1677 Parliament voted nearly £600,000 to build one first rate of 1400 tons, nine second rates of 1100 tons each, and 20 third-rates of 900 tons each, all to be built within two years. The only disappointment for the king was that the Act ordered the ships to be somewhat smaller than Pepys had argued for.[15]

INITIAL PLANNING

As Secretary to the Admiralty, Pepys immediately wrote to the Navy Board, the body responsible for implementing Admiralty decisions, to tell them the happy news. He went on to ask the Board to consider the supply of materials and men and to build the ships in as short a time as was possible.[16] However, Pepys's dealings with Parliament were not finished. He still had to steer the bill carefully through the Committee of Ways and Means. A storm blew up when the Lords insisted upon an amendment and, to a despairing Pepys, it looked for a short time as if his bill might founder. At the time the Navy Board sat in temporary accommodation at Mark House Mansion, as their offices at Seething Lane had burnt down in 1673. During this period, their members consisted of:

> Lord Brouncker, Comptroller of the Treasurer's Accounts
> Sir Thomas Allin, Comptroller of the Navy
> Sir John Tippetts, Surveyor of the Navy
> Sir Richard Haddock, Commissioner of the Navy
> Sir John Werden, Commissioner Extraordinary of the Navy
> Sir John Chicheley, Extra Commissioner of the Navy
> Sir Anthony Deane, Comptroller of the Victualling
> Mr Edward Seymour, Treasurer of the Navy
> Mr Thomas Hayter, Joint Clerk of the Acts
> Mr James Sotherne, Joint Clerk of the Acts

The Admiralty Commission met on 22 March 1677, attended as usual by the king and, on this occasion, the officers of the Navy Board. King Charles opened the meeting by considering the debate in Parliament for building thirty new ships and the Act for raising money. Time could not be wasted because Parliament had voted the ships to be built within two years. Charles suggested to the Navy Board officers that the season for felling trees was approaching and that they ought to enquire where the materials of all sorts needed for building may be had. They were also to consider where and how the building was to be distributed among the Royal yards and whether the finishing works could be completed after the ships were launched to allow the vacation of the stocks as soon as possible. They were to send out purveyors to look for timber, consider the guns, think about the dimensions and scantlings and take care that each builder's draught be viewed before a ship was started.[17]

By 16 April Pepys had managed to overcome the Ways and Means and the Act for raising money by a land tax was finally cleared. The coming spring must have been particularly satisfying for Pepys. He later acknowledged that his involvement in the thirty-ship programme was his greatest achievement. His success was not greeted with universal acclaim however, as he soon discovered when an attempt was made to stop him charging fees for making out ships' passes, an old right of his office. The political situation remained dangerous, with the Duke of York openly Catholic, although Charles achieved something of a religious balance by marrying off the Duke's daughter, Mary, to the Dutch Protestant William of Orange.

The Navy Board, driven by the Admiralty Commission, now began the huge task of setting the programme in motion. Lord Treasurer Danby, who also enthusiastically promoted the thirty ships, held a conference with the Navy Board at which he brought up further considerations, confirmed in a letter of 24 April. He informed them that Sir John Tippetts, the Surveyor of the Navy and an experienced Master Shipwright, had identified thirty-three ships that required urgent repair, and that berths needed to be found for them in the yards. When Sir Anthony Deane had first proposed a new shipbuilding programme he sensibly suggested building five ships a year for four years; thirty ships in two years tripled the building rate. Danby asked how many ships could realistically be built in the royal yards within the timescale and how many men would be required to build them. Other items for consideration were the guns, anchors, fittings, rigging and stores. Finally, and perhaps most importantly, Danby requested that an estimate of the cost of the ships and a method of keeping records of expenditure should be considered so that the money voted by Parliament could be accounted for and not used for other purposes.[18]

While the Navy Board deliberated Danby's comments, Charles and the Duke of York went down to Newmarket to enjoy the pleasures of the turf. Samuel Pepys joined them and found them "very solicitous about the building of the ships". Pepys returned to London before the royal brothers to meet the Navy Board and prepare them for the next Admiralty Commission meeting.[19] This meeting took place as soon as Charles returned from Newmarket on Tuesday 1 May. Danby's letter was read out and debated; no conclusions were drawn, although the Navy Board promised to provide some written answers soon.[20] It was clear, however, that a dramatic increase in manpower was needed in the dockyards and the following day Charles gave the Admiralty Commission power to issue press warrants, which in the seventeenth century could be issued for dock labour as well as seamen.[21] It was extremely unusual for press warrants to be issued in times of peace but shipwrights and dockyard labour were urgently required.

On 5 May the Admiralty Commission met again. Fears were expressed that the whole programme could not be completed in time and money would run out before all thirty ships were built. The possibility was raised of building some of the third-rates by contract. King Charles and those present were against this proposal as they considered it would always be to the contractor's advantage to build their ships as slight as possible. But, keeping options open, tentative enquiries had been made with merchant builders who were now asking £11 per ton against Pepys's earlier estimate of £9 per ton. In eager anticipation of orders they were already buying up timber. It appeared that the whole programme could lose direction.

THE CONTRIBUTION OF CHARLES II

At this crucial juncture King Charles made a vital contribution to the debate which would redirect the programme towards outstanding success. Firstly, noting the inconvenience of supplying different-sized equipment to each ship, he recommended the standardisation of gun ports, fittings, masts, blocks, etc. He then proceeded to debate the burthen of the ships, noting that they must be large enough to carry out the tasks for which they were designed. There was an expectation of war with France and he probably expected the area of conflict might be much greater than in previous wars. Instead of the North Sea the area of conflict could extend from the Channel to the Mediterranean, requiring much greater capacity to carry stores. Parliament approved the building of twenty third-rates of not less than 900 tons, much smaller than the *Royal Oak* and the last series of third-rates built in 1674. Although the Act stated 900 tons was a minimum, in fact there was not enough money to build larger. Pepys's minutes of the meeting then recorded: "His Majesty was pleased graciously to add that if such increase of charge should be expected against by Parliament (which he could not expect) he would rather choose to make it good out of his own purse than hazard the wronging the ships for want of it".[22]

In spite of his fine words Charles probably never had any intention of paying for even part of the building programme himself. In fact, in April 1679, after many of the ships had been built and with money running out, the Treasurer of the Navy reminded Charles of his promise to make good the deficit from his own purse. Charles, in his amusing, cynical way, "was pleased to answer that, as the state of his revenue now stands, he doubts his being able to make his word therein good".[23]

As a result of Charles's contribution to the meeting a long letter arrived at the Navy Board the next day, 6 May. They were instructed to compare the cost of building the ships to the smaller size stated by Parliament with

Above: *Samuel Pepys (1633–1703), diarist and naval administrator. This engraving is based on a portrait by Sir Godfrey Kneller painted in 1689, the year Pepys left his post as Secretary to the Admiralty Commission. (Author's collection)*

the greater size originally suggested by Mr Pepys and from which Charles judged it unfit to depart. To ignore Parliament's wishes could be dangerous, however, and so it was decided to look into the legal consequences of departing from the size of ships stated by Parliament. The Navy Board officers were also to examine the price of timber and the quantities and cost of materials from abroad such as pitch, tar, masts and timber plank.

Charles ordered that a table of the principal dimensions be drawn up for each rate of ship for his approval. He also ordered that a scantling list be made up for use in His Majesty's own yards and for any merchant builders who may be contracted. The members of the Navy Board were to make personal visits into the country to see what quantities of timber and plank could be had.[24]

Only two days later the Admiralty Commission met again and discussed the manning and arming of the thirty ships. They agreed that an establishment be drawn up of all the existing ships and new vessels to be built, to include the number of men and guns each ship should have. To help procure the vast quantity of materials necessary to build the new ships and repair the old, it was also agreed to publish invitations for tenders in the *London Gazette* newspaper and put up notices in the Exchange and Custom House.[25]

On 10 May, another two days later, the Admiralty put out an advertisement "to notify all persons, owners or traders for any English oak timber or plank or any other naval provisions or materials whether of English or foreign growth employed in the building of ships of war; such as pitch, tar, rozin, hemp, masts, fir timber, deals, sail cloth, oaken standards, knees or plank". The notice went on to say the Admiralty were ready to receive tenders at reasonable market prices on Wednesday and Friday every week at the Office of the Navy in Mark Lane, London.[26] Spurred on by the Admiralty, the Navy Board were soon able to report "we have bought considerable quantities of East Country timber and plank and encouraged the importing of more".[27]

Progress continued, with a full meeting of the Admiralty Commission held on 12 May, which the Navy Board officers also attended. They presented a table showing the principal dimensions of each rate of ship according to the larger dimensions as instructed by Charles on the 6th and first proposed by Mr Pepys to Parliament. Charles, however, was not satisfied. "His Majesty", Pepys reported, "took exceptions, as not coming up to the full length upon the gun deck, and other measures which he judged fit they should be of". He amended and returned the table with new dimensions to "answer all the ends of force and quality which such ships ought to have", and directed them to calculate anew. The Navy Board officers with whom Charles must have taken most exception were Sir John Tippetts, the Surveyor of the Navy and Sir Anthony Deane, the shipwright.

The dimensions that Charles had made up himself during the meeting and passed across the table to the Navy Board officers would be those to which the thirty ships would be built. Charles had ignored Parliamentary and monetary constraints to build the ships he knew the navy required. The new third-rates would now be about the same size as the largest third-rate yet built, the *Royal Oak*, built by Jonas Shish in 1674.[28] On the same day warrants were issued to Chatham and Deptford yards to impress 100 shipwrights. The warrants were left blank so that the names could be filled in as the men were pressed.[29]

Sir Anthony Deane sat down after the meeting and wrote a testimonial outlining his thoughts. He was understandably gloomy. The size of the twenty third-rate ships that he had proposed for building had been unceremoniously rejected by the king. Furthermore the arrears in producing a scantling list demanded by Charles fell on Deane's shoulders as a shipbuilder, as well as Sir John Tippetts. But Deane's darkest thoughts concerned Parliament: "It puts me in mind that we have another great work which the nation's eye is upon – but the doing what ever may contribute unto the advance of this great undertaking from whence there will spring many difficulties and more censure than may be proper for us to bear, notwithstanding all our endeavours." His pessimism was perfectly understandable in the political climate of the times.

Deane then proceeded in a more positive vein, emphasising that the Master Shipbuilders bring their plates into the Navy Board for approval. Plates in this context are ships' draughts mounted on wooden boards. Pepys referred to them when writing about draughts of ships and models: "neither the former, if done in the usual manner upon boards".[30] Until the plates were approved the builders could not start work making the moulds for the ships' frames. Although the thirty ships were built to an establishment regarding their size and equipment, the Master Shipwrights were allowed the freedom to draw their own draughts. Deane agreed with earlier suggestions that the season for felling and transporting timber would last only until the end of August and that members of the Navy Board go out into the country to secure as much as possible. He then suggested building third-rates at private yards in the country, at Hull, Harwich, Ipswich, Woodbridge, Albrough, Bristol and Shoreham. The royal yards would then be left free to build the three deckers and carry out repairs to the old ships. In the event two sites, Harwich and Bristol, were used to build third-rate ships as well as the private yards of Sir Henry Johnson and William Castle on the Thames.

Deane's next point concerned transportation costs, which he reasoned could be saved by converting (finishing) the timber in the woods rather than sending rough or rough-squared timber into the yards. Although shipwrights could work better in a shipyard with proper sawpit facilities, much of the timber could be brought to the yards sawn almost to their finished shape.[31]

At a meeting on 17 May it was agreed to go ahead with a plan to repair the old ships within twelve weeks. Progress was also made with drawing up the establishment for manning and arming ships.[32]

Shortly afterward, at the next meeting on 19 May, a protection against the press was applied for by the workmen employed on the rebuilding of St Paul's Cathedral, destroyed in the great fire of 1666. However, the building of the thirty ships was considered more important and protection was denied them. The Navy Board officers, deeply worried by Charles's insistence on larger ships, raised an old query. They wanted to know if they were liable in law for building ships different from that voted for by Parliament. Charles took full responsibility upon himself and confirmed that he would pay for any excess costs. Relieved and satisfied with the answer, the officers agreed to proceed with the building of ships to the dimensions directed by their monarch.[33]

A further query was raised concerning the calculation for obtaining the burthen of the new ships. It was very difficult to obtain the actual burthen by calculation. Because of this difficulty, a rough and simple formula for calculating burthen had long been in use:

$$\frac{\text{keel length x breadth x } \frac{1}{2} \text{ breadth}}{94}$$

For example, calculating the burthen of *Lenox* using her "touch" keel length of 131ft and a breadth of 39ft 10in results in 1106 tons. The touch length is the distance from the back of the keel to the point where the rabbet of the stem rises from the keel.

A major component in the formula was the length of keel. The trend in new ships had been to design them with longer keels in proportion to their gun decks. It was therefore agreed to alter the formula to take this into consideration.

The new method was the same as the old except a new theoretical keel length replaced the touch keel length. It was calculated by using the length of the gun deck between the rabbets of the stem and stern post, minus three fifths of the main breadth of the ship and a quarter of the perpendicular height of the stern post to the gun deck. The length of the gun deck of *Lenox* was 151ft 6in and the perpendicular height of the stern post to the gun deck about 21ft 0in giving a calculated keel length of 122ft 4in instead of a touch of 131ft 0in. The result was that the burthen of *Lenox* reduced from 1106 to 1033 tons under the new formula.

The new method of calculating the length of keel resulted in a supposed reduction in the burthen of *Lenox* of 73 tons. This saving would be very useful if Parliament ever asked for the burthen of the ships for which they had voted. To avoid further conflict with them it was agreed by the Admiralty Commission that "this question is a matter that cannot be thought likely to come within the comprehension of the Parliament – and therefore not fit to be taken notice of to them". In other words Parliament would simply not be told of the new method for calculating ships' burthen.[34]

Yet another meaning for the length of the keel complicates contemporary lists of ships' dimensions. This was the "tread", the distance from the back of the sternpost to the angle of the knee of the head. There were also different methods of resolving the "calculated" length of keel. Unfortunately, contemporary ship lists never state what keel length is used, rendering complete accuracy very difficult. Even Pepys, writing in his diary as Clerk of the Acts on 16 January 1668, found "My work this night

with my clerks till midnight at the office was to examine my lists of ships I am making for myself, and their dimensions, and to see how it agrees or differs from other lists; and I do find so great a difference between them all, that I am at a loss which to take."

In all his years as a great naval administrator, Pepys failed to create order from the chaos. Many years later in 1692 Edmund Dummer, the Surveyor of the Navy wrote, "Having obtained with some labour several catalogues or lists of the ships of their Majesties' navy, comparing the dimensions and burdens of them as they have been at different times and by different hands taken and calculated and being compared one with another it seems altogether doubtful from the vast disagreement in one and the same ship what a real and true dimensions and burden of any of them is".[35] *Lenox* is typical, her breadth is variously given as 39ft 8in, 39ft 10in and 40ft in different lists and the old formula for obtaining burthen continued to be used.

Following the meeting on 19 May a document was sent the same day from the Admiralty to the Navy Board confirming progress made in planning for the ships: "His Majesty having now been pleased to approve of the table of the principal dimensions last presented to him from your Board, to be observed in the building of the 30 ships of the first, second and third-rates, as being satisfied in the reasons of the difference between them as they are now offered from those tendered in your letter of the 11th instant". The normally easy-going Charles's insistence on ships larger than those voted for by Parliament and against the fears of a reluctant Navy Board does him great credit for understanding ship requirements and design.

The Navy Board was instructed to send a copy of the table to the Master Shipwrights who were to prepare draughts of their ships. These were then to be sent to the king himself for reviewing (see Appendix 1). The Navy Board was also reminded that the estimated costs requested on the 6th of the month and a scantling list for both the use of the royal and private building yards was still in arrears.[36]

PREPARATIONS AT DEPTFORD

The Navy Board sent a letter to the Admiralty Commission dated 25 May, broadly following Deane's proposals. The letter was then shown to the king, who agreed to the Navy Board members going into the woods to select timber.[37] In the Admiralty's reply they were encouraged to act with as little dispute as possible and to press any men necessary to carry out the work.[38] In fact Jonas Shish's second son, Thomas, was already in the woods negotiating the purchase of timber with landowners. Anthony Deane followed shortly and spent most of June scouring the woods of East Anglia for suitable timber.[39] While Thomas was procuring timber, Jonas's eldest son John, the Master Shipbuilder at Deptford, was busy repairing the old ships and preparing to build the first of the new. This lead ship would be *Lenox* and was the first large ship he had built. John received a letter from the Navy Board on the 25th requiring him to make draughts according to the list of "principal dimensions for the new ships" approved by the king nearly a week before. In his eagerness John had already made a draught for each of the rates of ship. Whether they conformed to the 978 tons as first proposed by Pepys to Parliament, or to the larger dimensions Charles passed across the table to the Navy Board on the 12th, is not known. However, on 29 May John Shish wrote to the Navy Board explaining that his draughts did not comply with their dimensions and there were no instructions concerning the length of the keel, the rake of the stem or the scantling of the timber. Also, he explained, if there were a contract drawn up for merchant builders, which would contain the scantling list, he would like a copy sent to him. On receipt of Shish's letter a Navy Board officer noted on it that the dimensions are not to be given him for the third-rate draught.[40] Indeed, the Navy Board could not send Shish the information he had requested as they themselves had

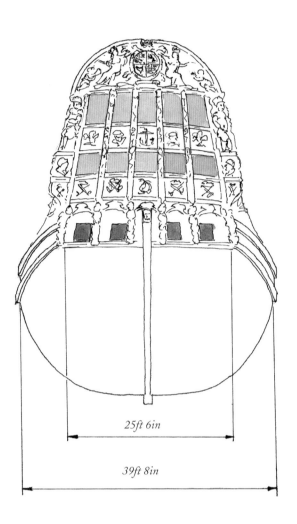

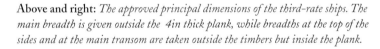

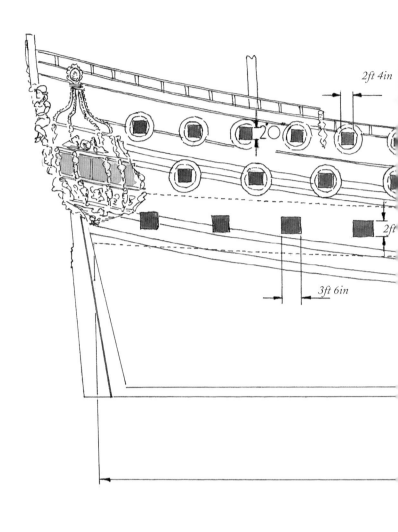

Above and right: *The approved principal dimensions of the third-rate ships. The main breadth is given outside the 4in thick plank, while breadths at the top of the sides and at the main transom are taken outside the timbers but inside the plank.*

not yet worked out what the dimensions should be.

The warrant for building *Lenox* would be issued only two weeks after Shish's letter and he started work on the ship immediately after receiving it. *Lenox* and the *Hampton Court* built alongside her, together with the *Captain* which was finished by Thomas Shish at Woolwich, did differ from the principal dimensions, mainly in having twelve upper-deck gun ports rather than thirteen. It is possible, therefore, that John Shish built his first two ships to an earlier, less complete dimension list than that finally approved, perhaps that which Charles himself had drawn up. At any rate the difference was small; the length of gun deck, breadth and depth of hold of *Lenox* when completed all agreed with the final principal dimension list. As a comparison, Phineas Pett II, Master Shipbuilder at Chatham, received his copy of the principal dimension list a few days earlier than Shish and replied that he had not yet started his draughts for the second and third-rate ships but had made some progress with that for the first rate.[41] His third-rates, in compliance with the principal dimension list, did have thirteen upper-deck gun ports but were not started or completed as quickly as *Lenox*.

The next meeting of the Admiralty Board occurred on 2 June, when the discussions moved away from the technical towards the practicalities of building the ships. Charles remained opposed to building by contract,

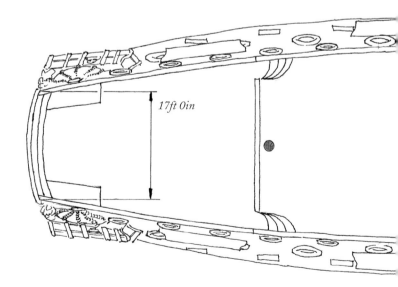

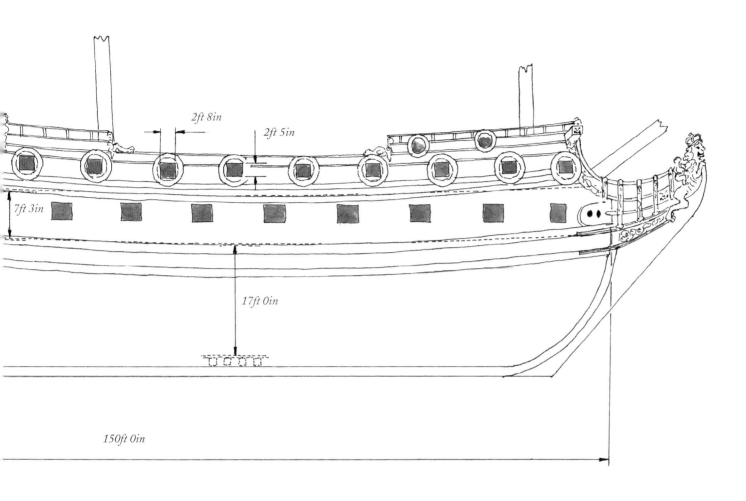

2ft 8in

2ft 5in

7ft 3in

17ft 0in

150ft 0in

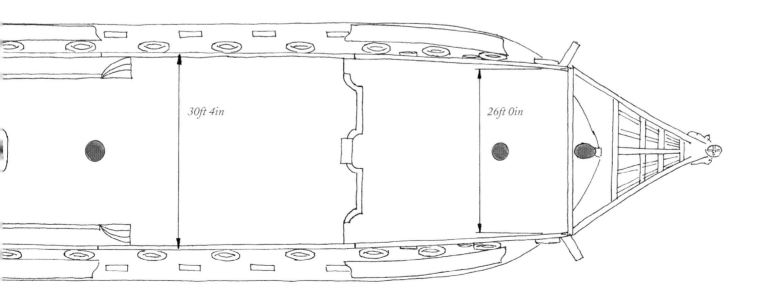

30ft 4in

26ft 0in

as the price for materials would increase as the contractors widened the market. The Navy Board officers agreed and also pointed out that all the shipwrights in England could be employed in His Majesty's yards and contractors would otherwise use men that could be employed by the king. The only exception would be Mr Baylie at Bristol whose remoteness would have little effect on the market near the Thames.[42] A week later the Admiralty wrote to the Navy Board, stating that orders would be sent for laying of keels at places where work was able to start.[43] Finally, on 12 June, a warrant was sent to Deptford ordering the building of the ship that would become *Lenox*.

THE SCANTLING LIST

As late as 7 July, when work began on *Lenox*, the Admiralty Commission was still discussing the design of ships. The king proposed that some of the ships not yet started be built with more upright stems. Pepys discussed the matter with Mr Pett who suggested bringing the draught of water to 2ft less in the second or third-rates. Sir John Tippetts and Sir Anthony Deane rounded on him, regarding Pett's idea as wholly impracticable. The Commission minutes record that having heard "the said officers, giving their reasons for this their opinion, His Majesty was pleased to determine that the depth of water appointed to the several rates of ships in his late table of dimensions shall be strictly kept to, and Mr Pett be directed to conform himself to do the same".[44]

By late July the scantling list had still not been compiled. The matter was brought up at an Admiralty meeting by Sir Anthony Deane who said that "a liberty is taken by the Master Shipwrights of His Majesty's yards determining variously according to their several private opinions the scantling of the principal timbers employed in the building of His Majesty's ships". Charles entered the debate, judging the matter of "too much moment to be left under so great uncertainty". It was agreed that timber scantling lists be prepared by Sir John Tippetts and Sir Anthony Deane with the advice of the King's Master Shipwrights.[45] The intention was that all the shipwrights in the king's yards and those working under contract were to observe them. The Navy Board immediately wrote to all the Master Shipwrights for a list of scantlings they recommended for the new ships. One of the King's Master Shipwrights, Isaac Betts at Harwich completed his scantling list four days later on 4 August.[46] Another, Daniel Furzer at Portsmouth, sent his the following day.[47]

A month went by before the Navy Board was able to respond. After receiving answers from all the Master Shipwrights, Deane and Tippetts studied the advice before writing their own list and sending a copy to the Admiralty Commission on 31 August.[48] However, it was not until 12 September that the king gave his approval,[49] and the Admiralty was finally able to issue the list.[50] It was too late to be used for *Lenox*, whose construction was already well under way.

Charles had first requested a scantling list four months before, immediately after Parliament's approval of the 30 ships programme. Deane had little reason to censure the Master Shipbuilders for proceeding according to their own opinions, as he was the man most responsible for providing the list for them to work to. John Shish had written to the Navy Board requesting one without success. Pepys makes it clear that Deane was responsible for drawing up the list and that it was Deane who eventually did so.[51] See Appendix 2 for details of the Scantling Lists.

COST RISES AND TAX COLLECTION

The cost of building a new third-rate ship rose as the project progressed. Lord Treasurer Danby requested an estimate on 24 April, to which the Navy Board responded with a price of £9,833 for the hull and masts for a ship of 900 tons.[52] This compared with £8,956 for the *Harwich* of 1674 of 995 tons.[53] Danby, "for his more regular proceedings", raised the matter of costs again two months later at an Admiralty Commission meeting on 23 June.[54] The request was passed to the Navy Board who estimated the hull, masts and yards completely built and fitted for a third-rate would now cost £11,000.[55] This estimate was based on the larger ships insisted upon by Charles. Danby attended the Privy Council meeting for passing the Privy Seal for fifteen of the thirty new ships. The total amount voted by Parliament was £584,978 2s 2 ½d. The estimate for all the 30 new ships, including all their stores and guns, amounted to £653,738 6s 2d.[56] At these estimates the programme would overspend by nearly £69,000; however, the true situation was even worse, as the actual cost for *Lenox* was not £11,000 as calculated by the Navy Board but £12,525.[57]

These were enormous costs for the time and the collection of the tax to pay for it was a monumental task in itself. The Act stipulated that seventeen monthly instalments each of £34,410 9s 6d were to be collected, starting on 25 March 1677.[58] Each county was assigned its fair share and Commissioners – namely, the local gentry – were given responsibility for collection. Norfolk, for example, was to raise £1,685 6s and the 175 commissioners included two lords and about forty knights. Among those lower down the list were Samuel Pepys and his cousin Roger[59], a lawyer and Member of Parliament.[60]

The Commissioners were charged with dividing the county into hundredths or other suitable divisions, with at least two commissioners to administer each area. In the town of Heacham, a total of fifty-nine citizens paid the tax. The wealthiest, John Crampe, paid £2 1s 2d for his land and £2 5s for his stock, while a much smaller landowner named Roger Thompson, paid only 1d for his land.[61]

DEPTFORD &
TIMBER SUPPLIES

THE DOCKYARD

Lenox was built at the old royal dockyard at Deptford, founded in 1513 during the reign of Henry VIII. It was situated about half a mile upstream from Greenwich where the ruins of the old Royal Palace of Placentia were slowly being cleared away. In the old palace's place now stood the Queen's House and nearer the river the King's House. Deptford was a small town surrounded by pleasant green fields whose Thames waterfront was dominated by the dockyard. The fourteenth century church of St Nicholas and Sayes Court, the gentlemanly home of John Evelyn, bounded the landward side. The grounds of Evelyn's home covered 100 acres and were laid out with gardens, walks, groves, enclosures and plantations.[1] Nearby lay a tightly packed community whose lives revolved around the dockyard.

Generations of tradesmen, passing their knowledge down through father to son, had built many famous ships at Deptford. By the 1670s the position of Deptford, situated as it was on the narrow Thames near London, was no longer an anchorage where fleets could be moored and victualled. It had become primarily a construction and repair yard. Its facilities included a great dry double dock, where *Lenox* was built, and two building slips or "launches", known as the Great Launch and the Lesser Launch. The double dock was used for the repair and rebuilding of old ships as

Below: *Deptford dock and the crowded streets in the vicinity. 'T' is the Master Shipwright's house and 'L' is the double dock. From* A Geometric Plan of Deptford *by Thomas Milton, 1753. (Author's collection)*

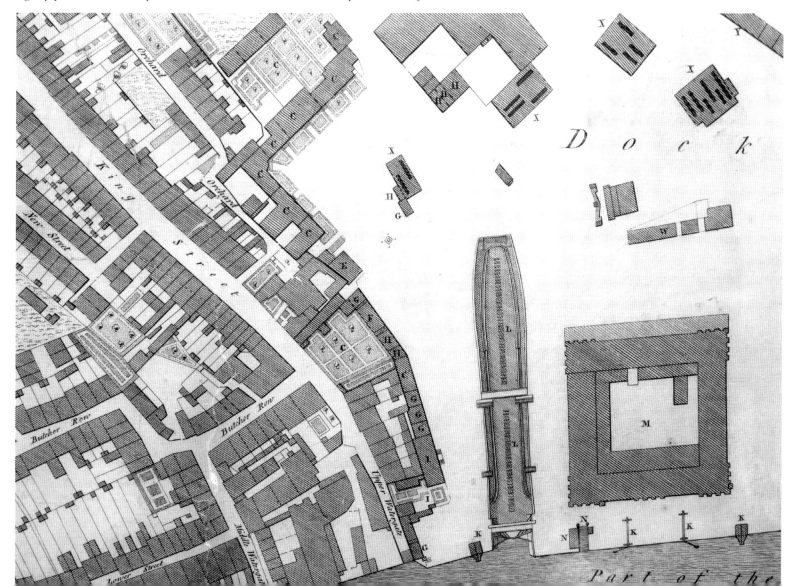

well as the building of new. It was of some considerable size; in 1734 it was measured as being about 320ft long and 53ft wide. At the river end it was 19ft deep and sloped upward to measure 8ft deep at the upper end. The sides were stepped with 'alters' for easy access of large timbers and provided secure locations for the shores that supported ships while they were in dock. At the water's edge was a large gate that kept the dock dry. The size of the dock enabled two large warships to be accommodated, placed one behind the other. For easy access around the ships a movable bridge was provided.[2]

In the spring of 1677, just before the new shipbuilding programme had begun, it was found that many of the dockyard facilities were in poor condition. The wharf around the dry dock near the great crane, and from the dry dock to the reed house – in fact, nearly the whole Thames frontage – was in urgent need of repair. It was thought that during the coming summer it could all collapse into the river. Many other buildings were in need of repair and an additional twelve house carpenters were requested to carry out the work.[3] The great storehouse, which made such an impressive elevation seen from the Thames, was used to store all shipbuilding materials except timber. It supplied not only Deptford but the other yards as well. It was in a poor state of repair and in 1686 a contract was made out for it to be pulled down and replaced.[4]

At the end of March 1677, after Parliament approved the building of the new ships, the number of workmen entered in the quarterly pay book increased from 203, of whom only sixty were shipwrights, to 318 men.[5] Many were engaged during the next eighteen months in the repair of old ships. Among the more interesting vessels under repair was the *Assistance*. Her chaplain, Henry Teonge, kept a diary during her most recent voyage to the Mediterranean and graphically described the appalling state she was in. After being battered and broken by storms she had many leaks patched by sheets of lead and strengthened with baulks of timber.[6] Another was the *Tiger*, an old fourth rate of 1647, which arrived at Deptford for repair during 1674.[7] Work started in February 1677, when she was taken into the dry dock and the planking stripped off above the waterline to the bottom of the top timbers.[8] Before repairs got very far it was decided to take her out of the dock and haul her up a launch to make way for the building of *Lenox*. Things went badly wrong as she was being taken out at half flood. She strained so much that holes had to be hastily cut in the bottom to sink her before her back was broken.[9] *Tiger* lay helpless at the stern of the double dock. With few options available it was decided to take her apart where she lay and, using a few of her serviceable timbers, build her again on the nearby launch. By October the Navy Board were told that her standing officers had been discharged and there was "no such thing as the *Tiger*".[10]

During January 1678, as *Lenox* neared completion, there was still a small part of the *Tiger* left in the dock.[11] Work on her took second place to the building of new ships and she was not completed until 1681. The re-built *Tiger* could be considered a new ship, for she ended up some 3ft broader than the original. While the double dock was being prepared for *Lenox* the great launch was being made ready for another new third-rate, which would become the *Hampton Court*.

Among the principal officers in the shipyard during the building of *Lenox* were:

Clerical Officers
Clerk of the Cheque William Fownes
Clerk of Control Frank Hosier
Clerk of the Survey John Shear
Keeper of Outstores Thomas Turner
Master Surgeon Sackville Wittle

Artificers
Jonas Shish
Master Shipwright John Shish
Assistant Shipwright Thomas Shish (until 13 October 1677)
Assistant Shipwright Fisher Harding (after 13 October 1677)
Purveyor of Timber Thomas Lewsly
Foreman Old Works Owen Bagwell
Foreman of the New Ships William Bond
Master Boatbuilder William Stygent
Master Caulker Peter Brunsden
Master Joiner Richard Dowson
Master Attendant John Kirk
Carpenter of the *Greyhound* Francis Bagwell

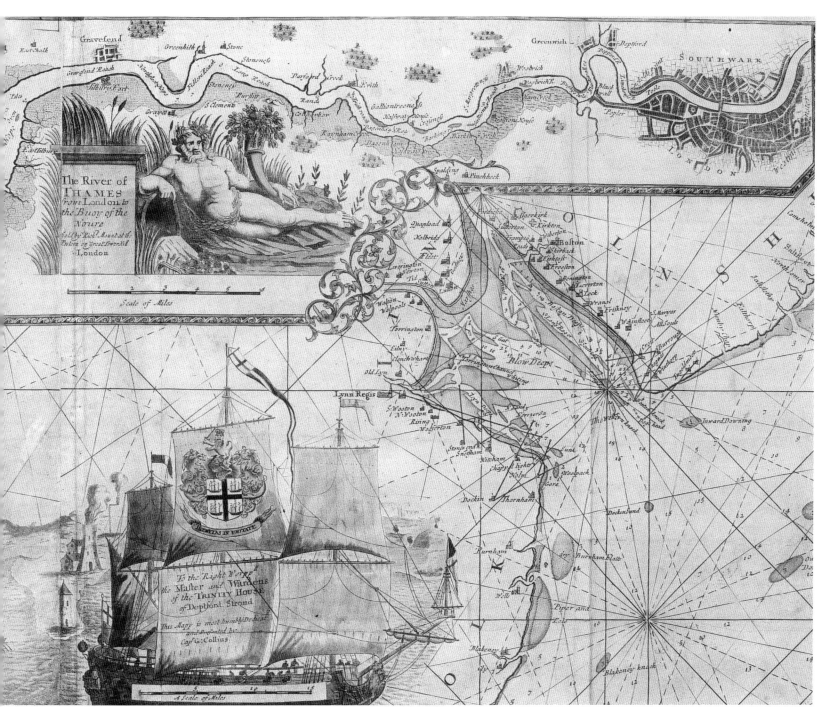

Above: *Detail of a chart by Captain Greenvile Collins, depicting "The River of Thames from London to the Buoy of the Noure", 1684. (Courtesy of the Worshipful Society of Apothecaries of London)*

Most of the royal dockyards had a Commissioner who was responsible for co-ordinating operations at the yard and reporting progress to the Navy Board. However, as Deptford was very near to London, the Navy Board exerted direct control. When he was a Navy Board officer, Pepys often travelled by boat or even walked to Deptford on official or personal business.

THE SHISH FAMILY OF SHIPWRIGHTS

The Shish family had been carpenters at Deptford for more than 300 years [12] before the birth of Jonas in 1605 marked the advent of their rise in status to prominent shipbuilders. Little is known of his early years but Jonas married Elizabeth Frauncis in 1636 [13] when he was thirty-one and she seventeen years old. During their long marriage they had eight children, of whom three sons and three daughters survived them. [14] He became Assistant Master Shipwright at Deptford and Woolwich during

the Commonwealth and retained his position at the Restoration in 1660. [15] During those years he established a reputation by building a number of successful ships.

Pepys was not so enthusiastic about Jonas Shish's abilities and said he discovered Shish's shortcomings in 1664, as he recorded in his diary: "I spent an hour in looking round the yard and putting Mr Shish to measure a piece or two of timber; which he did most cruelly wrong and to the King's loss 12 or 13 shillings in a piece of 28 (cubic) foot in contents". [16] Measuring the volume of timber should have been an easy matter using a scaled carpenter's rule, a device used by shipwrights since Tudor times. [17] Pepys also thought Jonas never pretended to draw a draught, but used

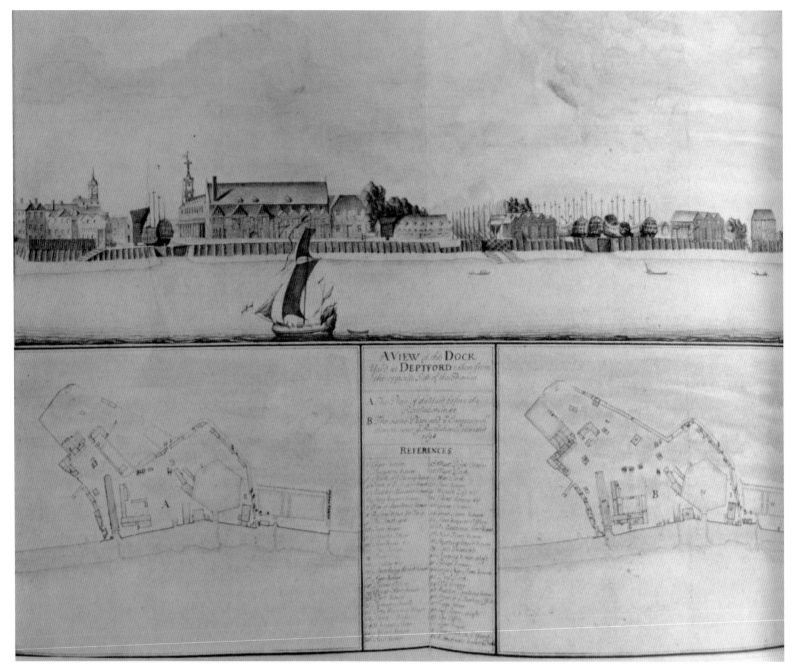

Above: *A view of Deptford dockyard, showing principal buildings. Two plans also illustrate changes and improvements made between 1688–1698. (BL, Kings MS 43)*

only his visual sense to build ships and that he talked so confusingly he could not pass his knowledge on to anybody else. Pepys's remarks are difficult to believe, considering the important ships Jonas built for the navy. Pepys did grudgingly acknowledge this, for he went on to say the truth of shipbuilding does not lie in the niceness of lines but in accommodating the shape so as best to take in all the variety of uses of and qualities required in a ship.[18] Indeed, ship design was an art in which years of practice, which Jonas had in abundance, counted more than seventeenth century science.

During the same year, 1664, an agitated Shish visited Pepys at the Navy Board offices. The problem, he explained, was that during the summer months the private shipbuilders worked from half past five in the morning until half past six in the evening, an hour longer than in the King's yards. This "great discouragement" of the private shipwrights was addressed by an order from the Duke of York to lengthen the day of the King's shipwrights to match the private yards.[19] Understandably the new order was not popular at Deptford and Shish complained to Pepys that the men would not work before six and worked only lazily thereafter.

If Shish hoped Pepys would be sympathetic toward his cause, he was badly disappointed. The ebullient young executive demanded from Shish the names of the shipwrights responsible so they might be punished. The surprised Assistant Master Shipwright refused, and Pepys applied his considerable rhetorical talents to extract the names. In spite of this pressure Shish could not bring himself to name his men. Pepys's assistant, William Hewer, had an idea and reached for a yard call-book and read out all the names, hesitating at each one, and obliging Shish to give an opinion of their work. At the end of the procedure Pepys ended up with the names of six men whom Shish had failed to endorse. Pepys immediately suspended them, but a few days later Shish returned to the Navy Board accompanied by the men concerned, and pleaded that they may be re-entered, as they were good and very able workmen. Pepys was not moved but Shish persisted and wrote to Mr Coventry, the Duke of York's secretary, saying the men were "sorry for their fault and do promise never to be so again". As a result the men were re-admitted after being suspended for less than two weeks.

Pepys found it very strange that Shish should concern himself so much with his men.[20] He had little concern for the shipwright's complaint and certainly had different ideas of how to motivate the men. However, by the time the thirty new ships were built the royal dockyards had reverted to their old summer working hours.

On 22 March 1668, Christopher Pett, the Master Shipwright of both Deptford and Woolwich died, and Shish once again visited Pepys, this time to ask if he would support his application for the vacancy. Pepys agreed, and immediately went up to Whitehall to the Duke of York's chamber where he heard that it was already decided that Shish should have the position.[21] Pepys's support for Shish at this juncture was at odds with some of his earlier remarks but acknowledges the skill he believed Shish had in building good ships. Shish built some of the best ships in the navy during the following years, including two first rates. He remained in his post until 1675 when, at the age of seventy, he filled the less demanding position of Master Shipwright at Sheerness. While *Lenox* was under construction Jonas, then seventy-two years old, and suffering from old age and infirmness, requested that he be discharged from his duties.[22]

Old Jonas did not fade into retirement but remained on the payroll at Deptford with four servants. He received a warrant to oversee the two contract-built ships at William Castle's private yard at Deptford.[23] As a member of Shipwrights Hall, Jonas formally surveyed Castle's ships and those built under contract by Sir Henry Johnson.[24] Those who had dealings with him usually treated Jonas with affection. His official letters to the Navy Board are often endorsed as coming from "Old Jonas".

Shish was not a university man in the mould of Pepys and Deane. Jonas could not have written and presented to Pepys a *Doctrine of Naval Architecture*, the beautifully written manuscript given to Pepys by Deane in 1670. Even his friend, John Evelyn, reckoned he was a plain, honest carpenter who could give little account of his art by discourse and was hardly capable of reading, yet of great ability in his calling.[25]

Apart from his honesty, Evelyn also thought Jonas remarkable for bringing up his children well and teaching them to be able shipwrights.[26] Jonas and Elizabeth's eldest surviving son, John, was born in 1643 and would have been educated in drawing and the spiling (marking out) of moulds.[27] During 1665 John became Master Carpenter of a new-built ship and three years later, at the age of twenty-five, filled his father's shoes as Assistant Master Shipwright at Deptford following the promotion of his father to Master Shipwright.[28] His personal happiness was completed on 22 November when he married Mary Lake[29] whom he later described as his "dear and loving wife".[30] A little more than ten months after their wedding their first son was christened at St Nicholas's church, Deptford.[31] Unfortunately the child, named Jonas, seemed to have followed the fate of so many children of the time and died in infancy.

During 1673 John Shish was appointed the first Master Shipwright of the small yard at Sheerness.[32] By that time he and his wife had another child, a two-year-old son called Kendrick. Two years later at the age of thirty-two, John exchanged positions with his father and became Master Shipwright at Deptford.

There followed a series of career moves for the Shish family. Thomas, John's twenty-seven year-old younger brother, became his assistant. Thomas's wife Rebecca had given birth to one child and would have at least seven more.[33] Two years later, on 13 October 1677, Samuel Pepys and the king signed a warrant making Thomas Master Shipwright at Sheerness in place of his father.[34] Fisher Harding – who had married one of Jonas's daughters – filled the position vacated by Thomas of Assistant

Below: *Although drawn in 1808, this view from what is now the intersection of the A2 and Church Street shows the town of Deptford surrounded by fields. St Nicholas' Church stands out as the prominent building in the town. (Author's collection)*

Master Shipwright at Deptford. Harding was at the time Carpenter of the first rate *Charles*.[35] The warrant took into account the long and faithful service of Jonas and the respective sobrieties and abilities of both Thomas Shish and Fisher Harding.[36]

Thomas Shish remained at Sheerness for less than six months before progressing to become Master Shipwright at Woolwich, following the death of a member of the Pett family.[37] The appointment of Thomas Shish was made by the king after the Admiralty recommended either Shish or Joseph Lawrence, the Assistant Master Shipwright at Chatham, as suitable candidates.[38]

Another of Jonas's sons, Jonas junior, was also a shipwright but he practised outside the navy as a private builder based at Reddriff, a yard owned by the Shish family.[39]

At the time *Lenox* was built, John Shish seems to have been able to work with his contemporaries without causing disputes or making enemies. Deptford by and large functioned smoothly, in marked contrast to the constant disputes taking place at Chatham. When some of his men were in trouble after being accused of theft, Shish, convinced of their innocence, intervened. When he launched the 90-gun *Neptune*, his final ship of the thirty ship programme in 1683, John Evelyn (who lived next door) described him as "my kind neighbour young Mr Shish"[40] and in 1685 mentioned "the great respect I have to Mr Shish".[41]

When John Shish's sister was dangerously ill in Kent she requested that he come and visit her. He immediately wrote to the Navy Board desiring leave from Deptford as soon as possible.[42] In 1679, in an act of public duty, some of the dockyard officers involved in building *Lenox*, including Shish, became "feoffees" – a kind of trustee – to the New School House, East Greenwich.[43]

In their professional capacities, John and Thomas Shish were consulted in 1680 about a new type of twin-hulled vessel designed for towing ships up the Medway from Chatham. A model and a scantling list were made but before she was built the Shish brothers were asked by the Navy Board for their expert opinion. They found the cross beams and other parts too large and suggested smaller sizes.[44] They were clearly well respected as naval architects, especially as the twin-hulled vessel probably evolved from the work of Sir William Petty and Henry Sheeres, both eminent engineers.[45] Pepys also wrote to Shish asking him for the details of a fourth-rate ship; the reply included the dimensions of the rising and narrowing lines of the frame timbers obtained by calculation.[46] It renders implausible Pepys's remark that Jonas built ships by eye when his sons were using Cartesian coordinates to position their ships' frames.

John Shish was not a favourite of Pepys. In 1686, when comparing his friend Sir Anthony Deane with other shipwrights for a position as a Special Commissioner, Pepys wrote of John Shish: "old Jonas Shish's son [is] as illiterate as he ... low-spirited, of little appearance of authority ... little frugality. His father was a great drinker." John Shish may well have been a good drinking companion but Pepys's malicious remarks must be treated with caution and are in marked contrast to all other evidence. Tellingly, in Pepys's biased comparison, none of the other Master Shipwrights fared any better than Shish.

John Shish was paid a salary of £200 a year and the wages of a skilled shipwright at 2s 1d per day. He lived in a large house situated in the dockyard, only yards away from the great double dock where *Lenox* was built. The structure of the house dated back to Tudor times and, rebuilt and modernised in 1705, it still stands today.

Shish's principal duties were laid down by the Duke of York's instructions of 1671. They included being in constant personal attendance in building new ships. He was also to assist the Storekeeper and the Surveyor in their duties. He was to take charge and check mauls, wedges, saws, crows, auger bits, axes and other tools and instruments of the King's against the Storekeeper's books. With regard to the workmen in his yard,

he decided their rate of pay, which for a Shipwright could be anything between 1s 1d and 2s 1d per day. He was further tasked to observe the appearance of all the men in the morning and their departure in the evening and give notice to the Clerk of the Cheque to "prick" the names in the pay book of those men who were found to be at fault. As the men left the yard in the evening at the ringing of the bell, he was to look for timber and plank being carried out which could be used in shipbuilding and reclaim it.[47] John Shish had four servants to assist him in his duties; they must have been very busy.

By 1678 old Jonas was able to reflect on the good fortune life had dealt him. His two sons were Master Shipwrights and he himself was respected as old men of his time naturally were. He wrote to the Navy Board:

"Sir's [sic], I have been a servant to this Honourable Board 29 years in which time there hath passed through my hands business of great concernments. And I bless God I have discharged a good conscience, which with great comfort I shall carry with me to my grave. Sir's [sic], I have nothing else but am thankful to your Honourables for your kind love to me and to my sons".[48]

Plain-minded and puritanical,[49] Jonas daily reminded himself of his grave, as Evelyn recorded: "It was the custom of this good man to rise in the night and to pray kneeling in his own coffin, which he had lying by him many years."[50] Jonas was certainly not illiterate and neither were his sons. Letters written by Jonas in his own hand at the age of seventy-three are perfectly legible, if a little shaky and lacking the style of a clerk.[51] Longevity at Deptford was not confined to old Jonas, for Maudlin Auger was buried in December 1672 aged 106 while Catherine Perry was buried in December 1676, having lived to the grand old age of 110. She would have remembered, as a twenty-two year-old young woman, ships being fitted out at Deptford to oppose the Spanish Armada.[52]

THE WOMEN OF DEPTFORD

The Shish family had the confidence of the king but received little patronage from Samuel Pepys. However, Pepys did advance one shipwright from Deptford in exchange for pleasant favours offered by his wife. William Sutherland, author of *The Ship-Builder's Assistant* of 1711, a valuable source for this present book, proudly wrote in his preface, "My uncle Mr (William) Bagwell died master builder of her Majesty's yard at Portsmouth". His uncle William was the son of Owen Bagwell, who in 1677 was foreman for repairing the old ships at Deptford. William Bagwell's wife, whom Pepys found pretty and seemingly virtuous and modest, found it an easy matter to seduce Pepys with sexual favours for the advancement of her husband. Pepys graphically illustrates the affair, which started in the 1660s, in his magnificently frank and now famous diary. Her first name has long remained a mystery but in 1684 there was an Anne Bagwell, wife of William, a Deptford shipwright who used the alias Bayley. At the time she is recorded as being involved in a dispute over a mariner's pay ticket belonging to Abraham Moody of the *Woolwich*.[53] There was almost certainly only one Mrs Bagwell with a shipwright husband called William living in Deptford at the time.

There is reason to suppose the Bagwells' happy triangular relationship with Pepys continued after the diary period (1660–1669) and into the time of the building of the thirty ships. Pepys wrote in October 1677 a testimonial concerning the advancement of Bagwell, then carpenter of the *Resolution*: "[I] do assure you that as well from the character you are pleased to give him as my own many years knowledge of him. I both have and shall endeavour to do him a good office ... Mr Bagwell hath had the opportunity of having his name and character by this occasion presented to His Majesty and will stand very fair for the first good turn that shall offer for him."[54] True to his word Pepys did a good turn shortly afterward and secured a warrant for the appointment of Bagwell to the position of Master Carpenter of the first rate *Prince*.[55] Then, as a Master Carpenter with experience of shipbuilding at Deptford, Bagwell was conveniently dispatched to Bristol to supervise the building by contract of another third-rate of the 1677 programme built by Francis Bayley. The alias used by the Bagwells was (strangely enough, or perhaps not so strangely) also Bayley. During the three-year building period Bagwell wrote regular reports as the works progressed. His reports are a valuable source for understanding shipbuilding practices of the period. While he was at Bristol his wife kept him informed of the great endeavours the Navy Board was making on his behalf.[56] At the same time he wrote letters requesting the advancement of his brother, Francis, from the position of Carpenter of the *Greyhound* to Master Carpenter of the ship he was supervising being built, later to become the *Northumberland*.[57] Mrs Bagwell was clearly still in a position of some influence.

Happily, William Sutherland had no way of knowing the reason for Bagwell's advancement as the family secrets were safely concealed in code in Pepys diary, not to be revealed until much later. Deb Willet, another woman who famously had an affair with Pepys during his diary period, later married a naval officer. She asked Pepys for help and soon her husband was conveniently found a position aboard a ship.[58]

The involvement of women in seventeenth-century warship construction would seem unlikely and limited to the sort of contribution made by Mrs Anne Bagwell. However, women do appear and did make a legitimate living. Some were rich widows who may have had little involvement in the running of their deceased husbands' affairs, such as Mary Brown who cast some of *Lenox*'s guns, but others were very active. One example in particular, who lived in Deptford and supplied much of the ironwork for *Lenox*, was Mrs Susana Beckford.[59] Her husband Thomas, a locksmith and ironmonger, had supplied ironwork to the yard but fell ill during December 1675. As he lay close to death, with his wife crying and fearing for her future, he told her "my dear, do not distress yourself for I give all that I have in the world to you."[60] He died shortly afterward, but luckily for Susan, as she was usually known, his verbal will had been witnessed by others. She duly inherited the business, and recovering from the loss of her husband, took the opportunity to run it herself.

While *Lenox* was being built, John Shish himself had dealings with her which he could not resolve, for he wrote to the Navy Board: "There is one brass plate lock belonging to His Majesty's Ship the *Greyhound* which wants a new key. The lock is very good, but Mrs Beckford refuses to make a new key to the said lock. The reason, as she informs me, is that she hath not a price answerable for such a key which I humbly leave to your Honourables consideration."[61] Mrs Beckford was clearly a sharp businesswoman, aware of the commercial advantage of supplying a new brass plate lock rather than just a key.

She was a thorn in the flesh of other officials at Deptford too. Frank Hosier, the Clerk of Control, wrote to the Navy Board complaining that she had demanded a copy of her rates of payment, but he did not give her one and wrote: "If your Honourables think fit to let her have one I may be excused, 'twas her extreme importunity has made me give you this trouble for which I humbly beg your Honourables pardon".[62] A few days later, on 22 June 1677, she wrote to the Navy Board herself complaining that she was not being paid enough.[63] Somehow she managed to persuade the Deptford officers to issue bills that did not correspond to her contract and receive money for them. This was soon noticed by the Navy Board Surveyor's office under Sir John Tippetts who wrote to William Fownes and other officers at Deptford, saying they were extremely displeased that the enclosed bills of Mrs Beckford should contain goods not agreeable to contract, and that they were never to receive any goods in the future unless agreeable to contract.[64]

The hardships suffered by the Deptford dockyard officers at the hands of Susan Beckford were nothing compared with those of their next-door neighbour, John Evelyn. During 1685 Evelyn was distraught and angry when his daughter disgraced him by eloping with the nephew of Sir John Tippetts, to the amusement of the whole of Deptford. Evelyn wrote to Samuel Pepys complaining that one Mrs Beckford was an assistant to the connivance.[65] In spite of her unfortunate attitude toward the menfolk of Deptford she delivered more than £461 worth of work for the new ships.[66] She seems to typify the brash, confident forwardness of many women of her time. Another woman who regularly visited Deptford at the time of *Lenox*'s construction was Mrs Ann Pearson, who was paid £28 a year for laying bane to rid His Majesty's stores at Deptford and Woolwich of rats – a tidy sum of money for the time.[67]

THE WORKMEN OF DEPTFORD

To become a shipwright required serving a seven-year apprenticeship acting as a servant to an experienced man. The apprentice would have lived in his master's house, receiving meals and tuition and perhaps almost becoming part of his family. On the downside the apprentice's wages would have been paid to the master who might retain some or all of them.

The work shipwrights performed was not only very skilled but also physically demanding. Ambrose Fellows was a qualified shipwright artificer from Deptford who worked on the building of *Lenox*.[68] At the time he was thirty-eight years old and lived with his family at Deptford.[69] In 1673 he had become a warrant officer as Carpenter of the *Constant Warwick*, at Deptford being repaired. He was still being paid 14s 3d for the 1677 summer quarter as her Carpenter even though he was, at the same time, working on *Lenox* and receiving shipwright's pay.[70] As an experienced shipwright he had two servants working under him, Thomas Vaughan aged twenty-three and John Williams, aged nineteen. Vaughan had served for more than three years and Williams eight months.[71] During 1685, it was written in an account of officers that Fellows, for loyalty, conformity, ability, sobriety, faithfulness and diligence was very well qualified and deserving preferment (promotion).[72]

Unfortunately the workmen were the least likely to write letters expressing their thoughts. It is no surprise to find that according to one old puritan, a Mr Kendall, men at Deptford indulged in "profane swearing, cursing, drunkenness and frequently practised other misdemeanours".[73] Among the more dangerous misdemeanours was smoking, a practice that had been forbidden for a number of years. The Admiralty heard that it was still often practised in Deptford yard and subsequently ordered that notices be put up stating that six days' wages would be lost for a first offence and dismissal for the second.[74]

There was a core of permanent workmen employed in the yard whose numbers could be increased as the workload demanded. They were well paid and secure and at their retirement could hand down their coveted positions to family members. Old Robert Gransden managed to do this at Deptford while *Lenox* was being built. Gransden was a "teamer", responsible for a team of four horses employed in moving timber about the yard. By reason of his great age he petitioned the officers at the yard that his position be surrendered to his son, Samuel. They made no objections and a warrant was duly issued. Nothing demonstrates more the detailed interest King Charles took in the navy than the fact that he and Samuel Pepys both signed the Gransden warrant on 24 November 1677.[75]

WORKING HOURS AND PAY

During summertime work began at six in the morning and finished at six in the evening. These long hours were evened out during the short winter days when work commenced at dawn and finished at dusk. Artificial light was used in dockyards, but only during times of war, when ships were being refitted and there was considerable strain on the limited number of docks available. Despite the great difference in the length of the working day between summer and winter, the men seem to have received the same daily rate of pay. A breakfast was allowed except during short winter days between 1 November and 2 February. Similarly, the standard one-hour dinnertime was increased during the long summer days to one and a half hours between 23 April and 24 August.[76] The working day would therefore last between about six hours during the winter and ten during the summer. It took about half an hour in the morning for the men to register their attendance or "to come to their call" before starting.[77]

Overtime consisted of a "night" for five extra hours' work or a "tide" for one and a half hours. The number of hours worked for a night or tide was not set in stone. At Chatham, where disputes often occurred, the men refused to change the time worked for a "tide" from one and a half hours to two hours during the building of the new ship *Anne*.[78] It was also proposed to make the men work through their dinnertime during the short winter days. Part of the justification was to end the normal working day at four o'clock in the afternoon rather than the usual five, because of the "roguery and villainy they commit when it begin to grow dark".[79]

For working a "night" a full day's extra pay was earned. As an example, a skilled shipwright who earned 2s 1d per day was paid 2s 1d extra for five hours' overtime or 6d extra for one and a half hours' overtime. In addition, an allowance for lodgings was made which amounted to about ½d per day.[80] A skilled shipwright would easily have earned, with moderate overtime and allowances, about 2s 6d per day.

The number of days worked in a year varied widely, but a rough average seems to have been something slightly more than 200. A skilled shipwright would therefore have an annual salary of about £25, which compared very favourably with the mean annual income per household of £7 2s.[81] To compare our typical shipwright's earning and spending habits with modern ones is difficult, but he was renowned for spending much of his income on beer, a pint of which cost him 1½d. In comparison, a loaf of bread cost 1d.

Daily Wage Rate during the Building of *Lenox* at Deptford.[82]

Trade	Skilled	Unskilled
Shipwrights	2s 1d	1s 1d
Caulkers	2s 0d	1s 0d
Joiners	2s 0d	1s 1d
House Carpenters	2s 0d	1s 4d
Wheelwrights	2s 0d	
Bricklayers	2s 0d	1s 2d
Seamen	1s 6d	
Scavelmen	1s 3d	
Labourers	1s 1d	
Plumbers	2s 6d	
Ocumboys	6d	
Teams 4 horse	6s 0d	
Sawyers	36d per 100' or 2s 6d per load breaking	

Wages were calculated by the Clerk of the Cheque, William Fownes, and his servants from records kept in pay books, each of which covered a three-month period. The pay books contained the name, trade, daily pay rate, days worked, nights worked, tides worked, wages earned and a lodging allowance. The wages, overtime and allowances were calculated at the end of each three-month period. At best, the men waited a further three months before the wages arrived at Deptford and were handed out.

After starting work, a man would have to wait about six months before receiving three months' money, even under the ideal conditions at the start of the thirty-ship programme when money was actually available. Their joy on these occasions can easily be imagined. Senior officers of the Navy Board, including Sir Thomas Allin, Comptroller of the Navy, and Edward Battine, Clerk to Sir John Tippetts, attended.[83] In the autumn of 1677 during the building of *Lenox*, Sir Thomas received £25 15s for the hiring of boat and coach and for his diet and entertainment when he attended the quarterly payment.[84]

Delays in paying wages caused understandable resentment against the navy, obliging many of the workmen and their families to rely on credit to tide them over. One unfortunate shipwright who worked on *Lenox* and fell into debt was Thomas Baker, who was arrested by a bailiff, Edmond Paddington, for allegedly owing 7s to one Anthony Moore. John Shish and Fisher Harding took up the case and wrote to the Admiralty to explain that they had credibly heard that the money had been offered to Moore since the shipwright's arrest but it had been refused. Before they could intervene, they heard Baker was to be carried off a prisoner to Canterbury, even though he had three quarters of a year's pay due for himself and his servant (apprentice).[85]

Later, in 1679, when much of the new shipbuilding work had been completed there was a general discharge of some workmen. One of those released was a shipwright named William Staines. He petitioned to be taken back into work as he was born and lived in Deptford and had served his apprenticeship in the yard. He also added that his creditors would deny further trust for food and necessities for his two children and his wife who was again with child. Furthermore, he added, there were "foreigners" (shipwrights from other yards) still working and he was long known by Mr Jonas Shish. He concluded that he must have been discharged by mistake.[86] By Michaelmas (29 September) 1679 the shipwrights employed on ship repairs were owed one year's wages, while those employed on building the new ships were owed six months' wages. They wrote a petition complaining they had wives and families to support and that they could no longer obtain credit in Deptford.[87]

THE PRESS

The resentment caused by delays in the payment of wages was further compounded by impressments, which could be enforced by the navy for dockyard workers as well as seamen in times of national need. The urgency to build the thirty new ships had already resulted in Charles issuing press warrants. Some men resented the loss of freedom to work where they liked for the best-paying jobs, while others were happy to work for long periods of time in the relative security of the king's dockyards. Not surprisingly some pressed men refused to appear at all, while others appeared at the wrong dockyard to suit themselves.[88] A few of the pressed men at Deptford deserted to find work elsewhere. Altogether eighteen pressed men never appeared at the yard and twelve volunteers had left by the end of January 1678.[89]

The Admiralty responded by issuing a warrant to the Navy Board for their arrest. All mayors, sheriffs, Justices of the Peace, bailiffs, constables, headboroughs and "all His Majesty's officers and loving subjects whom it may concern" were required to help. In the effort to find the absent shipwrights, three ships' commanders were given warrants to arrest the men.[90] The usual method of pressing shipwrights was for lists of names to be made out at the time the Admiralty issued press warrants. In the past these men often absented themselves from their usual habitation and proved very difficult to find. Now, to prevent the "abuse", the Press Masters were to leave twelve pennies at their abode with a notice in writing of when and where they were to appear. If they evaded the Press Masters, or refused to appear, a diligent search was to be made. If this failed, a warrant for their arrest was to be made with the Constable and the arrested man to be delivered to the Conductor or to a common gaol.[91]

For the dockyard workmen the system was more a form of conscription than our modern perception of press-ganging. A pregnant woman, Ann Maverly, whose husband had recently been pressed, wrote to the Navy Board during June 1677: "Please for to order that my husband John Maverly, a caulker, may have a months liberty before he goes into His Majesty's yard at Chatham being imprest last Wednesday in respect I am very big with child and look every day and having several things to do which I am not able to perform without my said husbands assistance. I humbly beg your Honourables compliance therein." After consideration the tolerant Navy Board granted John Maverly one week of paternity leave.[92]

Violence involving the press in recruiting shipwrights could come from unexpected directions. John Jackson, normally a ship's carpenter at troublesome Chatham, foolishly became an assistant in pressing shipwrights and caulkers, as "I best know where to find them". He must have deeply regretted the decision, for he later wrote, "They threaten me of my life that I go daily in danger about the streets". In fear Jackson petitioned the Navy Board to be moved to Woolwich and within a week he had orders to be entered there.[93]

CHIPS
One of the benefits workmen enjoyed were "chips", the pieces of timber not suitable or large enough for shipbuilding and unsuitable as fuel to boil pitch. Men were allowed to take them home, but the seemingly harmless allowance caused endless problems. Shipwrights might reduce large timber to render it useless except as a chip. At the least it did not encourage them to be economical in their use of timber. It had long been a problem; in 1664, one sawyer at Deptford had his wife bring him a breakfast two or three times a morning, and she was observed to leave carrying out chips every time. The enterprising couple then sold off the timber for between £2 and £5 each month.[94] Sir John Tippetts, the Surveyor of the Navy, discharged Richard Lawrence, a labourer, for splitting good timber into chips. Unfortunately for him he had worked in the yard for a year and a quarter and was owed most of his wages, which would now be forfeit. He appealed to the Navy Board for leniency, saying that he had a wife and child and that his creditors would have him thrown in prison. Showing commendable mercy, the Board directed that he should be included amongst the men who were generally discharged, and was therefore allowed to receive his wages.[95]

The Portsmouth officers questioned the practice of allowing chips but King Charles, mindful of the advantages of remaining popular, ruled that according to the ancient custom, lawful chips should be allowed.[96] It was also the custom at Deptford to allow the poor people of the town to enter the yard on Wednesdays and Saturdays to gather small chips. Over time the scheme gradually fell into abuse as the wives, children and friends of the workmen joined the poor people carrying out good timber along with chips. The situation deteriorated to the extent that eventually the workmen, ever willing to take maximum advantage, cut up useful timber and hid it for collection later. In 1698 a search revealed many such abuses, including lengths of good timber hidden in holes cut into the earthen sawpits. The authorities at last took action by discharging two men caught converting good timber into chips. The boatmaker, who had observed the abuses but did not report them, was deducted a month's wages from his pay. For the future, no wives, children or friends of workmen were to be allowed into the yard and the sawpits were ordered to be brick lined.[97] Workmen leaving dockyards with chips were always looked at by their superiors with suspicion, hence the saying "a chip on his shoulder".

Another way the poor people of Deptford were able to earn money was by unpicking old navy rope or "junk" into ocum (commonly called oakum in modern parlance) for use in caulking. They were paid directly by Thomas Turner, Keeper of the Outstores, who was later reimbursed by the Navy Board.[98]

DEPTFORD ADMINISTRATION FOR THE THIRTY SHIPS
Parliament had voted a fixed amount of money for building the thirty ships, and this money had to be kept separate from other dockyard accounts. The Navy Board issued orders on 24 May 1677 that distinct

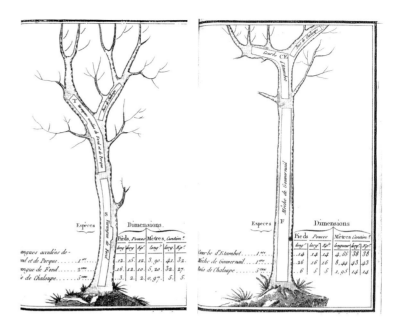

Above: *A great deal of skill was required to make the most efficient use of trees when selecting them for ships' parts. Natural growth and curvature could be utilised for specific timber pieces. From* De l'exploitation, Des Bois. *(Author's collection)*

accounts of the received and issued stores and the charges for workmanship for the new ships were to be recorded. Apart from separate book keeping, John Shish and the clerical officials discussed what other measures they would need to take. After viewing the yard and storehouses for suitable places they proposed building partitions in which to keep small stores. For timber, plank and masts a distinctive mark would be made on them for identification and a separate pay book kept for wages.[99] A small problem arose one month later when Thomas Turner fell behind with his weekly returns for the stores and goods. He soon received a reminder from the Surveyor's office to send them promptly in the future.[100]

The Clerk of the Cheque, William Fownes, kept meticulous records of costs and produced a book for each of the new ships built at Deptford. The books that survive are for the *Lenox*,[101] *Hampton Court*[102] and the *Duchess*[103]; they contain details of all the items bought for each ship. Further orders came from the Navy Board in September instructing yards to keep a distinct account of the repairs of the old ships and the building of the new.[104]

On 17 April 1683, with money for the new ships running out, Samuel Pepys attended the launching of the second rate *Neptune* at Deptford. He saw several hundred workmen lounging idly around for lack of materials, even though there remained a good quantity of small timber in the backyard. The workmen dared not meddle with as it was left over from the thirty new ships, of which the *Neptune* was the last to be finished there.[105]

THE QUANTITY AND COST OF TIMBER
The success of the shipbuilding programme relied on vast quantities of timber being supplied to the various dockyards. The problem had been identified early by the Admiralty Commission and instructions were issued for Timber Purveyors to be sent into the countryside, as it was the time of year for felling. The best oak was considered to come from the southeastern counties of England where the clay soil and moderate climate were ideal for its growth. Timber had to be hauled from the place it was felled to a convenient river where it was loaded into barges and taken to the dockyard. This was expensive; as an example, the cost of a load of timber felled eight miles from Rainham in Kent was £2 0s 0d. The further cost of hauling the timber overland by ox or horse to the Thames was

between 12s and 13s. Once afloat the cost of transportation dramatically reduced, costing only 3s to reach Deptford.[106]

The Navy Board bought timber either by means of a contract drawn up with a timber merchant or by a warrant for individual parcels of timber. Some of the small parcels purchased consisted of only twenty trees. The dual system of supply favoured competition and price comparison. It helped keep the Navy Board officers and merchants honest, although Pepys reveals in his diary that when he was Clerk of the Acts during the 1660s, he received gold and gloves among other gifts from the timber merchant William Warren. Warren was still a major supplier when *Lenox* was being built and supplied much of her timber.

The quantities and types of timber used on *Lenox* are listed in her Building List[107] (see Appendix 5). They are presented below as loads of timber; typically each load represented a small tree of fifty cubic feet.[108]

Loads of Timber used building *Lenox*:

New England deal 34.4
Spruce 352.0
Ordinary 38.46
Slit 1.89
Battens whole 3.1
Slit 0.05

Sawn oak plank up to 4in thick

1½in 1.55
2in 8.84
3in 50.14
4in 138.82

Elm timber 74.3
Fir timber 99.08

Oak stuff of 8in 42.9
Straight oak timber 733.56
Compass oak timber 657.48
Raking knees 41.6
Square knees 32.24

Total 2310.41 loads

Above: *(Left) A seventeenth-century squaring axe. The head is at an angle to the shaft to give clearance for the hands of the workman. The cutting edge is bevelled on one side only, allowing the blade to cut a flat surface. (Right) A seventeenth-century hatchet. (Author's collection).*

THE QUALITY OF OAK

Oak was by far the most important wood used in English ship construction in the late seventeenth century. Apart from its strength it was the best timber available to resist the effects of fungal rot caused by being periodically wet or dry. In common with other hardwoods, oak was much easier to cut when it was green. After it had seasoned for some three years it became very hard and difficult to work. This was a valid consideration, as the tools used during the seventeenth century were comparatively soft, although easy to sharpen. It was customary to fell between mid-winter and the spring when the sap was not rising[109,] as the summer heat could split the quickly drying oak. It would also be more susceptible to rot in summer, as the timber stripped of its bark would be exposed to the fruit spores of fungi in late summer.[110]

The amount of timber already in stores before the building programme started was minimal, resulting in most of the timber for the new ships coming straight from the woods. Only a few months after being felled much of it was afloat as part of a ship. For *Lenox*, which was built in a dry dock under great demand, the pressure was even greater to complete her quickly and vacate the space she occupied for the next ship. It was common practice for ships to season afloat, but for this process to be effective, careful attention was required if vessels were to remain in good condition.

For oak trees to reach the required size for shipbuilding they would have started life at least eighty years prior, during the reign of Queen Elizabeth I. Large timber necessary for some of the important ship's structure may have been more than 150 years old. It was difficult for landowners to plant and watch oak trees grow in the knowledge that they personally would not benefit from their investment. However, many managed estates were passed down through generations and contained oaks of all ages. In ideal conditions, saplings replaced the felled mature trees so that the estate supplied a regular quantity of timber.

STRAIGHT, COMPASS AND KNEE OAK

The bulk of oak timber used for ship construction was categorised as either straight or compass. Straight timber came from high forests where the trees grew tall and straight so that their branches could reach the sunlight. Curved compass timber generally came from places where the oaks did not need to reach to their maximum height and were found in copses and hedgerows. Compass timber cost £2 18s 4d a load, and was marginally more expensive than straight timber, which cost £2 16s 3d a load.[111] The ship frames were made from compass timber, the vast majority of which was English growth, while straight plank and thick stuff formed the majority of the fore and aft structure. The natural curve of the wood grain was matched as closely as possible to the shape of the frame timbers to maximise its natural strength.

Also made of oak were the knees, the majority of which were used to connect the deck beams to the side of the ship. They counted for less than five per cent of the oak in a ship but the difficulty in obtaining them from suitable branch stems increased the price to £3 10s 0d a load for raking knees and £4 10s 0d a load for square knees. Square knees, as would be expected, formed about a ninety-degree angle, while raking knees formed a more obtuse, or less sharp angle.

The definition of timber was vague. During 1677, Phineas Pett II, the Master Shipwright at Chatham, wrote, "We never yet had any rules set or direction how much a piece must be round to be called a compass timber, nor the extent a piece must spread to be called a raking knee".[112] Elm, which grew in long lengths and was resistant to rot as long as it remained wet, was used principally for the keel. It was slightly cheaper than oak, costing £2 15s 0d a load. Cheaper still was beech, which like elm was only suitable for use where it remained wet. It was often used for planking deep underwater near the keel.[113] Fir or pine softwood deals were used for the decks of the quarterdeck, forecastle and the upper gun deck away from the guns. It was also used for the partitions, storerooms and for use in the upper works of the ship, as it was both light and cheap.

Below: *A typical skilled shipwright, such as Ambrose Fellows, squaring timber for use on* Lenox. *He became her carpenter and had a very long association with the ship.*

Right: *Diagram showing proportions for squaring timber. From William Sutherland's* The Shipbuilder's Assistant, *1711, p29.*

OAK PLANK AND THICK STUFF

A great amount of oak was bought in as sawn plank. Plank is timber sawn to between 1½ and 4in thick. Anything thicker was unimaginatively known as thick stuff while anything thinner was called board. For ships the size of *Lenox*, 4in plank was used to "birth up" from the keel to the lower wale. Most plank was imported from the East, usually the Baltic areas around Danzig, Riga, Hamburg or Prussia. It was found to be more flexible than most English oak, a very desirable quality towards the ship's ends. However, when planking towards the buttocks and bows where the bend increased, English oak that had become pliable through old age was preferred because of its more durable qualities.[114]

During 1686 Samuel Pepys held a conference concerning plank for shipbuilding.[115] In attendance were a number of respected shipwrights, including John Shish. First, it was observed that for many years foreign supplies had been used to such an extent that only one fifth of plank now used was English. It was also agreed that planks of between 26ft and 40ft long, meeting at 14in average breadth at the top end, could be obtained that equalled or exceeded English plank in durability. It was further added that foreign oak trees did not taper so much near their top end so that the plank was of more constant width for its full length. Finally the poor method of sawing English plank, known as conversion, resulted in fewer loads of foreign timber being used for the same amount of plank. Sawing plank in straight lengths was performed using wind-powered mills in the East (the Baltic regions), whilst conservative English shipwrights used manual two-handed saws. It may seem somewhat surprising that sawmills

were not introduced into English yards, but the most likely explanation for this is that most of the conversion was performed on compass timber, with which it would have been difficult to employ a sawmill effectively.

SQUARED TIMBER

Timber was normally supplied to dockyards in squared condition: that is, with four flats cut at ninety degrees to each other along the trunk, leaving it partially round at the corners. The breadth of the flat was about twice that of the remaining round corner. The cross-section was cut at a constant size along its length. This custom provided a good basis for calculating the volume of the timber, known as square measure, as well as exposing it for inspection of defects. During May 1678, the Navy Board asked the officers at Deptford what basis they used for cutting the flats. Never having given it much consideration, they replied that there was no calculation to provide an exact account, but nevertheless went on to enclose a sketch of their practice.[116] Shortly afterward the Navy Board issued a rule stating that timber up to 4ft in girth (15¼in diameter) after hewing, was to have the side, or flat, 5in wide. Accurate measuring of timber was vital in controlling the cost of building the new ships and Master Shipwrights

Below: *An early photograph (c. 1880) of the south side of Flagon Row, Deptford, once owned by the Shish family. The view looks west towards Deptford High Street. The houses were demolished in 1896 to make way for an extension of Creek Road east to the end of Evelyn Street. (Lewisham Council)*

and Yard Commissioners often checked the timber themselves.[117]

It was not always desirable to hew timber square, as the swell near the butt, or root end, would be removed. This swell was often useful when converting trees to compass timber. In these instances the timber was left in its natural condition, but the price was adjusted to account for the additional waste. In this condition it was measured according to the girth measure. The frames of ships during the seventeenth century relied on great pieces of timber that were becoming increasingly difficult to find. During the next century, the shortage of these large timbers would eventually result in a new method of frame construction that employed smaller individual pieces. Great care and ingenuity was necessary for shipwrights to make the best use of the natural curves and sizes of the timber brought into their yards. It was essential to convert the timber with the minimum amount of waste. Some merchant shipbuilders were known to have gone out of business for their inefficient use of timber.[118]

THE TIMBER PURVEYOR

Even before Parliament approved the thirty-ship programme the problem of timber supply was being addressed. Thomas Lewsley was the Purveyor of Timber and, although based at Deptford, he also worked on behalf of Chatham and Woolwich yards. He was extremely knowledgeable and Pepys had consulted him about the quality of supplies since at least 1664.[119] He now received instructions from the Navy Board to go into Suffolk to view plank and negotiate prices and terms. He stopped first

near the coast at Walberswick, where he dealt with four merchants who between them had 384 loads of 2in, 3in and 4in sawn plank, probably imported from the East. The biggest 4in plank, meeting at 40ft long, was 16in broad in the middle and tapered down at the top end to 14in broad. Because of the great length and breadth, the price that he negotiated was a few shillings higher than the average of £4 10s 0d a load for delivered plank. Lewsley then rode south a further twenty miles through the leafy countryside to Woodbridge, where he met three more timber merchants with similar quality plank to that he had seen at Walberswick. Two of the merchants, the widows Gutteridge and Gurling, proved typically tough businesswomen. They had generally good wood but Lewsley could not persuade them to sell it on any other terms than ready money.[120]

He returned to London but was soon off again, all the way to Sheffield, riding the Navy Board's gelding, evidently a very good horse, for he "was never better carried of a journey in my life". The main purpose of his long journey was to view 1400 trees in the Earl of Arundel's parks. The massive trees, containing nearly five loads of timber each, were not from a managed estate and many proved to be very old, dead or decayed. Lewsley counted only about 200 trees that could be used for shipbuilding, which were of about equal quality to Danzig crown oak.[121] Danzig crown oak was considered the best timber available from the East but not the equal of oak from the south-east of England. Also out looking for timber was Sir Anthony Deane, who lamented that the greatest part of the wood available had been bought by house carpenters at expensive rates.[122]

In June, despite the normal time for felling being over, the Timber Purveyor Thomas Lewsley rode south into Sussex to view more timber for the navy. He stopped first at Limpsfield where he marked 100 loads of spire oak averaging 75cu ft, most of which would be serviceable for frame pieces. Further on he stopped at Slaugham, Woodmancote and Wineham where he viewed 500 loads of very useful timber for Deptford belonging to Mr Yoakhurst and Mr Barber. On his way back home he stopped at Lingfield and Chislehurst where he viewed some eight great elm trees, each averaging eight loads and suitable for keel pieces.[123] The Navy Board agreed to buy Yoakhurst and Barber's timber and Lewsley returned to Slaugham, Woodmancote and Wineham three weeks later to mark out the timber he had chosen with the broad arrow. He also visited Sir Charles Bickerstaff at Boldbrook where he saw another 800 loads of good timber, for which a contract was agreed for delivery to Chatham and Deptford.[124] However, to make up for the slow progress of timber supply to Chatham it was all delivered there.

As the programme progressed Lewsley made other journeys; during September he rode to Epping to oversee the transport of some great elm pieces for a keel. They duly arrived at Deptford but the keels for *Lenox* and *Hampton Court* were already laid and so the elm pieces were sent on to Chatham.[125] The individual yards co-operated a great deal and loads of timber were often allocated to the yard with the greatest need.

Lewsley's duties included viewing and inspecting many different supplies for shipbuilding, including deals brought in from the Baltic, masts from New England, brush and broom bavins (burnt as fuel to heat planks during bending), oars and capstan bars. He went on board two pinks lying in St Saviours Dock and viewed their cargoes of oak plank from Hamburg. The first belonged to Mr Redmand, carrying a cargo of only 3in plank of indifferent length measuring 18ft to 34ft long. The second pink belonged to Mr Johnson and contained generally very sound white wood plank of 3in and 4in thickness that he could recommend.[126]

THOMAS SHISH PROCURES TIMBER

Thomas Lewsley was not the only officer from Deptford to visit the woods. On 18 May Thomas Shish, the Assistant Master Shipwright, rode out toward Essex. He first went to Marks House to view a small parcel of timber before moving on to Romford, where he heard of another parcel at Maylors Greene near Hornchurch. There he found about twenty loads but they were not squared so he could not accurately measure them. That evening he went to Brentwood to stay for the night where he wrote, in a neat hand, the first of his daily reports to the Navy Board.[127] While there he heard that Lord Petre had a large amount of timber still standing between Brentwood and Ingatestone. In the morning he went and viewed the standing timber, which contained nearly 3000 loads suitable for deck beams, wale pieces and plank. Such a large amount of timber was enough to build a ship and therefore of considerable interest. Unfortunately he could not agree a price with Lord Petre and the matter was later discussed at the Admiralty meeting of 10 July.[128]

Thomas then met a Captain Henry Southcott who told him that very little timber had been felled in the area during the year because the price of bark was too low. Bark was a valuable by-product of the oak trade, used in the leather tanning industry. Thomas then heard of more timber he could view at nearby Hutton, even though the month of May had arrived and the usual time for felling had passed. He rode on through Fyfield to see yet more timber; unfortunately it was judged to be too small for shipbuilding, so he continued to Chelmsford where he spent the night.[129]

Shish rode on deeper into the countryside, visiting Totham and other small villages without finding any usable timber. By 21 May he passed through Witham [130] on his way to Colchester, visiting Cressing and its environs, but found it increasingly difficult to find timber large enough for shipbuilding. Once again he heard that the season for felling timber in the county was past.[131] The following morning he started on his way back towards Deptford. At Rayleigh he went to view some recently felled timber and found about 120 trees averaging about one load each belonging to Thomas Holland of Stifford. Shish judged they would be suitable for floor timbers, top timbers, and upper gun-deck beams, at a cost of £2 10s 0d per load delivered to Woolwich, the nearest dockyard.[132]

On reaching Deptford he wrote out a report summarising his tour of Essex.[133] As he wrote, work was starting in the dock upon the laying of the ways for *Lenox*. Thomas Shish did not stay long enough to see much building work, for he was sent back into Essex to negotiate prices with some of the timber-owners he had recently met. He found their demands unreasonable but persuaded at least two of them, including Holland, to travel to London to negotiate terms with the Navy Board itself.[134] Holland, like so many others, preferred ready cash to waiting for the Treasury to eventually pay out. He managed to negotiate successfully for a £100 imprest as soon as his timber was marked out, and terms for a contract were signed on 13 June. His contract is typical for a parcel of timber of the size he was supplying (for Holland's contract see Appendix 7).

A week later Shish was back in Essex marking out and stamping trees that Thomas Holland had sold to the navy.[135] He stamped broad arrows on the butts and marked identification numbers on the trunks to signify the navy's ownership. Some of Holland's timber soon arrived at Deptford, so soon in fact, that the Keeper of Outstores, Thomas Turner, had not yet received a copy of the contract to check the quality and volume. The storekeeper laid the timber up by itself while he waited for a copy of the contract to arrive.[136] Only when the dockyard officers were satisfied would Holland receive an "imprest" bill, which could be redeemed against the Navy Treasury. In the past, the state of finances usually resulted in suppliers suffering a considerable wait, often years, before it was their turn to be paid. But, with a specific amount granted by Parliament for building the ships, the delay was now very short in comparison. Holland must have been satisfied in his dealings with Thomas Shish and the Navy Board for he entered into another contract during September to supply Deptford with 67 loads of mixed straight, compass and knee timber. For his journey into Essex, Thomas Shish was paid £11 3s 4d travelling expenses, followed by a further £2 for marking the wood.[137]

For John Shish, the most immediate and fundamental timber problem holding up work on *Lenox* was the absence of suitable keel pieces. Luckily help was at hand in the nearby private yard of Robert Castle who had two great pieces sawn and converted to dimensions suitable for a second or third-rate ship.[138]

THE QUANTITY OF TIMBER IN STOCK

As work began on the new ship programme the Navy Board realised it would be necessary to record the amount of timber already at Deptford dockyard so that its value could be separated from the new shipbuilding. It was found that there were 276 loads, 44cu ft of straight oak averaging 78cu ft each suitable for footwaling and floor timbers. There were also 119 ends of compass oak containing 141 loads, averaging 59cu ft each. They were suitable for making futtocks or top timbers for the frames. Finally there were sixty-four knee timbers containing fifteen loads, averaging 12cu ft each. In total, there were about 433 loads of timber ready to be used.

As *Lenox* was the first ship to be started much of this ready timber must have been used in her construction. Although the amount of timber in the yard would seem to be vast, altogether it would amount to only a fifth of that needed to built a third-rate ship.[139]

THE QUANTITY OF TIMBER ORDERED

A report was made out by the Navy Board giving an account of the timber for which contracts had already been made with merchant suppliers:

Loads of timber ordered by 23 June 1677

	Deptford	Woolwich	Chatham	Portsmouth	Harwich
Oak Timber	2850	3854	1572	250	820
Elm Timber	23	23	12	105	–
Fir Timber	200	100	–	–	–
Oak Plank	350	450	400	–	72
Treenails	5	15	50	–	–
Ordinary Deals	4200ft	4200ft	5600ft	–	–
No. of places being prepared to build ship	2	4	3	2	2

In addition a contract was in the making for Francis Bayley to build one ship at Bristol.[140] In terms of timber supply for each ship, the report shows that Deptford was better placed than any other yard. The only shortage was treenails, a situation that would remain long unresolved.

The Deptford officers put enormous effort into acquiring sufficient quantities of very large timber to build their ships of war. The Master Shipwright himself, John Shish, and his brother Thomas both went out into the countryside looking for timber. They left Deptford at a time when the king and Parliament were expecting the new ships to be built in the shortest possible time. But ensuring the supply of large timber was the most important task undertaken by them and they were generally successful in their efforts. Their correspondence is to the point and much shorter than that of dockyard officers from other yards. They worked as a team and were particularly well served by the Purveyor, Thomas Lewsley. Later, John Shish made other trips to view timber belonging to Robert Castle at Albury, Ditton and the wharves at Guildford, but found most of it too small for his use.[141]

It was very different at Chatham. Richard Beach the Commissioner wrote concerning Phineas Pett II, " ... the Shipwright is so infirm (or at

least pretends to be) that it is not to be expected that he can travel the country to look for timber".[142] Needless to say Chatham's first ship, the *Anne*, was launched seven months after *Lenox*.

TIMBER MERCHANTS

In spite of the time and effort spent by the dockyard officers at Deptford in obtaining timber from the Home Counties the results were inadequate. In total they probably secured about 1000 loads that could have been delivered in time for use on *Lenox*. However, some of this timber went to Woolwich or Chatham. The timber that did find its way to Deptford was not used only on *Lenox*. The *Hampton Court* was under construction at the same time and the second rate *Duchess* was started when *Lenox* was half built. After all the depletions only a small proportion of the timber acquired by the Deptford officers would become part of *Lenox*. During the same period more than 7000 loads of timber, much of it foreign, was contracted with timber merchants for delivery to the Thames dockyards.

It is difficult to be precise about the quantities of timber that found their way to Deptford, as many of the contracts included delivery to more than one yard. (See Appendix 7 for an example of the contract made with Peter Causton). It seems that during the building of *Lenox* and *Hampton Court* more foreign than English timber went to Deptford. During the early days of the building, the contracts were often for large pieces of timber needed to construct the frame, supplemented by those for plank later on. Almost all the plank came from "the East Country or Hamburg", of which the majority was 4in thick for the underwater part of the hull.

The most important contributor to the timbers of *Lenox* was Sir William Warren, the timber merchant with whom Pepys had many dubious dealings in the past. Warren read the invitation to tender for shipbuilding materials in the *Gazette* on 10 May 1677. A week later he wrote to the Navy Board, offering enough timber to build two entire ships.[143] Negotiations began and Warren signed a contract on 6 June for 4000 loads of timber at a somewhat lower price than that originally asked for. The timber was to be delivered in equal proportions to Deptford and to Woolwich. The vast majority of the timber supplied was to be straight and compass oak spire and to be between 40cu ft and two loads (100cu ft) apiece, suitable for making frame pieces and beams (See Appendix 7 for the contract made with Sir William Warren).

By 12 July a report from Deptford to the Navy Board showed that nearly 700 loads of timber had been delivered since 24 May, fully half of which came from Sir William Warren.[144] This deluge of timber was one of the major factors in the rapid progression of *Lenox*, but there was a similarly major problem. John Shish wrote to the Navy Board explaining that although the timber arrived in quantity, little of it proved large enough to be made into the ship's frames. He requested that his brother, Thomas, be sent out to mark up the timber before it was delivered to make sure it was of sufficient size.[145] Within a month, and eager that nothing should hold up the progress of *Lenox*, John Shish went out to visit five of Warren's sites. He marked out about 200 loads of compass timber suitable for frame pieces to be delivered to his yard.[146] The actions of the Shish brothers in picking out the best of Warren's timber left an increasing amount of the smaller-sized timbers that would be difficult to use later. Warren also complained that his barges were beginning to wait a long time at the dockyards before they were unloaded.[147] Moreover, Warren experienced various other difficulties. After he procured 4in plank from the East, his ships the *Black Cock* and *Jacob* could not sail to fetch it until they had received passes from the Navy Board.[148] In spite of all his problems, Warren was paid £36,000 – a huge sum of money for timber for the thirty new ships.[149]

BUILDING LENOX

SOURCES

In the seventeenth century there were very few works published on ship-building, leaving huge gaps in our understanding of the art and practices of that time. This account tries to close that gap by relying on contemporary manuscript sources, uninfluenced as far as possible by the many works of the eighteenth century. Some interpretation was occasionally necessary and this author has tried to differentiate between fact and probability in the text. Although we are exploring the building of one ship, *Lenox*, it follows that other ships of the period were built in a similar manner. But there were differences; Phineas Pett II at Chatham for example, often used smaller sized but more numerous pieces of timber. As regards language, I have used contemporary, often more descriptive, shipbuilding terms than the specialist words employed later. For instance rather than use "sheer plan" I have used the term "broadside".[1] All ships mentioned along with *Lenox* are from the thirty-ship building programme, unless otherwise stated.

Thanks to the surprisingly good quality of surviving manuscript material, it has been possible to follow closely the building of the ship as time progressed and to illustrate each stage of her construction. Of primary importance are the building progress reports sent by John Shish to the Navy Board (see Appendix 4). He and the other Master Shipwrights, were ordered by the Admiralty to submit their reports in writing on a weekly basis.[2] Not unexpectedly the Admiralty did not always receive them in a timely fashion and the matter was raised more than once at Admiralty Board meetings.[3] However, most of the building reports for *Lenox* still survive, although they are somewhat scattered around in different locations within the National Archives. The reports started on Friday 13 July 1677 at the outset of building and continued every Friday for thirty-one weeks. The weekly reports then abruptly stopped, a couple of months before the ship's completion, but by then all the important structures were in place.

Of primary importance for the appearance of *Lenox* are six surviving Van de Velde drawings, whose clarity in showing her decoration and gun-port arrangement from all viewpoints, makes her one of the most well-illustrated ships of her period. A further painting by Van de Velde the Younger shows *Lenox* under sail and is probably based on the drawings made by his father. Another Dutch artist, Van Beecq, also painted *Lenox*, but he clearly did not have the help of the Elder Van de Velde's drawings. Comparing Van de Velde drawings of the same ship reveals some discrepancies and some careful interpretation is necessary when using them.

There are also limitations. As useful as the Van de Velde drawings may be, they do not help with the dimensions of the ship. These are provided by official sources. They are the actual primary dimensions of *Lenox*, recorded by Pepys in his Register of Ships,[4] The Principal Dimensions list of the thirty ships[5], Scantlings of the Thrity Sail[6] and the contract for

building a similar ship, the *Yarmouth* of 1694 to the same size and specification as those of 1677[7] (see Appendix 3).

Another book in the Pepysian Library, Cambridge, records all the materials and costs of building *Lenox*.[8] Every block and rope is listed with the dimensions and quantity. The masts and yards and all other details such as capstan bars, stern lanterns and ship's boats are similarly mentioned.

Another valuable primary source would have been a draught of the ship. Unfortunately none survives, but research into the nearest evidence relating to *Lenox* has led to a reliable reconstruction. A draught in the Danish Archives in Copenhagen[9] shows a 70-gun ship. Although its flags indicate it could date from 1677, the end view shows the ship has a vertical line between the upper and lower breadth sweeps, a feature not seen on draughts until much later than 1677. Another draught, which is supposed to be of the *Captain* of 1678, is printed in Charnock's *A History of Marine Architecture* published shortly after 1800. Unfortunately it is clearly not the same ship shown in Van de Velde drawings of *Captain*, having several differences, including four extra upper deck gun-ports and, more worryingly, an almost circular hull section.

THE WORK OF EDMUND DUMMER

A valuable source resides at Wilton House, Wiltshire, where there is a model and accompanying draught of a third-rate ship. The model has been known as the *Old Hampton Court* since at least 1751, when it was mentioned in an inventory.[10] The initials JS or IS on the cradle could be those of John Shish. The draught and model would have been excellent source material for *Lenox*, as the *Hampton Court* was built alongside *Lenox* at Deptford. But as is so common with ship models of the period, there are major doubts concerning her identification. The date on the cradle near the initials is 92, by which time John Shish was dead. The decoration and number of upper-deck gun ports do not match known Van de Velde drawings of the ship. The model, however, is strikingly similar to Van de Velde drawings of the *Essex*, a contract ship built in 1678 by Sir Henry Johnson at Blackwall. Perhaps the JS initials are those of Johnson.[11] The draught is a thing of beauty in its own right and was evidently drawn by somebody with artistic skills, a good knowledge of ships and access to official navy draughts.

A possible candidate for the draughtsman is Edmund Dummer, who in 1678 was an Extra Clerk in the Surveyor's Office working under the patronage of Sir John Tippetts, the Surveyor of the Navy.[12] He is well known for a number of drawings in the Pepysian Library known as "Mr Dummer's draughts of the body of an English man of war".[13] These show deck plans and sections through a first-rate ship and are excellent source material in their own right. The ink and wash escutcheons, which bear the titles and descriptions of both the Wilton House and Pepysian draughts, are similar and could have been executed by the same hand.

Dummer is also responsible for another valuable and reliable source, the partial hull lines he made from actual ships while they were still on the stocks. Tippetts took Dummer to Harwich in April 1678, when he recorded the bodies of two ships building there, the *Restoration* and the *Sandwich*.[14] After a meeting between Pepys and Anthony Deane on 11 February 1679, Dummer went to Deane's home where he was shown the experiments and rules to which his host worked and was given some useful   advice.[15] He developed an idea and wrote a letter to Pepys in which he professed himself belonging to the shipbuilding school of truth and reason rather than custom and tradition, and that ships' bodies be recorded and exact models made that could be tested in water.[16] His plan to make draughts of the actual ships, which could be compared with the builder's draughts, is in keeping with the philosophy of the king, Pepys and Deane.

The project would give valuable insight into the effects of hull shape when the sailing performance of the actual ships was known. The plan seems to have been adopted, and as the ships neared completion, Dummer visited all the dockyards and made drawings of the hull lines and compiled lists of dimensions.[17] He visited Chatham between 17 and 24 May 1679,[18] to take the lines of the *Berwick*, finishing only five days before she was launched on the 29th.[19] During the same period he measured the second rate *Duchess* at Deptford; meanwhile John Shish received a letter from the Navy Board ordering him to provide men and materials for the work. Unfortunately, by that time *Lenox* was already launched and her hull lines were not recorded.[20]

Dummer made his initial rough work on printed, foot-squared paper, with sub-divisions of 4in, making it possible to reproduce accurately to scale without fear of distortion. He was later reimbursed for paying John Seller to engrave the necessary copper plate and for 200 prints from Robert Harford the stationer.[21] For the finished draughts, twenty-two skins of writing vellum and four quires of royal printing paper were purchased.[22] The work would have resulted in accurate ships' draughts, matching the quality of the written naval records left to us. Dummer seems to have concerned himself with the lines only and not the general appearance or decoration of the ships. However, the famous father and son Dutch marine artists, the Van de Veldes, lived at Greenwich only a few hundred yards from Deptford. Van de Velde the Elder produced a series of large-scale drawings, which in the vertical measure ¼in to the foot, the standard scale of draughts, making it a possibility they were produced to aid Dummer in his work. Fortunately *Lenox* was one of those drawn at a large scale and being built so near to the Van de Velde home was the subject of more drawings than any of the other new ships.

As Dummer carried out his seemingly innocent work, disaster struck in the unlikely form of the Popish Plot, the bogus Catholic attempt to overthrow the king. As the political storm gathered, Dummer became worried by accusations made against the navy. He managed to speak to Mr Secretary Pepys, who gave him reassurance of his favour. This reassurance did not last long, however, for both Pepys and Deane became embroiled and found themselves accused, among other things, of selling navy secrets. These events must have badly shaken Dummer, but worse followed shortly afterwards, for as he walked along the Strand, he heard that four Members of Parliament had been "discoursing particularly of the misfortune of Sir Anthony Deane and Mr Secretary Pepys and among the crimes laid to them it was aggravated that they had employed a man to take the bodies of the kings [sic] ships supposed to be of no good intention", i.e. selling them to the French.[23] As can be imagined, poor Dummer was "extremely fearful"; there was no telling where it would end. He must have imagined being accused of treason and ending up on the scaffold. Luckily for him, he managed to be sent away to Bristol, but on his return was told by the navy that he would be severely whipped if he had disposed of any of his draughts. Dummer, aware of the political sensitivity of his work, managed to keep out of further danger as the furore surrounding the

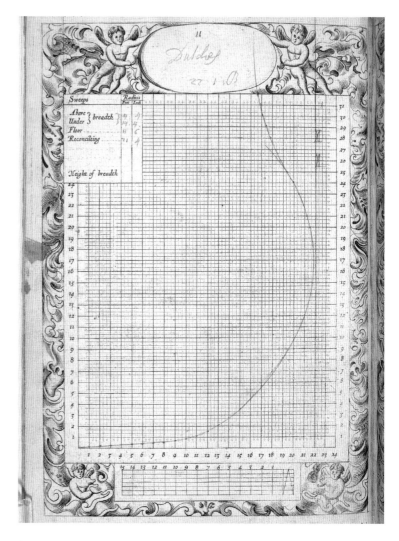

Above: *One of the sections recorded at 8ft intervals by Edmund Dummer of the Second Rate* Duchess, *built by John Shish. The plan is drawn on paper printed by Robert Harford the stationer, with the engraved plate supplied by John Seller. (British Library)*

Popish Plot gradually died away. The result of all Dummer's efforts for the navy was only five draughts.[24] Unfortunately none seems to have survived, with the possible exception of that in Wilton House. His work was ordered to be brought to the Navy Board "as fast as done" then locked under three different locks, kept by three different officers.[25] Quite probably, in the interests of misplaced "national security", his work was deliberately destroyed. All that seems to have survived is a small book with preliminary rough pencil draughts and various lists of dimensions for six of the new ships, on the pages printed by Robert Harford.[26]

One drawing that is of particular interest in Dummer's work is the clearly drawn sections of the *Duchess*. John Shish built her in the double dock at Deptford in *Lenox*'s place, immediately after the latter was launched. Unfortunately for a reconstruction of *Lenox*, she was a second-rate, three-decked ship and proportionally wider than a third-rate two-decker. However, Phineas Pett II wrote that his draught of a second rate was proportioned from draughts of the first and third-rates he had already drawn.[27] If the depth of *Duchess* is scaled down to suit that of *Lenox*, the breadth works out to be about 2ft too wide. Nevertheless, Pett did proportion his draughts and must have used different scales for the breadth and depth. Knowing the breadth, depth and length of both *Lenox* and *Duchess*, it is possible to redraw the lines of *Lenox* from those of *Duchess*. This method was used to recreate the lines of *Lenox* and they turned out to be very similar to Dummer's lines of the *Berwick* and the Wilton House draught of what is likely the *Essex*.

THE DRAUGHT

John Shish prepared his draughts for the proposed new ships some time between 12 and 25 May 1677, in the old Tudor Master Shipwrights' House, next to the great double dock at Deptford. This is the period between the king giving the Admiralty Commission a list of principal dimensions and Shish writing that he had already drawn them.[28] John Shish must have been taught how to make draughts by his father, "Old Jonas", and probably received his help in preparing those for the new ships, especially as they were the first he had made. Jonas's last ship had been the *Royal Oak* of 1674; her dimensions were very similar to the new third-rates and the draughts John Shish now made may well have been based on his father's final project.

There is every reason to believe Shish had no time, nor indeed authority, to add decorative details to his draught. Certainly, the draughts of new ships built at Chatham and Bristol did not have decoration, for their builders wrote to the Navy Board, as the ships neared completion, asking for details of the carved works[29] and of balconies.[30] These men were following orders of September 1677, that carved works could only be prepared with Navy Board authority.[31] The shipwrights barely had time to make a draught; it is extremely unlikely, if not impossible, that a dockyard model could have been built before the ship itself was started.

Shish would have made his draught to the usual scale of ¼in to 1ft. The broadside view was started first by drawing in the length of the keel. The stem and sternposts were then added followed by the waterline, wales, decks, gun ports, head and other details.[32] The waterline was not drawn parallel to the keel, but 18in deeper abaft than afore to give the ship better steerage.[33] The curved rising lines of the floor, breadth and top of the side were then drawn; they would play a vital role in controlling the hull shape when Shish came to draw the end views. A half-breadth plan view was then drawn below the broadside view and the narrowing lines of the floor, breadth and top of the side added.

The position of the frame stations was then drawn on both views, starting with the largest frame, the midship flat, about two-thirds the length of the keel from the aft end of the stern post. Every third frame was rather special; these were the "frame bends" and they would be the only ones drawn on the end views of the ship. During ship construction, the futtock timbers that formed the frame bends would be assembled and set

up first, before the other timbers were added. A timber, in shipbuilding terms of the time, was a piece of frame or rib. For smaller ships, sometimes every fourth frame became the frame bend. To avoid confusion, frame bends were identified by letters forward of midships and numbers aft. Because of their importance, the only numbers and letters shown on the draught were those belonging to the frame bends, i.e. 2, 5, 8, etc. It was normal practice in the seventeenth century to use the Latin alphabet with the letters J and U omitted, a practice seen in samplers of the period.[34] The shape of the hull in midships was more constant than anywhere else along its length; its moulds would be valid for more frames than normal.[35] On the draught, the lettering and number system of these midship frames, called flats, were marked inside a circle to avoid confusion with all the other frame bends. The Purveyor of Timber for Portsmouth, Thachery Medbury, mentions "between frames 2 and 11 and B through K",[36] suggesting nine frames formed the midship flats. This number and lettering system is used in the reconstructed draught. The Wilton House draught shows a similar method, as does a draught in *Deane's Doctrine*.[37] The frame timbers that crossed the keel – the floor timbers – were 1ft 1¼in fore and aft and spaced at every 2ft 3in, a distance known as the "space of timber and room".[38] The frame bends, spaced at three times the space and room, were therefore a distance of 6ft 9in apart.

Looking at the broadside view forward of midships, the floor timbers lay to the forward side of the frame bend stations, where the moulded outside profiles were drawn. This moulded profile was marked out on the floor timber before it was cut. The floor timber could then be cut square to the moulded shape, in the safe knowledge that an excess of material would be left on the forward side toward the narrowing bow. This condition was called "bevelled under".[39] However, if the floor timbers lay to the forward side of the frame bend station aft of midships, then the bevel allowance would have to be added to the moulded exterior profile, a difficult and time-consuming task. In order for the "bevelled under" condition to be maintained aft the midship flat, the frame bend station was moved.[40] There were a number of ways in which this could be done and no doubt individual seventeenth-century shipwrights had their own particular method. The distance between the midship flat and the first frame bend station forward, or aft, could be increased, or decreased by a timber width,[41] or two futtocks placed next to each other in midships.[42] Having two futtocks next to each other looks untidy and loses the strength of having overlapping timbers; however, it does mean the frame bend stations are evenly spaced.

With the broadside view finished, Shish would have turned his attention to the end views. They were drawn inside a box, the same width as the ship to the outside of the frame. The shape of the hull frames consisted of a series of tangential arcs and a straight line. It was usual practice to draw the fore body on the right hand side of the draught and the after body on the left. It was only necessary to show one side of the ship, as the

Below: *Frame letter and number system. The floor timbers at stations 5, 2, (3), (X), (C), B and E are the frame bends and are the only ones drawn on the end views of the draught. The floor timber was always marked out to the mould to leave an allowance for bevelling under. Note the reduction in distance of one timber between the midship flat (X) and the frame station at (C) to allow this.*

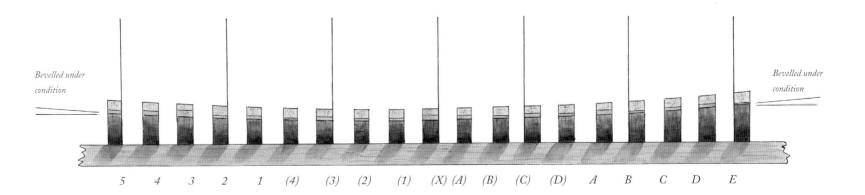

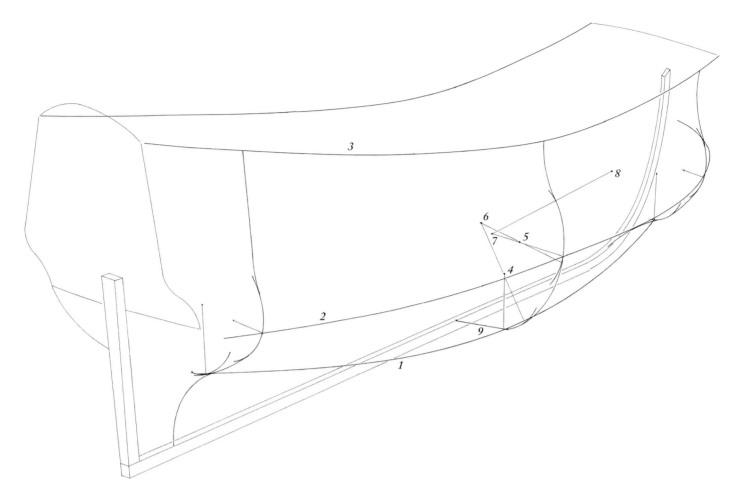

Above: *Construction lines. Three rising and narrowing lines controlled the shape of the hull at the floor, the breadth and the top of the side. Viewed from above they were called the narrowing lines and from the side the rising lines. The floor rising and narrowing line touches the bottom of the sweep and the breadth rising and narrowing line touches the side at the widest part of the hull. Greater control of shape was achieved by altering the radius of the breadth sweep along the length of the ship. Toward the bow and stern the straight floor line extending from the keel became a reverse sweep. Above the breadth line toward the stern the tumble home sweep increased in radius to become a straight line. For clarity the diagram shows only 3 frame bends instead of about 20.*

(1) Rising and narrowing line of the floor
(2) Rising and narrowing line of the breadth
(3) Rising and narrowing line of the top of the side
(4) Centre of the floor sweep
(5) Centre of the breadth sweep downward
(6) Centre of the reconciling sweep
(7) Centre of the above sweep
(8) Centre of the reverse sweep generally called the tumblehome
(9) Floor line

opposite side was of course a mirror image. The two most important arcs were at the maximum breadth, called the breadth sweep, and another lower down near the keel, known as the floor sweep. As the hull became smaller towards the ends of the ship, the position of the two sweeps was changed by theoretical rising and narrowing lines, previously drawn on the broadside and half breadth plan. The radius of the breadth sweep, which only struck downwards, also changed in accordance with the desires of the shipwright. Towards the stern, the breadth sweep was reduced in radius to create a narrow run aft leading toward the rudder, while at the

same time maintaining a broad stern above for a wide gun deck.[43] The breadth and floor sweeps were faired together by a large sweep, with a radius a little more than half the breadth of the ship.[44] In the midships a straight line extended from the keel to touch the lower edge of the floor sweep. Towards the ship's ends, this straight line was replaced by a reverse curve. Above the waterline, at the breadth, another sweep, somewhat larger than the breadth sweep, struck upwards.[45] Touching the rising and narrowing line at the top of the ship's sides and tangential to the upwards breadth sweep was a reverse sweep called the tumblehome. The radius of the tumblehome grew larger toward the stern until it gradually became a straight line.

The curved rising and narrowing lines, often arcs, and the change in the diameters of the sweeps, followed geometric forms that could be calculated by the use of arithmetic, a practice some shipwrights may have been employing since at least the mid-sixteenth century. It was therefore possible when marking out the full-size moulds to find the precise positioning of the floor and breadth sweeps without resorting to scaling from the ship's draught.

In 1674, John Shish gave Pepys the dimensions of rising and narrowing lines for all the frame stations of a fourth rate. They are a combination of arcs and curves that had been calculated using formulae based, for the most part, on Pythagoras' theorem – that is, that the square of the hypotenuse is equal to the sum of the square on the other two sides. To complete his draught, Shish would have plotted the waterlines onto the half plan view. This was a recent innovation which gave the shipwright a better understanding of the hull lines than could be seen on either the broadside or end views.[46]

Great secrecy surrounded the design of ships' bodies and being accused of taking unauthorised copies, as Edmund Dummer discovered, could place one in great jeopardy. However, the principles of ship design had evolved over a considerable time and were probably known by a great many people. Henry Sheeres, the prominent engineer, noted that every workman had access to the rising and narrowing dimensions, as they were all marked out on the moulds, marking-out rods and on the ship timbers themselves.[47]

By using the same well-known principles, it is possible to design a ship that sails either very well or very badly. It was probably the hard-won practical lessons of using the principles to the best advantage that were the secrets of the master shipwrights. They were used by Anthony Deane as a series of proportions based on the keel length.[48]

If ship design was something of a secret art, then putting the timbers together was not. At Deptford alone, more than 100 apprentice-trained shipwrights knew exactly how this was done. The relationship between frames and plank is fundamental in understanding seventeenth-century ship construction. The frames were not solid structures that carried the plank; in fact, most of the nine individual overlapping timbers that formed each frame of *Lenox* were not joined together at all and would fall to the ground if not held in place by the planking. The frame timbers were held in place by the plank, as much as the plank was held in place by the frames.

The ship's structure was highly stressed by the weight it carried, making it necessary for ship components to be made as light as possible.[49] The main gun-deck beams, for example, had to be strong enough to span 37ft across the inside of *Lenox* and to carry the main gun battery. The beams of the upper deck could be shorter and were reduced to 31ft by the tumblehome. The width and depth of these beams, the knees that secured them and all other associated timber, were reduced in proportion to its lighter duty. Many other practical considerations included making sure the guns did not interfere with the whipstaff and bitts on the main gun deck and cookroom furnaces and masts on the upper deck. Some ships were fitted with catheads made from a straight piece that went athwart the forecastle, interfering with the chase guns going forward. Surprisingly, some ships had chainplates fitted so that the main rigging supporting the masts and the shrouds covered the gun ports.[50]

Below: *The keel scarph. The tabled scarph of the keel based on the wreck of* Dartmouth *whose keel probably dates from 1678.* *(Courtesy of Colin Martin)*

THE KEEL

About 13 June 1677, only two weeks after Shish completed his draught, a warrant arrived at Deptford ordering the building of two new third-rate ships.[51] By that time he had already started preparing the dock floor and negotiating the purchase of keel pieces[52], presumably having heard that King Charles commanded the laying of keels on the 9th.[53] The great double dock had already been cleared of two old ships, the *Assistance* and *Tiger* to make space for one of them. Large wooden foundations, called groundways, spread the weight on the floor of the dock. On top of the groundways were blocks of hard knotty stuff, raising the hull off the ground so that men could work under the bilges. On top of the blocks were the splitting blocks, pieces of straight timber recessed by 1½in to secure and locate the keel.[54] This straight timber would eventually be cut away when the ship was ready for launching.

John Shish heard that close by, in the merchant yard at Deptford, William Castle had three large keel pieces suitable for the new ships. After first obtaining permission from the Navy Board, he went round to see Castle and inspect his great elm pieces.[55] Although oak could be used, elm was the usual timber for making keels, for not only did it grow very tall and straight, it did not rot when permanently wet. Even beech, a timber that rotted quickly if sometimes wet or dry could be used for the keel.[56] For a ship the size of *Lenox*, keel pieces were extraordinarily difficult to find, being 15in up and down and 16in broad in midships[57] and around 39ft long each.[58] The length of Castle's pieces would indicate that four such pieces would be used on the new ship.[59] This was not universal as one of the ships at Chatham, probably the *Anne*, had five keel pieces.[60] Keel pieces had to be free of defects, as they were very difficult and costly to replace after a ship was completed. To make matters worse, it could not be seen if elm had defects until it was cut down and squared.[61] When keel pieces were cut to size by the sawyers at Chatham, the fear of finding defects prompted the Commissioner to write to the Navy Board: "It will be the late end of next week before they be cut, God send they prove well".[62] Indeed, Castle's elms were no exception, as one of the pieces was found to have defects and was rejected. The remaining "two are very fit for His Majesty's service, one a forefoot [the foremost piece that joins the stem] and the other, which is sawn and converted to his scantlings, for a midship piece for a second or third-rate ship".[63]

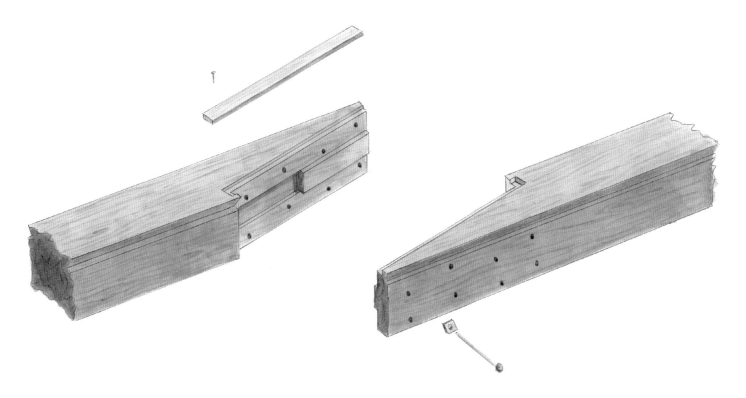

Shish was desperate for Castle's keel pieces and went to see him again a few days later, but had to wait until 11 o'clock at night before he turned up. Afterward he reported that they were the best pieces in the river, or anywhere else that he had heard of. One great elm would make a midship piece 50ft long, and the other 40ft.[64]

The importance of the keel pieces meant that a great deal of trouble was taken over them. Probably the most skilled, or perhaps the most careful shipwrights would have been entrusted with the work. The keel pieces were sawn[65] to their finished scantling and tapered a few inches athwart ship toward the bow and stern. The scarphs were 4ft 6in long, cut in the vertical plane and "tabled" to lock them together.[66] The forefoot keel piece had a greater depth at its forward end as it led into the upward curve of the stem, making its tabled scarph even more complex than normal. As well as the scarphs, a rebate or "rabbet" was made to take the "garboard strake", the plank nearest the keel. Finally, three mortises were cut at the aft end of the keel to take the tenons of the sternpost.

After the careful carpentry, the keel pieces were scarped together.[67] The tables were caulked with resin, the vertical flat faces with hair and tar and the butts at the extremes of the joint caulked with ocum.[68] Each scarph was bolted together from side to side with eight 1¼in diameter bolts,[69] staggered slightly to prevent splitting along the grain.[70] The bolts would probably have been driven from alternate sides to even out any weakness, then riveted or "clenched" over washers or "roves". This type of bolt was commonly used in ship building, but it could not be tightened nearly so much as a threaded bolt, which in those days was far too expensive to use. In ships built of green timber, as *Lenox* was, the problem became much worse as the timbers dried and shrank, leaving the bolts loose to protrude up to ¼in beyond the wood.[71] To prevent water seeping upward, a sealing board was nailed into the top of the scarph caulked with tar and hair.[72] The underside of the keel was protected by a false keel made from 4in thick[73] elm plank,[74] fastened to the keel by treenails (wooden pins).[75] The false keel protected the keel against damage during docking, grounding, from shipworm, and preserved the caulking of the keel scarphs.[76] Surprisingly, it was only after the scarphs were bolted together

and the false keel fitted that the assembly was lifted onto the blocks.[77] Finally, the positions of the frames stations were carefully marked on the upper side.

THE STEM AND STERNPOST

While the keel was being constructed, some shipwrights were busy hewing *Lenox*'s frames, stem and sternpost. Up to the time of building *Lenox*, ships had a circular stem with sharp bows and consequently little forward buoyancy. This tended to make ships pitch excessively and to overcome the problem, the stem was made more upright and the keel lengthened. Under the old system, *Lenox* would have had a rake and stem radius of 30ft.[78] Now there were two radii leading one into another, the lower of which was tangential to the keel and about 11ft diameter.[79] The stem was constructed from compass timber,[80] and consisted of two parts, the upper and lower pieces.[81] They were very large, being 26ft long and rounded to 2ft 7in.[82] The lower stem piece had a tabled scarph that fitted into the keel forefoot. After hewing to size it was erected by 13 July and the upper piece added a day or two later.[83] The two pieces of stem were scarped together with a 4ft long[84] plain scarph and fastened with treenails and clench bolts.[85] The false stem was then closely joined, or "fayed" to its inside surface. It was not often commented upon by Master Shipwrights and probably consisted of three easily obtained shorter pieces, even though they were 6in wider[86] each side than the stem itself. Its purpose was to overlap the stem scarphs and provided support for the butt-ends of the plank, properly called the hooding ends, which fitted into the rabbets of the stem.[87]

As the stem was being erected at the fore end of the keel, the sternpost was simultaneously erected aft. It consisted of a straight, single great piece of oak tenonned into the keel. According to William Sutherland, it was the opinion and the practice of Jonas Shish and his sons to rake the

Below: *Building progress, 14 July 1677*
The four pieces of keel are assembled and the stem and sternpost erected. The frame of the stern is fitted to the sternpost and two floor timbers are across the keel.

Inset: *Wooden staffs used to check the rising and narrowing lines of the ship during construction. From William Sutherland,* The Shipbuilder's Assistant, *1711, p 82.*

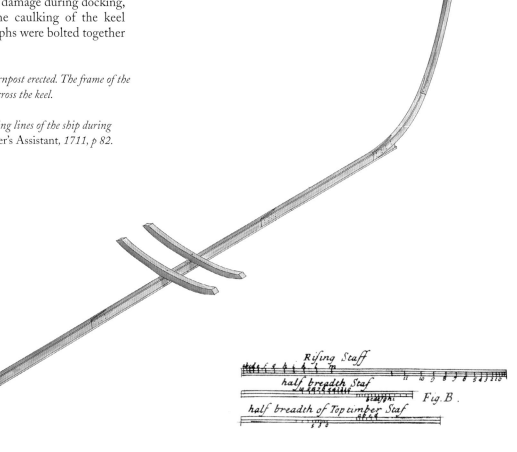

sternpost aft 2in for every foot of height.[88] It was vital to make sure the stem and sternpost were not twisted and lay truly in line with the keel. To ensure this alignment, a ram line was set up between them and a centre line marked along the middle of the keel. The structures now had to be secured against any movement and this was accomplished by setting up shores. The bases were "nogged and trigged"[89] by wedging and fixing with treenails into the groundways. Many of the shores used during the construction of *Lenox* were Gothenburg masts, but in order to make use of old materials, a decayed bowsprit and the mast of the *Cleveland* yacht were used as well.[90]

THE FRAME OF THE STERN

The day after the sternpost was fixed safely in position, the frame of the stern was raised and bolted to it.[91] The frame of the stern assembly consisted of the false post within and the four lower transoms fitted to it. Another transom, the helm port transom, was then fitted; it was rebated to take hold of the sternpost and formed the upper sill of the four chase gun ports.[92] Below this, at the widest part of the stern, the biggest transom of all, the wing transom, was fitted. It formed the lower sill of the chase gun ports and was scored over the sternpost.[93] The wing and helm port transoms were at the same level as the upper and lower main wales, so that later in construction, horizontal knees could be bolted to join them all securely together.[94] To further secure the structure, chocks or half transoms[95] were fitted between the full transoms. Months later, as the ship neared completion, the false post without[96] was fayed to the aft end of the sternpost. It probably consisted of 4in plank and performed the same function as the false keel.[97]

During the raising of these heavy structures great care had to be taken, not only for the safety of the shipwrights but for the timbers themselves. The lifting gear consisted of sheers at each end of the ship, from which blocks and tackles were suspended.[98] During August, Shish requested a supply of eighty small masts of nine, ten and eleven hands in size for standards[99]. These were used as vertical supports around the ship for stages; in effect a type of wooden scaffolding.

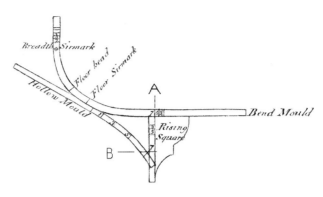

Above: *Although this whole mould is for a boat, the same principle for moulding the floors of large warships was still used in the late eighteenth century. The whole mould for the floor timbers consists of three parts, the bend mould, the hollow mould and the rising square. It is set up for marking out the floor timber at station 7. The rising square is laid on the vertical centre line of the floor at 'A' with the horizontal line 'B' level where the top of the keel would be at the 7 mark. The bend mould is then laid against the rising square at position 7 and the floor sweep, floor sirmark and floor head marked on the floor timber. The rising square and bend mould can then be removed and the hollow mould placed with the position 7 at the keel intersection and tangential to the already marked out floor sweep. The hollow mould is then marked on the timber. The moulds are then turned over and the process repeated for the opposite side of the floor timber or perhaps a left and right hand mould was made. From* Encyclopaedia Britannica, *1797. (Author's collection)*

MOULDING THE FRAME

By this time a total of 119 men were working on *Lenox* in the double dock and *Hampton Court* on the launch. These included fifty shipwrights, three house carpenters, thirteen pairs of sawyers and forty labourers.[100] Many

Below: *Fore and Aft Body. Note the positions of the sirmarks and heads and heels of timbers. Prints taken from* The Bends of a Ship, *by Thomas Fagge, c. 1680. His work almost certainly relates to the thirty ships of 1677. (Private collection)*

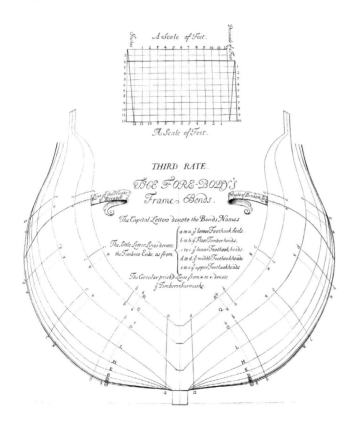

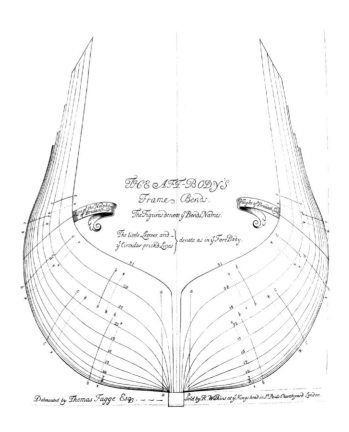

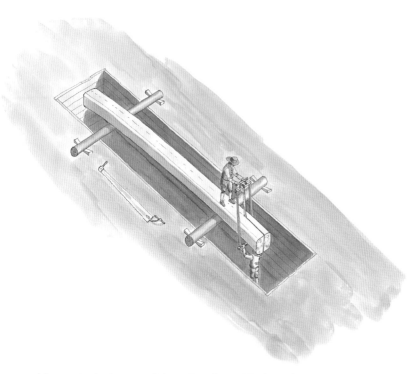

Above: *A pair of sawyers siding a floor timber. The 'iron' pit saw shown in use required a frame for support, but during* Lenox's *construction John Shish requested steel saws, shown alongside the pit, as they did not dull in such a short time.*

would have been making frame timbers according to the frame and floor timber moulds. Each frame was made up of a number of individual timbers, consisting of the floor, two lower futtocks, two middle futtocks, two upper futtocks and two top timbers. The lower and upper futtocks overlapped (more properly known as scarphed) the butt joints between the other timbers.

It was the responsibility of the Master Shipwright to order the "bends of moulds".[101] They were produced from Shish's draught in the mould loft, a building on the opposite side of the double dock to the Master Shipwright's house and just about long and wide enough for the draught to be copied full size. Mould lofts had many windows for light and a large flat floor washed with black size so the lines could be marked out in chalk. The drawing process started with broadside and half plan views, both the

full length of the ship and probably overlapping to save space. The rising and narrowing lines in particular would be carefully drawn according to calculations, if they had been made, and any small corrections taken in that could not be seen on the 1:48 scale drawing. The dimensions of the rising and narrowing lines were then carefully recorded on staffs (rods). Although there is little evidence for moulds being made for the rising wood, stem or other parts of the ship in the broadside view, it is quite probable they were. They would have been made from dry, seasoned fir or pine, generally called deal, about 3in wide, planed very smooth and nailed together.

The chalk drawing on the mould-loft floor was then rubbed out and the end views drawn, according to the draught, but using the staffs to position the rising and narrowing lines at full size. The first part of the frame that required moulding was the floor and rising floor timbers. The mould was almost certainly universal and would suffice for every floor. It would have consisted of a two-part frame and a square, with the risings and narrowing of the floors, at frame bend stations marked on them. By moulding every floor timber, the accuracy of the foundation of the ship was assured.

Reference marks known as "sirmarks" were placed near the heads and heels of every timber, so that on assembly, the sirmarks on one timber could be lined up with those of the next to ensure their correct positioning. Marks were also made at bevel positions, where the fore and aft ribbons were temporarily nailed to the frame timbers. The ribbons ensured the frames lined up properly and stayed in position before they were planked.[102] Looking on the end view, the sirmarks were generally all at an angled straight line, following the planking.

The bevel allowance at the sirmarks could be determined on a drawing in the mould loft, similar to the waterlines on the half-breadth plan, the difference being an angled plane passing through the bevel positions. This replaced the horizontal plane intersecting the hull surface. During the eighteenth century the planes seen on the end view were called "diagonals" and were always straight lines. Predictably, seventeenth-century practice did not always follow such rules and curved lines were sometimes used.[103]

Below: *Floor timber assembly. The stages of fitting and bolting floor timbers: (1) Every other floor timber is placed in position and bolted through the keel. (2) The ribbon is brought about at the sirmark. (3) The remaining floor timbers are placed between those already in position. (4) The keelson is scored down over all the floor timbers and alternate floor timbers bolted.*

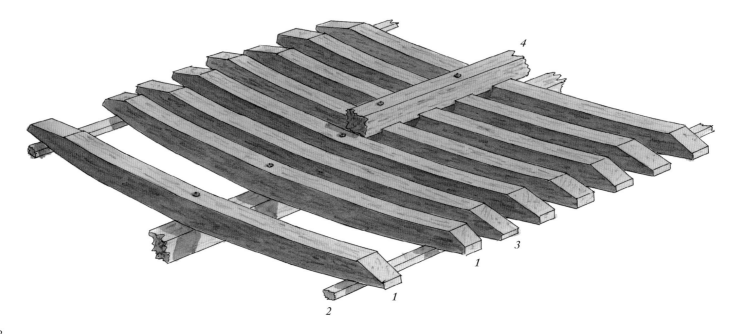

The in-and-out dimension of the timber was usually obtained by drawing two straight lines, the length of which was the same as the distance along the hull profile from the keel to the top of the side. At one end the distance between the two lines was the thickness of the frame at the keel and the other the same as the thickness at the top of the side.[104] The in-and-out dimension along the straight lines would be recorded at known distances from one end and marked at similar distances on the timbers round the hull profile.

The mould for the first futtocks may have followed the same principle as the floor timber mould. Indeed, as the floor sweep and the reconciling sweep were constant radii, a universal mould may have served for most of the first futtocks. The only variable was the hollow mould that could have been produced following the principle of the hollow mould for the floors. It may be that simple moulds were made to serve two futtocks; the outside edge one and the inside edge another.[105] In 1711, William Sutherland noted that he once observed a mould made for every timber from keel to breadth,[106] indicating this was unusual and there must have been a variety of forms the moulds might have taken. The radii of the downward breadth sweep changed towards the ends of the ship, affecting the middle and upper futtocks moulds. This must have resulted in more individual moulds, possibly a mould was made for each frame station which was long enough to incorporate both futtocks. It is a possibility that moulds were made for the frame bends only and the fill-in timbers each side of it were made to the same mould. As the ribbons located the fill-in frames, any error would have been small. Any mould, however it worked, would have been thoroughly checked against the chalk drawing in the mould loft before being used to mark out the ship's timbers.

FLOOR TIMBERS

To avoid waste, great care was taken in selecting the logs from which the floor timbers were made. They were chosen so the grain matched as closely as possible the moulded profile and were no longer than necessary. Rather, a shorter log was used and any deficiency made up with a chock.[107] After selection, the logs were taken to the sawpits and sawn to their fore and aft "sided" dimension. The sided timber was then marked out according to the floor mould. As we have seen, whether the floor timbers were fore or aft the midship flat, the outside profile would be bevelled under, allowing the shipwright to cut accurately along the marked line.

Below: *Moulding squared timber with an adze.*

Below (top to bottom): *(a) The rising timber at frame H. (b) The rising timber at frame O. (c) Floor timber at frame 8. (d) Rising timber at frame 20. (e) The aft most rising timber at frame 26, showing how it may have been made from more than one piece of timber. (f) The rising timber at frame 14. (g) Half timber at frame 29. (h) The adjoining futtock at frame 29. (inset) Frames from above, showing how the timbers scored into the rising wood.*

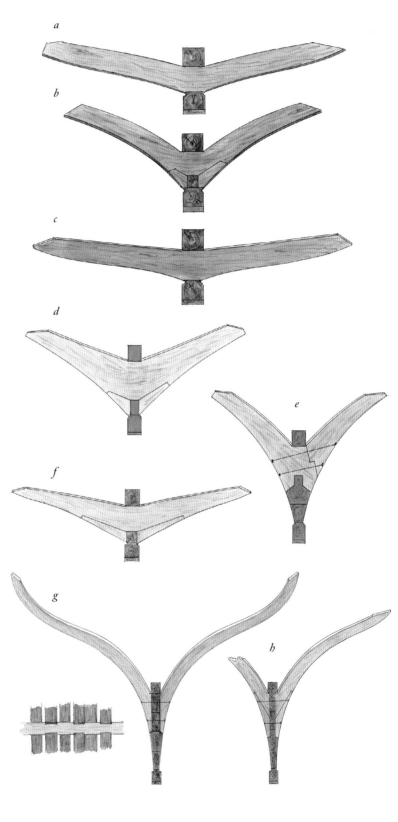

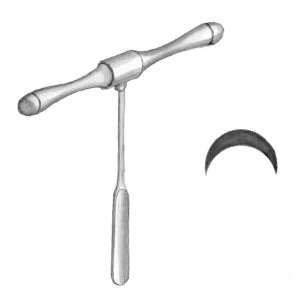

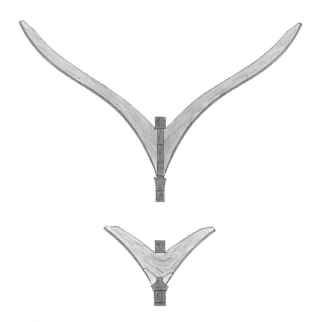

Above: *An auger for drilling treenail holes. It consisted of the handle and detachable bit. During the construction of* Lenox *4 extraordinary bits and 120 ordinary bits were used. Extraordinary bits were 6in long and ordinary 5in long.*

Above: *(Top) Half timber at frame V. (Bottom) Rising timber at frame R.*

Even if he did not cut the bevel angle but left the profile square, he would always leave a safe trimming or "dubbing" allowance. The situation for the inside profile was rather more complicated, for the shipwrights had to take account of the bevel allowance and the moulded thickness. This was not a serious problem, because, of course, the inside profile was not as critical as the outside. Typically, the bevel for the floors was 2in, but ranged from nothing in midships to about 4in near the bow and stern. The position of the sirmarks was taken from the mould and marked on the floor timbers. After the inner and outer moulded profiles were sawn, the bevels at the sirmarks for the ribbons were accurately cut.[108] This accuracy was only necessary where the ribbons were attached to the timbers.[109] Two limber holes were also cut on the underside of the floor timber, each side of the keel, for water to pass through to the pumps.[110]

The first two floor timbers were laid across the keel at the time the stem and sternpost were raised.[111] At this stage only every other floor timber was fitted and bolted through the keel[112] with bolts of 1¼in diameter.[113] Holes for the bolts would have been augured through the floor, keel and false keel from above. The false keel may have needed replacing at a later date and so it was probably rebated through to the underside of the keel. The bolts were then driven in and clenched. The largest floor timbers were 23 ft long and weighed the best part of a ton. When handling this weight in position, great care had to be taken not to tear the rabbet of the keel.[114] When shipwrights used the term "floor timbers", they generally meant all the lowermost timbers that crossed the keel, but floor timbers proper covered less than a third the length in the midship area, the rest being called the rising timbers.

RISING TIMBERS

As the hull changed shape towards the bow and stern, the floors began to rise more steeply and compass timber was needed to make them.[115] At Chatham, these "rising timbers", as they were known, would serve at least between frames 2 and 11 aft and frames B and K forward.[116]

RISING WOOD AND KNEE

A week after the frame of the stern was raised; the junction between the keel and sternpost was strengthened by a huge knee with 8ft long arms[117] and the first piece of rising wood.[118] The knee was bolted with one arm against the sternpost and the other directly on top of the keel.[119] During

the same week, the fashion pieces were fitted to the outer forward edges of the transoms and, despite having double curves, were probably made in one piece and joined by treenails alone,[120] a practice that continued until at least 1737.[121] Although not strictly necessary, the knee was sometimes joined directly to the rising wood by a small length of timber.[122] Two weeks later a second piece of rising wood was fitted above the knee and another piece at the forward end of the keel.[123] At the aft end, near the sternpost, the rising wood was made wider and shaped to follow the outside profile of the hull up to a height of about 6ft so the outside planking would be in direct contact with it. Above that, for approximately the last ten frames, the rising wood continued upwards vertically for another 4ft. In its wake, the hull profile rose so steeply that rising timbers crossing the keel could not be used and no frame timbers were fitted for the moment. The solid rising wood ensured that if the aft part of the keel were torn away in a mishap, the ship would remain watertight.[124] A similar manner of construction was used at the forward end of the keel, but did not rise so high.

With the rising wood in position, the more extreme rising timbers were placed across it, about as far as the last ten frames from the stern. Between these rising timbers and the keel it became necessary to fit chocks against the rising wood[125] as filling pieces to take the hull profile down toward the keel.

FLOOR RIBBON

At this stage a ribbon, often made from an old mast,[126] was nailed to the bevelled heads of the floor timbers near the sirmarks, to fair the hull and provide support for the frame timbers.[127] The temporary ribbons would eventually be taken off when the outside plank was in place. For this reason care was taken that they followed the line that would later be taken by the planking. They were securely shored with "very able" pieces of timber because of the weight they would have to bear.[128]

HALF TIMBERS

With the ribbon in position, attention turned to the aft vertical rising wood where it was impractical to have timbers crossing the keel. Here, half timbers were fayed to each side,[129] scored down by about an in. This arrangement left a step in the rising wood at the transition; it also meant the half timbers could reach higher up the side of the hull. The half timbers were put in after the ribbon, making it probable they were made to fair in with them as well as be answerable to a mould.[130] There was no universal practice of making half timbers. Indeed, after the old third-rate *Dunkirk* was rebuilt in 1697, it was reported that she had no half timbers at all. Her "after floor being whole timbers standing on chocks on tiptoe upon the dead wood (having no half timbers)".[131] When in place, the half timbers were secured to the rising wood with bolts from one side to the other. It was only seven weeks since Shish had received the warrant to build his two ships and yet surprisingly, only thirty-seven shipwrights and sixty labourers were employed upon them at the time.[132] Progress was all the more commendable as August was extremely wet, which hampered progress at Chatham.[133]

THE KEELSON AND REMAINING FLOOR TIMBERS

As soon as the floor ribbon was secured, the remaining floor timbers were placed between those already bolted down through the keel.[134] They were kept in position by the ribbons and probably wedges, but not bolted down until the keelson was fayed upon them. The timber required to make the keelson was 16in up and down and 17in broad,[135] about the same size in section as the keel, but because it was made from five, rather than four pieces, they were much easier to find. The scarph was set in the horizontal plane, in opposition to the keel itself and for strength's sake the scarphs were set between those of the keel scarphs.[136] The keelson was scored down onto the floors by about 1in to prevent any movement.[137] On 27 July 1677, John Shish was able to report that two thirds of the floor was in and two pieces of keelson placed upon it.[138] Once the keelson was in place, the alternate floor timbers not yet bolted down were augured through and secured with long bolts to the keelson and keel.[139]

FRAME BEND FUTTOCKS

Before all the half timbers were fayed to the rising wood, John Shish had already assembled the frame bend futtocks and erected them on the floor,[140] a feat accomplished within two weeks. The timbers that made up

Below: *Fig A shows the floor timbers placed across the keel. O is a ram-line. D are standers for the stages. a is the splitting blocks. Fig B shows the frame bends and ribbons supposrted by shores. Sutherland,* The Shipbuilder's Assistant, *1711, p27.*

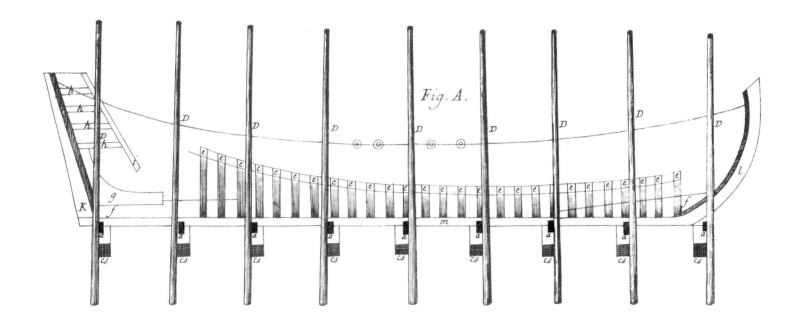

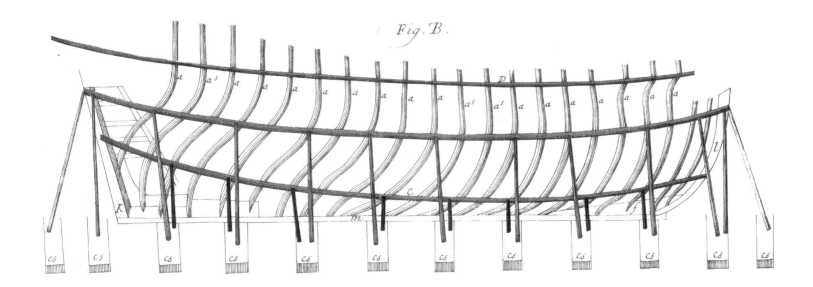

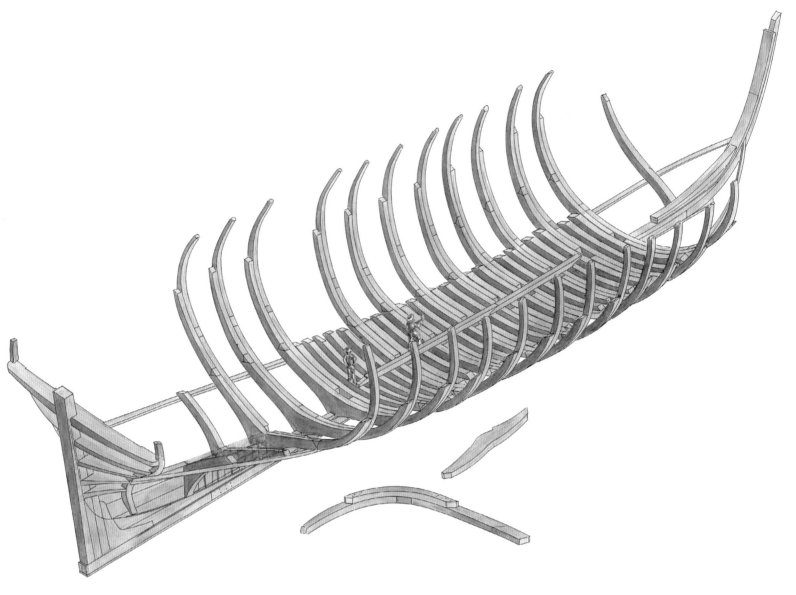

Above: *Building progress, 10 August 1677*
The floor timbers are in and the floor ribbon is about. For clarity some of the rising timbers are not shown and one is on the ground. The rising wood on top of the keel is completed and the frame bends, except one, are raised. Note how the arrangement of frame bends fore and aft the midship flat are reversed so all the floors and middle futtocks are bevelled under.

Right: *Floors across, first frame, cross spawls. This view shows the sheerlegs for lifting the frame bends in position and the cross spawls for securing them at the correct distance apart. Shores are placed under the ribbon supporting the floor. Note the crabb capstans providing power to raise the frame bends. From Thomas Milton's* Geometric Plan of Deptford Dockyard, *1753.*

the futtocks were "sided and moulded" from compass oak trees, some of which were large enough to make "four futtocks and four top timbers in a piece".[141] The frame bend futtocks were made as pairs, one for each side of the ship and consisted of lower, middle and upper futtocks. They took the framing on both sides of the hull up to its maximum breadth, leaving only the toptimbers to go in above them. As with the floors, the futtocks were bevelled accurately at the sirmarks in the wake of the ribbons. The middle futtocks, which would stand on the head of the floor timbers, were

8in

4¼in

TOPTIMBER

1ft ¼in

*UPPER FUTTOCK
(ALSO KNOWN AS
GUNDECK TIMBER
OR THIRD FUTTOCK)*

11½in

10¼in

6ft 6in

*MIDDLE FUTTOCK
(ALSO KNOWN AS
SECOND FUTTOCK
OR SIMPLY FUTTOCK)*

6ft 8in

CHOCK

1ft 1¼in

*LOWER FUTTOCK
(ALSO KNOWN AS
NAVAL TIMBER OR
FIRST FUTTOCK)*

FLOOR TIMBER

1ft 1¼in

1ft 4½in

KEEL

1ft ¼in

2ft 3in

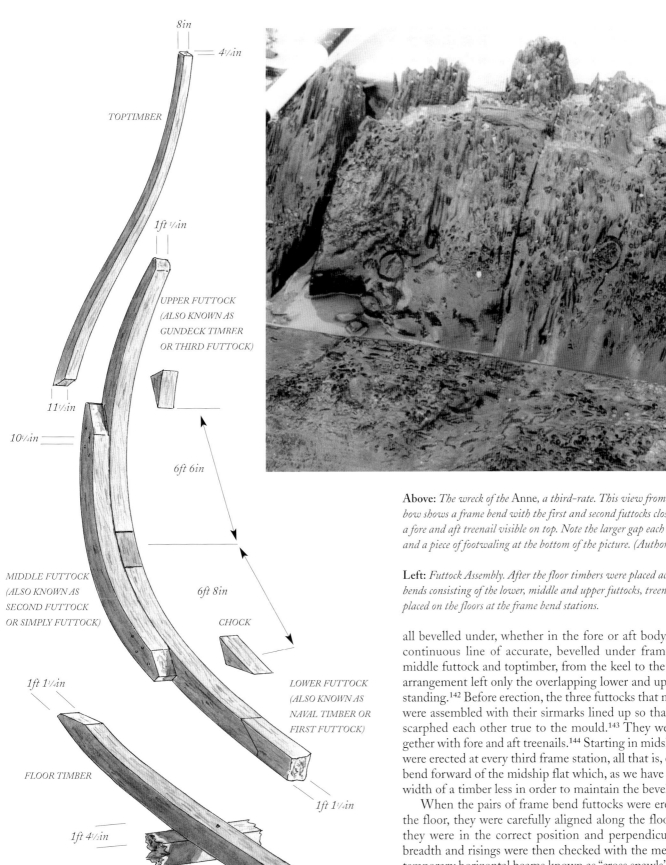

Above: *The wreck of the Anne, a third-rate. This view from the inside of the port bow shows a frame bend with the first and second futtocks closely joined together with a fore and aft treenail visible on top. Note the larger gap each side of the fill-in futtocks and a piece of footwaling at the bottom of the picture. (Author's collection)*

Left: *Futtock Assembly. After the floor timbers were placed across the keel, the frame bends consisting of the lower, middle and upper futtocks, treenailed together, were placed on the floors at the frame bend stations.*

all bevelled under, whether in the fore or aft body. This would make a continuous line of accurate, bevelled under frame through the floor, middle futtock and toptimber, from the keel to the top of the side. This arrangement left only the overlapping lower and upper futtocks bevelled standing.[142] Before erection, the three futtocks that made up a frame bend were assembled with their sirmarks lined up so that they overlapped, or scarphed each other true to the mould.[143] They were then fastened together with fore and aft treenails.[144] Starting in midships, the frame bends were erected at every third frame station, all that is, except the first frame bend forward of the midship flat which, as we have already seen, was the width of a timber less in order to maintain the bevelled under condition.

When the pairs of frame bend futtocks were erected on each side of the floor, they were carefully aligned along the floor sirmarks to ensure they were in the correct position and perpendicular to the keel. The breadth and risings were then checked with the measuring staffs, before temporary horizontal beams known as "cross spawls"[145] were nailed across to secure them. The breadth ribbon was then nailed in place at the position of the upper wale, under which shores were placed to prevent any movement.[146] To help with the positioning of timbers around the gun deck ports, their positions were marked out on the ribbon.[147] The semicircular curve of the ribbon at the bow, known as the upper false harping, was probably too severe to bend a piece of fir around, and compass oak,

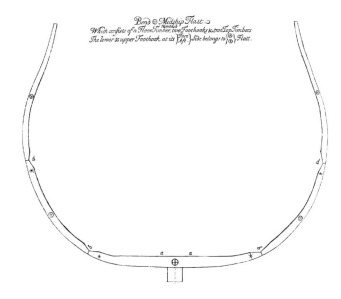

Bend ⊙ Midship Flatt
Which confists of a Floor Timber, two Foothooks to two Top Timbers
The lower to upper Foothook, at its Fore Side belongs to Flatt.

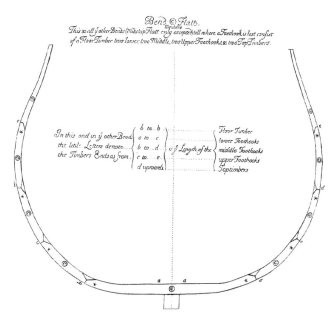

Bend ⊙ Flatt.
This is all y other Bends (Midship Flatt only excepted) till where a Foothook is left confist
of a Floor Timber two lower, two Middle, two Upper Foothooks to two Top Timbers.

In this and in y other Bends
the little Letters denote
the Timbers Ends as from
b to b
a to c
b to d
c to e
d upwards
y Length of the
Floor Timber
lower Foothooks
middle Foothooks
upper Foothooks
Toptimbers

Above: *Frame timbers up. The entire floor is in and the frame bends up and shored. Some loose futtocks are on the ground waiting to be placed in the frame. From Thomas Milton's* Geometric Plan of Deptford Dockyard, *1753.*

Left: *(Top) The Floor, Middle Futtock and Toptimber make a continuous line of bevelled under timbers. (Bottom) Midship Flat. Both taken from* The Bends of a Ship, *Thomas Fagge, c. 1680. (Private Collection)*

on the middle tier higher up the frame. Stages were erected to enable the shipwrights to work in safety further from the ground.[151] The middle futtocks were placed directly above the floor timbers and their heads reached as high as the widest part of the ship, just below the upper main wale.[152] At their heads they measured 11½in fore and aft, leaving 4in gaps between the timbers, and scarphed the upper and lower futtocks by a nominal 6ft 8in.[153] After the middle futtocks were in place the final tier of futtocks, the upper futtocks, were put up into the frame. Progress on the ship went very smoothly and it took only four weeks between the erection of the frame bends and filling in with all the other futtock timbers. This orderly progression of work was exceptional; at Chatham the fitting of the tiers of futtocks was not so even and they were put in as they became available; it also took place over a longer period of time.[154]

HARPING TIMBERS
The half timbers, not having to cross the keel, reached higher up the ship's side than the rising timbers. This made it unnecessary to have so many timbers to make a frame and as a result one futtock was eliminated on each side of the ship. The remaining timbers were called harping timbers by John Shish, although at the stern where the uppermost were very long and straight, they seem simply to have been called long timbers.[155] At Chatham, Phineas Pett II called the harping timbers "foreship double futtocks",[156] later they were all known as long timbers.[157] The bevelling of the harping timbers and half timbers at the bow was considerable and a great deal of timber and effort was required to produce them. The heels of the harping timbers fitted against the rising wood and were bolted to it in the same manner as the half timbers.[158]

CHOCKS
The frame timbers came in all manner of lengths and if a timber was longer than that given by the scantling list, it would not be cut shorter.

sawn to shape may have been used instead.[148] At this time the term harping meant a severe curve, usually at the bow. The harping timbers were frame timbers at the bow and the harping of the wale was the foremost piece of the wale. The curved lower counter timbers were also known as the harping timbers.[149]

LOWER, MIDDLE AND UPPER FUTTOCKS
With the frame bends rigidly secured, the lower tier of fill-in futtocks that went between them was laid in place. They were not fastened to any other frame timbers and only when the outside planking and the ceiling on the inside were treenailed to them would they add strength to the structure. The frame timbers were reduced in thickness, both in sided and moulded dimensions, the higher up they were in the frame. The lower futtocks were ¼in narrower than the gap between the floor timbers to allow air to circulate and dry any dampness. In order to maintain this gap until the frame was planked, the futtocks were held firmly in place by elm wedges. These were ordered in huge quantities; during August 1677, 800 elm wedges were ordered for Deptford. They were generally 4in square at the head and 18in long.[150]

As soon as the lower tiers of fill-in futtocks were in place, work started

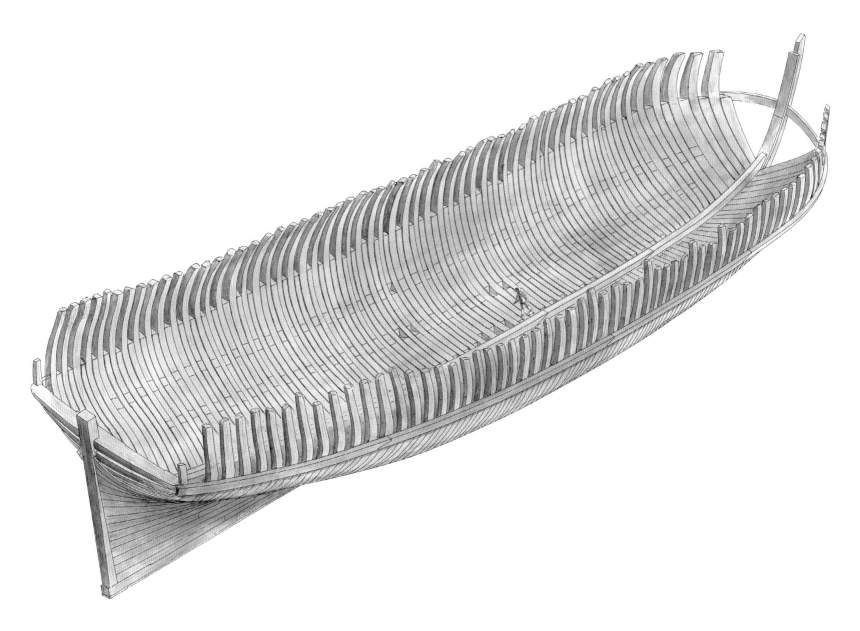

Above: *Building progress, 14 September 1677*
With the exception of the toptimbers the frame is complete with the keelson securing the floor. On the outside the planking has reached the lower futtock heads and the lower main wale is complete.

Instead, its extra length would be made use of to allow a shorter piece to be used above or below it. Alternatively, where a timber was short or not curved enough, the gap would be filled later with a large chock held in place by treenails.[159] In fact, chocks filled all the gaps between the heels and heads of timbers[160], making a solid frame from the maximum breadth down to the keel.[161] The sole exception was the heel of the first futtock, which finished a number of ft short of the keel. Consequently a chock was fitted to reduce the distance to about 20in[162]. This would still allow a gap for bilge water to pass through to the limber holes.[163] Phineas Pett I had introduced the practice of chocking the heel of the first futtocks when he built the *Royal Prince* in 1610.[164] The chock allowed the planking to be treenailed in the same manner as the rest of the hull underwater; the only exception being the garboard strake next to the keel, which could only be treenailed to the floor timbers. In midships, a number of chocks at the heels of the lower futtocks were left short to form a chamber so the chain pumps could reach down almost to the outside planking.[165] The chocks

were of irregular shape and made individually to suit gaps in the frame; it is doubtful if they were ideally shaped as shown in drawings of the eighteenth century.[166] Chocks were also fitted between frames in the wake of the chainplate bolts to take the thrust of the shrouds.[167]

PLANKING

On 17 August, as soon as the floor timbers were in and the frame bends up, John Shish began strengthening the outside of the hull with planking. Amongst the many things he had to consider for the sake of strength was keeping the butts clear of the scarphs of the keel. It was also important to keep them clear of the pumps, as there was a danger a butt may be started by the water flow or the ocum sucked out of the seam.[168] The process started with the outside surface of the floor being faired and trimmed smooth with the hand adze, a process known as "dubbing".[169] The frames of some ships were tarred, but this was not recorded as being done to *Lenox*, probably because her timbers were very green and would not dry out if sealed. The 4in thick planking near the keel and for 10ft upwards, which would always remain wet, was often made from elm or beech.[170] Although no beech plank appears to have been purchased for use on *Lenox*, a large amount of elm timber was, and this may well have been sawn into plank at Deptford. Planking started with the garboard strake

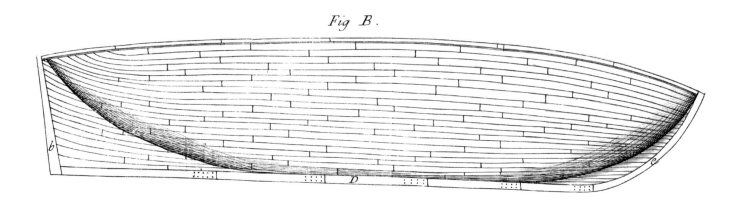

Fig B.

Above: *The underwater planking below the wale. The position of the butts are staggered to prevent weakness. From William Sutherland,* The Shipbuilder's Assistant, *1711, p49.*

that fitted into the rabbet of the keel[171], a strake being a number of planks that ran in line from one end of the ship to the other. The strake next to the garboard strake was left off at this stage to allow for the easy removal of wood chips.[172] A week later "two strakes were about and fast" and by the end of August Shish was able to inform the Navy Board that six planks aside were completed.[173] As the planking reached upwards toward the mainwales, the floor ribbons became redundant and were subsequently removed.

For a ship the size of *Lenox*, somewhere between twenty-three and twenty-seven strakes each side, would be required to birth her up to the underside of the lower wale.[174] Although the planks were all 4in thick they were not all parallel in width. The shape of the hull changed greatly along its length and the distance along the surface between the keel and the underside of the lower wale ranged from about 21ft at the bows, 25ft in midships and 30ft at the stern. If the same numbers of strakes were used at the stern as the bows, the planks would be unmanageably broad at the

stern, or unnecessarily narrow at the bows. In order to overcome the problem, two planks were merged into one, or one into two, as the situation required, a process known as to "steal" or "drop" a strake. The edges of the plank were made perpendicular to the hull surface, except for a small taper as an allowance for caulking. The ends of the planks were only butted together, but to prevent any weakness they were staggered by "shifting the butts" a minimum of 6ft away from the nearest in an adjoining strake, creating a pattern that tended to repeat every four or five strakes.[175] Many of *Lenox*'s planks were in the order of 30ft long but in order to shift the butt, at least one strake at the bow is known to have been only 11ft long. This plank, which fitted into the rabbet of the stem, was termed a hooding plank.[176]

Towards the stem and stern, plank needed to be made pliable by heat for it to bend to the desired shape.[177] To heat the planks for bending they were placed over an open fire of brush and broom bavins[178] and kept from burning by the occasional soaking with water. The hot planks were hauled up on the hull, then forced round the curve of the hull with wrain bolts and treenailed in position. A surprising number of eighty-three wrain bolts were purchased for the construction of *Lenox*.[179] When heated, East country 4in thick straight plank could bend a great deal in and out, but because of its great breadth was difficult to bend up and down. Instead, the plank edges were trimmed to shape in a process that remained unchanged for many years. In order to determine the developed shape, a thin staff, about 1/8in thick, 4 in wide and 20ft or 30ft long was bent round the frame of the hull where it was intended the plank would be fitted. Making sure it was not deflected up or down, the offsets to the required position of the plank were measured and marked on the staff. The offsets were then transferred to the plank.[180] Straight east-country oak plank was used for most of the ship, but for the most severe curves at the buttocks and bows, plank was sawn to shape from compass timber.[181] For this particular use John Shish ordered forty loads of English oak at the end of August 1677.[182]

TREENAILS

Exactly 32,801 treenails were used on *Lenox*, ranging in length from between 1ft and 3ft long.[183] The treenails were made from well-seasoned oak from the top of the tree, straight grained and free from knots.[184] They were first roughly cut to shape, then passed through a die to make them

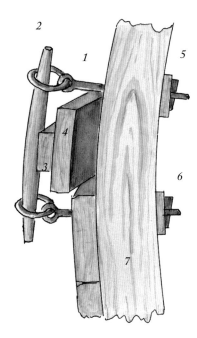

Left: *Wrain bolts and their associated staves were used to bend hot planks to the frame timbers. After setting up the bolts through open treenail holes the stave was inserted through the rings and wedges driven between stave and plank. (1)Wrain bolt. (2) Wrain stave. (3) Wedge. (4) Hot plank set against frame. (5) Forelock. (6) Chock to suit length of bolt. (7) Ship's frame.*

Abovet: *Buttock and Bow Plank cut from compass timber. Sutherland notes that it is better to cut according to the required shape of the hull even if it cut across the grain. William Sutherland,* Shipbuilding Unveiled, *1717, Part 1 p126.*

smooth and circular, a process known as "mooting". The important planking underwater was secured with treenails of 1¾in diameter.[185] Two treenails were used in every frame timber and to prevent splits, or rents as they were known occurring between them, their positioning was slightly staggered along the grain.[186] Before the inside planking was brought on, the two holes were bored through, although only one treenail was driven through the plank and frame, leaving the second hole open. Treenails were tallowed before use and after being driven home were trimmed and secured by being cut with a cross, which was then wedged with oakum. Finally the treenails were cut off level with the plank, a process known as "to cut off and wedge". Later, when the inboard ceiling was brought on, the open hole was bored through and the inboard ceiling, outside planking and frame all treenailed together. In the upper works, where the timbers were much lighter, only every other frame needed to have an open hole left. Alternatively, one treenail and one iron nail was an acceptable practice.[187] A relatively small number of 4000 1ft long treenails were used, in their rough condition, on the upper works of *Lenox*.[188] The vast quantities of 2ft, 2½ft and 3ft long treenails consumed in planking the bottom soon resulted in supplies running short. There had been a shortage of treenails at the start of the building, a situation from which Deptford never fully recovered. During October 1677, Shish wrote to the Navy Board that there was "a great want of 3 and 2 ½ foot treenails in His Majesty's stores at Deptford, we have but very few left to carry on the works of the new ships, I have formerly acquainted your Honourables of it and doubt not but that your Honourables will give directions that we may be speedily supplied." The Surveyor of the Navy, Sir John Tippetts, endorsed the letter, ordering 3000 treenails for Deptford.[189] However, 12,000 treenails of this length would be required and as a result just 100 were left in the stores a month later. Shish himself located a good parcel at Guildford belonging to Sir John Shorter and he immediately wrote to the Navy Board requesting them.[190] Nothing seems to have happened, for a week later Shish was still trying to get permission for his timber purveyor, Thomas Lewsley, to acquire them.[191] Happily, Lewsley managed to find some supplies in the nearby Thames merchant yards, but disappointment followed shortly afterwards, when an expected large delivery to Deptford was sent to Harwich instead. In despair Shish wrote that if he was not immediately supplied his caulkers would soon be standing still.[192] The treenail crisis often came close to causing delays during the building of *Lenox* but the yard always managed to find enough to keep their men busy.

At the end of August 1677, progress was proceeding at a prodigious pace, and many more shipwrights were working on the ship. With six strakes completed on each side, the first two caulkers were taken on to prepare the gaps between the planks.[193] Amongst the other considerations Shish had to deal with was obtaining permission from the Navy Board to order eight steel two-handed saws, as the iron ones in use went very dull in a short time.[194]

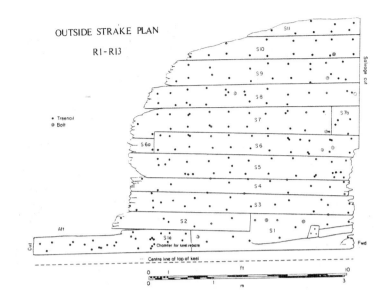

Above: *A drawing by Colin Martin of planking from the wreck of the* Dartmouth *(wrecked 1690) illustrates the reality of planking in an old ship, showing how untidy such planking typically became after repairs. (Courtesy of Colin Martin)*

MAIN WALES

Only a week later, the bilge was planked and at the same time all the frame timbers were in place up to the widest part of the ship. It was now time to strengthen the structure by fitting the lower main wales.[195] They were the largest fore and aft thick stuff on the outside and measured 14in up and down and 9½in thick.[196] The lower wale was brought on first, as the breadth ribbon was probably placed in the position of the upper wale. Most of the wale was fashioned from straight timber, but the foremost piece, the harping of the lower wale, was cut to shape from compass.[197] Because of its width, the top edge of the wale projected further outboard than the next piece of thick stuff above. This projection was made parallel to the horizon, creating an angle with the up-and-down face in contact with the hull. These angles, and others like it, were measured with a simple instrument using the principle of a plumbline and a pivoting limb called a "level and bevel". The bevellings were measured at between three and five positions along the place where the wale piece would fit on the frames and recorded. The recorded bevels were then marked along the wale piece and the excess trimmed away with an adze. To prevent waste on the under side of the wale, it was bevelled at a compromise angle between being perpendicular to the hull and parallel to the horizon.[198]

Once in place the lower wales were fastened to the frame timbers with treenails.[199] In spite of their complexity and size, the lower wales on both sides of the ship, were brought about in only one week between 7 and 14 September 1677. Work then started on the next strake of 6in thick stuff[200] under the lower wale.[201] One week later, on 5 October a second strake of 6in thick stuff was brought about on each side and the 4in plank brought up to meet it. Shish was then able to report that the hull was "shut in with plank under the lower wale on both sides".[202]

CEILING

Work inside the hold purposely lagged behind the progress of the outside and was only started when the outside planking had reached the lower futtock heads. The ceiling was fitted for structural reasons only and was not caulked to make it watertight. The frames were first dubbed down[203] following the practice on the outside. The waste was swept to the bottom of the hold and out through the space left by the missing strake next to the garboard strake. The ceiling strakes were heated to make them pliable

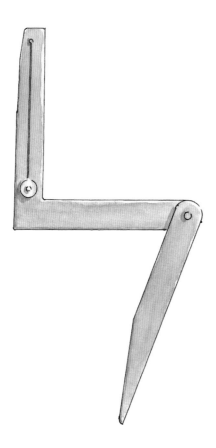

Above: *The level and bevel instrument used for measuring angles against the vertical when forming the sides of wales and similar structural pieces.*

Above: *(Top) The ceiling and part of a plain scarph. Forward of the scarph is a rider or hanging knee. (Bottom) A section through the ship's side showing the frame timbers, ceiling and outside plank. Both the wreck of the third-rate* Anne. *(Author's collection)*

then pushed into shape by props supported from the opposite side of the ship. At the head of the floor timbers, the wrongheads, a band of footwaling consisting of five timbers 7½in thick and two of 5in thick were brought on. The footwaling narrowed in proportion from 1ft 4in wide in midships following the shape of the hold towards the ends.[204] The individual lengths of thick stuff were joined by plain scarphs rather than butt joints.[205] Further up the inside of the hull, two even larger pieces of thick stuff, the middlebands, measuring 8½in thick and 18in broad[206] were brought on, one above the other and tabled 1½in together. They not only supported the frame but also acted as orlop deck supports.[207] The great bands of footwaling covered about two-thirds the area of the hold, the rest being made up with 4in thick plank. One plank, next to the keelson, was non-structural and was known as the limber board. It was loose fitting, resting at an angle between the footwaling and the upper edge of the keelson. Its purpose was to create a passage free from ballast so the bilge water could flow to the pumps. By 2 November John Shish was able to report "all the footwaling underneath the orlop was perfected".[208]

TOPTIMBERS

The sequence of building events did not neatly follow one after another; tasks were started as soon as possible and overlapped each other. As soon as the lower wales were in place, preparations were made to raise the toptimbers. Further stages were erected so the workmen could work high up on the ship, a distance of between 20ft and 50ft above the floor of the dock. The frame-bend toptimbers were first fitted into the frame at the head of the frame-bend middle futtocks.[209] They were secured from one side to the other with cross spawls and the ribbon at the top of the side,[210] then the other fill-in toptimbers were put up in their places. The toptimbers were of much lighter construction than the futtocks and being above the waterline did not have to withstand the pressure of the sea. They

measured only 8in fore and aft and 4¼in in and out at the head to make them as light as possible.[211] The frame in the wake of the toptimbers was not a solid structure and because of the diminishing size of the timbers there were many gaps of about 19in between them.[212]

Because the toptimber heels came down as far as the lower wale, work on the outside thick stuff over them could not resume until the toptimbers were all in place. As soon as they were, work proceeded upwards on the two 6in thick strakes between the wales on each side. When they were completed, the ribbon that occupied the place of the upper main wale was removed and the upper main wale, the same size as the lower wale, brought on.

HAWSE PIECES

Just before the footwaling at the bow was finished, the foremost frame timbers, the hawse pieces, were put in place. These were similar to frame timbers except they were made in one piece 2ft 4in wide. Around the hawseholes they were about 1ft 6in thick,[213] being the thickness of the frame, plus the inside and outside planking. This arrangement, although expensive in timber, meant the anchor cables passing through the hawseholes could not catch on any joints.[214] There were probably two hawse pieces, secured with ragged bolts, each side of the stem that reached up to about 18in above the level of the upper deck. Other timbers close against the stem rose even higher to form the partners for the bowsprit.[215] The hawse pieces and the footwaling beneath the orlop were all finished in the week before 2 November 1677.[216]

Below: *Building progress, 12 October 1677*
The toptimbers are all in place except 6 aft between the larboard frame bend toptimbers. The outside planking has reached the lower main wale and 10 strakes of footwaling have been fitted on the inside.

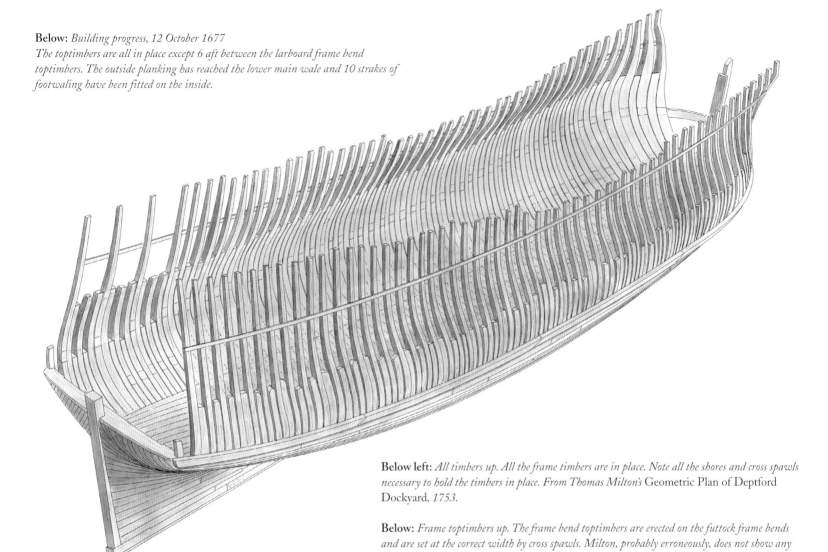

Below left: *All timbers up. All the frame timbers are in place. Note all the shores and cross spawls necessary to hold the timbers in place. From Thomas Milton's* Geometric Plan of Deptford Dockyard, *1753.*

Below: *Frame toptimbers up. The frame bend toptimbers are erected on the futtock frame bends and are set at the correct width by cross spawls. Milton, probably erroneously, does not show any planking. From Thomas Milton's* Geometric Plan of Deptford Dockyard, *1753.*

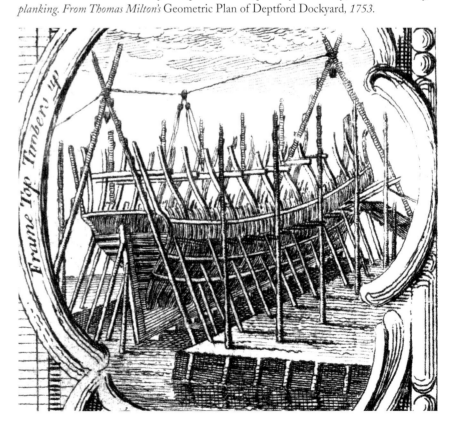

Left: Planking. Note the stages that have been erected. These were necessary to allow the shipwrights good access for planking. From Thomas Milton's Geometric Plan of Deptford Dockyard, *1753.*

CAULKING

At the start of November, John Shish reported to the Navy Board that he was "now ready to go in hand to caulk the new ship building in the dry dock, there is wanting to carry on that service four dozen reeming irons and two dozen beetles to have new iron hoops on. There is a smith in Deptford ... and if it be your Honorables pleasure that we shall have them off him I will give order for them."[217] Reeming irons were wedge-shaped iron tools for opening the seams of the plank and beetles were wooden mallets reinforced with iron hoops used for driving the irons.[218] In total, seventy-two reeming irons and twenty-four reeming beetles were used during the construction of *Lenox*.[219] The seams were caulked with ocum made by unpicking old, but not rotten, rope. It was made up to consist of three double threads of white (untarred) ocum, one thread of hair and two double threads of black-tarred ocum. Tar was extracted from pine trees and is not to be confused with what is now known as tar. Once the ocum was driven home, it was sealed with boiled tar called pitch.[220] Caulkers could drive 30ft of 4in plank in one day[221] and were paid almost the same rate as shipwrights. Over 200 shipwrights and thirteen caulkers were now working on Shish's two ships.[222]

Below: *Building progress, 16 November 1677*
The outside planking is finished up to the chainwale. The hawse timbers forward and the counter timbers aft are in place. All the orlop beams are fitted and kneed and the gun-deck clamps started.

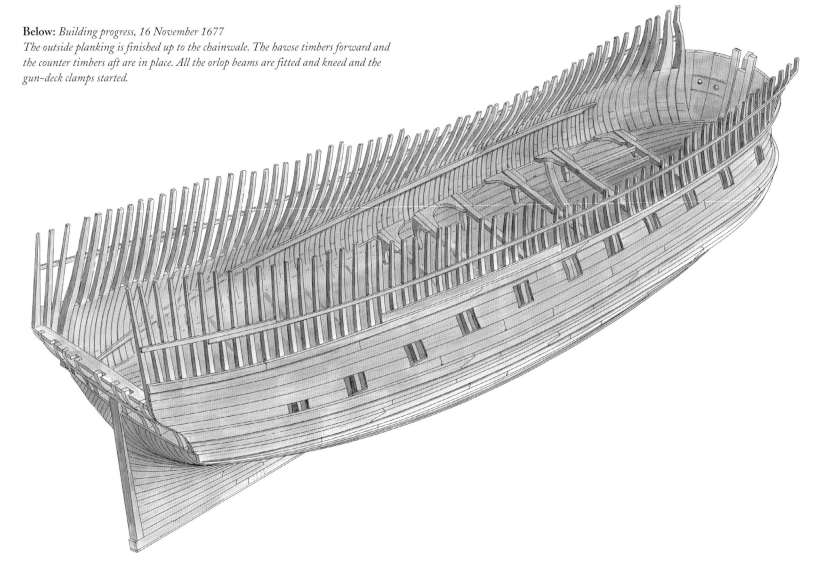

ORLOP

Now that the middlebands were completed on the inside, ten orlop beams, five each side of the mainmast were let down upon them. The beams of all the decks not only had to bear great weight, but also support the two sides of the hull in heavy weather. It was common practice to construct the beams from two pieces and scarph them together, and it is possible the only one-piece beams were those each side of the main hatchway.[223] Large beams had a scarph 12ft long, about a third the length of the beam and were secured with eight bolts.[224]

The end of every beam, not just the orlop but the gun-deck and upper deck, were secured to the side of the ship by two knees secured by three bolts in each arm. One knee, the lodging knee, laid horizontally, the other, the hanging knee, lay vertically. Knees were cut from the junction of a tree trunk and a large branch so the wood grain followed its shape. The sharper the angle of the knee, the weaker it became and the more difficult it was to find. The cheaper, more open raking knees were made use of by ingenious shipwrights wherever possible. The preference for raking lodging knees meant that forward of midships they lay aft the beam, a situation that was reversed aft, where the knees lay forward of the beam.[225] Similarly, the tumblehome helped open the angle of the upper deck hanging knee. After the orlop beams were kneed, fore and aft carlines were set down between them.[226] One row of carlines lay each side of the main hatchway and another 8ft outboard of these. Between the carlines, and parallel to the beams, smaller pieces of timber known as ledges were let in. Eventually a 1½in thick deal platform was laid "convenient for the coiling of the cables".[227] The platform was not laid over the beams but let into rebates so the plank was level with the top of the beam. This allowed the plank to be removed when stowing or rummaging for stores in the hold.

THE HEAD

During early November work started on the knee of the head, the structure at the bow that supported the lion figurehead.[228] Its main purpose was to tack the weather clew of the foresail, secure the bowsprit and act as a toilet for the crew, usually and politely called a house of ease. However, it was not just a practical structure; during the seventeenth century and indeed throughout the age of sail, the head was an ornament that often attracted critical comment from observers.[229] The drawings of Van de Velde the Elder show that the head of *Lenox* appears a little old fashioned, being a little lower and longer than most of her sisters. Because it secured the bowsprit by the gammoning ropes, its failure could result in the loss of the bowsprit and possibly other masts as well. The size of the knee meant it could not be made from a single piece of timber, but made up of several pieces, the arrangement of which could take many forms

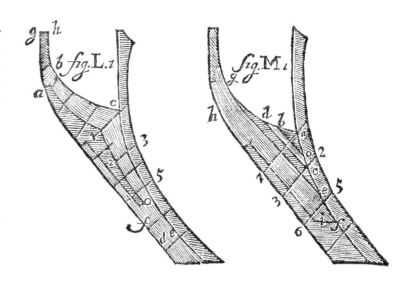

Above: *Two forms of knee of head from William Sutherland's* Shipbuilding Unveiled, *1717, Part 1, p128. Fig L1 shows the preferred timbers with full contact between the stem and knee. Fig M1 shows alternative construction if such large knee pieces could not be found.*

Right: *Construction of the head, omitting the head rails for clarity. The standard was bolted to the stem through the keelson, lace and trailboard (Sutherland,* Shipbuilding Unveiled, *1717, p85). These bolts were often exposed as they passed through the trailboard carvings, making them liable to corrosion and failure (Sutherland, 1711, p64). (1) Stem. (2) Harping of the lower main wale. (3) Harping of the upper main wale. (4) Lower cheek. (5) Part of the upper cheek. (6) Knee of the head. (7) Trailboard. Only 1in thick at the fore end and 5in thick against the stem. It was carved and pierced from one side to the other. (8) Lacing. 10in thick aft and 8½in fore. (9) Brackets. It is doubtful if each could have been made in one piece and they were probably in two pieces scarphed together at the base. (10) Keelson. 10in thick aft and 6in fore. (11) Standard, often called the knee. It was 5ft 6in at the throat. (12) Lion backpiece. Evidence for this is sketchy but it is seen on Van de Velde drawings of ships that have lost their figurehead and shown on a true-framed model of a 50-gun ship. (13) Lion support. It was located by a flat scarph to the knee of the head. (14) Lion. Third rate ships such as Lenox almost invariably had a lion figurehead . (15) Part of the forefoot.*

Below: *A view of the stern without planking. The counter timbers formed the sides of the stern gun ports and were erected first. The lower counter was then planked before the timbers of the upright were fitted. The half, or false transoms between the transoms and the planking are not shown. Note how the main wales line up with the helm port and wing transoms so that large knees on the inside can bind them together.*

depending on the timber available. Ideally the principal piece was curved and would be in contact with the stem for its entire length. This situation appears to have been unusual and more often than not the knee was made up of more numerous straighter pieces.[230] At its aft end, the knee of the head was the same width as the stem but narrowed towards the cutwater.[231] Two months after the main timbers of the knee and gripe were ready, they were put in place against the stem. The bolts that secured them were up to 9ft long and were driven in from the outside. On the inside, tapered forelocks went through a slot in the bolt and secured it against a rove. This allowed a limited amount of tightening at a later date should the head work loose.[232] With the knee in place the lower cheeks, each side of the head, were fayed level with its upper edge and securely bolted to prevent sideways movement.[233] Looking from the broadside, the curve of the cheeks followed the curve of the wales. They were of such complex shape that a mould was made to ensure they were of the correct form.[234] To make the head as strong as possible, the cheeks were placed at the same level as the breasthooks on the inside and the wales on the outside.[235] On *Lenox* the lower cheeks were bolted through the lower main wale.

Once the lower cheeks were fast the trail board was brought on. *Lenox*'s trail board consisted of carved scrolled acanthus leaves "grooved in

every way", as Battine described them. It fitted in above the lower cheeks and against the stem. It was about 9ft 6in long, 1ft thick and tapered from 1ft 6in deep at the aft end to 1ft 4in afore.[236] On top of the trail board came the lacing, a beam approximately 12in by 8in,[237] which was tenoned into the stem.[238] Now that the knee of the head was strong enough to receive it, the magnificent carved lion was attached to the forward end.[239] To secure the assembly further, the upper cheeks on each side of the lacing were fayed and bolted in position.[240] So far, the work on the knee of the head had taken twelve weeks. This amount of time makes it reasonable to suppose a small team of specialised shipwrights carried out the work. If so, their next task was to fit the fore and aft head rails and the up-and-down kneed brackets. At the back of the lion's head was a piece of timber known as a "quafe",[241] made saddle fashion to which were located the forward end of the upper curved head rails. At the aft end they separated out and were attached to each side of the forecastle by two bolts.[242] Later on, a keelson was fitted over the kneed brackets and a knee or "stander", fayed over the keelson. This was bolted through the stem and the knee of the head timbers. To complete the structure, crosspieces were fitted athwart the rails and gratings fitted.[243]

THE STERN

Work on both the head and the stern of the ship started simultaneously. Progress at both ends proceeded at about the same pace. Work first began on the counter timbers that were rebated into the top of the wing transom and edge of the helm port transom. They swept in two arcs, first rearwards, then upwards, finishing at the level of the lower stern windows. They were evenly spaced and would form the sides of the four rear gun ports.[244] Before proceeding further, the lower counter was planked. This increased the overall strength of the structure.[245] The six straight pieces that formed the timbers of the upright of the stern were then put up and four transoms, which formed the sills of the stern windows and supported the decks, were added. Afterwards, four rails and some of the carved work were fayed in position.[246]

RIDERS

After the fore and aft timbers on the inside of the hold were finished, large thwartship timbers known as riders were fitted to reinforce the hull. Before the thirty-ship building programme, the arrangement consisted of floor and futtock riders that reached up to the height of the orlop deck. Shortage of knees had for some time been a problem, a situation made worse by the huge ship building programme. When the scantling list for the ships appeared at the end of August 1677, the problem was addressed in a novel way. The hanging knees of the orlop and the hanging knees of the gun deck between the ports, could be replaced with riders fayed and bolted to the beams.[247] The new method of construction was not compulsory, but offered as an alternative if there were a shortage of knees. The practice lasted about fifteen years[248] and as there was a multitude of different ways of fitting riders, may have taken a number of forms. *Lenox* was kneed in the traditional method, as Shish mentioned while building her,[249] as did a later report in 1693.[250] This is not surprising as work on *Lenox* was far advanced by the time the scantling list was issued. The floor riders were considerable pieces of oak, measuring 24ft long in midships,[251] 1ft 6in fore and aft[252] and deep enough to cover the keelson by 1ft. On 23 November 1677, John Shish reported that five floor riders were fayed,[253] this was almost certainly the same number as the *Warspite* of 1666, the *Edgar* of 1668 and the *Yarmouth* of 1691, all having five bends of floor and futtock riders.[254] Not specified, but usually fitted toward the bow, were one

Below: *Building progress, 14 December 1677*
The work on the gun deck is advanced and the upper deck clamps are in place as are the knee of the head and gripe.

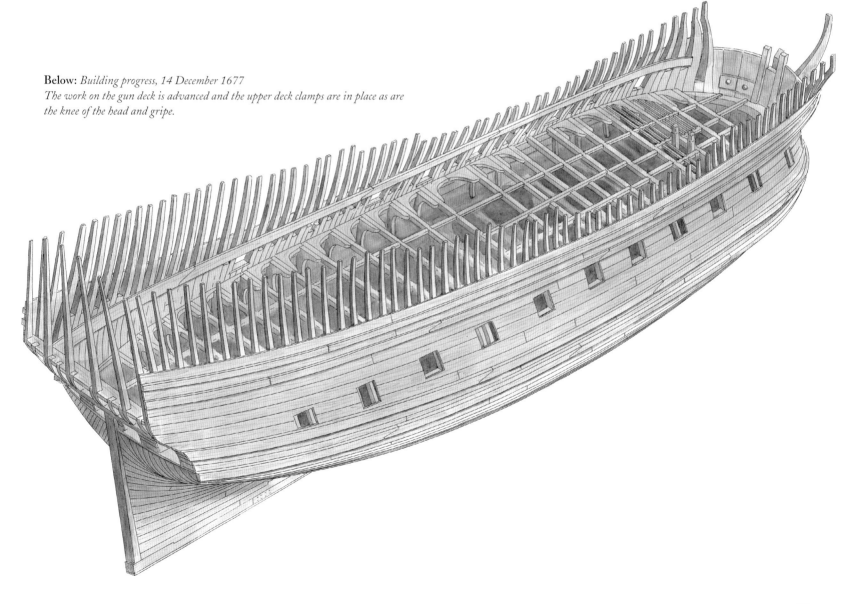

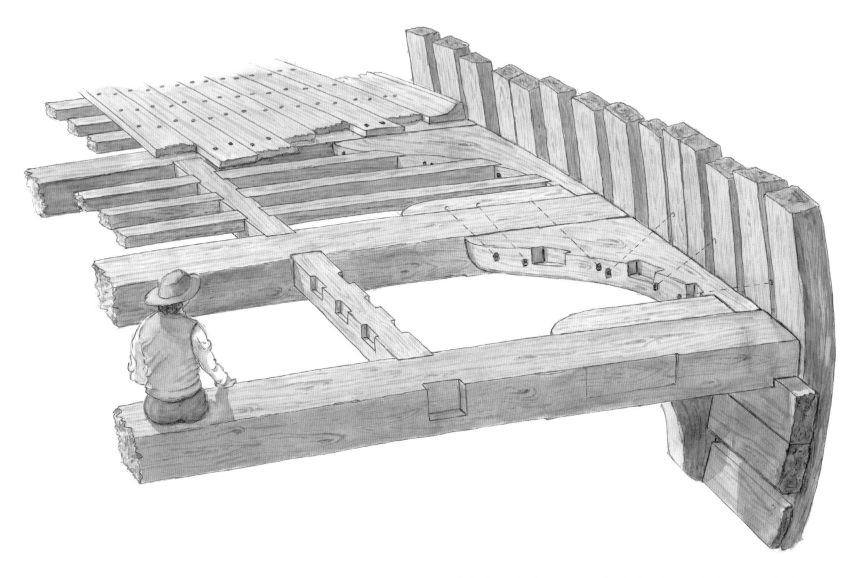

Above: *Gun deck construction. This view shows the arrangement of the gun deck beams, knees, carlines and ledges based on the Wilton House draught of* Essex, *a lodging knee from the wreck of* Dartmouth *and a draught of* Resolution, *1708. Note the two spikes in each beam and two treenails in each ledge.*

or two short riders that had no corresponding futtock riders.[255] Each floor rider was bolted through the ship from the inside with nine bolts, 3ft 6in long and clenched on the outside planking of the ship; they were also staggered to prevent splitting along the grain.[256] On the outside, the boltholes were countersunk and caps of lead nailed over the clench.[257]

Overlapping the floor riders by 7ft and laid beside them were the futtock riders. Slightly smaller in section than the floor riders, they reached up to the beams of the orlop. They were each bolted with seven bolts and because they were more accessible from the outside were clenched on the inside.[258] Although work started in mid-November, it was not until 25 January 1678 that they were all completed.[259]

BREASTHOOKS, TRANSOM KNEES AND CRUTCHES

Work began on the breasthooks as soon as work started on the floor riders. They were similar to the riders and bound the two sides of the ship together at the bows. In the hold, the uppermost were level with the gun deck clamp and the lower mainwale through which it was bolted. The rest of the breasthooks in the hold were equally spaced to the foremast step.[260] Two other breasthooks were fitted, one under the upper deck and another under the hawse holes.[261] Although it is not recorded, two crutches,[262]

small V-shaped riders, were probably fitted towards the stern, aft of the mizzen mast.[263]

Another series of reinforcing timbers in the hold were the transom knees. As their name suggests, they tied the transoms to the side of the ship. The knees at each end of the wing transom were huge, being 5ft 6in along the transom and 15ft along the inside. The long arm was bolted every 20in[264] through the spirketting of the gun deck and the lower wale.[265] All the other transoms below were likewise kneed, but of progressively smaller dimensions.

THE GUN DECK

As soon as the ceiling in the hold was complete up to the orlop deck, work began on the gun deck. To begin with, large fore and aft timbers known as clamps were fitted to the inside of the frame to act as supports for the huge gun deck beams, which measured 15¼in deep by 16½in sided.[266] The clamps were strengthened by the main wales on the outside of the hull that curved above and below them. The first strake of clamp on each side, was in place on 16 November, quickly followed by all the rest a week later. They were then, in one of John Shish's favourite words, "perfected". Before the end of the month all the gun-deck beams were in place and let down, about 1in, into the clamps.[267] Considerable work was involved in making the beams, a good proportion of which probably consisted of two pieces scarphed together.[268] They were also cambered to allow drainage to the sides, to help reduce the recoil of the guns and for strength. Structurally they need not be quite so deep near the sides and were consequently

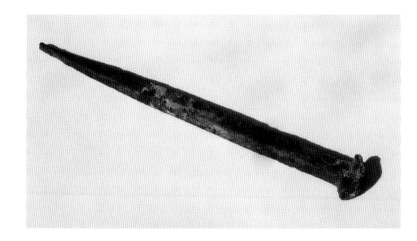

Left: *A seventeenth-century 8in spike, of the type and size used on the gun deck of* Lenox.

reduced by about 1–1½in.[269] Strength was vital, for each gun on the deck weighed 2½ tons, and there were twenty-six of them. One beam was placed under and one between every gun port, except in the wake of the main hatchway where the beams were 8ft apart.[270] Shish practised good forward planning, as work on the beams must have started long before they went into the ship. The only problem was a lack of knees, of which Shish requested ten loads in order to carry on the work.[271] The resulting

delay seems to have slowed the fitting of gun-deck knees and it was a month before they were all finished.

As soon as the deck beams were in place, two tiers of carlines were brought on. Ship building contracts suggest only short carlines spanned the distance between the beams; however, Shish made the inner tier around the hatches using long carlines.[272] They formed the sides of the hatchways and were scored to fit over the beams. The outer tiers of short carlines were placed equidistant between the long carlines and the lodging knees and were scored into the beams by 1½in.[273] The ledges were the final supporting structure of the gun deck and were scored into the carlines to form a flat surface for the deck. With the supporting structure finished, attention turned to the gun deck planking, although parts of it may have been left open for a while to allow access for fitting cross pillars and bitts in the hold. The planking was 4in thick in the wake of the guns up to the edges of the hatches and fixed by two iron spikes in each beam and two treenails in each ledge.[274] The guns did not run over the deck between the mast partners and the main bitts so the planking was reduced to 3in thick.[275]

Below: *Building Progress, 11 January 1678*
The upper deck beams are in and work has advanced on the gun deck. The gun deck ports are cut out and the upper deck port sills faying.

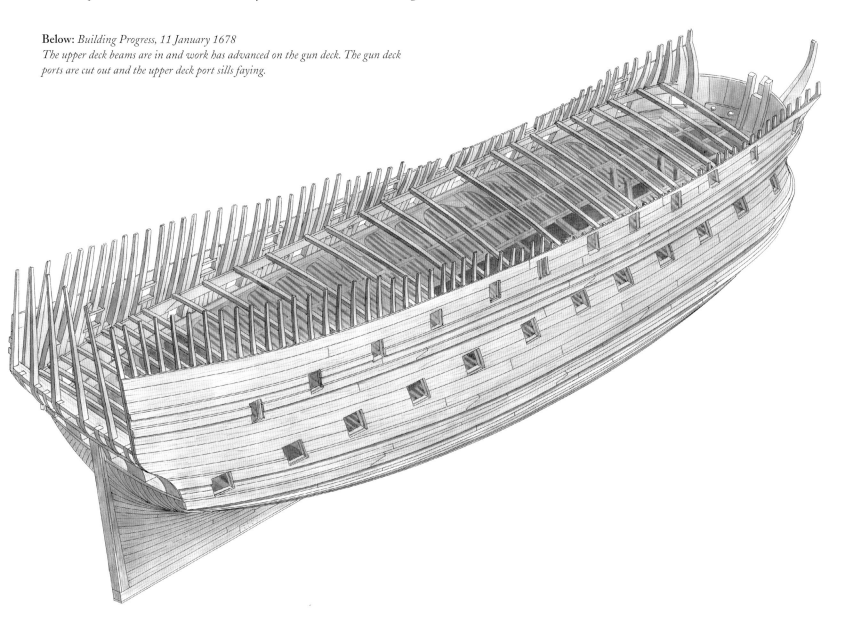

PLANKING THE TOPTIMBERS

During the first week of November 1677, at about the time the footwaling beneath the orlop was completed, work started on the outside planking above the upper wale. Below the wales the framing was almost solid, but above them the situation was very different, as there were only light toptimbers with large gaps between.[276] As the spacing of the gun ports was not co-ordinated with the spacing of the frame timbers, individual arrangements had to be made for the sides of the gun ports and in many cases a toptimber had to be cut through. At this stage, however, the toptimbers were only held in place by a ribbon at the top of the side and the upper wale below. To stabilise them before the gun ports were fashioned, the outside of the hull was first planked. By 4 December the outside planking had reached the channel wales.[277] There was no point in planking over a gun port and so gaps were probably left accordingly, with a trimming allowance. Before the port sills were fayed, fill-in pieces of frame timber must have been placed where there were no toptimbers to form the sides of the ports. Only then was Shish able to report that the port sills were fayed and "the gun-deck ports are cut out on both sides".[278] The framing arrangement may have been a little untidy, but it would never be seen except by the shipwrights working on the inside of the ship.[279]

Only after the gunport sides and sills were completed could the inside work proceed. In spite of the short, cold winter days, progress continued at pace. During January 1678, the waterways at the edge of the gun deck and the 6in thick spirket rising below the gun ports were "all finished", while the ceiling planking between the ports was "half completed".[280]

UPPER DECK

Work on the upper deck generally lagged behind that of the gun deck by about four weeks. Its structure was very similar, but proportionally smaller. A week after the gun deck beams were completed, the upper deck clamps were fayed and made fast.[281] Ten upper-deck beams were resting on the clamps by 21 December and these were quickly followed by the rest a week later.[282] The beams were placed above the gun-deck beams, with one between and one under every gun port. In the forecastle, where the cookroom would be situated, the beams were probably laid more closely to carry the extra weight.[283] During January the shipwrights continued their rapid progress on the upper deck as well as the gun deck and when the outside planking reached the height of the gunwale the port sills were fayed. During the same period, the fore-and-aft long carlines that formed the edge of the hatchways and rose 6in above the deck[284] were

Below: *Building progress, 1 February 1678*
The upper deck is partially laid and the quarter galleries are in hand.

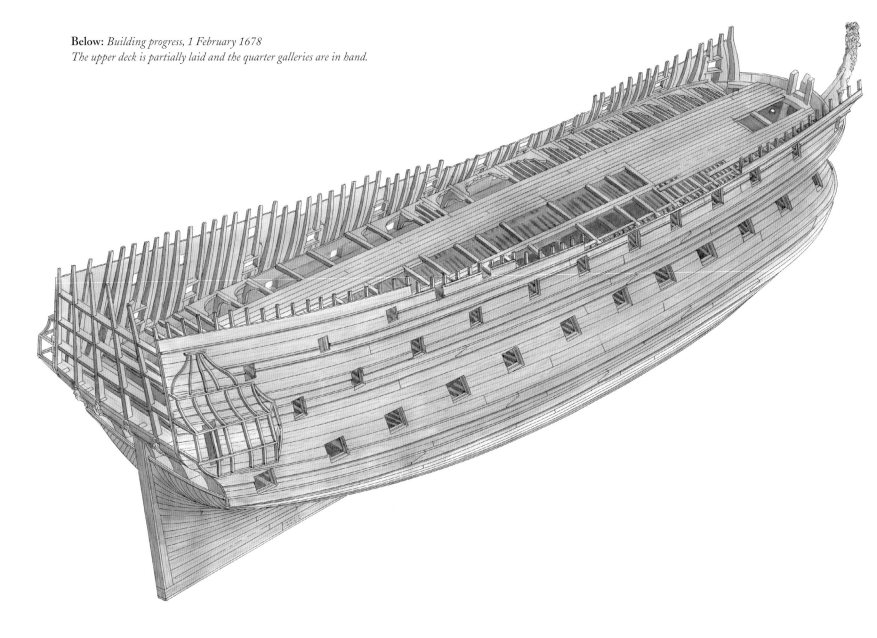

completed.[285] Soon all the carlines were fayed and the "major part of the upper deck knees and ledges fayed".[286] By the end of the month more than 100 knees were completed and part of the deck itself laid, a testimony to the organisation and workmen of Deptford.[287] The upper deck was laid with 3in oak plank in the wake of the guns and with "Prutia", or Prussian deal softwood elsewhere, of the same thickness.[288] Deal was much lighter than oak and was used wherever possible.

Now that the planking had reached as high as the gunwale in the waist some of the decorative features so essential to seventeenth-century ships were started, including the moulded great rail and upper rail on each side. Work also started on the quarter galleries that fitted against the outside planking at the stern.[289]

Above the upper-deck ports, on the inside of the waist, ran a fore-and-aft 8in by 10in string of oak.[290] It was scored down between the 4in deep toptimbers, enclosing them to form solid wood above the ports. This strength helped to stop the ship hogging (bending downwards at the bow and stern) and prevented damage as stores were taken over the side.

Above: *A view looking down the main hatch. The gallows just forward of the fore mast were used to stow spare spars and fishes for repairing masts and yards, the forward end of which rested on the forecastle. Note the end of the pump axletrees onto which the winches for turning them fitted.*

FINISHING THE HOLD

The positions of the masts, measured from the rabbet of the stem, was recorded for fifteen of the twenty new third-rate ships.[291] Unfortunately *Lenox* was not one of them. However the *Stirling Castle*, built by John Shish after *Lenox*, was, and thus her dimensions were used in the reconstructed drawings. They are about 1½ft to 2ft further forward than most of the other new ships, but fit very well in the drawing.

With the construction of the decks far advanced, it was time to start work on some of the ship's fittings. The step for the foremast was put in place by the end of January 1678[292] and probably at about the same time, the cross pillars in the hold. Their purpose was to strengthen the ship's structure when it grounded, either by accident, or for cleaning. They reached from the head of the floor riders to the gun-deck beams where they were attached by knees. The main bits, used for securing the ship's cables while at anchor, were situated at the fore end of the gun deck. They were vertical timbers bolted to the gun deck and orlop beams with the heels stepped on the footwaling in the bottom of the hold. They were arranged in two pairs, with each pair braced by a cross piece and knees.[293] Two of *Lenox*'s bits were in place by the middle of December and were finished a week later.[294]

PUMPS

Fundamental to the safety of the ship were the chain pumps, and they would also have been started about this time. All ships leaked to a greater or lesser extent and required regular pumping to keep them dry. During her career *Lenox* would experience storms so severe that the working of the

seams caused leaks likely to sink her if the pumps were not continually manned. A pump using the same principle as the chain pump had been used in Roman London, and an advanced Tudor version was found in Hampton Court Palace. It was further developed by the time *Lenox*'s pumps were made and development would continue further. Tests were conducted at Deptford in the 1680s between a standard chain pump and Sir Robert Gourden's pump. It moved water at thee times the rate of the standard pump, but whether it was adopted or how it worked is unknown.[308]

Lenox was fitted with two chain pumps[309] just aft the mainmast. They were situated in the well and ran from the bottom of the hold to a cistern on the gun deck, just above the waterline. In case of failure many spare parts, including a complete chain, and the tools necessary to maintain them were provided in the carpenter's stores.[310] The principle of the pump

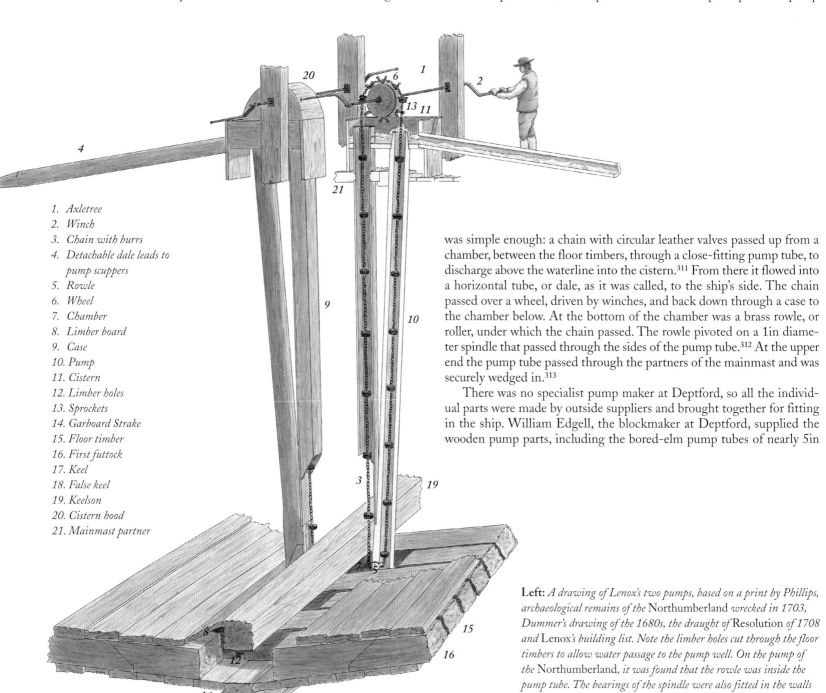

1. Axletree
2. Winch
3. Chain with burrs
4. Detachable dale leads to pump scuppers
5. Rowle
6. Wheel
7. Chamber
8. Limber board
9. Case
10. Pump
11. Cistern
12. Limber holes
13. Sprockets
14. Garboard Strake
15. Floor timber
16. First futtock
17. Keel
18. False keel
19. Keelson
20. Cistern hood
21. Mainmast partner

was simple enough: a chain with circular leather valves passed up from a chamber, between the floor timbers, through a close-fitting pump tube, to discharge above the waterline into the cistern.[311] From there it flowed into a horizontal tube, or dale, as it was called, to the ship's side. The chain passed over a wheel, driven by winches, and back down through a case to the chamber below. At the bottom of the chamber was a brass rowle, or roller, under which the chain passed. The rowle pivoted on a 1in diameter spindle that passed through the sides of the pump tube.[312] At the upper end the pump tube passed through the partners of the mainmast and was securely wedged in.[313]

There was no specialist pump maker at Deptford, so all the individual parts were made by outside suppliers and brought together for fitting in the ship. William Edgell, the blockmaker at Deptford, supplied the wooden pump parts, including the bored-elm pump tubes of nearly 5in

Left: *A drawing of Lenox's two pumps, based on a print by Phillips, archaeological remains of the* Northumberland *wrecked in 1703, Dummer's drawing of the 1680s, the draught of* Resolution *of 1708 and* Lenox's *building list. Note the limber holes cut through the floor timbers to allow water passage to the pump well. On the pump of the* Northumberland, *it was found that the rowle was inside the pump tube. The bearings of the spindle were also fitted in the walls of the pump tube.*

Right: *Steering mechanism, general arrangement. (1) Quarter-deck.*
(2) Upper deck. (3) Mizzen mast. (4) Roundhouse bulkhead.
(5) Rudder. (6) Tiller. (7) Sweep. (8) Whipstaff. (9) Rowle secured in
its scuttle. (10) Gooseneck, called a nagshead at Deptford. (11) Shoe.
(12) Binnacle or bittacle, based on the Essex *model at Wilton house and*
a print from Blanckley. (13) Whipstaff companion for communication
between an officer 'conning' the ship and the helmsman, who had limited
vision up the stairs to the quarterdeck. It is raised to provide clearance for
the whipstaff.

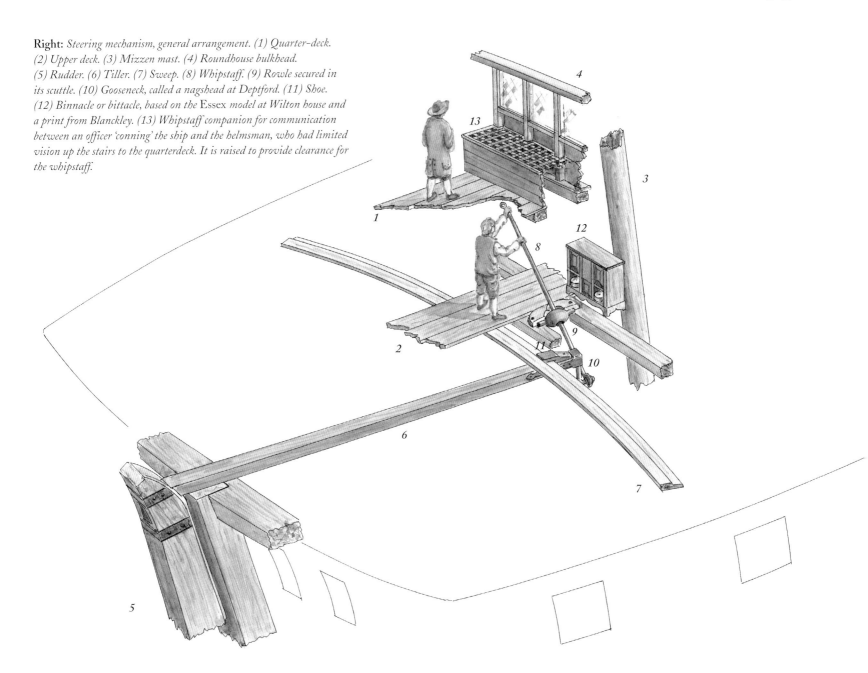

diameter[314] and the elm wheels,[315] which were about 18in diameter.[316] He also supplied the dales and the square cases through which the chain passed to the chambers below. The cases were cut away near the bottom[317] to allow access to the chains for repairs. The dales were hollowed-out elm logs that took the water from the cisterns at the top of the pumps, to the ship's side where the largest lead scuppers in the ship, measuring 4¼in diameter,[318] discharged it into the sea. The dales would have to be removed for action, as they would interfere with the training of nearby guns. Henry Loader, the anchor smith at Deptford, supplied the ironwork for the pumps, including the chain, swivels and hooks[319]. The valve assemblies were attached at every 30in along the chain.[320]

There is no mention of burrs in the building list of *Lenox*, the thin metal plates that supported the five leathers[321] in each valve, and it may be that they were made in Deptford dockyard. Loader also supplied the fittings for the solid elm wheels at the top of the pumps. Each had eight sprockets driven into it that engaged the back of the valves and to prevent splitting, were reinforced with iron hoops. Since at least 1688, hoods were fitted over the wheels of many ships to stop splashing.[322] A square hole was cut in the centre of the pump wheel, through which a shaft, known as an

axletree, was fitted. The axletree was also square in section where it was in contact with the wheel, and was secured by iron wedges driven between the two parts. At each end of the axletree was a location for detachable winch handles, by which means the men worked the pump. At the bottom of the pump, a gap was left between the floor timbers to create the chamber, or sump.[323] The limber holes through the floor timbers sometimes became stoaked (blocked) by ballast or other debris[324] and a special tool called a limber iron was used to clear it.[325] One was purchased in 1678 for the *Hampton Court* at a cost of four pence.[326] It is difficult to find evidence for hand pumps aboard *Lenox*, but it is probable they were used as portable items for fire-fighting or cleaning decks and provision may have been made for them to be fitted in the well.

STEERING GEAR

Another vital fitting was the steering gear. In the days before the introduction of the steering wheel in about 1703, ship's tillers were moved by a whipstaff. The 25ft long tiller of *Lenox* was fitted to the top of the rudder and ran forward underneath the beams of the upper deck to finish just aft the mizzen mast. The tiller was supported at its forward end by a

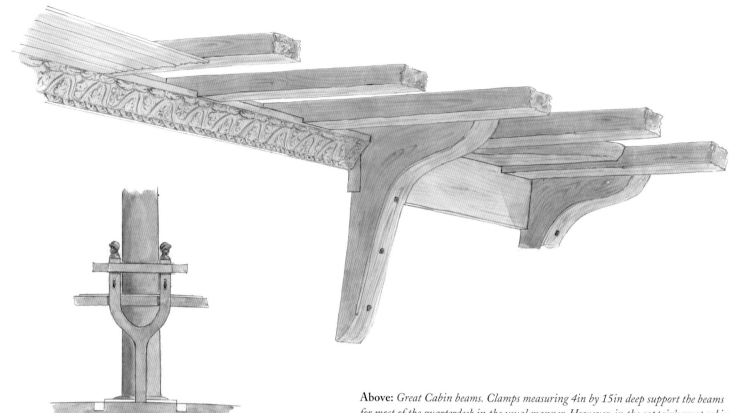

Above: *Great Cabin beams. Clamps measuring 4in by 15in deep support the beams for most of the quarterdeck in the usual manner. However, in the captain's great cabin the beams are dovetailed into a carved string, thus eliminating the necessity for knees. The ceiling under the beams was boarded and the sides were panelled.*

Above: *Fore Sheet Bitt. The awkward arrangement of the fore sheet bitts inside the forecastle was necessary to allow room for the foremost guns on the upper deck.*

curved beam called a sweep, which was bolted to the underneath of the beams. A shoe on top of the tiller fitted into a rebate in the sweep, connecting the two and allowing sideways movement.[327] At the forward end of the tiller was a gooseneck, which at Deptford seems to have been called a nagshead, supplied by Henry Loader, which was connected to the lower end of the whipstaff. The whipstaff was a simple lever that moved the tiller from side to side[328] and pivoted in a rowle with a large clearance hole. The hole allowed the whipstaff to slide in and out with a necessary small amount of movement fore and aft. This allowed the whipstaff to move the tiller over at least 9ft each way, or about twenty degrees, before it became impractical to use.[329]

The whipstaff was made and supplied by William Edgell, the blockmaker, and was 12ft long.[330] The whipstaff had a ring at each end, the lower of which fitted onto the gooseneck,[331] while the ring at the upper end may well have been intended to be hooked by convenient upper-deck tackles when the forces on the whipstaff became too great for the helmsman to handle. Edgell also supplied the *lignum vitae* rowle[332] that was held in place by a plate bolted over it.[333]

The whipstaff was so long that it projected through the quarterdeck above and into the coach. It was covered by the whipstaff companion that had a grating on top, making it possible for instructions to be shouted below. These were given by an officer who "conned" the ship, aided by a view of the sails and weather through the coach bulkhead windows.[334]

As the helmsman in *Lenox* would have little view of either the sea or the ship's sails, a binnacle was fitted for him to steer by.[335] This was a type of wooden locker that held navigational equipment: two compasses, a sand glass and candles.[336] The binnacle had both left- and right-hand

compartments, each of which was provided with a compass. This arrangement allowed the helmsman to steer his course, whichever side of the whipstaff he happened to be.[337] A second binnacle was located on the quarterdeck for the use of the ship's officers.

CAPSTANS

Ships the size of *Lenox* were fitted with two capstans, the main capstan on the gun deck and the jeer capstan in the waist. During January 1678 the main capstan, situated on the gun deck abaft the mainmast, "was in hand".[295] The capstans fitted to *Lenox* are probably the first "drum fashion" type fitted to a new ship.[296] The drum-fashion capstan had half-capstan bars at the same height, where previously capstan bars were at awkwardly different heights. The spindle of the main capstan projected below the deck and was stepped in the orlop,[297] on a flat iron bearing[298] supplied by Henry Loader, the anchor smith of Deptford,[299] who had been supplying the navy since at least 1652.[300] The main drum capstan was 30in in diameter[301] and capable of receiving the anchor cable itself. Vertical pieces called whelps, which tapered upwards, were fitted to the sides of the capstan, preventing the cable sliding down and rubbing on the deck. The main capstan was intended to raise the anchors but was seldom used except in extremity.[302]

The jeer capstan was a double capstan with a common spindle; one part was situated on the upper deck in the waist and the lower part on the gun deck. It was the everyday work capstan and was used to haul in the anchor cable along the gun deck to the hatches, or raise the yards from the upper deck to the masthead. It had one drum only, on the upper deck, and was somewhat smaller than the main capstan, being 22in in diameter. The lower part could not take the anchor cable; instead a looped rope called a vyol was attached to the anchor cable, by which means it was pulled to the hatch. Here the vyol was detached and run further aft to run through the huge 38in vyol block[303] secured to the mainmast. The vyol rope then ran forward again to the jeer capstan to repeat the cycle. *Lenox* was supplied with twenty-four half-capstan bars, indicating that each capstan had twelve bars.[304] To stop the capstan turning uncontrollably, iron pawls were fitted to the deck that could be pivoted to jam against the whelps. The pawls were locked in position with chained pins supplied by Susan Beckford, the favoured supplier at Deptford of small metal fittings such as these.[305] At the beginning of February 1678, when *Lenox* was a little more than two months away from launching, John Shish's progress reports suddenly stop. He also ceased sending reports for the other ships he was building, including the *Hampton Court*, launched on 10 July 1678 and the *Duchess*. The second-rate, three decker *Duchess* was built in the space occupied by *Lenox* at the head of the dry dock and was launched on the

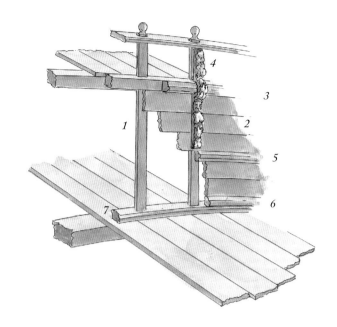

Right: *Bulkhead construction. Based on a model of the* Bedford *of 1698 in the National Maritime Museum and the works of William Keltridge and Edward Battine. (1) Stanchions, 5½in wide by 4in thick. (2) Prussia deal plank, 2in thick with rabbeted edges. (3) Rail Aloft, 5in deep by 4½in thick. (4) Brackets, 5in wide. (5) Middle rail, 4½in deep by 3½in thick. (6) Rail alow, 3½in by 3½in. (7) Oak plank, 4in thick and 10in broad, laid with tar and hair. The seams were sealed with lead.*

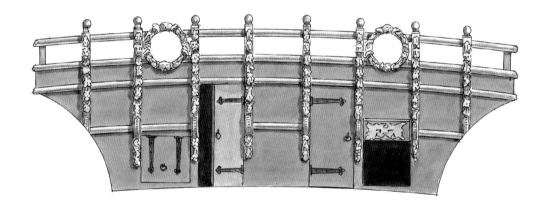

Above: *The beakhead bulkhead.*

Below: *The roundhouse bulkhead.*

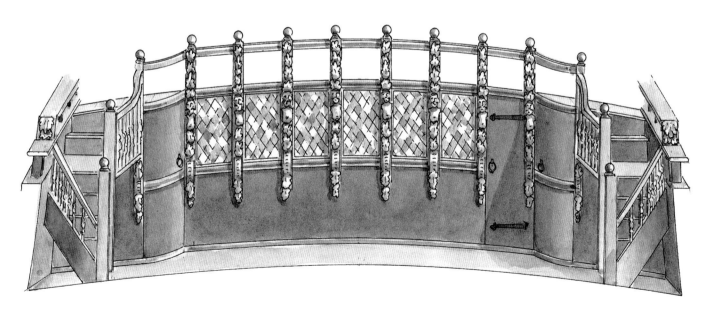

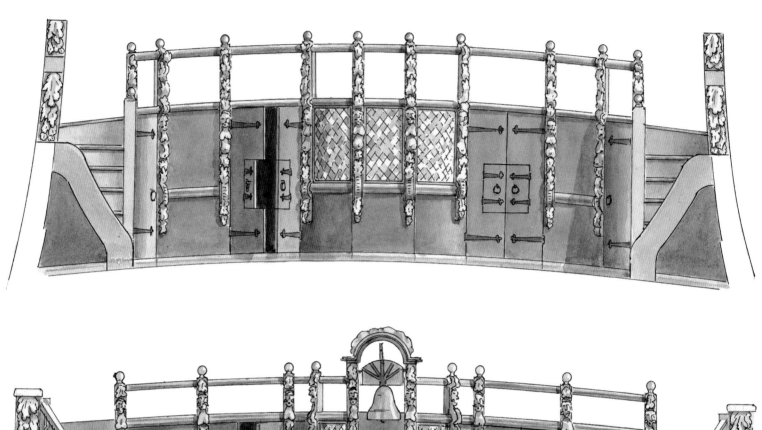

Above: *Lenox's bulkheads continued. (Top) The steerage bulkhead. (Bottom) The forecastle bulkhead.*

15 May 1679. Already the shipwrights had hewed the foremost keel piece, main post, scarped the stem and the wing transom was in hand.[306] After *Hampton Court* was finished, Shish built another third-rate, the *Stirling Castle* launched on 29 July 1679.[307]

FORECASTLE, QUARTERDECK AND POOP

In the final weeks of building, the last decks were put in place. Above the upper deck was the forecastle and quarterdeck. These two half decks were at the same level but were separated by the open space in midships known as the waist. The beams for both decks measured 6in deep by 8in sided[338] with only every other beam kneed by a hanging knee alone.[339] An exception to this arrangement occurred at the waft end of the quarterdeck, in the captain's great cabin, where the beams were only 4in deep and no knees whatsoever fitted.[340] Instead of knees to secure the ends of the beams, they were dovetailed and bolted into a large fore-and-aft string, 8in thick and

11in deep. This arrangement left a high, open, uncluttered space ready to be embellished by the joiners. There were three large bulkheads along the upper gun deck, one in the quarterdeck and two in the forecastle. These bulkheads were very substantial and the two that faced each other in the waist had rounded projections. To strengthen them, the beams in their wake were double kneed. At the base of the bulkheads was a 10in broad by 4in thick plank rebated to take vertical stanchions. To make a watertight seal the plank was laid with tar and hair and the joints sealed with lead. At the ship's sides the bulkheads were further strengthened by 9in sided standards.[341]

The uppermost deck at the stern was the poop under which the roundhouse was situated. One drawing, by Van de Velde the Elder,[342] appears to show the roundhouse bulkhead forward of the third port from the stern, placing it forward of the mizzen mast. The same drawing also shows the bulkhead with the usual rounded cabins at the sides. If this were so, then the cabins would be impossibly small due to the proximity of the gun port. It would also mean the mizzen mast passing through the poop deck, a layout unknown in any other third-rate of the period. Moreover, other

Van de Velde drawings[343] seem to show the bulkhead in the normal position, further aft. It must be concluded, that in all probability, the roundhouse bulkhead of *Lenox* was in the normal position.

The poop was very lightly built with beams 4 in square.[344] As with the quarterdeck and forecastle, only every other beam was kneed, with one hanging knee at each end.[345] An interesting little detail is the fitting for the ensign staff that allowed it to lay either flat on the deck, or pivoted upright. It consisted of a cylindrical rowl, pivoting between two small cheeks fitted to the stern timbers at the aft end of the poop.[346]

THE COOKROOM

After the forecastle was completed, work could start inside on the cookroom, where the boiling and roasting of provisions for captain and crew was undertaken. It was situated on the centre line of the ship towards the aft end of the forecastle. Much of the crew's diet consisted of substantial quantities of salted cask beef and pork. To boil the meat and cook it, *Lenox* was provided with two copper furnaces - at least one measured 36in in diameter.[347] They were made from copper plates riveted together and at the forward end were spouts, to which cocks were soldered to let the boiled liquid run out.[348] The two furnaces, their associated covers, funnel

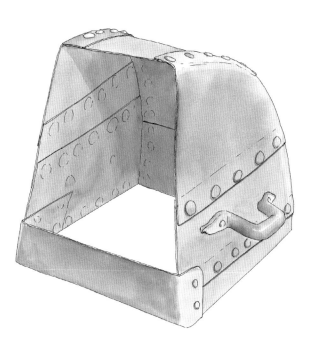

and hood were supplied to Deptford by Henry Wilks, a brazier of London, at a cost of more than £60 in a contract made on 22 March 1678.[349] The furnaces were set up upon a brick hearth with space between for the fire. The sides of the furnaces were totally enclosed by brickwork that was doubtless of the highest seventeenth-century standard with little gap between. At the aft end was a fireback, for a range, where meat could be roasted. The wages earned by the bricklayers indicate that two men would have spent about six weeks making the hearth, during which time they used 2000 bricks.[350] These bricks were probably specially made and somewhat smaller than the seventeenth-century norm.[351] Between the range and bulkhead was an enclosed area with a bench or dresser,[352] where the food was prepared. The deck in *Lenox*'s cookroom was tiled with twelve large paving tiles. To prevent fire, 500 plain tiles and a number of tin plates were used; the tiles probably covered the range and the tin plates probably lined the wooden bulkhead. Not all the cookrooms of the new ships were tiled, for some were lined with lead sheet.[353] Cooking utensils were provided, including tongs, spits, fire forks, a shovel, fenders, a fish kettle and a smaller kettle.[354] Later, during 1684, a double copper kettle measuring 2ft 2in long, 1ft 8in wide and 1ft 3in deep was supplied to third-rate ships.[355] Altogether, *Lenox*'s cookroom cost a little more than £87.[356]

SCUPPERS

The camber of the decks allowed water to run into waterways at the deck sides, where holes were bored through to the outside of the hull, into which lead scuppers were inserted. They were made from milled lead sheet, formed into shape and soldered at the joints. During 1678 and later in 1691, trials were carried out using cast lead scuppers but difficulty in casting an even wall thickness prevented their widespread use.[357] In all fifty scuppers were fitted to the decks above the waterline of the new third-rate ships.

Number, diameter and length of scuppers fitted to a third-rate of 1677:

Deck	Number	Inside Diameter	Length
Gun deck	20	3¼in	2ft 7in
Pump and Manger	6	4¼in	2ft 7in
Upper Deck	14	3¼in	1ft 10in
Steerage	6	3in	1ft 10in
Cuddy	4	2in	1ft 2in

List by Daniel Furzer, Master Shipwright at Portsmouth, to the Navy Board on 23rd April 1678.[358]

The spacing of ten scuppers on either side of the main gun deck appears to agree with those shown on a drawing of *Lenox* by Van de Velde the Elder shortly after she was launched.[359] The scuppers of the gun deck were very close to the sea and when the ship heeled in rough weather, sea water was liable to wash back into the ship. To prevent this happening, twenty leather spouts, called leather scuppers, were purchased[360] and nailed to the outlet of the gun-deck lead scuppers. When immersed, the leather would collapse and close, preventing water returning back into the ship. The leather scuppers were cut according to a pattern from the necks, flanks and shanks of bucks and measured 14in at their upper end, 4½in at the lower and were 24in long. The leather scuppers were bought by contract from tanners[361] at thirty-two shillings a dozen,[362] the design of which was

Left: (Top) The furnace for boiling meat. From the wreck of Stirling Castle, *lost in the Great Storm of 1703. (Bottom) The chimney hood, made from riveted copper plate panels, could be moved into one of four positions depending on the wind direction. From the wreck of the* Stirling Castle, *lost in the Great Storm of 1703.*

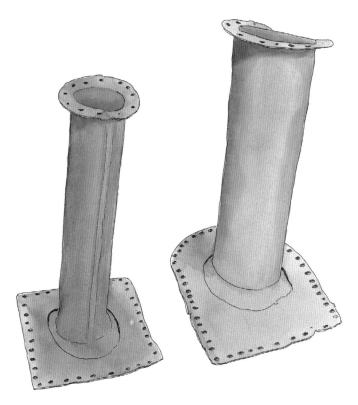

Above: *Lead scuppers, drawings based on archaeological remains from the wreck of the* Stirling Castle, *lost in the Great Storm of 1703.*

Right: *Roman numeral draft marks. These marks, similar to those found on the wreck of* Dartmouth, *are fitted to the* Essex *model, a third-rate, ranging from XII to XX on the starboard side of the sternpost and from XII to XVII on the port side of the stem.*

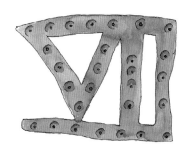

Right: *Cabin and storeroom arrangement in the Hold. The layout of cabins was compiled using the establishment of cabins of 1673 and 1686, Dummer's drawings of a first rate,* Deane's Doctrine, *the draught of* Resolution *of 1706 and letters written during the construction of the thirty ships. Tables are shown as allocated in the 1686 cabin establishment. (1) Sail Room. (2) Boatswain's storeroom. (3) Scuttle to Boatswain's lower storeroom. (4) Gunner's storeroom. (5) Filling room. (6) Carpenter's storeroom. (7) Scuttle to Carpenter's lower storeroom. (8) Scuttle to powder room in hold. (9) Wing bulkheads probably 2 or 3 ft high. (10) Cable tier. (11) Main Hatch. (12) Well. (13) Purser's cabin. (14) Steward's room and cabin. (15) Open platform to breadroom. (16) Breadroom. (17) Cockpit. (18) Scuttle to after powder room below. (19) Hatch and scuttle to fish room below. (20) Captain's storeroom. (21) Surgeon's Mate's cabin. (22) Slop clothes room.*

similar to those found on the Tudor warship, *Mary Rose*.[363] Sometimes there were problems with lead scuppers; during 1684, a survey of the *Grafton*, taken after her voyage to Tangier with Samuel Pepys aboard, found that all the lead scuppers were broken because of the working of the ship.[364]

JOINERS' WORK

Joiners fitted staircases, cabins, doors, panelling and many other small works in the ship. The deal cabins, although of enormous interest to sea officers, were of minor importance to shipbuilders and were rarely worthy of comment. Usually ships were taken into the navy, with many, if not all of the cabins missing. This was not necessarily a fault by the builders, but intentional, for the cabin bulkheads stopped air circulating and prevented new ships from drying out. After delivering two of the thirty new ships, Daniel Furzer received a letter from the Navy Board, complaining that his joiner's work was excessive. He was told that the third-rate ships should have no partition between the bulkhead of the great cabin and the bulkhead of the quarterdeck and joiner's work should not line the deck beams above. The long table and pillars should not be provided as formerly performed.[365]

For use in service, the great cabin was specially treated, the underside of the beams was lined[366] and the absence of knees left it clear to be decorated in the classical seventeenth-century manner, including a cornice fashioned from the rising, or string, as it was also known.[367] The sides would have wainscot panels with the captain's private arms and paintings in the panels above. Even the deck of the great cabin was often treated differently and left off for finishing later.[368] The *Restoration*, built at Harwich and launched a month after *Lenox*, had no flooring in the great cabin or roundhouse, was short of two cabins in the gunroom, had no aft powder room, no shot lockers or well, four cabins lacking between decks, the great cabin missing a bulkhead thwartships and a bulkhead alongship for the captain's bedroom.[369] In a survey of 1684, seven of the twenty new third-rates lacked roundhouse and great cabin floors.[370]

Lenox was delivered with most of her cabins finished, which must have been broadly in accordance with an establishment set up in 1673.[371] The only joiners' work missing were the shot lockers[372] and probably the powder and filling rooms. Certainly the *Hampton Court*, the ship John Shish started building while *Lenox* was under construction, did not have them until 1688.[373] The cabin arrangement changed little over the years and was confirmed by the Pepys Special Commission in 1686.[374] A typical seventeenth-century joiner was capable of producing work of the highest quality. Sympson, the joiner on *Lenox*, made the first of Samuel Pepys's bookcases, now preserved at Magdalene College, Cambridge,[375] and the type of carved cornice and workmanship found on it would also have been found in *Lenox*'s captain's great cabin.[376] The great cabin was of ample size, measuring about 22ft by 12ft, but in a crowded ship only the captain was afforded such luxury. More junior officers' cabins, for example the carpenter's mates, measured only 5ft 9in by 4ft 6in.[377] Many cabins, not mentioned in the establishments, were erected when the ship was being prepared for sea, with as many as ten being "canted" on the gun deck alone.[378] They would have been temporary structures, made from old canvas or if none was available, slit deals.[379] Even worse off were the ordinary seamen, who had barely enough space to hang a hammock.

Some cabins were fitted with scuttles and lights through the ship's side to provide daylight for the more senior officers.[380] Lights were also provided for the cabins in the coach, steerage, cookroom and similar places. These were glazed with leaded, diamond-shaped mica, or muscovy glass, about 3in by 2¾in. They can be seen in Van de Velde the Elder's drawings of *Lenox*.[381] Muscovy glass is a natural silicate mineral and was supplied by John Hardwin,[382] a supplier since at least 1651.[383] From 1690 onward Muscovy glass was gradually replaced by ground glass that could be made in larger sections. To withstand the heavy duty of shipboard life and the shock of gunfire, it was nearly ¼in thick.[384]

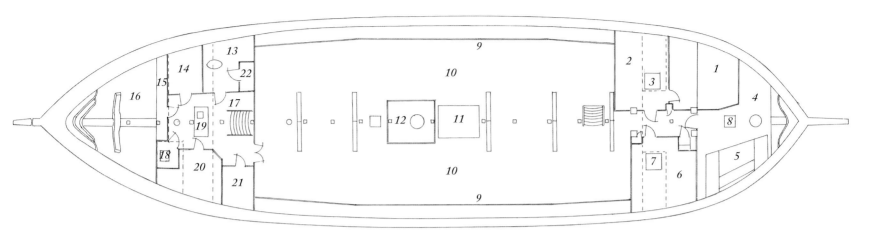

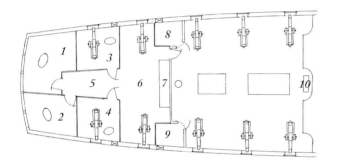

Immediate left: *Quarterdeck cabin arrangement. (1)First lieutenant's cabin. (2) Master's cabin. (3) Minister's cabin. (4) Second lieutenant's cabin. (5) Cuddy to roundhouse. (6) Coach. (7) Helm companion. (8) Land officer's cabin. (9) Chief mate's cabin. (10) Binnacle.*

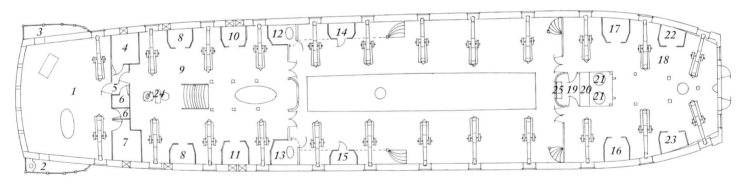

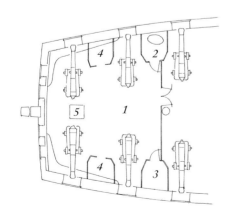

Above: *Upper deck cabin arrangement. The precise arrangement of the captain's quarters is speculative. (1) Captain's great cabin. (2) House of office. (3) Wardrobe. (4) Captain's bedplace. (5) Lobby. (6) Pantries. (7) Dressing room. (8) Servants' half cabins, about 3ft high. (9) Steerage. (10) Pilot's cabin. (11) Second mate's cabin. (12) Boatswain's cabin. (13) Carpenter's cabin. (14) Coxswain's cabin. (15) Midshipman's cabin. (16) Cook's cabin. (17) Boatswain mate's cabin. (18) Forecastle. (19) Cook room. (20) Range for roasting meat. (21) Furnaces for boiling meat. (22) Midshipmen's cabin. (23) Carpenter mate's cabin. (24) Binnacle. (25) Cook's food prepartion table.*

Bottom left : *Gun deck cabin arrangement. (1) Gunroom. (2) Gunner's cabin. (3) Surgeon's cabin. (4) Mate's cabin. (5) Scuttle to bread room.*

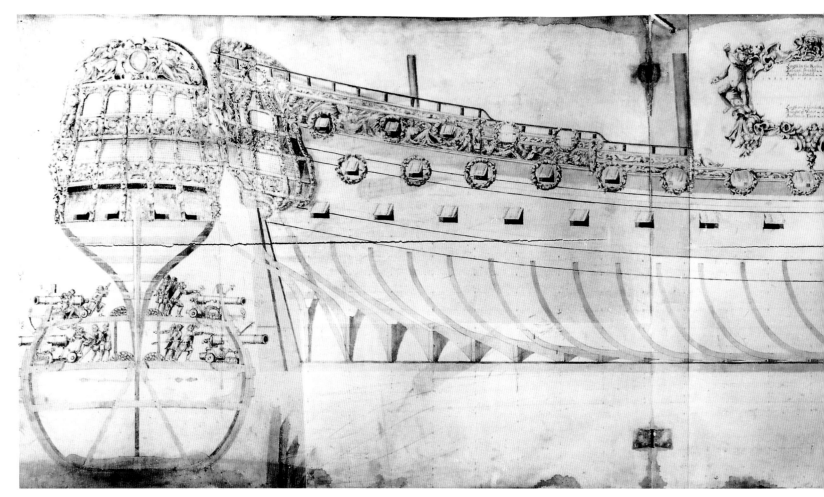

Left: *A draught of* Essex, *a two-decker third-rate of the thirty ships building programme. This was one of the key sources consulted when researching* Lenox's *construction. (By kind permission of the Earl of Pembroke, Wilton House).*

Below left: *A drawing of* Lenox *by Willem Van de Velde the Elder. Present owner unknown, although the mark 'PS' is inscribed on the lower left corner, indicating it was once in the Paul Sandby collection. (Author's collection)*

Below: Lenox *viewed from the larboard bow by Willem Van de Velde the Elder. This particular drawing is an offset of the Van de Velde drawing of* Lenox *(left). (Scheepvaart Museum, Amsterdam)*

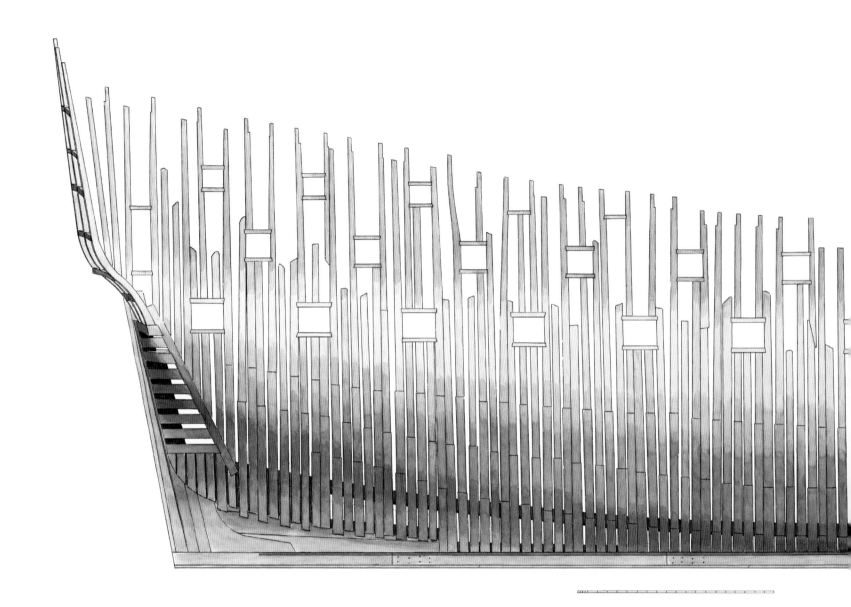

Above: *Broadside view of frames. Notice that only every third pair of frames, the
frame bends, are joined together. The rest of the frames could not stand in the ship
unless supported by the planking, making this view theoretical only. The spacing of the
frames was not co-ordinated with the spacing of the gunports so that individual
arrangements were made for each one as they were cut out.*

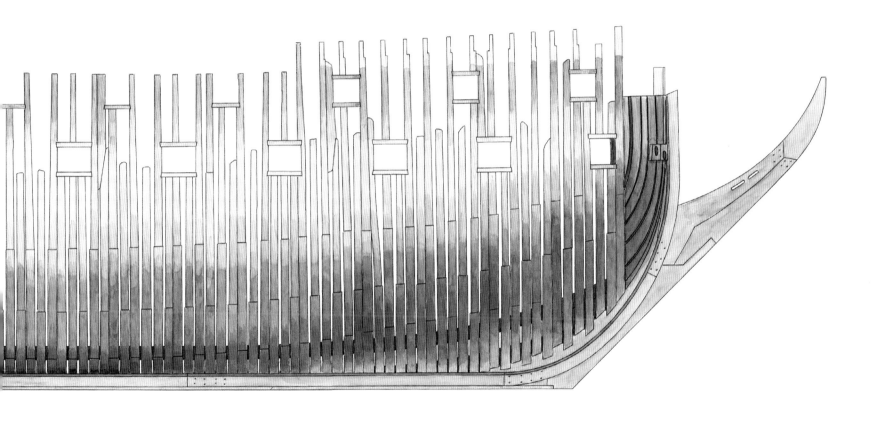

Below: *The table found in the gunroom of the* Stirling Castle *lost in the Great Storm of 1703 belonged to John Laws, the Gunner. The tabletop was made of deal, while the rails and turned bobbin legs were made of oak.*

Inches

DEAL TABLES

Deal tables were provided for the five standing warrant officers, the lieutenants and the minister.[385] In 2002, an oval swingleg table was found on the 1703 wreck of the *Stirling Castle* on the Goodwin Sands. It was found on the port side of the gun deck, just forward of the third gun port from the stern.[386] The tabletop is 3ft 6in long and rather crudely made of deal, while the rails and turned bobbin legs are beautifully made of oak.

Judging by the wear and its position close to the gunroom, it must have belonged to John Laws, the Gunner, and therefore dated from about 1690. This is the period when *Lenox* and her sisters, including *Stirling Castle*, were commissioned at Chatham and received furniture such as this. A contract exists for supplying such tables to Chatham in 1690 (see Appendix 10).[387]

IV

CARVINGS & FINISHING

ART AND THE NAVY

During the seventeenth century aesthetic quality was as important on the king's ships as it was in a great house or palace. The transient beauty of ships was recognised by King Charles, who engaged the famous Dutch artists, the Van de Veldes, to record them for all time. He retained both the elder Van de Velde and his son with a yearly pension of £100 each,[1] paid in regular quarterly payments of £25.[2]

The Van de Veldes were sometimes directly involved with the decoration of the king's vessels. During the early part of 1677 a new yacht, the *Charlotte*, was nearing completion at Woolwich. Charles, taking his customary interest in his ships and especially his yachts, gave personal instructions to the Master Shipwright, Phineas Pett III, as to how it should be finished. Charles asked Pett to see the great Dutch master at Greenwich, who was to prepare paintings on panels of "the posture of ships sailing several ways". Pett duly obliged and agreed a price of £74 with the artist.[3]

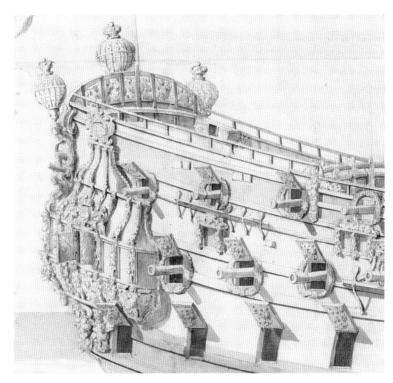

Below and right: Lenox *viewed from the starboard beam by Willem Van de Velde the Elder, in pencil and wash. Inscribed at the bottom "60 stux", confirming the ship was built with no provision for guns on the forecastle or poop. The drawing (excluding the ensign) measures 12½in by 42³/₈in and, interestingly, scales 1:48 in the vertical. The detail right is an enlargement of her 4 aft ports and 2 aft broadside gunports. (Museum Boijmans van Beuningen).*

Given the freedom to negotiate with Van de Velde, it is not surprising to find the subjects of the paintings were ships recently built by Pett. These still survive and are easily identifiable; of the hundreds of Van de Velde paintings extant, only three are of a suitable size, on panel and date from the correct period. The paintings are almost certainly of the *Charles Galley* of 1676, the *Woolwich* of 1675 and possibly the *Portsmouth* yacht of 1674.[4] On receipt of the paintings from Van de Velde, Pett would have mounted them in the great cabin of *Charlotte* in a similar way to those shown in a contemporary print of the longitudinal section of a yacht.[5] The print is clearly not the *Charlotte* but certainly a yacht of the period.

Charles did not confine himself to works by Van de Velde to grace his yacht. The King's Master Carvers, Joseph Leadman and Joseph Helby, were commanded not only to provide carved work for the yachts and boats but also provide models for his own use. Their finished works were surveyed by Pett on 25 April 1677 and altogether valued at £61 8s 6d.[6] Joseph Helby not only supplied prestigious carvings and models for yachts, he was also the carver for Deptford and Woolwich dockyards and was responsible for the carvings on *Lenox*.

Few would question Charles's impeccable good taste but not everyone shared his enthusiasm for the arts; or, rather, their cost. At a meeting of the Admiralty Commission on 3 November 1677, a letter from Sir Richard Beach, the Commissioner at Chatham, was read out, criticising the lavish expenditure upon the yachts. Charles agreed, and expressed his concern regarding shipwrights adorning their ships with carvings, painting and gilding without any limitation but at their own pleasure. In agreement the Commission resolved that costs must be kept within the limits of the builders' estimates and that Sir John Tippetts, the Surveyor of the Navy, and Sir Anthony Deane were to prepare a standing settlement for carving, gilding and painting for all His Majesty's ships, in effect an establishment of decoration. Samuel Pepys, to his credit, noticed the shipwrights were being rather unfairly criticised and pointed out that in

general the Master Shipwrights, when questioned about a warrant for their proceedings "do generally pretend His Majesty's verbal command". With commendable honesty Charles agreed that it was his pleasure to give verbal commands, but let it be known as a standing rule for the future that the Master Shipwrights must also have approval in writing from the Lords of the Admiralty. It was further resolved that any unauthorised work would be paid for out of the wages and salary of the Master Shipwright who was responsible.

Charles then remembered the Van de Velde paintings he had asked Phineas Pett III to supply for the *Charlotte*, and agreed with the Commission that he should be speedily paid his expenses.[7]

CARVINGS

Specialists working under contract had for some time made ships carvings. The carvings of the last third-rate built, the *Defiance* of 1675, cost £100 according to contract. Her carvings were the equivalent of an earlier ship, the *Resolution* of 1667.[8] By late October 1677 it had been decided that the carvings for the new ships should also be made by contract.[9] At Portsmouth Lewis Allin, the carver, made an estimate for carvings that paralleled those of the *Rupert* of 1666. He estimated that to match earlier ships in quality and quantity would cost £160, claiming the considerable increase was because the new ships were much larger than the old.[10]

Up to 1675 the sterns of two-deck English ships had hardly changed since the 1640s. They consisted of a large Stuart or, for a short time, Commonwealth Arms situated above a single row of windows for the captain's great cabin. During the Third Dutch War, French ships, with two rows of stern windows, visited Spithead where they were observed by Charles, James and Anthony Deane. Among them was the *Superbe*,[11] a two-deck ship with two rows of stern windows, an aspect of her design that was greatly admired by the English.[12] As a result the new fashion was subsequently adopted in English ships, including the fourth rate *Woolwich*, built by Phineas Pett III in 1675, and the recently repaired *St David*.[13]

Lenox and the other new ships were similarly styled with two rows of windows, the lower for the captain's great cabin, and the upper for the coach, containing the master and first lieutenant's cabins. This change

Below: *Drawing of* Lenox *viewed from the larboard quarter, in pencil and wash. The drawing, by Willem Van de Velde the Elder, measuring 16¼in by 31¾in , was probably made at the launch of the ship. (Museum Boijmans van Beuningen)*

Above: *Detail from Van de Velde's drawing of* Lenox's *larboard quarter (pictured opposite), showing her stern carvings. (Museum Boijmans van Beuningen)*

reduced the size of the Stuart arms and pushed them into the area above the upper windows. Females in Restoration costume replaced the lion and unicorn supporters of the arms. These females bore a strong resemblance to the figure of Britannia, modelled by Frances Teresa Stewart for a medal celebrating the Peace of Breda in 1667, and on coins ever since.[14] Charles had long pursued the beautiful Frances Stewart who became the Duchess of Richmond and Lennox in 1667. She may well be considered the model for one of the supporters. It is irresistible to speculate the other figure is Louise de Keroualle, Duchess of Portsmouth and Charles's French mistress, with whom he had a son named Charles Lennox. The previous Duke of Richmond and Lennox, husband of Frances Stewart, died without issue and King Charles recreated the title to make his son the Duke of Richmond of the third creation, and among his other titles, Earl of Lennox. It is generally accepted the new ship was named after him.[15]

The carved badges on the stern that filled the panels followed typical Stuart devices. The rose and thistle beneath a crown, the crossed sceptres, and intertwined "C" had also been used for a long time. Although difficult to define from Van de Velde drawings, the badges below the quarter gallery windows may be the Lennox arms or a copy of the Darnley jewel.

Between the windows and panels and following the upright stern timbers were bold brackets in the form of figures playing flutes. In contrast, the horizontal rails were restrained mouldings. Craftsmen made the

mouldings using a number of combination moulding planes, each with a particular concave or convex shape. The taffrail, which formerly consisted of heavy carvings, now followed the example of the rails to form a simple smooth curve. The strong vertical emphasis, and strange mix of traditional and neo-classical carvings were a unique architectural style peculiar to English ships and immediately distinguishable from either Dutch or French practice.

Third-rate ships had long been decorated with an emaciated lion figurehead with a whorl above its head. This was retained for *Lenox,* but a crown replaced the whorl above its head with a cherubic supporter blowing a trumpet worked in just abaft on either side. However, the designs for the rest of the decoration, the ring ports, quarter galleries and bulkheads were unaltered by the new order and probably remained in the hands of the Master Shipwright and Master Carver.[16] Only the number and situation of these carvings were subject to control.

Although Sir John Tippetts and Sir Anthony Deane were to prepare the standing settlement for decoration, credit for the new designs probably belongs to King Charles. He retained an enormous interest in the subject and continued his old practice of giving verbal directions. During the following year he met William Bagwell who was acting as overseer for one of the new third-rate ships building at Bristol. Charles gave Bagwell verbal orders to build two balconies in the ship's stern. Bagwell, aware of the new resolutions, wrote to the Navy Board informing them of Charles's order and requested confirmation of them.[17]

Apart from Charles, another person known to be involved with the designs was Edmund Dummer, an Extra Clerk in the Surveyor's office. It was he who drew the designs of the sterns and sent them to each yard.[18] Dummer may have taken orders directly from Charles or indirectly through Sir John Tippetts, the Surveyor. The designs for the ornamentation of the new ships were formally presented to the king and Admiralty Commission on 12 January 1678, "to prevent the extravagant and uncertain expensefulness of builders in their adorning of their ships". His Majesty approved and confirmed the designs presented to him.[19]

A week later a contract was signed between the Admiralty Commissioners and Joseph Helby for the carvings for the third-rate ships that were currently building at Deptford and Woolwich. The carvings were to be equal in goodness to the carved work of the *Defiance.*[20] (See Appendix 6 for a transcription of the contract for the carvings).

It may be thought the forward planning and principles of the "standing settlement" leading to the contract would have saved the king money or at least prevented any cost increase. This proved not to be the case. The original estimate of £160 for the new ships' carvings was confirmed in the new contracts, a huge increase over the £100 of the earlier carvings of the *Defiance.* The difference does not seem justified by the relatively small increase in the size of the new ships. Much has been made, then and ever since, of the lavish cost of carvings. Bearing in mind *Lenox* cost £16,139 to build, rig and store, the carvings represent just one per cent of the total.[21] The cost represents about twenty carvers working for three months; this indicates a huge productivity rate considering the number of carvings on a seventeenth-century warship. A standardised design did not mean that all the ships would be identical; nothing would be more abhorrent to seventeenth-century eyes. Each ship would have its individual characteristics and unique details.

In order to produce the carvings, a copy of the contract, a drawing of the stern and another of the lion was sent from the Navy Board to the Master Shipwright, who then passed them on to the Master Carver.[22] The largest individual carving was the lion which measured 14ft in length.[23] The arms and taffrail at the top of the stern were made from large fir timber[24] 10in thick[25], although elm was also used in other places.[26]

The contract states the king's timber was to be used to produce the carvings.[27] For the quarter pieces, for example, it was roughly cut to usable

Above: *Drawing of* Lenox *from the larboard bow, in pencil and wash, measuring 21¼in by 36½in by Willem Van de Velde the Elder, probably made at the launch of the ship. Inscribed "Lenocx". (National Maritime Museum, PT 2427)*

shape in the dockyard, then sent to the carvers for finishing. When finished, carvings such as the brackets on the stern and the gun-port decoration were hollowed out and fastened in position.[28] It is difficult to know exactly how good the quality of carving was, but as those for the *Charlotte* yacht cost £507 8s 6d and *Lenox* cost £160, there must have been a considerable difference, especially as *Lenox* was much bigger.[29] It is probable that *Lenox*'s carvings were more direct and coarse, with facets made by the

carvers' chisel still visible, while those for the yacht were smooth and better defined.

PAINTING AND GILDING

Once the carvings were in place they were painted. This was not an ideal situation for *Lenox*, as she was a new ship and most of her timbers were unseasoned. Damp wood does not allow paint to enter the pores and paint is liable to peel off in the heat of the sun.[30] The head, stern, bulkheads, brackets, port rings and galleries were primed three times, stopped with putty and painted yellow (ochre).[31] The insides of the gun-ports were similarly prepared but finished with red lead.[32] The planksheers, gunwales,

thoroughly dried.[36] Some yards employed contractors to carry out the painting, most notably at Portsmouth where Mary Harrison was responsible for a period of at least twenty years. The painting at Deptford may well have been undertaken by the dockyard workers, as there is little evidence of contractors being employed. If a fixed price for painting was not agreed beforehand, the yardage was calculated by running a measure on the outside of the work but not into the hollows of the carvings.[37] The cost of painting *Lenox* was £98 6s 7d,[38] more than half the cost of the carvings. It is reasonable to suppose the quality of the workmanship followed other seventeenth-century crafts and was exceptional. However, during the time William Sutherland was employed measuring painters' work at Portsmouth, he saw painting being performed over dirt, and in the rain, which washed the paint off, making the ship "strangely besmeared".[39]

No additional allowance was made for gilding the ships. However, such austerity did not mean she was completely devoid of gold, for the lion of the head and the king's arms in the stern were expected to be gilt[40] to the value of about £10.[41] The only vessels that were lavishly gilded were royal yachts. To gild with "rich gold" the cabins of the *Isabella* yacht cost more than the total cost of painting *Lenox*.[42]

SHEATHING

The purpose of underwater protection was to prevent damage caused by shipworm and to reduce the growth of marine weeds and barnacles that hindered sailing performance. The protection applied to *Lenox* was organic and would deteriorate with time. When this happened the old, decayed materials were removed and new ones applied. All the materials necessary for her protection were recorded in her building list[43] and are the same as those found on the remains of the *Dartmouth*, wrecked in 1690.[44] As protection against shipworm, deal sheathing to the underwater regions of her hull was nailed in place. It was hoped that shipworm would not eat its way through the sheath before it was replaced. The sheathing extended from the waterline to cover the wales and the space between them.[45]

A layer of tar and hair was applied by mops made from thrums[46] that were then covered with 1in thick board nailed to the ship plank.[47] Some three hundredweight of sheathing nails[48] were used to hold it in place. When ships were commissioned to serve in southern waters where shipworm was more prevalent, sheathing was always applied and frequently renewed. If not, then the sheathing could be almost completely eaten away.[49] Another problem was that sheathing nails caused almost as much damage to ship plank as the worms.[50]

Sheathing was an inadequate solution and not satisfactorily solved until the introduction of copper-bottomed hulls in the eighteenth century. Experiments had been carried out during the 1670s using milled lead sheet. Unfortunately the electrolytic action of different materials was not understood at the time, causing bewildered consternation among those who witnessed it. A typical result of these experiments was the considerable damage done to the ironwork of the *Henrietta* on a voyage to the Mediterranean.[51]

Some 2½ tons of milled lead were used on the *Lenox*. Some of this may have been used to sheath the area near the keel and to line the bread room and for other similar applications.

PAYING

Van de Velde drawings seem to show that the hull of *Lenox* up to the water line was covered, or payed, with "white stuff". White stuff was made from train oil (whale oil), rozin (pine resin), and usually brimstone (sulphur),[52] boiled together. The brimstone offered some protection as poisonous antifouling. White stuff, as its name suggests, turned white when wet and was considered of pleasing appearance. One of the problems with white stuff was that the surface finish was rough and increased water resistance. It lasted for a period of about three years depending upon the usage of the

bulkhead timbers, bitts, cross pieces and capstans were again similarly prepared but finished in black. Internally, the roundhouse, galleries and captain's bedstead and other areas, amounting to about 180 square yards, were finished in grained walnut colour[33] or veined stone.[34] Presumably the grained walnut imitated wood and veined stone marble. Other painted cabins included the steerage, coach, gunners', boatswain's, carpenter's, surgeon's, purser's and two cabins at the bulkhead of the steerage.[35] It is almost certain the rest of the interior was left natural wood, a practice that would certainly be beneficial in helping wood dry out (season). However, John Shish did write in 1685 that he recommended priming the beams, knees and spirketting between decks with oil and colour if they were

Abovet: *Drawing of* Lenox *viewed from starboard quarter, in pencil and wash and measuring 15¾in by 20½in, by Willem Van de Velde the Elder, probably made at the launch of the ship. (National Maritime Museum, PY 4119)*

ship. Large quantities of soap and tallow (animal fat) were ordered for the ship to be tallowed. Mixed together and payed over the white stuff it formed a very smooth underwater surface. As one would expect, it lasted only a couple of months but during this time it certainly increased sailing performance.

The Van de Velde drawings show the main wales, the thick stuff between them, the first plank above and the planking from the main wale to the waterline payed with black stuff; a practice that had taken place for a very long time.[53] Black stuff was a mixture of tar and pitch and when applied hot would penetrate the wood grain. When cool it would harden to a durable finish that would not wear off as quickly as white stuff.

It was usual practice to pay ships' sides above the black stuff with rozin (resin) and tallow to leave the natural finish seen on Van de Velde paintings. When left in ordinary, it was thought better if the upper deck, quarterdeck, forecastle, bulkheads within and without, and the planksheers be paid with a mixture of one third pitch and two thirds tar, a mixture called soft stuff.[54]

STERN LANTHORNS

Lenox was equipped with three raking poop lanthorns at the top of her stern. The larger middle lanthorn was 6½ft high and 4ft diameter while the outer two were 1ft shorter and 6in narrower. They were supplied by John Hardwin, a specialist tradesman.[55] He was the established supplier to Deptford, but over the years his stock had been reduced by emergencies and wars, for which he was still awaiting payment. Within a week of *Lenox* being launched, Hardwin was complaining that the last of his stock of lanthorns had been exhausted by his recent delivery to Deptford; a delivery that must have been for *Lenox*. In the same letter he tendered for new lanthorns for the thirty new ships, which were slightly bigger and more expensive than those supplied to *Lenox*.[56]

SHIP'S BOATS

The newly built *Lenox* was supplied with two boats; a longboat and a pinnace, the usual complement of boats for a third-rate ship.[1] William Stygent, the Master Boatbuilder at Deptford, was responsible for their construction.[2] However, the pinnace intended for *Lenox* appears not to have been ready in time and was left behind when she sailed for the fleet anchorage at Chatham. Thomas Shish subsequently sought this pinnace for the new ship *Captain* he was building at Woolwich.[3] In the absence of the pinnace, another boat, measuring 33ft long, was found. This boat was about 3ft longer than that normally given to a third rate. In fact, it was the same size as the pinnace of the first rate *Royal Sovereign* in 1688.[4] It cost £24 15s, calculated at the rate of 15s a foot. The 32ft longboat was wider and of much heavier construction than the pinnace. It cost £28 16s at the rate of 18s a foot.[5]

The two boats did not stay with *Lenox* during the years she was laid up in ordinary but were kept safely ashore in storage.[6] During the period leading up to war in 1688 most of the fleet was commissioned before *Lenox* and many ships were short of boats. As a result *Lenox* lost hers to them. In November 1688 she was recorded as having a 30ft pinnace but no longboat.[7] Six months later, as she remained idle in ordinary at Chatham, she was found to have no boats at all.[8]

By 1690 a third, smaller boat was added to the complement of boats for third-rate ships. Those aboard *Lenox* referred to this boat in later years as either the jolly boat or yawl.[9] Three-deck ships always carried three boats and the *Duchess*, built immediately after *Lenox* at Deptford, had a longboat, pinnace and a 20ft skiff. When *Lenox* was commissioned in 1690 she would have had, in common with her sisters, a 31ft longboat,[10] a yawl of about 23ft and a 30ft pinnace.

In common with all ship's boats of the period, those belonging to *Lenox* would have been made from oak. The planks were laid edge to edge in carvel rather than the overlapping clinker manner. The carvel method of planking mirrored that used on ships and was preferred and well understood by ship's carpenters who had invariably started their careers as dockyard shipwrights. *Lenox*'s long serving carpenter, Ambrose Fellows, was a typical example, having been a shipwright at Deptford dockyard. All of *Lenox*'s boats were constructed in a simplified method of ship construction. The frame timbers were spaced at a distance known as the "room and space". Each frame was made up of a floor and two futtocks that scarphed, or overlapped, each other. The boats, including the small yawl, were planked with 1in thick oak board.[11]

It is generally thought that the sturdily constructed longboat was used to carry anchors and water casks while the lighter, narrower pinnace carried officers about their business. This simplified image is somewhat misleading, for *Lenox*'s boats were used for a great variety of roles. The pinnace and longboat raided merchantmen to press men; the pinnace also took sick men ashore. All the boats on occasion towed *Lenox* into harbour,

her boats saved a man from drowning who had fallen into the sea, and the longboat brought ballast aboard and went on firewood expeditions. If the boats were essential in servicing *Lenox*, their use in action was even more profound. At the Battle of Beachy Head, the longboat passed a hawser for towing a disabled Dutch ship out of danger and prevented her capture. In the attack on Cork, the boats towed *Lenox* into the harbour and landed men for an amphibious assault. French ships were attacked and either taken or burned by the boats. The Battle of La Hogue was perhaps the most famous attack ever made by ship's boats. Those from *Lenox* played their full role in helping burn a large part of the French fleet. It could even be claimed the boats were responsible for more damage to the enemy than *Lenox* herself.

During her years of service up to the end of the seventeenth century, *Lenox* lost a number of longboats. One was lost at the Battle of Beachy Head, another was cut away while being towed, and another given to the *Edgar* when she left the Mediterranean fleet for England. In a tragic accident a pinnace was sunk after an accident involving the yawl.

The longboat was a sturdy, seaworthy boat of much heavier construction than the pinnace. It was equipped with a 10in windlass for handling the buoy rope and other anchor tackle.[12] Boats of this size were equipped for sixteen oarsmen double-banked – that is, two oarsmen, each with an oar on every thwart. In 1677 *Lenox* was supplied with forty-two boat oars and eighteen barge oars. The boat oars were cheaper than the barge oars, probably because they were smaller and used by double-banked oarsmen in the longboat. Four of the thwarts were secured with knees, while the rest were removable to allow space for large stores such as casks.

The boat could be rigged with a fore and mainsail for longer voyages.[13] The seaworthy longboat was normally towed behind and only hoisted aboard *Lenox* when she ventured far out into the Atlantic. Hoisting the heavy longboat aboard was a major operation, involving bracing the main and fore yards towards each other and rigging tackles from the yards and mast tops.

The pinnace was primarily used to transport men, particularly the officers, and had a separate compartment at the stern to accommodate a helmsman. It was rowed by eight oarsmen, sitting single-banked on the opposite side of the boat from where the oar entered the water. Long barge oars necessary for this arrangement were supplied to *Lenox* in 1677. The pinnace was normally hoisted into the waist when *Lenox* went to sea and stowed on top of spare topmasts that lay fore and aft from the gallows to the forecastle.

The yawl, of about 23ft, was a useful complement to the longboat and pinnace. In design it was related more closely to the pinnace than the longboat. Like the pinnace it had a separate compartment for a helmsman and was rowed single-banked. It was used as a complement to the larger boats and performed tasks that did not require a large carrying capacity.[14]

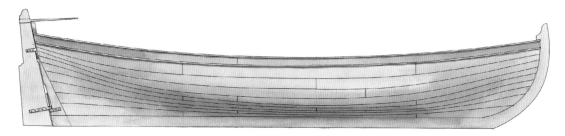

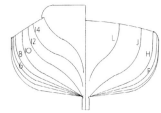

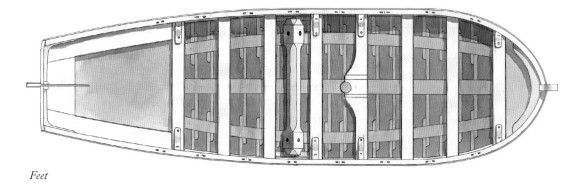

Feet

Left: Lenox's *31ft longboat, c. 1690. Based on a boats contract (NA ADM 106/3069, see Appendix 8) and contemporary models.*

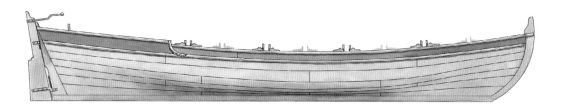

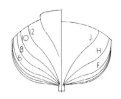

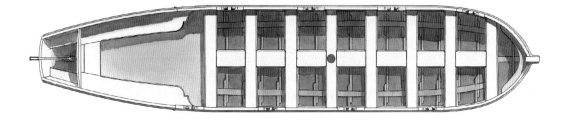

Left: Lenox's *30ft pinnace, based on a contract from 1690 and contemporary models.*

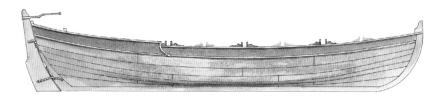

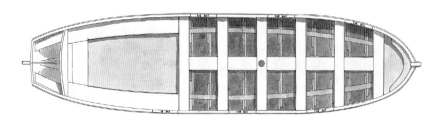

Left: Lenox's *23ft yawl, based on a contract from 1690 and contemporary models.*

ANCHORS

Lenox was supplied with seven anchors; four of these were very large, ranging in size from between 40cwt and 48cwt.[1] The two forward main anchors were called bowers because of their position near the bows of the ship. The best bower, weighing 46cwt, was situated forward on the port bow. Opposite, on the starboard bow, was the small bower weighing 44cwt. Behind the small bower was the largest anchor carried, the 48cwt sheet anchor.[2] The smallest of the four main anchors was the 40cwt spare anchor, which was normally carried opposite the sheet anchor but would sometimes have been stowed in the hold. Van de Velde drawings and paintings seem to show that bowers were always carried to sea but that the sheet and spare may have been interchanged. Moreover, the spare is often not visible, suggesting it was stowed within the ship.

The anchor most often used was the best bower situated on the port bow. This anchor was the first to be cast off, followed if necessary by the small bower on the starboard side. This arrangement ensured the cables would not cross and foul as the wind in the northern hemisphere usually veered in a clockwise direction. The sheet anchor was the seaman's last hope and was never used except in extreme necessity. *Lenox*, in common with all other ships of her type, had four hawseholes in the bows so that in theory she could ride by four anchors. However, it had long been known that it was better to ride with one anchor on a "shot" of two cables spliced together rather than by three anchors on short cables.[3] Admiral Leake in the *Prince George* survived the great storm of 1703 in the Downs riding snug on two and two thirds cables, while many other ships were destroyed on the Goodwin Sands.[4]

Below: *Lenox's best bower anchor. The best bower anchor was about 2½in smaller than the sheet anchor and 2½in bigger than the small bower. The spare anchor, the fourth large anchor was 8½in smaller than the best bower.*

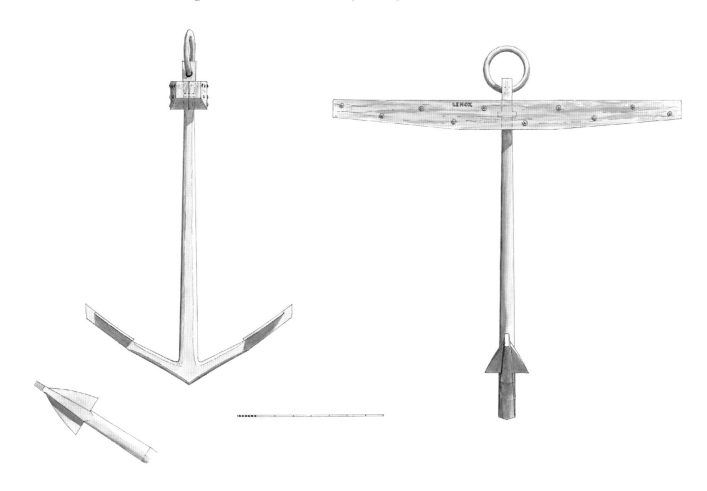

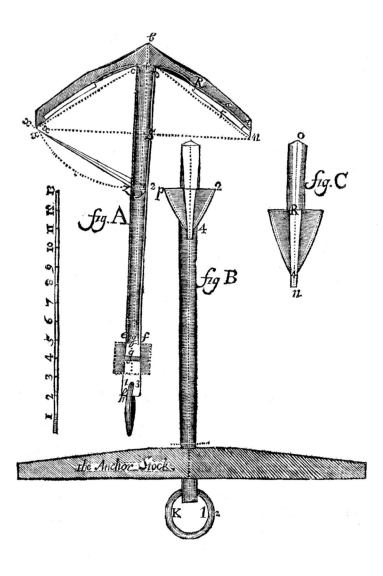

Above: *Print of an anchor from William Sutherland,* Shipbuilding Unveiled, *Part 1, 1717, p16. (Author's collection)*

The largest of *Lenox*'s three small anchors was the stream anchor. It weighed 11cwt and was just a quarter of the size of the main anchors. It was used, as its name suggests, to anchor in an easy stream or tide.[5] The next smallest anchor, the kedge or kedger, weighed only 4cwt. It could easily be carried in a boat and dropped ahead of *Lenox* in narrow streams. By winding in her cable with the capstan *Lenox* could position herself to take advantage of the tide or to prevent herself running ashore.[6] A yet smaller kedger weighing only 2cwt was also used for similar work in very light currents.

Making an anchor was a skilled and expensive exercise and cost twice as much as a gun of the same weight. The anchors cost so much more because the manufacturing process was very different. A gun was cast by pouring molten iron into a mould. As the iron cooled it formed regular shaped crystals. This structure was far too brittle to withstand the tension to which an anchor was subjected. To increase the strength the grain structure was elongated by forging iron consisting of two-thirds Spanish iron into long bars.[7] The iron bars were then heated till soft then forged together into the shape of the anchor. At the crucial joint of the crown, where the arms meet the shank, the bars were bent so that the elongated crystals followed the bend in a manner similar to grain in wood. The temperature of the bars during forging was critical. Too cool and they would not properly weld together, too hot and the bars would melt, destroying the elongated crystal structure.

To make sure the fluke, or palm, of the anchor embedded into the sea bottom, a large cross-piece of timber called the stock was fitted at right angles to the arms. The stock was made in two halves and bolted together.[8] The name of the ship to which it belonged was painted or carved into it.[9]

Henry Loader, the anchorsmith of Deptford, supplied the anchors for *Lenox*. At the end of July 1677 he contracted to deliver fifty-two large and sixty-six small anchors for the thirty new ships.[10] Some of these anchors are the same or close enough in size to be those given to *Lenox*. The cost of the anchors and cables was enormous, being rather more than an entire lower gun-deck battery of 32pdr demi-cannon.[11]

As important as the anchors were the anchor cables. They were specially made for underwater use and consisted of three right-hand laid hemp ropes laid together left handed. When under tension the fibres became tightly compressed, making the cable virtually watertight. *Lenox* was supplied with seven 18in circumference (5¾in diameter) cables for her four main anchors and six smaller ones for her lesser anchors.[12] Each of the 18in cables was ninety-five fathoms in length.[13] Seven main cables were the standard issue for third-rate ships.[14] Two of the cables were spliced together for the sheet anchor and called the sheet shot; three for the best bower; one for the small bower and one for the spare.[15]

LAUNCHING

Preparations for the launch of *Lenox* began more than a month before the event, when requests were made to the Navy Board to supply items not made at Deptford, such as launching blocks and cables.[1] Even though *Lenox* was built in a dry double dock she had to be slid down ways into the Thames, and the construction of the ways and cradle involved by far the greatest work.[2] The slope of the floor of the dock did not allow high tides to reach the depth of water necessary to float out a ship the size of *Lenox*. Bilge ways, in effect, wooden rails nearly 3ft wide, had to be laid each side of the ship leading down into the water. Huge quantities of timber were used in their construction, including nearly sixty loads of straight oak timber, twenty-one loads of 7in oak thick stuff and hundreds of feet of plank. The hawsers ordered to move and restrain *Lenox* included 136

fathoms of 8in hawser and 630 fathoms of 6in white hawser. Altogether the total cost of the materials was more than £430, too much for the materials to be discarded after the launch. Once used, the timbers and hawsers were set aside for use on other new ships that would follow.[3] The sliding cradle was attached to *Lenox* by upright, vertical timbers called clamps[4] at the bow and stern. They were made to fit closely the shape of the hull and were bolted to it.

The considerable weight of *Lenox* could now be transferred from the blocks under the keel onto the cradle. The splitting blocks under the keel were then cut away with twenty-eight splitting wedges supplied by Henry Loader.[5] *Lenox* was, as it was termed, struck down upon the ways.[6] As work neared completion, a launch date agreeable to the king was finalised. On Friday, 5 April 1678, John Shish told the Navy Commissioners that if the king so directed, the new ship could be launched the following Thursday.[7] The matter was discussed the next day at an Admiralty Commission meeting, attended as usual by Charles, where it was decided the

Below: *The flags are flying as this ship is launched and celebrations begin. From* A Geometric Plan of Deptford Dockyard, *Thomas Milton, 1753. (Author's collection)*

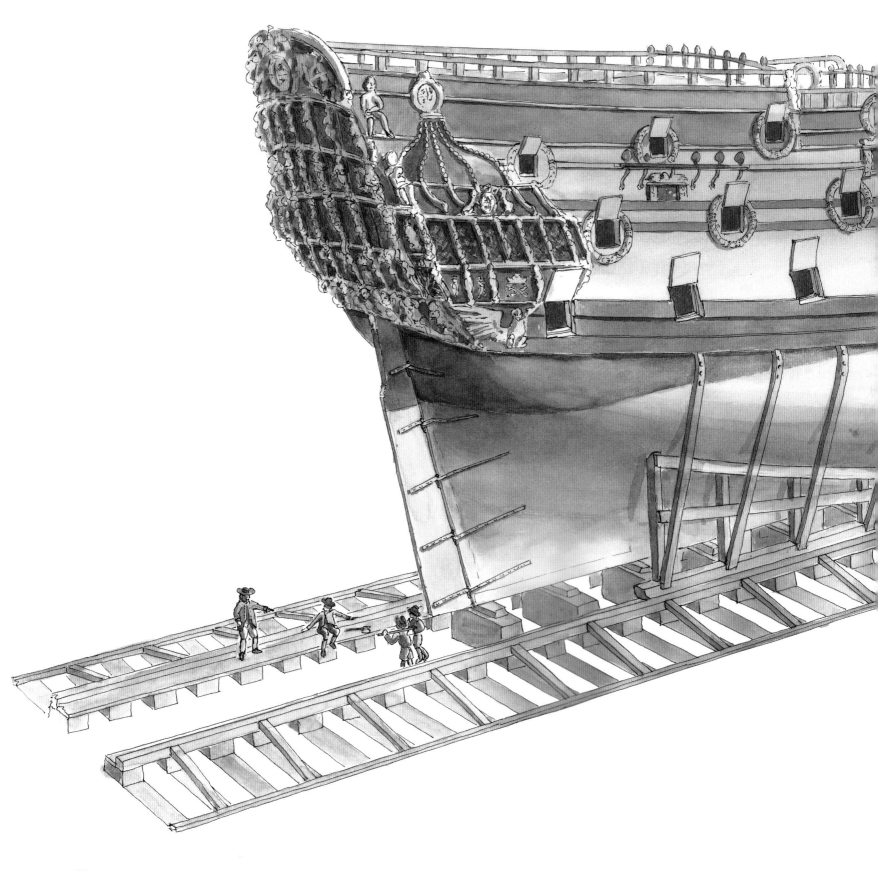

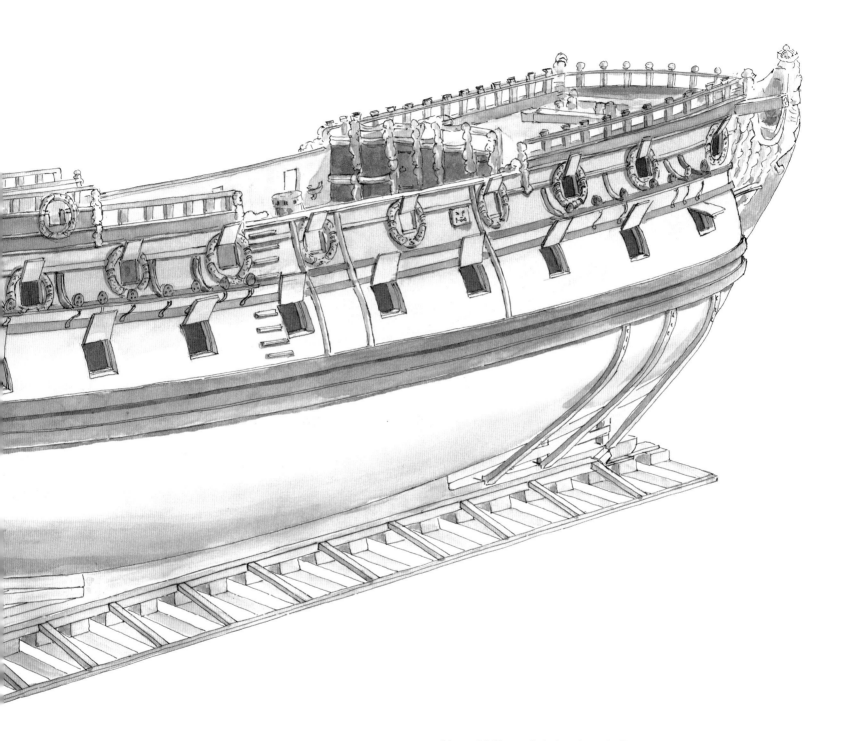

Above: *Making ready for launch, 12 April 1678*
The bilge ways and cradle are in place. All that now remains are for the splitting blocks under the keel to be removed, thus transferring the weight of Lenox *onto the cradle. The cost of the materials used to construct the bilge ways and hawsers was the sizeable sum of £431 7s 2d, although this cost could be offset as, following* Lenox's *launch, they could subsequently be re-used for launching other new ships.*

new ship should be launched the following Friday, and not Thursday as suggested by Shish. Charles probably changed the date so that he could make arrangements convenient to his own diary commitments.[8]

During the week left to him, Shish made preparations for the ritual that attended a launching. Poles were set up in the position of the masts and flags. These included the royal standard, the flag of the Lord High Admiral, union flags and a red ensign, ready to be flown as if the ship were a flagship with the king aboard.[9]

The new ship could not be allowed to slide down into the Thames without restraint. The Thames was so shallow that the stern of the ship would need lifting by lighters to prevent it hitting the bottom. To commence the launch the workmen at the crab capstans heaved the hawsers taught and tightened the screws at the top of the ways against the cradle[10] to slide the ship down the tallowed bilge ways to the stern of the dock. Once there, shores were set up to prevent any mishap.[11]

Friday 12 April 1678 dawned at Deptford with a fresh northeasterly wind blowing. Although breezy it was described aboard the *Dreadnought*, anchored in the Downs, as "fine mild weather".[12] King Charles, probably with much of the court in attendance, came down the River Thames from the Palace of Whitehall in his barges.

It is probable the King's mistress, Louise de Keroualle, Duchess of Portsmouth, and her five-year-old son by the King, Charles Lenox, were among them. Louise lived in "glorious apartments at Whitehall"[13] but as a French Catholic was unpopular among the general populace at a time when fears of Popish plots were rife.

Louise met King Charles years before when she first came to England as maid-of-honour to his sister, Princess Henrietta, who had married Philippe d' Orleans, the younger brother of Louis XIV. After the tragic, early death of Princess Henrietta, Louise returned to England in 1670 at the request of Charles.[14] She became his mistress, but in 1674 contracted a venereal disease from him, suffered a miscarriage and became very ill for a considerable period.[15] Understandably their relationship suffered badly, and Charles embarked on a number of relationships with other women. However, they had recently reconciled and thereafter remained very close until the end of the King's reign.[16] Naming the first of the new ships after their son was a gesture of Charles's re-discovered love for Louise. The event seems to have been a private naval affair that was not mentioned outside naval records. Others attending the launch included principal officers and administrators of the navy, who were provided with accommodation and three dinners during their stay, at a cost of £72 8s.[17]

As high tide approached, everyone, including all the workmen at Deptford, would have waited with great expectation as the uncertain process of launching the new and as-yet unnamed ship began. As well as being the first of the thirty new ships, she was the first big ship to be launched at Deptford since the *Royal Oak* four years earlier. To prevent the stern of the new ship hitting the bottom of the river, lighters belonging to Giles Thorpe, Francis Wyborne and John Baker were ready to lift the stern until the ship was floating in the river on an even keel.[18] Launching

could be hazardous and dangerous; some years later Fisher Harding, the Assistant Master Shipwright and brother-in-law of John Shish, was badly hurt when a prop being hewed away fell on his head.[19]

On this occasion, however, the operation proceeded smoothly and the ship entered the water and was secured to her moorings. Charles and his party probably chose this time to come alongside the new ship and go aboard. They were not the only ones: soon the ship would be packed with

Right: *The launch of* Stirling Castle *at Deptford on 28 July 1679, drawing by Willem Van de Velde the Elder. She was the third and last 70 gun third-rate ship John Shish built for the thirty ship building programme, the earlier two being* Lenox *and* Hampton Court. *Although probably built to the same draught as* Lenox, *Shish has added another upper deck port, crowded in toward the bow, to suit the official principal dimensions issued after building work on* Lenox *had commenced. This picture clearly shows the lavish celebrations that accompanied the launch of the ships. Shish was subsequently ordered by the Admiralty to account for the large expense. The king's trumpeters and kettle drummers can be made out on the poop, making it a distinct possibility that the figure with the outstretched hand at the forward end is King Charles II himself, in the act of officially naming the ship by throwing wine over the half-deck below. (National Maritime Museum, PY3920)*

all sorts of people associated with her who wanted to witness the ceremony of naming the ship. Those who could not come aboard, or chose to celebrate in their own boats, soon covered the river in craft, ranging from great barges to small wherries.[20] On the poop of the ship a company of His Majesty's trumpeters and kettle drummers under the command of the sergeant trumpeter, Traverse Price, played "according to their former custom".[21] Standing at the forward end of the poop, Charles performed the ancient ceremony of drinking part of a cup full of wine, throwing the rest forwards over the half deck and giving the ship her name, *Lenox*.

The name, always spelt *Lenox*, may be thought a misspelling of the name of the Scottish Lennox family dating back to the eleventh century. However, the King's son was called Charles Lenox and is referred to as Prince Charles Lenox on contemporary engravings; leading one to the conclusion that the ship must therefore have been named after him.[22]

Above: *A large third rate becalmed off Greenwich, by Willem Van de Velde the Elder. Large third rates such as this were only normally seen so high up the Thames after their launch or rebuilding at nearby Deptford dockyard. The ship has the same gunport arrangement as* Lenox *and thus may be her or her sister ship the* Hampton Court. *(National Maritime Museum, PZ7278)*

Charles, if he followed the custom, then heaved the standing cup overboard.[23] As a gratuity for building *Lenox*, John Shish then received from the king a customary piece of silver. Among the items given during previous years were flagons, cups and covers, state plates and cups, cans, tankards, chased bowls, basins and pieces of plate.[24] They were much prized, and when Anthony Deane requested a silver dish for building the *Resolution* in 1667 he was told it must be a bowl or some kind of drinking vessel, as it was the custom to drink His Majesty's and the Lord Admirals' health from it.[25] The piece given to Jonas Shish for building the *London* in 1670 was truly magnificent[26], and cost £22 4s from Humphrey Stoaks, a goldsmith of London (see plate section).[27]

To commemorate the building of *Lenox*, Stoaks received a bill to make a flagon to the value of £20. It was recorded as having the Duke of York's arms on it with an anchor through and the inscription "Att the launching of his Ma:ts ship *Lenox* ye 12th day of April 1678: Built at Deptford by John Shish his Ma:ts Master Shipwright there, Burthen 1096 Tunns, Menn 440 Gunns 70".[28] Shish knelt before the king to receive the flagon, then, following the old practice, drank His Majesty's health.[29] With his official duties over, Charles and his court then enjoyed the entertainment laid on by John Shish.

The celebrations were not confined to the royal party. The common workmen were also generously provided with food and drink.[30] They were entitled to celebrate with their king. Their ship had been completed in only ten months and finished before any of those building at Chatham, Woolwich, Harwich or Portsmouth. It must have been on occasions such as this that Charles mixed with ordinary shipwrights, such as William Bagwell, "encouraging all men of that trade, beginners as well as old practisers, and even assistants and foremen as well as master builders, nay, down to the very barge maker and boat maker".[31] After the celebrations Charles returned to Whitehall and the following day gave an audience to an envoy of the King of Denmark.[32]

Normally after a launching, the Master Shipwright immediately wrote a letter to the Navy Board, informing them of the depth of water the new ship drew. However, for easily imaginable reasons in the aftermath of the celebrations, John Shish was unable do this; instead his younger brother Thomas, the newly appointed Master Shipwright at Woolwich,[33] measured *Lenox* for him. "I went on board His Majesty's new ship the *Lenox* and when all the people were out, the ship drew fourteen feet eleven inches abaft and ten feet and a half afore and I judge there might be about eight tons of ballast in the midships at that time".[34]

Charles must have thoroughly enjoyed the company and hospitality provided by Shish, for he continued to attend the launchings of the next four ships at Deptford over the course of the next two and a half years.[35]

Encouraged by Charles, merry launching celebrations reached new heights during the building of the thirty new ships. They were, however, rather costly. During 1680 Shish applied to the Navy Treasurer for the reimbursement of his expenditure of £176 for launching ceremonies. Unfortunately for Shish, the Admiralty officials who attended the launchings had nearly all been replaced following the political upheaval of the Popish Plot. The new Admiralty Commissioners were not at all sympathetic to the old regime and ordered the Navy Board to examine Shish concerning the particulars of his expenses. At his examination Shish gave details and offered to swear they were true. Satisfied his account was genuine, the Navy Board wrote to the Admiralty Commission with their findings but to prevent such extravagance in the future they recommended that no such disbursement should be paid again without particular orders in writing.[36] The Admiralty Commission immediately issued orders in agreement with the Navy Board and further ordered that bills be forthwith made out to John Shish for the cost of entertaining His Majesty and for the diet and drink for the workmen.[37]

After the launch, the responsibility of *Lenox* passed forever from John Shish. Although his father Old Jonas must have been very proud of his two sons; both Master Shipwrights with growing reputations, he was deeply saddened by the loss of his wife Elizabeth, who died only one month before *Lenox*'s launch, on 5 March, aged fifty-nine years. Yet with twelve grand-children, Old Jonas must have been confident that the dynasty of shipbuilders he had created would last into the future, perhaps becoming remembered as the famous Petts were.

Jonas himself lived a few more years before dying on 7 May 1680 at the age of seventy-five. During his last sickness he characteristically wrote his own epitaph, part of which read "Once I was strong but am intombed now / To be dissolved to dust and so must you". He was buried at St Nicholas Church, Deptford, in the coffin in which he had prayed for many years. He had gained the respect of many people and at the funeral John Evelyn and three knights acted as pallbearers.[38]

Jonas was spared the tragedy that befell his sons. Thomas died in 1685 aged thirty-seven, the same year that Kendrick, the only surviving child of John and his wife Mary, died, at the age of fourteen. John fell ill during October 1686. On Sunday the 10th he sent for his close friend, Stephen Holland, an attorney. Holland arrived in the early afternoon about one or two o'clock and found John sick and weak but fully alert with his memory unaffected. John was anxious to settle his affairs and make a will. Among those present were his younger brother Jonas and his cousin John Chamberlain. Fortunately Holland could write in shorthand and was able to take instructions as John dictated them. John started off revealing his religion with a typical Puritan preamble worthy of his father. After he had finished writing his will, Holland read it back to him, he approved the contents and asked for it to be drawn up accordingly. Holland left to carry out his instructions and returned about one o'clock on Monday morning with four and a half pages of the drawn-up will. He found John much worse: "the violence of the ... distemper being a fever had so seized his head that he was not then of perfect mind nor in a condition to sign and seal." John Shish never recovered and died shortly afterward at four o'clock that morning.[39] News of his death swept throughout the royal dockyards and was recorded the same day in the journal of Edward Gregory, an officer at Chatham who would soon become Commissioner.[40]

Fortunately for those concerned the will was approved and John's final wishes carried out. His "dear and loving wife" Mary was sole executrix and he left her a considerable fortune. There were tenements and premises in Canterbury; tenements, gardens, closes, lands and premises in Erith; tenements, yards and premises in Deptford; and freehold lands, closes, meadows, pastures, pasture woods, underwoods, other singular premises and a number of acres lying in Hatfield Broad Oak. Mary was also to have all his goods, chattels, and household stuff including his plate, rings and jewels. Apart from all this there was Flagon Row, a street in Deptford, which he had bought jointly with Rebecca Shish, his sister-in-law and wife of his younger brother Jonas. His share was left to Rebecca. He left smaller amounts to other members of his family and to the poor of Deptford. Perhaps the most poignant request in his will was for his wife to look after the welfare of the infant son of his dead brother Thomas, and take as much care as if he were her own. He also left him £300 put in trust until he was twenty-one years old. The infant, named Thomas after his father, was presumably an orphan. The will also advised Mary as to who should be left property in her will.[41]

It was a sad, sudden and early end to the Shish family of Master Shipwrights. Only the youngest brother, Jonas, survived into old age, as a private shipbuilder based at Reddriff. A fitting memorial honouring the family survives to this day in St Nicholas Church, erected by the local Deptford community.[42]

Section One Plans

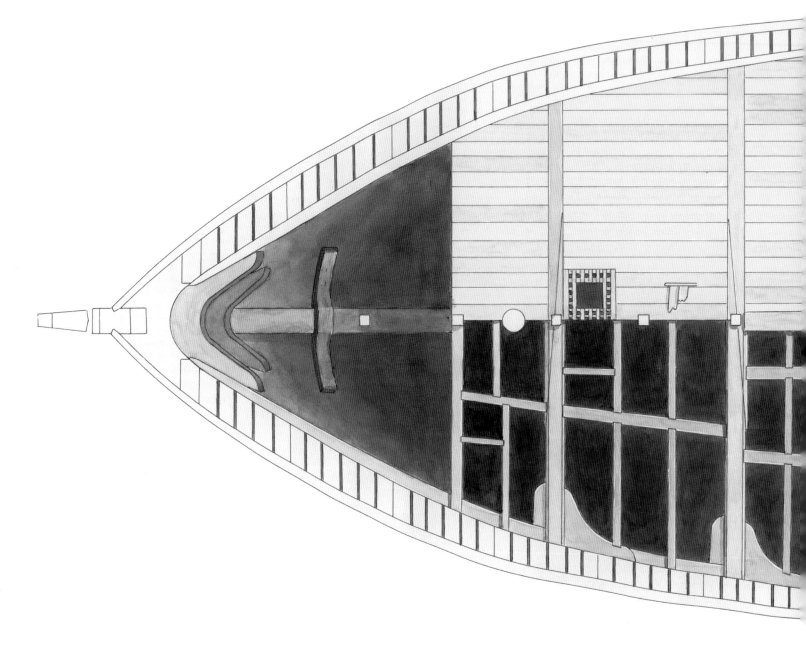

Feet

ORLOP DECK

Above: *Orlop deck (1:72). Note how the deck planks do not go over the beams but are instead laid into rebates so that they can easily be removed for 'rummaging' in the hold below to access stores.*

LONGITUDINAL
SECTION

*ght: Longitudinal Section (1:72). For clarity the cabins and internal bulkheads are
* shown. Ships were usually left in "ordinary" in this condition to allow the free
rculation of air.

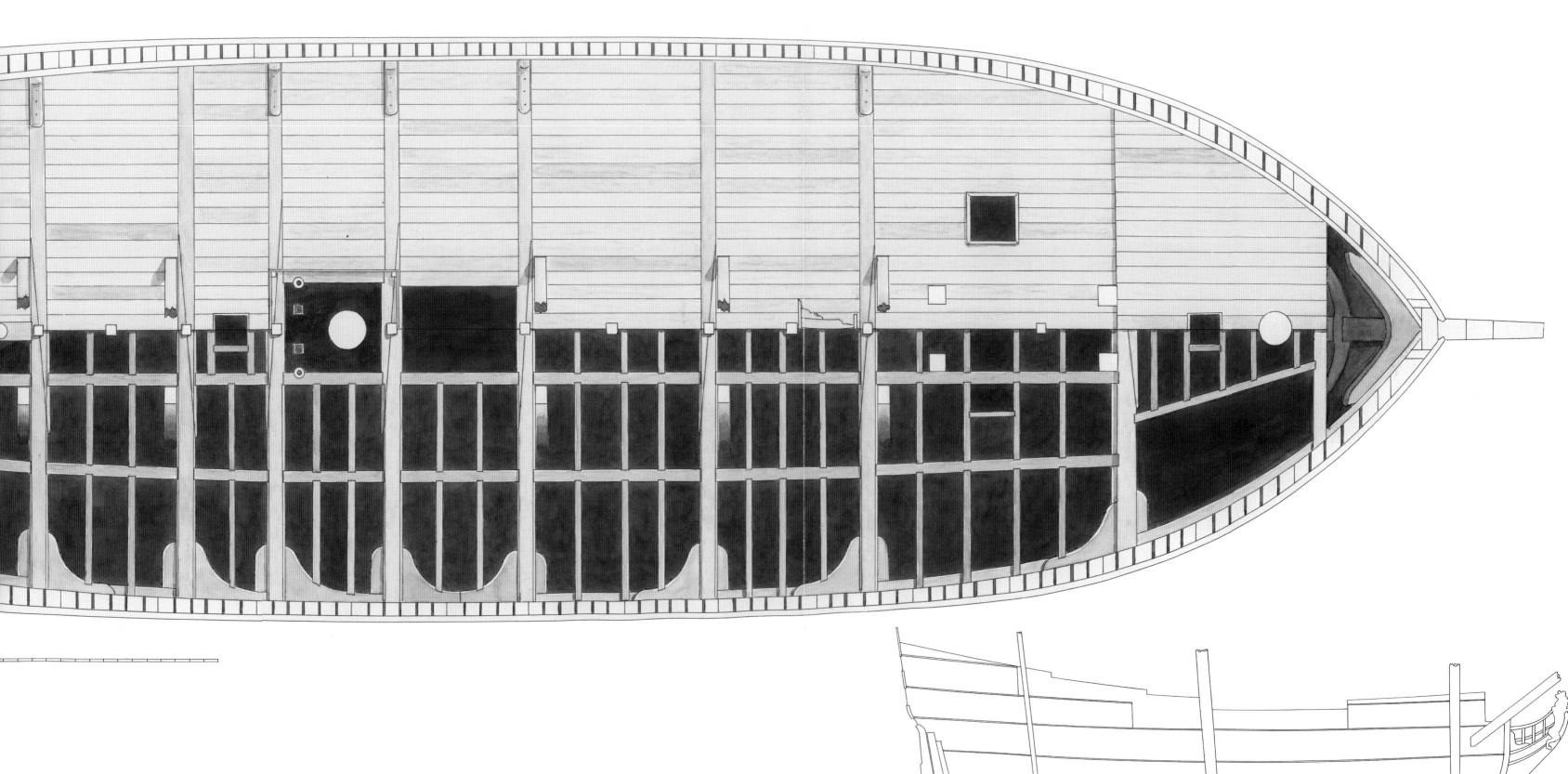

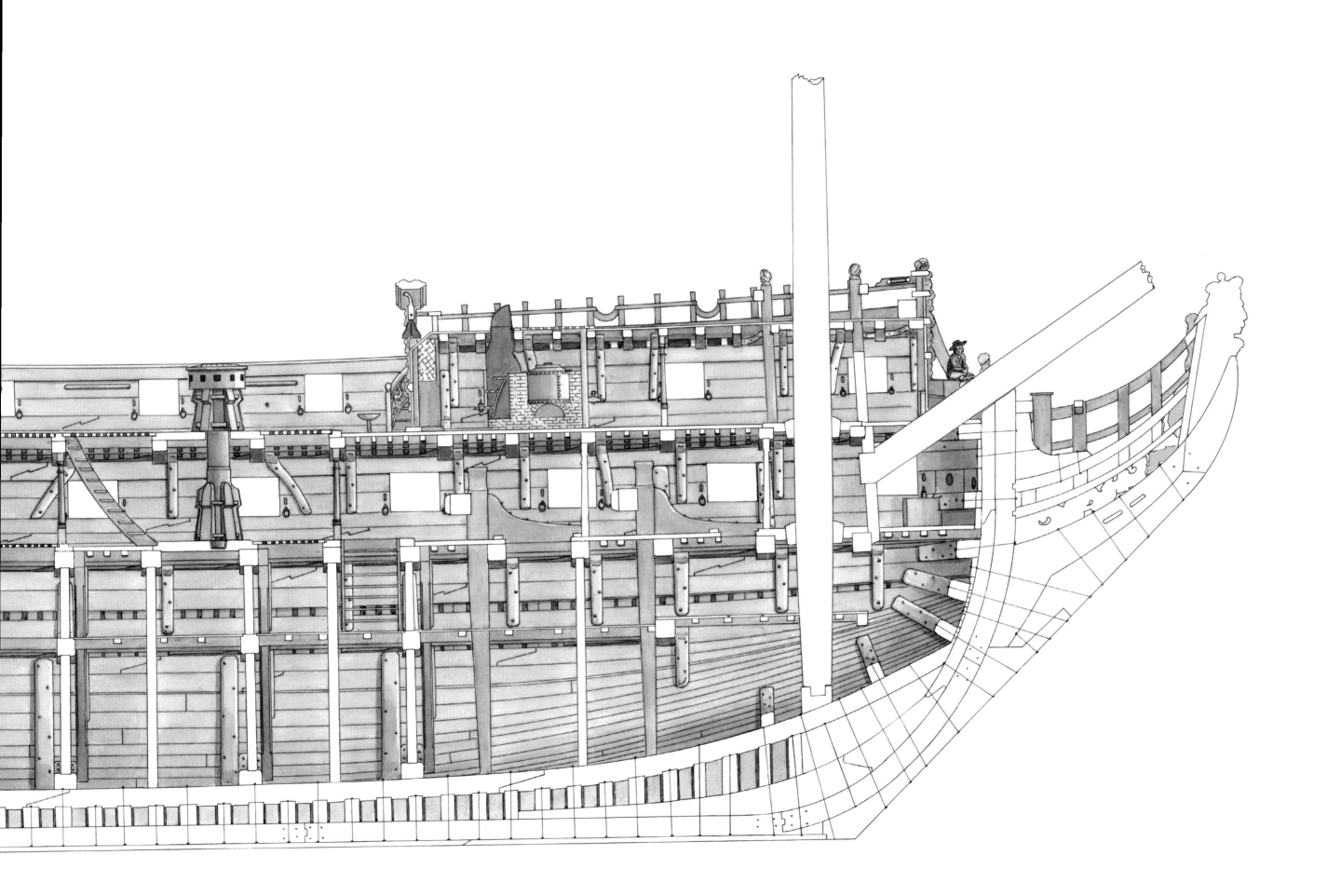

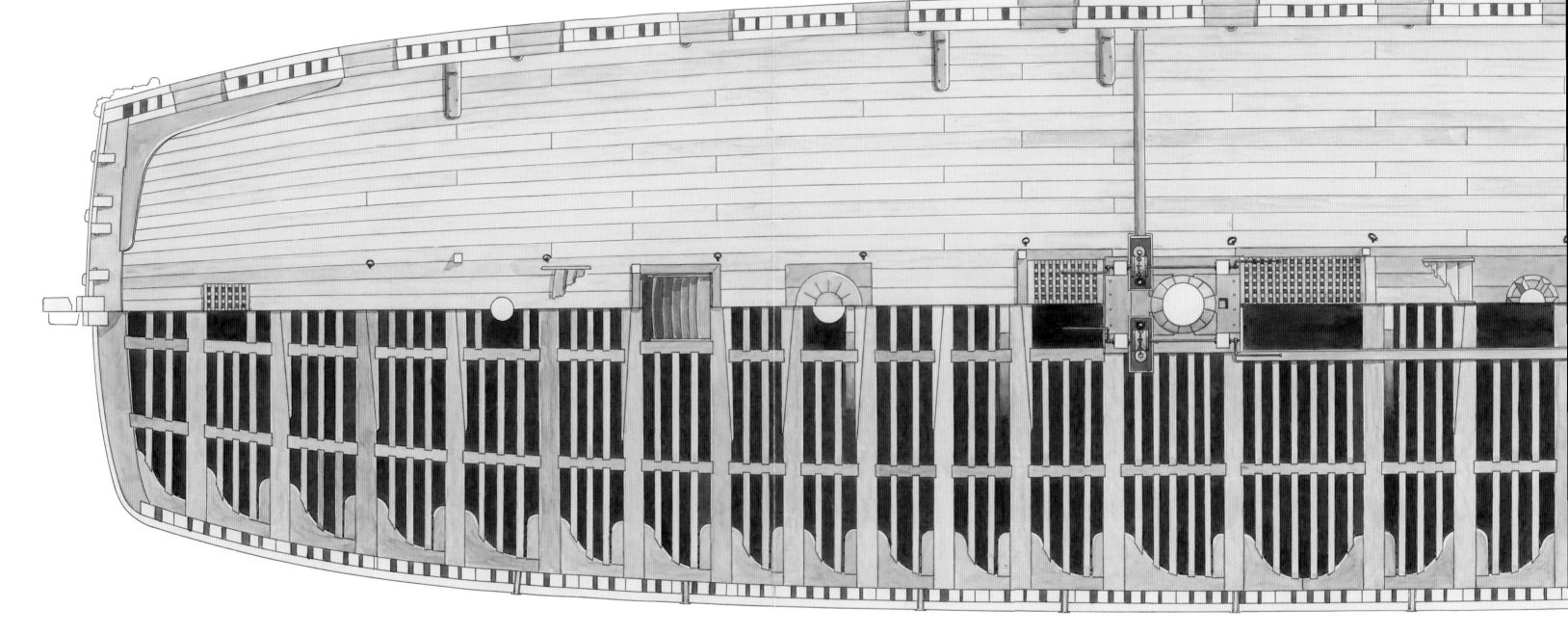

GUN DECK

Above: *Gun deck (1:72).*

MIDSHIP SECTION

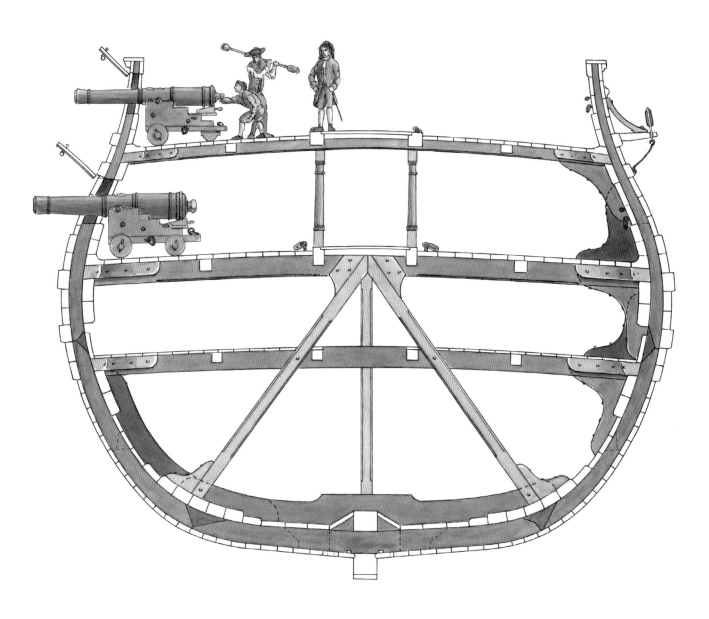

Above: *Midship section through waist (1:72). For clarity, the pointers and lodging knees are shown on the same side of the beams.*

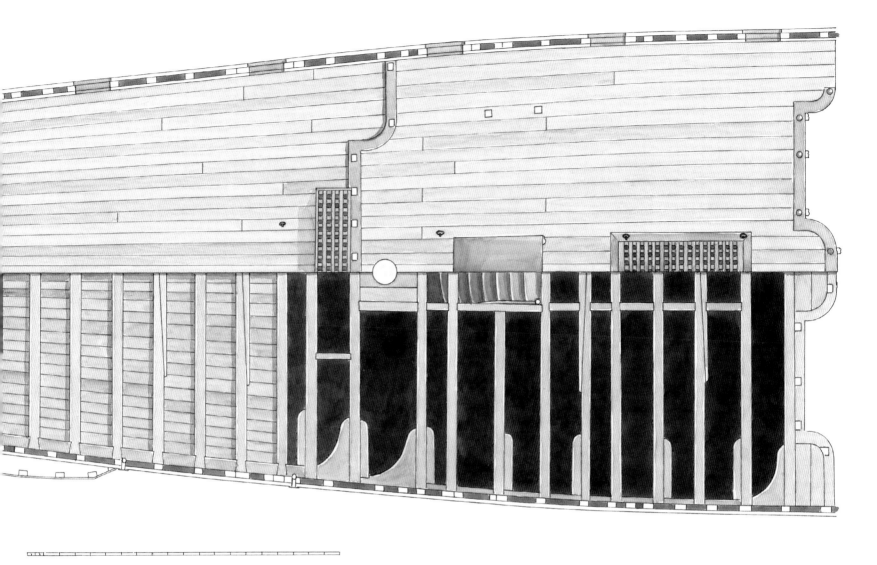

QUARTERDECK

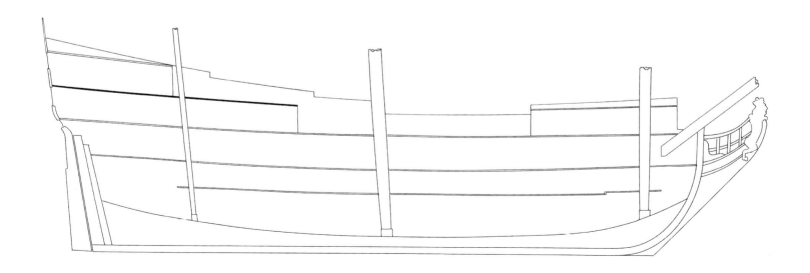

SHIPWRIGHT PRODUCTIVITY

In November 1675, Samuel Pepys estimated that just ninety men could build a third-rate ship in one year.[1] It seems almost unbelievable that so few men could achieve such productivity in the days when every piece of timber was hewn or sawn by hand. It is readily apparent that Pepys's estimate applied only to those men employed in the dockyard engaged as shipwrights or their supporting trades. All manner of bought-in goods produced by others such as sawn plank from the East, ironwork, carving and painting work were not included. Altogether this work accounted for approximately ninety more men. Considerable numbers of records survive, enabling us to make reasonable estimates of the productivity of the men employed at Deptford. These records include the Deptford pay books[2] and *Lenox*'s Building List,[3] which has an account of the total amounts paid to the dockyard trades. A rough estimate of the number of men required to build *Lenox* can be obtained simply by dividing the sum of the wages earned during the building, which totalled £3972 14s 8d, by the daily wage of 2s 1d for a qualified shipwright. This works out at 38,138 days' work for one man or 424 days for ninety men as suggested by Pepys. However, *Lenox* was about a fifth bigger by burden than the third rate Pepys originally had in mind, so that it is fair to make an estimate assuming 110 men were employed for one year on her construction. In this case 347 days would be required. By this rough estimate *Lenox* could have been built in a little under a year by 110 men, a productivity rate very close to that originally proposed by Pepys in 1675.

Samuel Pepys's estimate may have been accurate as a financial exercise for calculating the cost of building a third-rate ship. However, his calculation assumed all the men worked every day of the week and earned the maximum wage as shipwrights. In practice the real situation may have been very different and many factors need to be considered if the true productivity of the dockyard workers is to be understood.

Any calculation must take into account the average daily wage earned by tradesmen. For example, a shipwright having served his seven-year apprenticeship earned 2s 1d a day but many apprentices or "servants" earned far less. It will be remembered that a working day ranged rather imprecisely from six hours during deep winter to ten in high summer. A great deal of overtime was worked which ranged from a "night" for five hours and a "tide" for one and half hours. A day's wages were paid for working a night and, for a qualified shipwright, sixpence for a tide. On top of this a lodging allowance was often made. It is also interesting to examine how many days a week the men worked on average, for they must have had some days off.

In order to obtain statistics from the pay books, a representative cross section of the characters for each trade were selected for study. It was not recorded which ship the men worked on, but it must have been either *Lenox* or *Hampton Court* in the early months, as well as *Duchess* towards the end of the period. The pay books are divided into "Ordinary" and "Extraordinary" accounts. The Ordinary are payments made for the keeping of ships laid up and not in service and these are not included. The Extraordinary are payments made for building new, repairing old ships and, for a period, getting ships ready for an abortive war with France. The pay books are meticulous records for these activities. Although it is unknown which ship the men worked on it is reasonable to assume their working practices and productivity would have been the same whichever ship it was. The Deptford accounts were made up every quarter year before the workmen were paid.[4]

One of the characters chosen for study was Samuel Stephens, a nineteen year-old servant, or apprentice, to Nicholas Carryer, the quartermaster to the ship in the dock (*Lenox*). He was a raw recruit, whose wages would have been paid to Nicholas Carryer; his indenture was dated 25 March 1677.

At the other extreme William Bond was the foreman of the building of the new ships.[5] Many of the other characters were qualified shipwrights. Of these Ambrose Fellows was the "master" of two servants, and, as events turned out, would have a very long association with *Lenox*.

It was not possible to calculate the attendance of sawyers as their pay was calculated on how much timber they had sawn. Similarly, the teams of horses were also excluded, for although we know the teamer was paid six shillings per day, we do not know the expense of keeping four horses. In any case the number of teams was only between two at the beginning and three later on and would have no great influence in the calculations.

The following tables were extracted from the Deptford Extraordinary accounts. From them the average number of days worked per week and the average number of hours worked per day was calculated. The calculations for hours worked each day do not include breakfast and dinnertime. I have assumed a ten-hour working day for the Midsummer quarter, eight working hours a day for the Michaelmas and Lady quarters and six hours a day for the Christmas quarter.

Midsummer Quarter between 1 April and 30 June 1677

Shipwrights	Days	Nights	Tides	Daily Rate s d	Wages £ s d	Lodging s d
William Bond						
For 30 Ships	12	0	0	2- 1	1- 5- 0	0- 4
Repair Old Ships	61	7	4	2- 1	7- 3- 8	2- 2
Total Days Worked	73	7	4			
Days in Period	91					

Average number of days worked per week 5.6 | Average number of hours worked per day 10.6 | Average daily earnings 2s- 4.1d

	Days	Nights	Tides	Daily Rate s d	Wages £ s d	Lodging s d
Ambrose Fellows						
Repair Old Ships	65	6	14	2- 1	7- 14- 11	2- 4
Total Days Worked	65	6	14			
Days in Period	91					

Average number of days worked per week 5.0 | Average number of hours worked per day 11.0 | Average daily earnings 2s- 5d

	Days	Nights	Tides	Daily Rate s d	Wages £ s d	Lodging s d
Samuel Stephens						
For 30 Ships	5	0	0	1- 2	0- 5- 10	0- 2
Repair Old Ships	55 ½	3	3	1- 2	3- 8- 9	1- 11
Total Days Worked	60 ½	3	3			
Days in Period	91					

Average number of days worked per week 4.6 | Average number of hours worked per day 10.4 | Average daily earnings 1s- 2.1d

	Days	Nights	Tides	Daily Rate s d	Wages £ s d	Lodging s d
Robert Hughes						
For 30 Ships	11	0	0	2- 1	1- 2- 11	0- 4
Repair Old Ships	11 ½	1	1	2- 1	1- 6- 6	0- 6
Total Days Worked	22 ½	1	1			
Days in Period	91					

He only worked 11 ½ days, probably because he was recently pressed.
Average number of hours worked per day 10.3 | Average daily earnings 2s- 2.8d

	Days	Nights	Tides	Daily Rate s d	Wages £ s d	Lodging s d
John Bass						
Repair Old Ships	2	0	0	2- 1	0- 4- 2	0- 2
Total Days Worked	2	0	0			
Days in Period	91					

Bass was a pressed man for he had RUN against his name. But he returned, since his name reappears in the following quarterly returns. Average number of hours worked per day 10.0 | Average daily earnings 2s- 1.0d

	Days	Nights	Tides	Daily Rate s d	Wages £ s d	Lodging s d
Thomas Sherman						
For 30 Ships	6	0	0	2- 1	0- 12- 6	0- 2
Repair Old Ships	21	3	3	2- 1	2- 11- 6	0- 9
Total Days Worked	27	3	3			
Days in Period	91					

Average number of hours worked per day 10.7 | Average daily earnings 2s- 4.9d

	Days	Nights	Tides	Daily Rate s d	Wages £ s d	Lodging s d
Edward Shish						
For 30 Ships	3 ½	0	0	2- 1	0- 7- 3	0- 1
Repair Old Ships	11 ½	1	1	2- 1	1- 4- 6	0- 3
Total Days Worked	15	1	1			
Days in Period	91					

Average number of hours worked per day 10.4 | Average daily earnings 2s- 1.7d

William Bayley
For 30 Ships	6	0	0		1- 5	0- 8- 6	
Repair Old Ships	65	17	14		1- 5	6- 0- 3	1- 11

Total Days Worked	71	17	14				
Days in Period	91						

Average number of days worked per week 5.5 | Average number of hours worked per day 11.5 | Average daily earnings 1s-10.1d

William Burnley
Repair Old Ships	71 ½	10	10		1- 3	5- 3-11	2- 6

Total Days Worked	71 ½	10	10				
Days in Period	91						

Average number of days worked per week 5.5 | Average number of hours worked per day 10.9 | Average daily earnings 1s- 5.9d

Samuel Wall
Repair Old Ships	74 ½	6	6		1- 6	6- 2-9	2- 7

Total Days Worked	74 ½	6	6				
Days in Period	91						

Average number of days worked per week 5.5 | Average number of hours worked per day 10.5 | Average daily earnings 1s- 8.2d

James Seabrook
For 30 Ships	1	0	0		1- 8	0- 1- 8	
Repair Old Ships	40	3	3		1- 8	3- 12-11	1- 5

Total Days Worked	41	3	3				
Days in Period	91						

Average number of days worked per week 3.2 | Average number of hours worked per day 10.5 | Average daily earnings 1s-10.2d

Caulkers

Thomas Webb
Repair Old Ships	67 ½	11	16		2- 1	8- 11- 6	2- 4

Total Days Worked	67 ½	11	16				
Days in Period	91						

Average number of days worked per week 5.2 | Average number of hours worked per day 11.2 | Average daily earnings 2s- 6.9d

Richard Wise
Repair Old Ships	62 ½	11	16		1- 5	5- 8- 5	2- 1

Total Days Worked	67 ½	11	16				
Days in Period	91						

Average number of days worked per week 5.2 | Average number of hours worked per day 11.2 | Average daily earnings 1s- 7.6d

Joiners

John Haile
Repair Old Ships	69 ½	0	25		2- 0	7- 11- 6	2- 5

Total Days Worked	69 ½	0	25				
Days in Period	91						

Average number of days worked per week 5.4 | Average number of hours worked per day 10.5 | Average daily earnings 2s- 2.6d

Scavelmen

	Days	Nights	Tides		Daily Rate s d	Wages £ s d
Stephen Butler						
Repair Old Ships	69 ½	2	5		1- 3	4- 11- 0
Total Days Worked	69 ½	2	5			
Days in Period	9					

Average number of days worked per week 5.3 | Average number of hours worked per day 10.3 | Average daily earnings 1s- 3.7d

Labourers

	Days	Nights	Tides		Daily Rate s d	Wages £ s d
Francis Carpenter						
For 30 Ships		1	1		1- 1	0- 1- 5
Repair Old Ships	69 ½	2	14		1- 1	4- 1- 2
Total Days Worked	69 ½	3	15			
Days in Period	91					

Average number of days worked per week 5.3 | Average number of hours worked per day 10.5 | Average daily earnings 1s- 2.3d

Michaelmas Quarter between 1 July and 30 Sept 1677

Shipwrights	Days	Nights	Tides	Daily Rate s d	Wages £ s d	Lodging s d
William Bond						
For 30 Ships	77	0	63	2- 2	9- 18- 8	2- 8
Repair Old Ships	1	2	2	2- 2	0- 7- 6	0- 0
Total Days Worked	78	2	65			
Days in Period	92					

Average number of days worked per week 5.9 | Average number of hours worked per day 9.4 | Average daily earnings 2s- 8.1d

	Days	Nights	Tides	Daily Rate s d	Wages £ s d	Lodging s d
Ambrose Fellows						
For 30 Ships	35	0	34	2- 1	4- 10- 0	1- 3
Repair Old Ships	41	10	30	2- 1	6- 1- 3	1- 5
Total Days Worked	76	10	64			
Days in Period	92					

Average number of days worked per week 5.8 | Average number of hours worked per day 9.9 | Average daily earnings 2s- 9.8d

	Days	Nights	Tides	Daily Rate s d	Wages £ s d	Lodging s d
Samuel Stephens						
For 30 Ships	58 ½	0	45	1- 2	3- 15- 9	2- 0
Repair Old Ships	19	2	18	1- 2	1- 7- 6	0- 8
Total Days Worked	77 ½	2	63			
Days in Period	92					

Average number of days worked per week 5.9 | Average number of hours worked per day 9.3 | Average daily earnings 1s- 4.4d

	Days	Nights	Tides	Daily Rate s d	Wages £ s d	Lodging s d
Robert Hughes						
For 30 Ships	46 ½	0	33	2- 1	5- 13- 4	1- 8
Repair Old Ships	20 ½	2	21	2- 1	2- 17- 4	0- 8
Total Days Worked	67	2	54			
Days in Period	92					

Average number of days worked per week 5.1 | Average number of hours worked per day 9.4 | Average daily earnings 2s- 7.0d

John Bass

For 30 Ships	58	0	54	2- 1	7- 7-10	2- 0
Repair Old Ships	2	0	2	2- 1	0- 5- 2	0- 1
Total Days Worked	60	0	56			
Days in Period	92					

Average number of days worked per week 4.6 | Average number of hours worked per day 9.4 | Average daily earnings 2s- 7.0d

Thomas Sherman

For 30 Ships	16 ½	0	17	2- 1	2- 2-10	0- 7
Repair Old Ships	52	7	37	2- 1	7- 1- 5	1- 9
Total Days Worked	68 ½	7	54			
Days in Period	92					

Average number of days worked per week 5.2 | Average number of hours worked per day 9.7 | Average daily earnings 2s- 8.7d

Edward Shish

For 30 Ships	39	0	33	2- 1	4- 7- 9	1- 4
Repair Old Ships	23 ½	2	20	2- 1	3- 3- 1	0- 10
Total Days Worked	62 ½	2	53			
Days in Period	92					

Average number of days worked per week 4.8 | Average number of hours worked per day 9.4 | Average daily earnings 2s- 5.4d

William Bayley

For 30 Ships	49	0	40	1- 5	4- 1- 1
Repair Old Ships	22	15	39	1- 5	3- 3- 9
Total Days Worked	71	15	79		
Days in Period	92				

Average number of days worked per week 5.4 | Average number of hours worked per day 10.7 | Average daily earnings 2s- 0.5d

William Burnley

For 30 Ships	61 ½	0	48	1- 3	4- 6-10	2- 2
Repair Old Ships	13 ½	3	14	1- 3	1- 3- 6	0- 5
Total Days Worked	75	3	62			
Days in Period	92					

Average number of days worked per week 5.7 | Average number of hours worked per day 9.4 | Average daily earnings 1s- 6.1d

Samuel Wall

For 30 Ships	25 ½	0	25	1- 6	2- 6- 7	0- 11
Repair Old Ships	49 ½	10	36	1- 6	5- 1- 3	1- 8
Total Days Worked	75	10	61			
Days in Period	92					

Average number of days worked per week 5.7 | Average number of hours worked per day 9.9 | Average daily earnings 2s- 0.1d

James Seabrook

For 30 Ships	18 ½	0	19	1- 8	1- 18- 9	0- 8
Total Days Worked	18 ½	0	19		
Days in Period	92				

Average number of days worked per week 1.4 | Average number of hours worked per day 9.5 | Average daily earnings 2s- 1.6d

Caulkers

Thomas Webb

For 30 Ships	28	0	29	2- 1	3- 12-10	1- 0
Repair Old Ships	45	12	56	2- 1	7- 6- 9	1- 7

	Days	Nights	Tides	Daily Rate s d	Wages £ s d	Lodging s d
Total Days Worked	73	12	85			
Days in Period	92					

Average number of days worked per week 5.6 | Average number of hours worked per day 10.6 | Average daily earnings 3s- 0.5d

	Days	Nights	Tides	Daily Rate s d	Wages £ s d	Lodging s d
Richard Wise						
Repair Old Ships	73	15	89	1- 5	7- 6-11	2- 6
Total Days Worked	73	15	89			
Days in Period	92					

Average number of days worked per week 5.6 | Average number of hours worked per day 10.8 | Average daily earnings 2s- 0.6d

Joiners

	Days	Nights	Tides	Daily Rate s d	Wages £ s d	Lodging s d
John Haile						
For 30 Ships	15 ½	0	15	2- 0	1- 18- 6	0- 7
Repair Old Ships	44	0	21	2- 0	5- 2- 2	1- 6
Total Days Worked	59 ½	0	36			
Days in Period	92					

Average number of days worked per week 4.6 | Average number of hours worked per day 8.9 | Average daily earnings 2s- 4.8d

Scavelmen

	Days	Nights	Tides	Daily Rate s d	Wages £ s d	Lodging s d
Stephen Butler						
Repair Old Ship	69	1	21	1- 3	4- 14- 6	
Total Days Worked	69	1	21			
Days in Period	92					

Average number of days worked per week 5.3 | Average number of hours worked per day 8.5 | Average daily earnings 1s- 4.4d

Labourers

	Days	Nights	Tides	Daily Rate s d	Wages £ s d	Lodging s d
Francis Carpenter						
For 30 Ships	75	2	60	1- 1	5- 3- 5	
Total Days Worked	75	2	60			
Days in Period	92					

Average number of days worked per week 5.7 | Average number of hours worked per day 8.5 | Average daily earnings 1s- 4.5d

	Days	Nights	Tides	Daily Rate s d	Wages £ s d	Lodging s d
John Hetherington						
For 30 Ships		1	1	1- 1	0- 1- 5	
Repair Old Ships	69 ½	2	14	1- 1	4- 1- 2	
Total Days Worked	69 ½	3	15			
Days in Period	92					

Average number of days worked per week 5.3 | Average number of hours worked per day 8.5 | Average daily earnings 1s- 2.3d

Christmas Quarter between 1 October and 31 December 1677

	Days	Nights	Tides	Daily Rate s d	Wages £ s d	Lodging s d
William Bond						
For 30 Ships	72 ½	0	0	2- 2	7- 17- 1	2- 6
Repair Old Ships	3	4	11	2- 2	1- 0- 1	0- 1
Total Days Worked	75 ½	4	11			
Days in Period	92					

Average number of days worked per week 5.7 | Average number of hours worked per day 6.5 | Average daily earnings 2s- 4.6d

Ambrose Fellows
For 30 Ships   77   0   0   2- 1   8- 0- 5   2- 8

Total Days Worked   77   0   0
Days in Period   92
Average number of days worked per week 5.9 | Average number of hours worked per day 6.0 | Average daily earnings 2s- 1.4d

Samuel Stephens
For 30 Ships   77   0   0   1- 2   4- 9-10   2- 8

Total Days Worked   77   0   0
Days in Period   92
Average number of days worked per week 5.9 | Average number of hours worked per day 6.0 | Average daily earnings 2s- 1.4d

Robert Hughes
For 30 Ships   56 ½   0   0   2- 1   5- 17- 8   2- 0

Total Days Worked   56 ½   0   0
Days in Period   92
Average number of days worked per week 4.3 | Average number of hours worked per day 6.0 | Average daily earnings 2s- 1.4d

John Bass
For 30 Ships   75   0   0   2- 1   7- 16- 3   2- 7

Total Days Worked   75   0   0
Days in Period   92
Average number of days worked per week 5.7 | Average number of hours worked per day 6.0 | Average daily earnings 2s- 1.4d

Thomas Sherman
For 30 Ships   62   0   0   2- 1   6- 9- 2   2- 1

Total Days Worked   62   0   0
Days in Period   92
Average number of days worked per week 4.7 | Average number of hours worked per day 6.0 | Average daily earnings 2s- 1.4d

Edward Shish
For 30 Ships   61 ½   0   0   2- 1   6- 8- 1   2- 2

Total Days Worked   61 ½   0   0
Days in Period   92
Average number of days worked per week 4.7 | Average number of hours worked per day 6.0 | Average daily earnings 2s- 1.4d

William Bayley
For 30 Ships   68   0   0   1- 6   5- 2- 0
Repair Old Ships   6 ½   4   7   1- 6   0- 18- 1

Total Days Worked   74 ½   4   7
Days in Period   92
Average number of days worked per week 5.7 | Average number of hours worked per day 6.4 | Average daily earnings 1s- 7.3d

William Burnley
For 30 Ships   73   0   0   1- 3   4- 11- 3   2- 6

Total Days Worked   73   0   0
Days in Period   92
Average number of days worked per week 5.6 | Average number of hours worked per day 6.0 | Average daily earnings 1s- 3.4d

Samuel Wall
| For 30 Ships | 75 | 0 | 0 | | 1- 6 | 5- 12- 6 | 2- 7 |

| Total Days Worked | 75 | 0 | 0 |
| Days in Period | 92 | | | |

Average number of days worked per week 5.7 | Average number of hours worked per day 6.0 | Average daily earnings 1s- 6.4d

James Seabrook
| For 30 Ships | 69 | 0 | 0 | | 1- 8 | 5- 15- 0 | 2- 5 |
| Repair Old Ships | 6 | 0 | 0 | | 1- 8 | 0- 10- 0 | 0- 2 |

| Total Days Worked | 75 | 0 | 0 |
| Days in Period | 92 | | | |

Average number of days worked per week 5.7 | Average number of hours worked per day 6.0 | Average daily earnings 1s- 8.4d

Caulkers

Thomas Webb
| For 30 Ships | 65 | 0 | 0 | | 2- 1 | 6- 15- 5 | 2- 3 |
| Repair Old Ships | 11 | 2 | 14 | | 2- 1 | 1- 14- 1 | 0- 5 |

| Total Days Worked | 76 | 2 | 14 |
| Days in Period | 92 | | | |

Average number of days worked per week 5.8 | Average number of hours worked per day 6.4 | Average daily earnings 2s- 3.2d

Richard Wise
| For 30 Ships | 8 | 0 | 0 | | 1- 5 | 0- 11- 4 | 0- 3 |
| Repair Old Ships | 67 ½ | 3 | 15 | | 1- 5 | 5- 4- 3 | 2- 4 |

| Total Days Worked | 75 ½ | 3 | 15 |
| Days in Period | 92 | | | |

Average number of days worked per week 5.7 | Average number of hours worked per day 6.5 | Average daily earnings 1s- 6.8d

Joiners

John Haile
| For 30 Ships | 73 ½ | 0 | 0 | | 2- 0 | 7- 7- 0 | 2- 6 |

| Total Days Worked | 73 ½ | 0 | 0 |
| Days in Period | 92 | | | |

Average number of days worked per week 5.6 | Average number of hours worked per day 6.0 | Average daily earnings 2s- 0.4d

Scavelmen

Stephen Butler
| For 30 Ships | 76 ½ | 7 | 0 | | 1- 3 | 5- 4- 4 | |

| Total Days Worked | 76 ½ | 7 | 0 |
| Days in Period | 92 | | | |

Average number of days worked per week 5.8 | Average number of hours worked per day 6.5 | Average daily earnings 1s- 4.4d

Labourers

Francis Carpenter
| For 30 Ships | 71 | 0 | 0 | | 1- 1 | 3- 16-11 | |

| Total Days Worked | 71 | 0 | 0 |
| Days in Period | 92 | | | |

Average number of days worked per week 5.4 | Average number of hours worked per day 6.0 | Average daily earnings 1s- 1.0d

John Hetherington
For 30 Ships   75 ½   3   6   1- 1   4- 7- 0

Total Days Worked   75 ½   3   6
Days in Period   92
Average number of days worked per week 5.7 | Average number of hours worked per day 6.3 | Average daily earnings 1s- 1.8d

Will Billinghurst
For 30 Ships   57 ½   0   0   1- 1   3- 2- 3

Total Days Worked   57 ½   0   0
Days in Period   92
Average number of days worked per week 4.4 | Average number of hours worked per day 6.0 | Average daily earnings 1s- 1.0d

Ocum Boys

James Woodward
For 30 Ships   11 ½   0   0   0- 6   0- 5- 9   0- 5
Repair Old Ships   62 ½   2   11   0- 6   1- 13- 2   2- 2

Total Days Worked   74   2   11
Days in Period   92
Average number of days worked per week 5.6 | Average number of hours worked per day 6.4 | Average daily earnings 0s- 6.7d

Team of four horses

Robert Gransden
For 30 Ships   77   0   0   6- 0   23- 2- 0

Total Days Worked   77   0   0
Days in Period   92
Average number of days worked per week 5.9 | Average number of hours worked per day 6.0 | Average daily earnings 6s- 0.0d

Lady Quarter between 1 January and 31 March 1677/8

	Days	Nights	Tides	Daily Rate s d	Wages £ s d	Lodging s d
William Bond						
For 30 Ships	69 ½	3	18	2- 2	8- 0- 0	2- 5
For French War	3	8	10	2- 2	1- 17-11	0- 1
Total Days Worked	72 ½	11	28			
Days in Period	90					

Average number of days worked per week 5.6 | Average number of hours worked per day 9.3 | Average daily earnings 2s- 9.2d

	Days	Nights	Tides	Daily Rate	Wages	Lodging
Ambrose Fellows						
For 30 Ships	73 ½	1	5	2- 1	7- 17- 8	2- 6
Total Days Worked	73 ½	1	5			
Days in Period	90					

Average number of days worked per week 5.7 | Average number of hours worked per day 8.2 | Average daily earnings 2s- 2.1d

	Days	Nights	Tides	Daily Rate	Wages	Lodging
Samuel Stephens						
For 30 Ships	75	0	5	1- 4	5- 1- 3	2- 7
Total Days Worked	75	0	5			
Days in Period	90					

Average number of days worked per week 5.8 | Average number of hours worked per day 8.1 | Average daily earnings 1s- 4.6d

Robert Hughes

For 30 Ships	58	0	5	2- 1	6- 3- 4	2- 1
Total Days Worked	58	0	5			
Days in Period	90					

Average number of days worked per week 4.5 | Average number of hours worked per day 8.1 | Average daily earnings 2s- 1.9d

John Bass

For 30 Ships	73	1	5	2- 1	7- 16- 8	2- 6
Total Days Worked	73	1	5			
Days in Period	90					

Average number of days worked per week 5.7 | Average number of hours worked per day 8.2 | Average daily earnings 2s- 2.2d

Thomas Sherman

For 30 Ships	55	1	1	2- 1	5- 17- 2	2- 0
For French War	13 ½	13	27	2- 1	3- 8- 8	0- 5
Total Days Worked	68 ½	14	28			
Days in Period	90					

Average number of days worked per week 5.3 | Average number of hours worked per day 9.6 | Average daily earnings 2s- 9.0d

Edward Shish

For 30 Ships	55	0	6	2- 1	5- 17- 7	1- 11
Total Days Worked	55	0	6			
Days in Period	90					

Average number of days worked per week 4.3 | Average number of hours worked per day 8.2 | Average daily earnings 2s- 2.1d

William Bayley

For 30 Ships	39 ½	0	3	1- 7	3- 3- 8	
Repair Old Ships	30	13	43	1- 7	4- 4- 2	
Total Days Worked	69 ½	13	46			
Days in Period	90					

Average number of days worked per week 5.4 | Average number of hours worked per day 9.9 | Average daily earnings 2s- 1.5d

William Burnley

For 30 Ships	63 ½	0	3	1- 5	4- 10- 8	2- 3
Repair Old Ships	6	9	13	1- 5	1- 5- 1	0- 2
Total Days Worked	69 ½	9	13			
Days in Period	90					

Average number of days worked per week 4.9 | Average number of hours worked per day 9.0 | Average daily earnings 1s- 8.4d

Samuel Wall

For 30 Ships	75	0	5	1- 8	6- 7- 1	2- 7
Total Days Worked	75	0	5			
Days in Period	90					

Average number of days worked per week 5.8 | Average number of hours worked per day 8.1 | Average daily earnings 1s- 8.7d

James Seabrook

For 30 Ships	6	0	0	1- 8	0- 10- 0	0- 3
Repair Old Ships	37 ½	2	2	1- 8	3- 6- 8	1- 4
Total Days Worked	43 ½	2	2			
Days in Period	90					

Average number of days worked per week 2.9 | Average number of hours worked per day 8.3 | Average daily earnings 1s- 9.6d

Caulkers

Thomas Webb

For 30 Ships	5	1	3	2- 1	0- 14- 0	0- 2
Repair Old Ships	64	12	21	2- 1	8- 8-10	2- 3
Total Days Worked	69	13	24			
Days in Period	90					

Average number of days worked per week 5.4 | Average number of hours worked per day 9.5 | Average daily earnings 2s- 8.2d

Richard Wise

For 30 Ships	29 ½	1	5	1- 6	2- 7- 5	1- 0
Repair Old Ships	40	15	27	1- 6	4- 11- 6	1- 5
Total Days Worked	69 ½	16	32			
Days in Period	90					

Average number of days worked per week 5.4 | Average number of hours worked per day 9.8 | Average daily earnings 2s- 0.4d

Joiners

John Haile

For 30 Ships	55 ½	0	5	2- 0	5- 13- 6	1- 11
Repair Old Ships	18	0	0	2- 0	1- 16- 0	0- 7
Total Days Worked	73 ½	0	5			
Days in Period	90					

Average number of days worked per week 5.7 | Average number of hours worked per day 8.1 | Average daily earnings 2s- 0.8d

Scavelmen

Stephen Butler

For 30 Ships	69	0	6	1- 3	4- 8- 3
Total Days Worked	69	0	6		
Days in Period	90				

Average number of days worked per week 5.4 | Average number of hours worked per day 8.1 | Average daily earnings 1s- 3.3d

Labourers

Francis Carpenter

For 30 Ships	73 ½	0	10	1- 1	4- 3- 0
Total Days Worked	73 ½	0	10		
Days in Period	90				

Average number of days worked per week 5.7 | Average number of hours worked per day 8.2 | Average daily earnings 1s- 1.6d

John Hetherington

For 30 Ships	74	5	22	1- 1	4- 12-11
Total Days Worked	74	5	22		
Days in Period	90				

Average number of days worked per week 5.8 | Average number of hours worked per day 8.8 | Average daily earnings 1s- 3.1d

Will Billinghurst

For 30 Ships	54 ½	1	4	1- 1	3- 1- 5
Total Days Worked	54 ½	1	4		
Days in Period	90				

Average number of days worked per week 4.2 | Average number of hours worked per day 8.2 | Average daily earnings 1s- 1.5d

Ocum Boys

	Days	Nights	Tides	Daily Rate s d	Wages £ s d	Lodging s d
James Woodward						
For 30 Ships	13 ½	0	5	0- 6	0- 7- 2	0- 5
Repair Old Ships	56	13	22	0- 6	1- 16- 4	2- 0
Total Days Worked	69 ½	13	27			
Days in Period	90					

Average number of days worked per week 5.4 | Average number of hours worked per day 9.5 | Average daily earnings 0s- 7.9d

Midsummer Quarter between 1 April and 30 June 1678

	Days	Nights	Tides	Daily Rate s d	Wages £ s d	Lodging s d
William Bond						
For 30 Ships	64	6	11	2- 2	7- 17- 2	2- 4
For French War	7 ½	15	29	2- 2	3- 3- 3	0- 2
Total Days Worked	71 ½	21	40			
Days in Period	91					

Average number of days worked per week 5.5 | Average number of hours worked per day 12.3 | Average daily earnings 3s- 1.4d

	Days	Nights	Tides	Daily Rate s d	Wages £ s d	Lodging s d
Ambrose Fellows						
For 30 Ships	22	2	3	2- 1	2- 11- 6	0- 10
Total Days Worked	22					
Days in Period	91					

Average number of days worked per week 1.7 | Average number of hours worked per day 10.7 | Average daily earnings 2s- 4.5d

	Days	Nights	Tides	Daily Rate s d	Wages £ s d	Lodging s d
Samuel Stephens						
For 30 Ships	74	4	5	1- 4	5- 5- 3	2- 7
Total Days Worked	74	4	5			
Days in Period	91					

Average number of days worked per week 5.7 | Average number of hours worked per day 10.4 | Average daily earnings 1s- 5.5d

	Days	Nights	Tides	Daily Rate s d	Wages £ s d	Lodging s d
John Bass						
For 30 Ships	72 ½	6	6	2- 1	8- 6- 6	2- 6
Total Days Worked	72 ½	6	6			
Days in Period	91					

Average number of days worked per week 5.6 | Average number of hours worked per day 10.5 | Average daily earnings 2s- 4.0d

	Days	Nights	Tides	Daily Rate s d	Wages £ s d	Lodging s d
Thomas Sherman						
For 30 Ships	52	1	0	2- 1	5- 10- 5	1- 9
For French War	7	9	5	2- 1	2- 0-10	0- 3
Total Days Worked	59	10	5			
Days in Period	91					

Average number of days worked per week 4.5 | Average number of hours worked per day 11.0 | Average daily earnings 2s- 7.2d

	Days	Nights	Tides	Daily Rate s d	Wages £ s d	Lodging s d
Edward Shish						
For 30 Ships	32 ½	2	1	2- 1	3- 12- 4	1- 1
Total Days Worked	32 ½	2	1			
Days in Period	91					

Average number of days worked per week 2.5 | Average number of hours worked per day 10.4 | Average daily earnings 2s- 3.1d

William Bayley
| For 30 Ships | 60 ½ | 4 | 2 | | 1- 7 | 5- 2-10 | |
| For French War | 7 | 14 | 23 | | 1- 7 | 2- 2-10 | |

| Total Days Worked | 67 ½ | 28 | 25 |
| Days in Period | 91 |

Average number of days worked per week 5.2 | Average number of hours worked per day 12.6 | Average daily earnings 2s- 1.9d

William Burnley
| For 30 Ships | 59 | 3 | 2 | | 1- 5 | 4- 8- 5 | 2- 2 |
| For French War | 6 ½ | 12 | 15 | | 1- 5 | 1- 10- 7 | 0- 2 |

| Total Days Worked | 65 ½ | 15 | 17 |
| Days in Period | 91 |

Average number of days worked per week 5.0 | Average number of hours worked per day 11.5 | Average daily earnings 1s-10.2d

Samuel Wall
| For 30 Ships | 73 ½ | 2 | 1 | | 1- 8 | 6- 6- 3 | 2- 6 |

| Total Days Worked | 73 ½ | 2 | 1 |
| Days in Period | 91 |

Average number of days worked per week 5.7 | Average number of hours worked per day 10.2 | Average daily earnings 1s- 9.0d

Caulkers

Thomas Webb
| For French War | 53 ½ | 12 | 23 | | 2- 1 | 7- 7-11 | 1- 10 |

| Total Days Worked | 53 ½ | 12 | 23 |
| Days in Period | 91 |

Average number of days worked per week 5.4 | Average number of hours worked per day 11.7 | Average daily earnings 2s- 9.6d

Joiners

John Haile
| For 30 Ships | 71 | 1 | 0 | | 2- 0 | 7- 4- 0 | 2- 6 |

| Total Days Worked | 71 | 1 | 0 |
| Days in Period | 91 |

Average number of days worked per week 5.5 | Average number of hours worked per day 10.1 | Average daily earnings 2s- 0.8d

Scavelmen

Stephen Butler
| For 30 Ships | 67 ½ | 2 | 1 | | 1- 3 | 4- 7- 2 |

| Total Days Worked | 67 ½ | 2 | 1 |
| Days in Period | 91 |

Average number of days worked per week 5.2 | Average number of hours worked per day 10.2 | Average daily earnings 1s- 3.5d

Labourers

Francis Carpenter
| For 30 Ships | 70 ½ | 2 | 36 | | 1- 1 | 4- 10- 6 |

| Total Days Worked | 70 ½ | 2 | 36 |
| Days in Period | 91 |

Average number of days worked per week 5.4 | Average number of hours worked per day 10.9 | Average daily earnings 1s- 3.4d

John Hetherington
For 30 Ships   71   13   41   1- 1   5- 4- 8

Total Days Worked   71   13   41
Days in Period   91
Average number of days worked per week 5.5 | Average number of hours worked per day 11.8 | Average daily earnings 1s- 5.7d

Ocum Boys

James Woodward
For 30 Ships   19   1   9   0- 6   0- 10- 9   0- 11
For French War   49   15   23   0- 6   1- 13-11   1- 8

Total Days Worked   68   16   32
Days in Period   91
Average number of days worked per week 5.2 | Average number of hours worked per day 11.9 | Average daily earnings 0s- 8.3d

From these examples it is clear that at the start of building in the first Midsummer quarter, between 1 April and 30 June 1677, many new men joined the yard. They therefore appear to have worked fewer days during this period and are not included in the calculations to find the average number of days worked per week. The men worked more overtime in the form of nights and tides when repairing the old ships and getting the fleet ready for an abortive French war than on the new ships. These periods are also eliminated from the calculations, as they are unrepresentative of working practice on the new ships. *Lenox*, in spite of her very rapid construction, could have been completed even more quickly if the same amount of overtime had been worked on her.

Using the remaining relevant data, an average daily wage, with and without days off, was calculated for each trade. From this a resultant total number of days worked during the construction of *Lenox* was obtained. The lodging allowance was included in their earnings.

Number of days worked during construction of *Lenox*

Trade	Account of wages earned upon *Lenox* £		s	d	Average daily wage from examples in pennies	No. days employed without days off	Average daily wage including days off in pennies	No. days employed including days off
Shipwrights	2517-	1-	7		23.52	25684	17.51	34500
Caulkers	125-	10-	8		23.2	1299	18.53	1626
Joiners	229-	3-	5		24.7	2226	19.76	2783
House Carpenters	5-	6-	0		24.0	53	18.0	71
Wheelwrights	10-	0-	0		24.0	100	18.0	133
Bricklayers	6-	9-	0		24.0	65	18.0	86
Seamen-Riggers	83-	17-	7		18.0	1118	15.0	1342
Scavelmen	59-	7-	1		15.7	907	12.26	1162
Labourers	316-	15-	2		14.6	5207	11.08	6861
Plumbers	1-	10-	0		30.0	12	25.0	14
Ocum Boys	19-	7-	2		7.0	663	5.2	893
Teams 4 horse	50-	18-	0		72.0	121	51.4	170
Sawyers	547-	9-	0		24.0	5474	17.8	7381
Total Wages	£3972- 14s- 8d					42929		57022

Lenox was built in about forty-two weeks, between 25 June 1677 and 12 April 1678. By dividing the 57,022 working days taken to build her by the number of days between these dates it can be concluded that about 194 men built her in that time.

If *Lenox* were built in one year instead of forty-two weeks then the number of men required would drop to 156. These estimates are hardly precise but they do show that Deptford dockyard, with its mixture of skilled and unskilled men, working a little more than five days a week with moderate overtime could produce a ship the size of *Lenox* in one year with a little more than 150 men. However, if the calculation is made using days worked without days off, then 118 men could have built her in one year, again very close to the estimate made by Pepys in 1675.

There is other evidence to support the number of 194 men required to build *Lenox*. William Fownes, Clerk of the Cheque at Deptford, mentions the number of men working on the new ships on 7 December 1677, about the midpoint of the ship's construction.[6]

If 429 men were working on new ships and 194 of them were working on *Lenox* then 235 must have been employed on *Hampton Court* and on repairing the old ships, which does seem very plausible.

The Deptford Extraordinary papers containing the quarterly pay books for the new ships are summarised below. The number of men's names on the pay books does not mean they were all working on the new ships at the same time. A man's name would be entered whether he worked for one or seventy days in a quarterly period.

Number of men working on the new ships on 7 December 1677

Trade	Number of men
Shipwrights	248
Caulkers	3
Joiners	12
Wheelwrights	2
Scavelmen	5
Labourers	114
Ocum Boys	3
Teams 4 horse	2
Sawyers	40
Total	429

Number of men entered on the quarterly pay books for building the new ships at Deptford

Trade	June 1677	1 July–30 Sept 1677	1 Oct–31 Dec 1677	1 Jan–31 Mar 1678	1 Apr–30 June 1678
Shipwrights	24	209	285	359	377
Caulkers		6	23	45	46
Joiners	3	15	16	47	51
House Carpenters		3			
Wheelwrights		1	2	2	
Bricklayers				8	10
Seamen-Riggers				10	39
Scavelmen			6	6	6
Labourers	52	85	119	135	92
Plumbers					2
Ocumboys		3	14	21	21
Teams 4 horse	2	2	3	3	3
Sawyers	16	32	40	42	39
Total	107	356	508	636	686
Wages Totals	£126-3s-5d	£1316-13s-10d	£2434-0s-4d	£2875-7s-1d	£2930-8s-7d

Judging from the pay books it is noticeable that at the start of work on *Lenox* there was a much higher proportion of skilled men than in later months, almost certainly caused by the shortage of skilled shipwrights as the huge building programme employed all those available across a number of new ships at several yards. As one would expect, the length of the working day was considerably longer during summer than winter. In addition, some men worked longer hours than others. During the short working days of the Christmas quarter William Bond, the Foreman, was one of the few who worked extra hours. These hours may well have been spent with the Master Shipwright discussing and poring over the draught

of *Lenox*. Of the qualified shipwrights examined, Thomas Sherman consistently worked longer than his colleagues.

Among the servants, William Bayley worked longer hours than anyone else examined. Moreover, as the servants became more skilled, their pay increased. During the first quarter Bayley earned 1s 5d per day but this increased to 1s 6d then to 1s 7d before *Lenox* was completed.

The lodging allowance was paid to skilled and unskilled shipwrights but was never paid to labourers and scavelmen. At the bottom of the pay scale were the ocum boys, and probably because of their youth they were paid an allowance.

SAILS & RIGGING

MASTS AND SPARS

Only two weeks after her launch it was reported from Deptford that *Lenox* was "nearly ready to sail".[1] To achieve such rapid progress, preparations had begun four months previously when John Shish wrote to the Navy Board asking for orders "to go in hand to make her masts".[2]

The Admiralty Commission had decided long before that the masts and yards of the thirty ships should be made to common dimensions to ease the supply of replacements after battles or storms.[3] These dimensions were determined by the use of a formula based on the size of the ship. The length of the mainmast was calculated first, then the diameters and lengths of all the other masts and yards proportioned from it. The formula itself took many forms according to the preference of the shipwright, or, in *Lenox*'s case, the Navy Board. Typically, the breadth and depth of the ship in feet were added together then multiplied by three and divided by five to give an answer in yards.[4] A list of dimensions for the masts and yards for "All the new 3rd rates" was issued just in time to be applied to *Lenox*.[5]

In spite of Shish's forethought in asking for orders to make the masts, initially nothing happened. Eventually, the Navy Board decided that Edward Grey, a merchant mastmaker from London, would supply the masts and yards. He signed a contract on 13 March 1678, which stated that they were to be made according to the authorised list. The contract applied not only to *Lenox,* but also to all six of the new third-rates built on the Thames at Deptford and at Woolwich.

Grey was contracted to deliver the masts and yards ready for use, with the main- and foremasts cheeked and headed with oak and supplied with caps and cross trees. The contract also stipulated that each suite of masts and yards would cost £480, but if the mainmast and bowsprit could not be supplied then the price would drop to £330. Supplying the mainmast and bowsprit was evidently going to be difficult, as they were each expected to be made from one Riga or Gothenburg tree each. In contrast the foremast could be "fished", or made up from more than one piece of timber.[6] It was very difficult to find trees from European sources big enough to make a mainmast or a bowsprit in one piece, although masts big enough for a first-rate were supplied from Danzig during 1679.[7] The only place such large trees were normally found in sufficient quantity was across the Atlantic in New England.

The types of fir timber used for making the masts and yards possessed different qualities and names. The European varieties are generally known as *pinus sylvestris* and the American varieties as *pinus strobus.*[8] The strength of the types of trees varied, or was considered to vary, significantly. For example, a 9in diameter Gothenburg mast was reckoned to be of equal strength to one of 10½in from Riga and one of 12in diameter from New England.[9] The relatively big but lighter New England trees were used for the mainmast and bowsprit. Further aloft, Riga or Gothenburg trees were

used for the topmasts, topgallant masts and yards.[10] Edward Grey, who had been contracted to supply the masts and yards for *Lenox*, managed to provide all the upper masts and yards but had to invoke the clause in his contract not to provide the mainmast and bowsprit.[11] He almost certainly had no means of obtaining the great New England trees from which they could be made in one piece. There were, however, men possessed of greater resources than Grey who would be able to deliver the requisite timber from New England. These were the great timber merchants Sir William Warren, Sir John Shorter, William Wood and Gregory Page.

A contract was drawn up with them on 27 August 1677, in which they agreed to dispatch ships to New England and collect a total of thirty-six masts, twenty bowsprits and twenty large yards for the thirty new ships.[12] The partners hired two great ships for the venture, the *Black Cock* and the *King Soloman*.[13] The masts were safely delivered, butthey arrived too late for use on *Lenox*.[14] Instead, a New England mainmast was found for her that was considerably bigger than the standard dimensions, being 33in instead of 30¼in diameter. This compares with 29½in diameter for the *Pendennis Castle*,[15] 30¼in for the *Eagle*[16] and 30½in for the *Hampton Court*.[17] The main mast of the three decker second-rate *Duchess* was only slightly bigger at 34½in.[18]

The New England masts were not delivered in their finished round condition but were "wrought into 16 squares" or sixteen flats to reveal any defects. The maximum diameter of the mast was to be at the partners, or supporting timbers, where the mast passed through the uppermost deck, but was only two thirds the maximum diameter at the hounds, the uppermost circular section of the mast. Above the hounds the mast overlapped the topmast and was cut square in section. Below the hounds, on each side of the mast, were cheeks made of oak.[19] It was important that masts retained their suppleness and were not left outside in the open air to dry out and become brittle. If they were to be stored for any length of time they were kept wet in a purpose-made mast pond at the northern edge of Deptford dockyard. This would not have been necessary for most of *Lenox*'s masts, as they were used almost as soon as they arrived.

After *Lenox* was safely launched, her lower masts were put in place. An old ship, the *Stateshouse* hulk, whose masts had been replaced by sheer legs acting as a crane, was brought alongside. The lower masts were held vertically by the hulk and then lowered down through *Lenox*'s decks to rest in position on the mast steps. At the position where the masts passed through the decks, they were secured from moving by partners 10in thick bolted into the nearby gun-deck beams.[20] The trestle trees, cross trees, and tops were then fitted to the mastheads and the paunch to the fore sides of the lower masts.

When the lower masts and bowsprit were in place, they were secured by rigging. Responsibility for the rigging of ships at Deptford fell to Captain John Kirk, the Master Attendant, a post (continued on page 120)

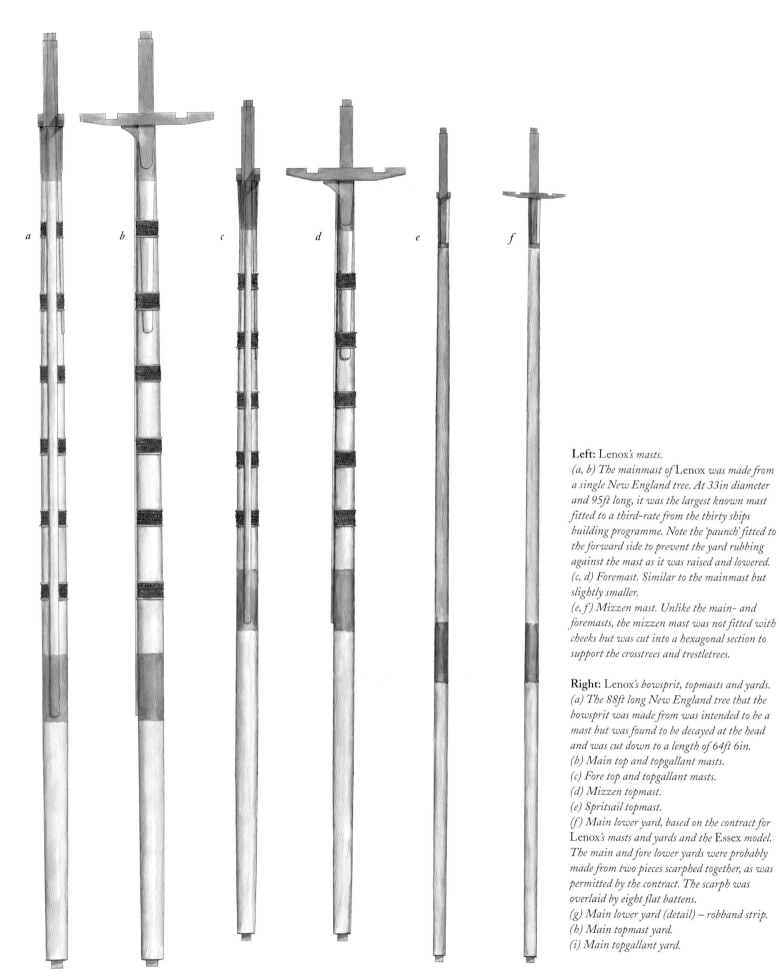

Left: Lenox's *masts.*

(a, b) The mainmast of Lenox *was made from a single New England tree. At 33in diameter and 95ft long, it was the largest known mast fitted to a third-rate from the thirty ships building programme. Note the 'paunch' fitted to the forward side to prevent the yard rubbing against the mast as it was raised and lowered.*

(c, d) Foremast. Similar to the mainmast but slightly smaller.

(e, f) Mizzen mast. Unlike the main- and foremasts, the mizzen mast was not fitted with cheeks but was cut into a hexagonal section to support the crosstrees and trestletrees.

Right: Lenox's *bowsprit, topmasts and yards.*

(a) The 88ft long New England tree that the bowsprit was made from was intended to be a mast but was found to be decayed at the head and was cut down to a length of 64ft 6in.

(b) Main top and topgallant masts.

(c) Fore top and topgallant masts.

(d) Mizzen topmast.

(e) Spritsail topmast.

(f) Main lower yard, based on the contract for Lenox's *masts and yards and the* Essex *model. The main and fore lower yards were probably made from two pieces scarphed together, as was permitted by the contract. The scarph was overlaid by eight flat battens.*

(g) Main lower yard (detail) – robband strip.

(h) Main topmast yard.

(i) Main topgallant yard.

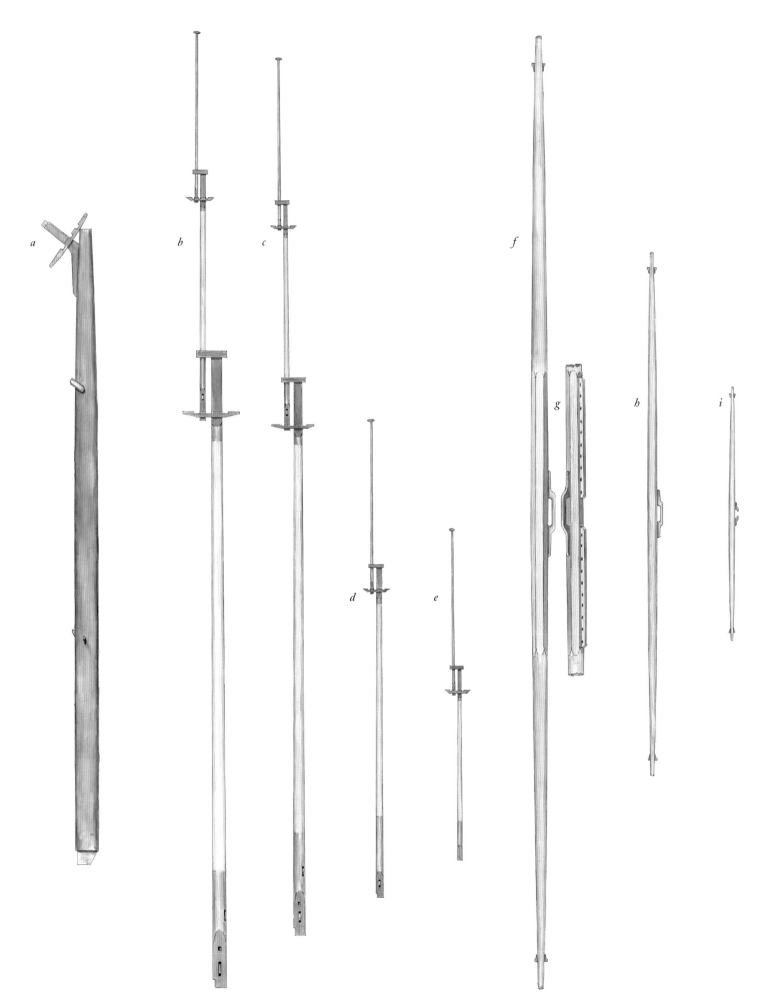

a

b

c

d

e

f

g

h

i

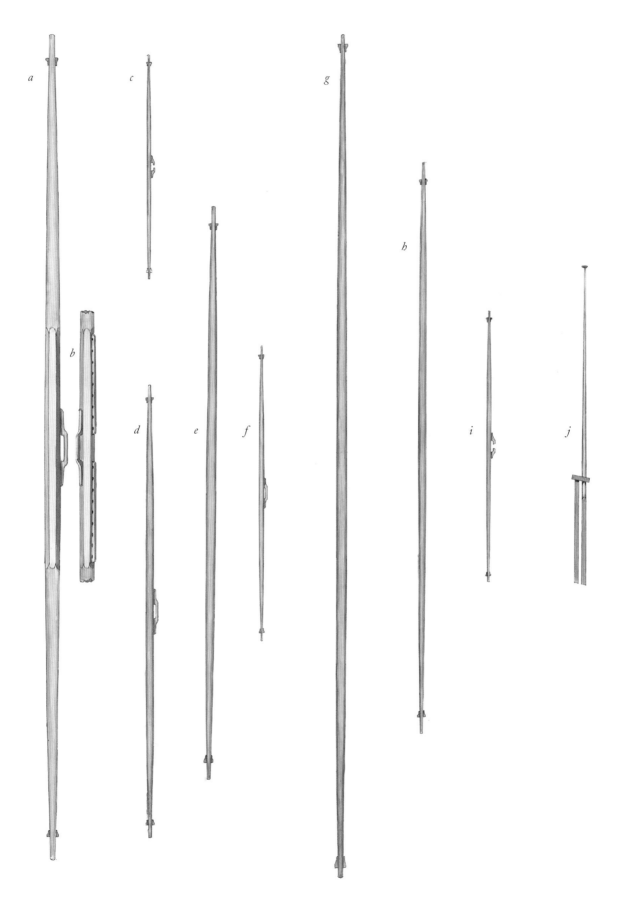

Above: Lenox's *topmasts and yards (continued).*
(a) Fore lower yard. (b) Fore lower yard (detail) – robband strip. (c) Fore topgallant yard. (d) Fore topmast yard. (e) Spritsail yard. (f) Spritsail topsail yard. (g) Mizzen yard. (h) Crossjack yard. (i) Mizzen topmast yard.(j) Jackstaff.

Right: *Rigging plan of a ship that could be* Lenox. *The ship has a bobstay, indicating that the drawing probably dates from the end of the seventeenth century. (Author's collection)*

Parts, Members &c of a Ship

Mizzen Mast *& Rigging*
1 Mizzen Mast
2 Yard & Sail
3 Sheet
4 Shrouds & Laniards
5 Bowlines
6 Brayles
7 Jeer
8 Peak Hallyards
9 Crossjack Yard
10 Lifts
11 Braces
12 Puttock shrouds
13 Mizzen Top
14 Top Armour
15 The Capp
16 Crowfoot
17 Stay & Sail
18 Halliards

Mizzen Topmast *& Rigging*
19 Topmast
20 Yard & Sail
21 Braces
22 Lifts
23 Shrouds
24 Halliards
25 Back Stays

26 Bowlines
27 Sheets
28 Clewlines
29 Stay
30 Crosstrees
31 Cap
32 Stump
33 Stay
34 Truck
35 Spindle
36 Vane
37 Slings of the Crossjack Yard

Mainmast *& Rigging*
38 Mainmast
39 Runners & Tackles
40 Tackle
41 Shrouds & Laniards
42 Stay & Sail
43 Staysail Halliards
44 Yard & Sail
45 Jeers
46 Sheets
47 Tacks
48 Buntlines
49 Bowlines
50 Braces
51 Leachlines
52 Puttock Shrouds
53 Crowfoot
54 Lifts

55 Top
56 Top Armour
57 Top Rope
58 Cap
59 Mainyard Tackles

Main Topmast *& Rigging*
Topsail
60 Main Top Mast
61 Tackles
62 Shrouds
63 Back Stays
64 Halliards
65 Stay & Sail*
66 Staysail Halliards*
67 Yard & Sail*
68 Braces*
69 Bowlines*
70 Sheets*
71 Clewlines*
72 Lifts*
73 Runner*
74 Buntlines*
75 Crosstrees
76 Cap
77 Stump
78 Stay
79 Truck
80 Pendant

Foremast *& Rigging*
81 Foremast
82 Runner & Tackles
83 Tackle
84 Shrouds & Laniards
85 Stay
86 Yard & Sail
87 Sheets
88 Tacks
89 Braces
90 Bowline
91 Buntlines
92 Leachlines
93 Yard Tackle
94 Jeers
95 Puttock Shrouds
96 Crowfoot
97 Top
98 Top Armour
99 Top Rope
100 Lifts
101 Cap

Fore Topmast *& Rigging*
102 Fore Top Mast
103 Tackles
104 Shrouds
105 Backstays
106 Halliards
107 Stay & Sail
108 Halliards
109 Yard & Sail

110 Runner
111 Lifts
112 Braces
113 Bowlines
114 Sheets
115 Clewlines
116 Buntlines
117 Crosstrees
118 Cap
119 Stump
120 Stay
121 Truck
122 Spindle
123 Vane

Bowsprit *& Rigging*
**Sprit Sail*
124 Bowsprit
125 Horse
126 Yard & Sail**
127 Lifts**
128 Sheets**
129 Clewlines**
130 Braces**
131 Bobstay**
132 Top
133 Top Armour

Sprit Sail,
Topsail *& Rigging*
134 Topmast
135 Shrouds

136 Halliards
137 Craneline
138 Yard & Sail
139 Braces
140 Lifts
141 Sheets
142 Crosstrees
143 Cap
144 Jackstaff
145 Truck
146 Jack
147 Best Bower Buoy
148 Cable

Hull
A The Cutwater
B Stem
C Hassholes
D Cathead
E Wastecloths
F Fore channel
G Main channel
H Mizzen Channel
I Chestree
K Entering Port
L Head
M Gallery
N Taffarell (Taffrail)
O Poop Lanterns
P Ensign Staff
Q Truck
R Ensign

Left: A mast hulk - a cut down old warship used to lift masts into ships. From Thomas Milton's A Geometric Plan of Deptford Dockyard, *1753.* (Author's collection)

ordered them to be delivered to Harwich, where other third-rates were nearing completion. Upon hearing this, on 25 April Kirk immediately wrote to the Navy Board, asking them to give orders that seven bower cables of 18in and one stream cable of 11in be sent from Chatham so that *Lenox* could sail on the next spring tide.[24] Kirk exerted heroic efforts, and within two weeks of his letter the ship was fully rigged and ready to sail, with all her stores and cables aboard .

SAILS

The yards carried sails made of canvas. Generally the lower and bigger yards carried sails of stronger construction than yards further up the masts. Each sail was made of breadths of canvas hung vertically and stitched together. The main and fore courses, topsails and mizzen course were made from the heaviest "Duck" canvas. Next in strength were "Noyals", used for the spritsail and the main staysail. "Ipswich" canvas was used for the spritsail topsail, mizzen topsail, mizzen staysail and the main studding sails. The lightest canvas used for *Lenox*'s sails was "Vittery" canvas from which the lightest sails; the fore and main topgallant, fore and main top staysails and main studding topsails were made.[25] As gales or sudden gusts of wind often split the sails at the seams, a second set of the more important sails – the fore, main, mizzen and spritsail courses and the fore and main topsails – were provided as spares. Any split or damaged sails were stitched and repaired on board.

Contractors supplied canvas to Deptford in lengths known as bolts or bails. A typical bail of Noyal delivered to Deptford in September 1677 consisted of 296 yards. The breadth of the canvas varied from between 27in[26] and about 30in.[27] Once in the King's storehouse, the sails were cut out to shape.[28] It was a simple matter to cut the canvas for square sails such as the fore and main courses. The main course of *Lenox* required thirty-six breadths of canvas "cloth" each cut into 14-yard lengths. This amounted to 504 yards of canvas; added to this was a small allowance of 4½ yards for the leech, gore and reef – amounting to 508½ yards of canvas in total.[29]

The shape of the topsail and topgallant sails were slightly more complex than the courses, as they were wider at the top than the bottom. A rectangle of canvas was cut, in which the number of breadths of cloth required was calculated by adding half the length of the upper yard to half the length of the lower yard. To form the correct shape between the yards, a triangle of canvas was removed from one side and the excess added to the other. This practice would ensure that the sail was symmetrical about its respective mast.[30]

Once the canvas had been cut out it was made up by sailmakers, often working by contract. The heavy Duck and Noyal canvases were sewn with a double flat seam, while the lighter Ipswich and Vittery canvases required a round seam with another round seam in the middle of the cloth. The canvas was sewn with the best English twine, and the number of stitches ranged from between 100 per yard for Duck and 90 for Vittery. The sail was reinforced by extra pieces of canvas in places where it was attached to the rigging. Around the edge of the sail, a three-stranded rope was sown in to strengthen it and form loops for securing the rigging. This rope, the boltrope, was 5in in circumference for *Lenox*'s main course[31] and had a gentle twist, making it more pliant.[32]

he had previously held at Chatham.[21] He was aided by Richard Glinn, *Lenox*'s boatswain, and the riggers at the yard. As with the masts and yards, the size and lengths of rigging were based on both relative proportions and many years of practical experience. The circumference and length of every piece of *Lenox*'s rigging and the length, type and material of the blocks was entered in her building list, a small book made up by the Clerk of the Cheque at Deptford, William Fownes. The form of the rigging list follows closely a well- established pattern that had been in practice for at least twenty years. As there was no ropehouse at Deptford, the rigging was supplied from Woolwich dockyard.[22]

The first rigging to be fitted were the shrouds and stays to support the lower masts. This rigging was generally known as the standing rigging and, except for small adjustments, it remained static. After the lower masts were secure, the topmasts were hoisted up by top ropes through sheaves at the lower end, without the need of a hulk or external crane. Shrouds and stays were then rigged to secure these masts.

Each mast carried a yard and each yard carried a sail. They were controlled by running rigging and were constantly being raised, lowered or in some way adjusted. The rigging of *Lenox* was such an everyday occurrence that it was not deemed worthy of comment in official records.

Unfortunately the cables had not arrived on time, so Captain Kirk enquired as to their whereabouts with Mr Bodham, the clerk at the ropehouse at Woolwich.[23] Kirk was informed that the Navy Board had

Right and top right: *Rigging plans of a late seventeenth-century seventy gun third-rate, showing broadside, stern and bow views of a fully-rigged ship. The vast amount of canvas a ship of this type carried is clearly visible. Both items courtesy of Cumbria Archives (Lons/L13/7/2/53)*

Types of Sailcloth by His Majesty's Sailmaker, Mr Blakeway, 1686[33]

Name of canvas	Breadth (Inches)	Price per Yard (pence)	Uses
Hollands Duck	30	26 ¼	For courses, single
Hollands 2nd Rate	30	15 ¼	Topsails, spritsails, mizzens, single
Great Noyal	25	10 ⅛	Topsails, main staysails, spritsails, single
Small Noyal	25	8	Double courses, mizzen, spritsail, topsail, mizzen topsail
Pertrees	32	8 ¾	Topgallant, top staysails, top studding sails, single
Suffolk	27	17	Courses, topsails, single
Buckt Ipswich	26	13 ½	For double courses, mizzen topsails, spritsail, topsail and mizzen staysails, main studding sails

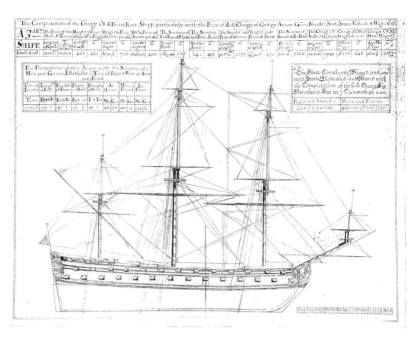

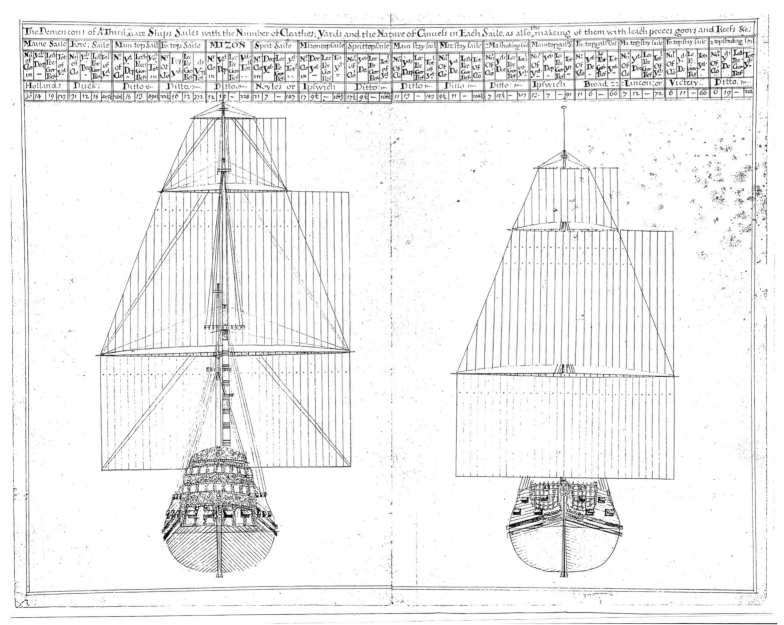

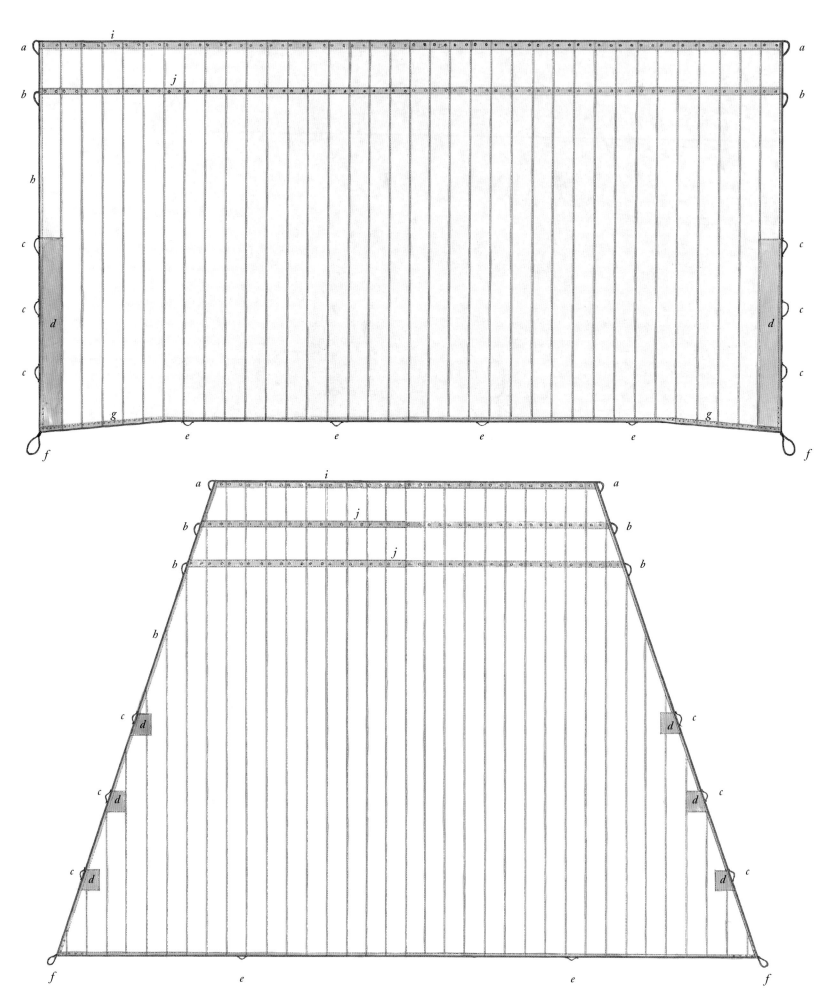

During January 1686, some eight years after *Lenox* had been launched and laid up at Chatham, a survey was made of the stores remaining on account of the thirty new ships.[34] The report shows that while the masts and yards were made to standard sizes, the sails varied slightly in size. The list of sails for the ships laid up at Chatham also reveals that some of the new third-rates had been issued with bonnets, the rather archaic additional sails that were laced to the bottom edge of the courses in light winds. *Lenox* was not supplied with them at the time of her building, although a painting of her believed to have been produced about 1680 by Van de Velde the Younger does show them.[35] As *Lenox* did not go to sea until 1690, the painting is probably an imaginary scene based on the drawings Van de Velde made at the time of her launching.

Left: *(Top) The main course of* Lenox, *36 breadths wide by 14 yards deep, based on the drawing of a sail of a first rate (BM ADD MS 9303) c. 1690 and the recorded details of* Lenox *in the building list PL 1339.*
(Bottom) Lenox's *main topsail, 27 breadths wide by 18 yards deep.*

(a) Earing, for securing the sail to the yardarm cleats. (b) Reef cringles for shortening sail. (c) Bowline cringles for holding the weather leach of the sail up to the wind when close hauled. (d) Leach piece sown on to locally double the strength of the sail. (e) Buntline cringles for gathering up the sail for furling. (f) Clews attached to clew garnets (known as clew lines on the upper sails) for hauling up the sail corners when shortening sail. They were also attached to the sheet for hauling the sail aft and downward and the tack for hauling it forward and down. The tack was only necessary on the lower sails. (g) Marlin holes for strengthening the fastening of the sail to the boltrope near the clew. (h) Boltrope sown to the sail edges. (i) Robbands for securing the sail to the yard. (j) Reef band for shortening sail.

Below: *(a) The spritsail top. This was 7ft 10in diameter, and was the same size as the mizzen top. It cost £2 14s 10p. Prices according to Edward Battine,* Methods of Building Ships of War, *1684, p52. (b) Overhead plan of the spritsail top. (c) Underside of the spritsail top.*

Above: *(a) Typical single block from the 1703 wreck of* Stirling Castle. *(b) Another single block from the same wreck - this example has an unusual shell. (c) A brass sheave, 14in diameter and 3in thick, also from the 1703 wreck of* Stirling Castle, *possibly used in the cathead or for some other heavy duty work.*

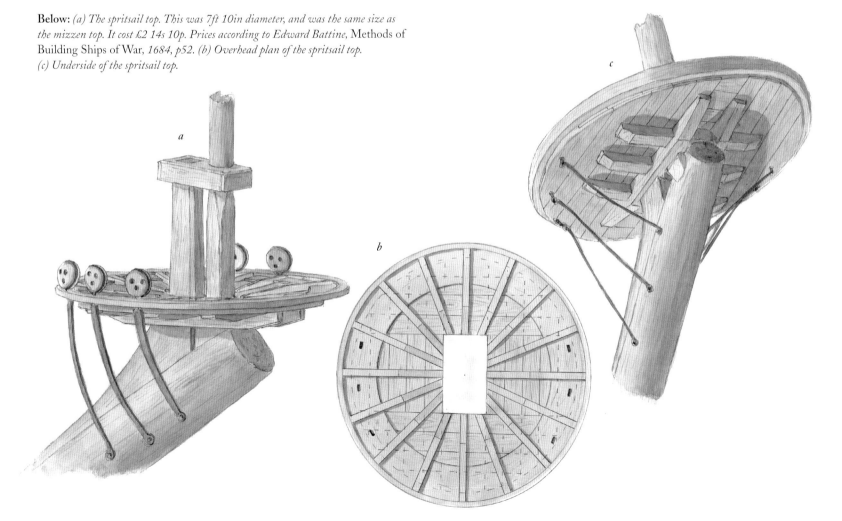

Below: *(a)The main top was 15ft 8in diameter and cost £8 12s 4p, while the fore top was similar but measured 14ft 8in diameter and cost £7 5s 8p, according to Edward Battine. (b) Underside of the main top, showing the trestle-tree and cross-tree that supported it. (c) The trestle-tree at the head of the main topmast.*

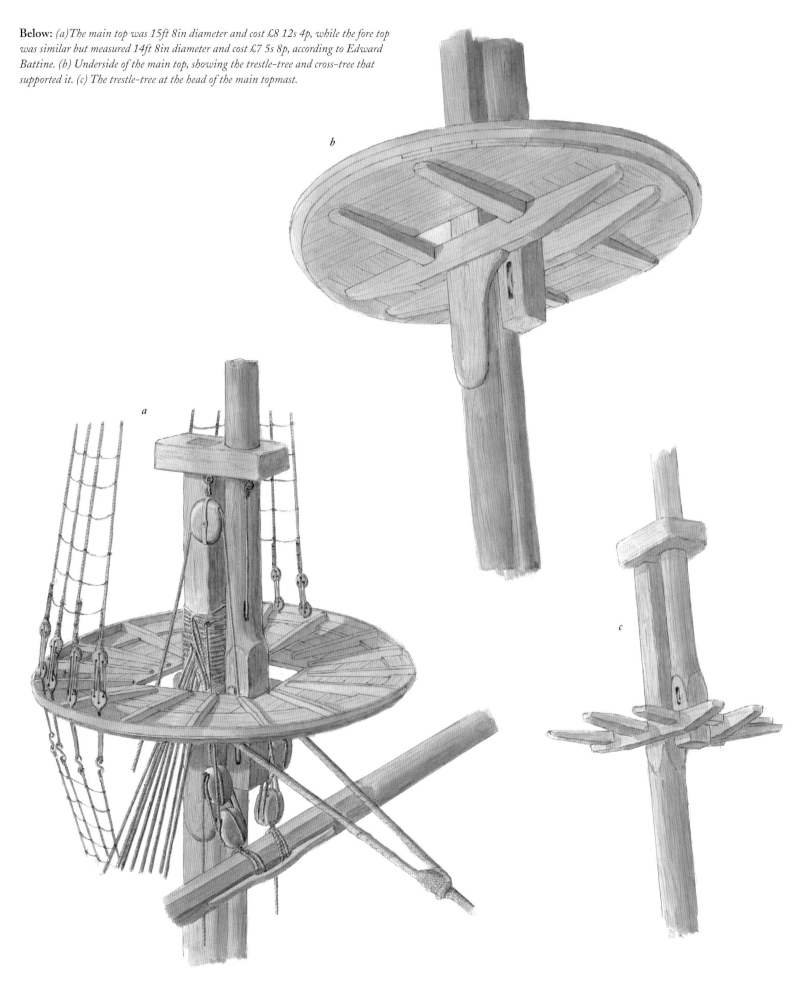

LAID UP
IN ORDINARY

During the preparations to leave Deptford, *Lenox*'s five warrant officers were appointed. Their status was different from commissioned officers who normally only came aboard at the start of war and left when it was over. The warrant officers were the cook, carpenter, gunner, boatswain and the purser. These permanent officers stayed with their ship even when it was laid up in ordinary, out of commission with no one else aboard. Their duties during these periods of inactivity do not appear to have been too onerous. Apart from lodging aboard every third night[1] they were required to look after their stores, keep out wet weather, air the hold on warm days, keep an eye on the moorings and keep out unwelcome visitors. Except for the cook, they all had a young servant,[2] or apprentice, of about seventeen years of age to help them.[3] The officers also had clerical responsibility that required them to keep a detailed account of all the stores under their jurisdiction. To save them the task of writing by hand, a standard printed list was available on which they had only to mark down the quantities. With no captain aboard they came under the orders of the dockyard resident senior officer, the Commissioner, who had considerable other duties ashore running the dockyard. It was therefore an advantage if the warrant officers were disposed towards a certain amount of self-discipline. Unfortunately, for all their other admirable qualities, Restoration men from the King downward were often sadly lacking in this respect. It was perhaps a mark of the small world in which they lived that both King Charles and the Secretary of the Admiralty Commission, Samuel Pepys, personally signed their warrants.

Below: *Navy Office protection against the press. (NA ADM 106/3541 Part II)*

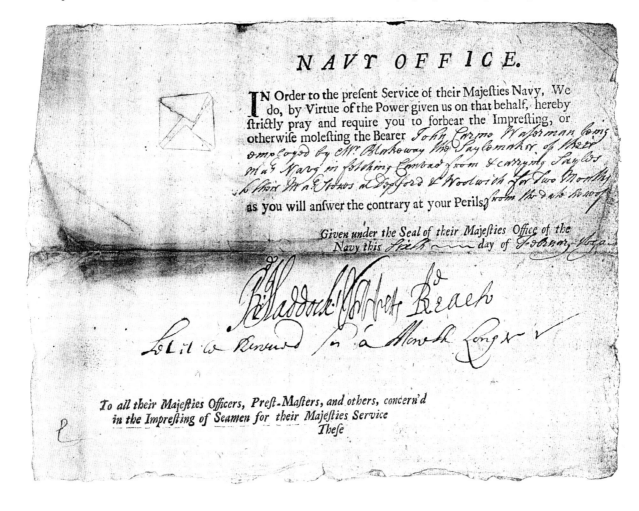

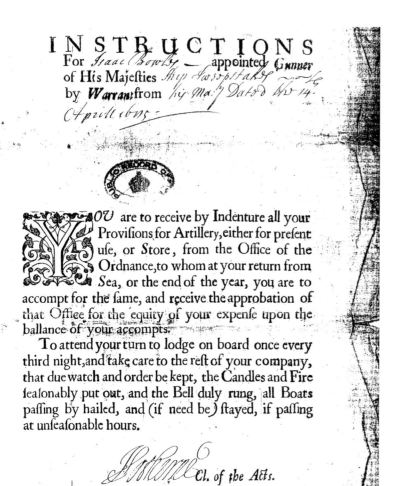

INSTRUCTIONS
For *Isaac Bowly* appointed *Gunner* of His Majesties *Ship Sweepstakes* by *Warrant* from *his Ma.ty Dated his 14 Aprill 1665.*

YOU are to receive by Indenture all your Provisions for Artillery, either for present use, or Store, from the Office of the Ordnance, to whom at your return from Sea, or the end of the year, you are to accompt for the same, and receive the approbation of that Office for the equity of your expense upon the ballance of your accompts.

To attend your turn to lodge on board once every third night, and take care to the rest of your company, that due watch and order be kept, the Candles and Fire seasonably put out, and the Bell duly rung, all Boats passing by hailed, and (if need be) stayed, if passing at unseasonable hours.

Cl. of the Acts.

Above: *Warrant officer's instructions for the gunner of the* Sweepstakes. *The five warrant officers and their servants were the only people aboard when the ships were laid up in ordinary and they found it difficult to follow their instructions and stay aboard ship. (NA ADM 106/3539 Part II).*

The appointment of officers for *Lenox* was first considered during February 1678 when the Navy Board informed the Admiralty that a list of recommended officers would shortly be sent to them.[4] By the first week of March nothing had happened, prompting Captain John Kirk, who would command her, to write to the Navy Board saying that a boatswain would shortly be needed to look after the fitting and rigging of the new ship building in the dock.[5] Kirk sent a reminder on 10 April[6] and five days later the first warrant officer was appointed. He was George Sewell, the cook, whose warrant was dated 15 April 1678. Four days later the King signed four other warrants. Ambrose Fellows, the shipwright artificer from Deptford who had worked on the building of *Lenox*,[7] became the carpenter.[8] Gabriel Walters was appointed gunner, promoted out of the smaller *Revenge*, a 58-gun third rate, and Charles Ashton of the yet smaller *Swallow*, a 48-gun fourth rate, took his place. Richard Glinn became boatswain and Joseph Phillips the purser; these two officers also came from the *Revenge* and their promotions resulted in officers from smaller ships being promoted into their places.[9]

While this was going on measures were taken to man *Lenox* for her first voyage. In order to take her down the narrow, twisting Thames, about 120 experienced seamen and watermen would be required rather quickly. The watermen, however, were difficult to recruit and could only be removed from their regular river trade by the press, from whom they normally gained official protection. However, shipwrights were pressed to build the new ships and so, now, the watermen were pressed too. But pressing the watermen and ordering them to report to *Lenox* was one

thing; getting them to turn up and stay aboard was to prove quite another. Altogether a surprising total of sixty-two watermen ignored their press notices and failed to report at Deptford.[10] Two watermen who did appear for their first entry in the pay book disappeared immediately afterwards.[11]

While the Deptford officers were busily engaged finding a crew, Kirk wrote to the Navy Board on 25 April informing them that *Lenox* was nearly ready to sail but wished to know "His Majesty's pleasure concerning her disposal as to whether she should be fitted for the sea or taken to the fleet anchorage at Chatham".[12] Orders from the king arrived five days later, signed by three members of the Admiralty Commission, the Earl of Anglesey, Lord Craven and Prince Rupert. "These are in pursuance of his Majesty's pleasure signified to us at our late attending him on the affairs of the Admiralty to pray and require you forthwith to cause his Majesty's ship the *Lenox* lately launched at Deptford to be carried about to Chatham and laid up there, His Majesty having been pleased to resolve that the said ship shall not at present be fitted out for the sea".[13]

There were fears of Catholic plots and during November inquiries were made in the dockyards to root out and suspend any officers, seamen or workmen who were "papists". At Portsmouth only one shipwright was found to have a Catholic connection; he confessed that during his childhood he was educated a Catholic but had ever since been a Protestant. Such was the fear of "papists" setting fire to ships that the man was suspended.[14] In this climate it was decided that provision must be made for the security of *Lenox* during her passage to Chatham and Captain Kirk requested "six or eight small guns together with gunners stores, ammunition and some small arms".[15] Shortly afterwards, eight guns arrived and were mounted at some of the gun ports. By 9 May *Lenox* was rigged and loaded with six months' carpenter's and boatswain's stores[16] and formally placed under the command of John Kirk.[17]

As a result of the shortage of watermen *Lenox* could not depart as ordered and the spring tide was lost.[18] In desperation to make up the numbers of absentee watermen, some of the shipwrights and labourers were recruited from Deptford dockyard. The ship's company then consisted of Captain John Kirk, five warrant officers and a mixed company of ninety-three seamen, watermen and dockyard workers.[19]

As permitted by the Admiralty, the motley crew started their generous allowances of sea victuals on the day Kirk became *Lenox*'s commander. The anchor was raised three days later, exactly one month after she had been launched. The sails appear to have been set for a northeasterly breeze that carried her half a mile to Greenwich Reach, where the Thames turns north. With the wind against her, progress was for the moment halted but it gave the Dutch artist Jan van Beecq, who had been living in England for about five years,[20] the opportunity to record the scene. Flags flew at every masthead in celebration as if the king himself were aboard, the only difference being that the union flag, instead of the royal standard, flew at the main masthead. The main yard was lowered to lie portlasted across the deck and the main topsail yard was lowered to the top with the sail clewed up. In van Beecq's scene, the small guns that Kirk requested earlier could be seen pointing through a number of gun ports. An old fourth rate lies at anchor near *Lenox*, either the *Falcon* or the *Constant Warwick*, Ambrose Fellows's old ship. Both ships were being fitted out for sea at Deptford at the time.[21] Two days after *Lenox* set sail, she reached the Hope, where seamen were borrowed from other warships to help continue the passage to Chatham where she arrived on 17 May.

Lenox was taken about a mile up river past the dockyard into Lyme Kiln Channel. She spent much of her time there during the next few years, at the seventh or eighth mooring down from Rochester Bridge.[22] Richard Beach, the Commissioner at Chatham, immediately wrote to the Navy Board to inform them of the event.[23] Beach was an old sea-serving officer with a long and distinguished career dating back to the early 1660s. As a fighting captain he had fought many actions against the Algerines.

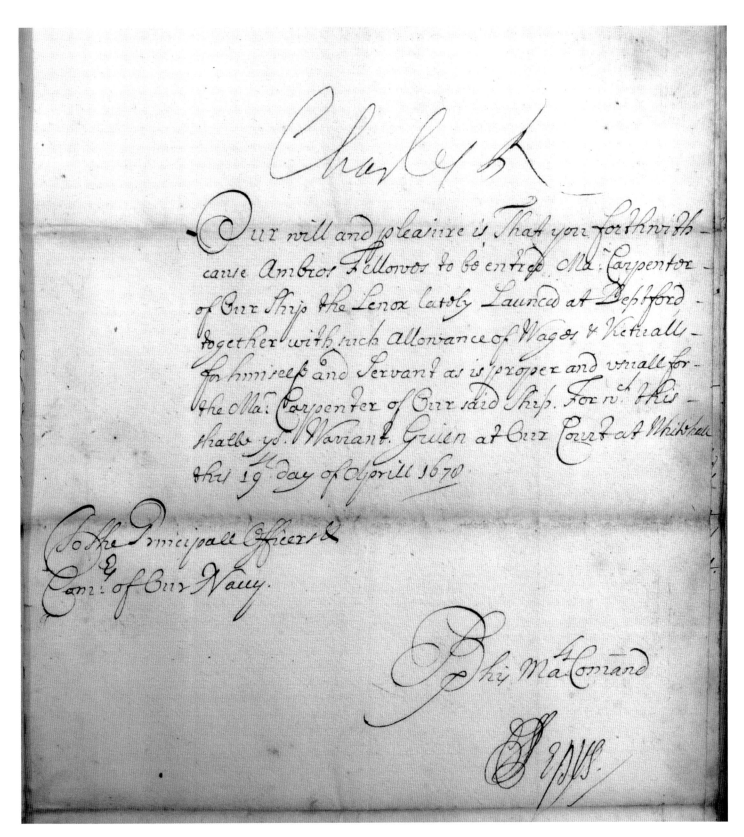

For the four-day voyage during a rather pleasant mid-May, most men aboard *Lenox* earned a very reasonable 4s 10d. For eight of the men, the ordeal proved too much to bear and they somehow deserted ashore. Not surprisingly they received no wages at all.[24] The problems caused by the absence of the pressed watermen were not forgotten by the navy and a list containing the names of sixty-one missing men was prepared for presentation to the Admiralty so that proceedings to punish them could be initiated. The Admiralty promptly issued a warrant ordering mayors,

Above: *Ambrose Fellows' warrant for the position of ship's carpenter on* Lenox, *signed by King Charles II and Samuel Pepys. Fellows was recommended by John Shish and had already worked on the building of* Lenox *as a shipwright at Deptford, earning 2s 1d a day, before becoming the ship's carpenter. He became the longest serving man aboard and during the 1690s, in recognition of his loyalty, his was the first name entered in her pay books before the Captain. He finally left* Lenox *in June 1695 'for preferment' after an association that had lasted a remarkable 18 years. (NA ADM 106/3)*

Above: *A workman caulking the seams of the replica ship* Batavia *at Lelystad, Holland. Rot has started to appear on the wale under the gunport. In 1681 caulkers began work on* Lenox *but found many planks too rotten to caulk. (Author's collection)*

sheriffs, Justices of the Peace, bailiffs, constables and all His Majesty's officers and loving subjects to apprehend the watermen.[25] The discontent aboard *Lenox* even spread to the pilot, one Mr Orrell, who had taken her from Deptford to Blackstakes. He complained he was undervalued after being allowed a fee of only 40 shillings.[26]

As *Lenox* was not at present to be fitted out for sea, Commissioner Beach's thoughts turned to the question of having the new ship laid up. She was for the moment moored with all her masts and yards up and fully rigged. Left to the weather they would soon deteriorate. The sails in particular were very vulnerable to rot in damp conditions and Beach had them brought ashore for storage, together with the colours (flags) and cables. However, there was not enough space in the storehouse ashore for the rigging of the new ships and the repaired old ones that would soon be arriving. Beach was worried because he had received no orders concerning the problem. Seeking guidance he wrote to the Navy Board on 26 May suggesting the rigging be made up, tallied and put into the breadroom or storerooms of the ships. He also pointed out that leaving the rigging in storage aboard the ships would save a great deal of trouble in transportation from ship to shore. The disadvantage, which he failed to mention, was that keeping the rigging on the orlop would reduce the circulation of air and hinder the drying of the fresh, damp timbers. For the future he suggested a new storehouse be built to hold all of the ships' stores.[27] Beach received no immediate reply to his letter and he wrote again before May was out, reminding the Navy Board that although *Lenox*'s sails were safely ashore, all her rigging was still standing.[28] By mid-July Beach seems to have taken the decision to take the running rigging down and store it

aboard, although he left the standing rigging in place as he thought she might be ordered suddenly to sea.[29]

At about this time it was noticed the ship was "very leaky", which Beach thought might have been caused by the removal of clamp bolts at her launching. To expose the leaks he proposed to the Navy Board the ship be hauled ashore, a suggestion to which they readily agreed.[30] As a result Beach was able to report on 14 June that "With the hauling of the *Lenox* we have found her leaks under water which were rents (splits) and a rotten knot in the plank and are now made tight".[31]

New warships required constant attention to keep them in good order and as part of this procedure an estimate was made during August for the charge of ordinary repairs "yet to be performed to His Majesty's ships in harbour to preserve their upper works to keep them tight above water, securing their glazing and tarring the masts". The new ships, as we have seen, were built with fresh, green timber and during the heat of summer the exposed work above the waterline began to shrink as it dried out. This was not an unexpected or unknown problem and was common in newly built ships. It was necessary to caulk the seams and pay the sides with stuff and tar to prevent moisture entering the opened seams. However, if no treatment was performed, water would enter the shrunken seams and the timber would soon rot and decay. The estimate for caulking the sides of *Lenox* above the water and for her decks was £85.[32] Ominously for the future, there is no record of the work being carried out.

On arrival at Chatham, Joseph Phillips, the purser of *Lenox*, made up his accounts of provisions received and expended during the passage from Deptford to Chatham in the stylish hand of a well-educated man. Phillips started his career as a warrant officer in 1666 when he was twenty-three years old and was now an experienced man having served in four previous ships.[33] He applied to Commissioner Beach for a week's leave to go to London to have his accounts passed. To ease the burden on his fellow warrant officers he proposed going during the time when it was not his turn to keep day- and night-watch duties. The officers were allowed leave on a rotational basis and proved ingenious at finding ways of staying ashore rather than afloat. Unfortunately for Phillips he had already incurred the displeasure of the Commissioner by absenting himself from duty and further leave was denied him.[34] The disgruntled but unbowed purser decided to bypass Beach, and on 6 June wrote directly to the Navy Board with a further request for leave. His application stated that the Commissioner had refused him permission to go up to London to pass his accounts, without which he could not claim the payments due to him.[35] The Navy Board must have thought his request reasonable and duly granted his request, and, satisfied, the purser made off to London.

Meanwhile, aboard *Lenox*, the boatswain, Richard Glinn, apparently fell so ill that he was unable to carry out his duties and went ashore. At the same time other officers took the leave due to them. Soon the only officer left on board to look after the ship was the lonely gunner. Beach, who must have been furious when he heard that Phillips had gone to London, wrote to the Navy Board to complain about his absence. He was also worried about future discipline, not just aboard *Lenox,* but all the ships in ordinary. He commented with knowing prophecy, "This will be such a way not only for Pursers but others to creep out at; that their attendance for the future will be far worse than it is now at present which I assure you is bad enough".[36] The absence of warrant officers was an old problem; Samuel Pepys described how, in 1663, he spent the whole night visiting the ships "in which I found for the most part, ne'er an officer aboard nor any men so much as awake".[37]

Commissioner Beach found his own way of dealing with Phillips and his fellow warrant officers, however, for six months after *Lenox* arrived at Chatham the officers and their servants had still not been paid their nine days' sea wages for *Lenox*'s passage from Deptford. Others who had been aboard, including John Kirk, the captain, and the watermen had all received their pay on 7 August.[38] The purser heard this news while he was enjoying his stay in London, from where he wrote to the Navy Board to make his complaint.[39] During the same week that the purser wrote his letter, Kirk, who had only been Master Attendant at Deptford for six months, unexpectedly died,[40] leaving his widow without income to support her family. Luckily the position of housekeeper at Chatham Dockyard became available and she immediately applied for the position. Her application was granted in consideration of her husband's long service to the navy.[41]

On 28 June, while Beach grappled with the problem of errant warrant officers, two hoys arrived alongside *Lenox* with her gun carriages. They were brought down from Tower Wharf to be stored on the gun decks. The guns themselves were very heavy and were only brought aboard ships when they were commissioned for sea. Robert Stanier, Master of the *Batchelors Adventure*, brought forty-nine carriages while Isaac Clerke of the *Dorothy of London* brought the remaining twenty-one. A mistake appears to have been made in the delivery, for during September the *Dorothy and Elizabeth* came alongside again to exchange a 3-pdr carriage for a demi-cannon carriage.[42]

The efforts of Commissioner Beach and the Navy Board to keep the warrant officers aboard their ships met with only limited success, especially in the depths of winter. Early in November Beach, at a muster of the warrant officers of the great ships, read them their orders and let them know: "the hazard they would run in losing their places if they were not more careful and diligent in their respective offices than they had been; being guilty of the breach of all orders that had proceeded either from the Right Honourables, the Lords of the Admiralty, your Honourables (Navy Board) or myself... that their duty was no longer to be dallied with, especially in these dangerous times." Beach was anxious to know the effect of his warning and the very next night, at half past ten, went out on the

Above: *A hoy, a small vessel used to ferry men and equipment between ship and shore. From William Sutherland,* Shipbuilding Unveiled, *1717, p16. (Author's collection)*

Medway among the ships in a very small boat so that no one would suspect a senior officer was aboard. Having heeded the warning many ships kept a good watch but six of them did not; when Beach came to the *Lenox* he found nobody but the purser's boy stirring, while his master remained in his cabin as was common practice. Beach heard the officers excuse themselves by blaming their deputies or pretending they were ill. Observing the "carelessness and contemptuous negligence of the officers", he wrote "Had I been one that designed to have destroyed any of the ships with two or three men, I might have set fire on several of them and have burnt the officers in their cabins."[43]

Three weeks later Beach was still troubled by the cook and purser of *Lenox.* Joseph Phillips approached Beach and demanded leave to go to London about the old business of making up his accounts. Beach was willing to let him go in a week or fortnight's time but for the moment all officers were required to do their duty aboard ship during the troubled times. Upon this refusal Phillips took the liberty to make off to London again and by the 24 November had already been gone eight or ten days. To make matters worse George Sewell, the cook, had gone with him. Beach decided not to suspend them but warned the Navy Board that if such behaviour were permitted, every officer would presume to do the like.[44] The Navy Board seem to have had enough of Phillips for the moment and halfway through December replaced him with a new purser, Moses Turner.[45] This was not the end of Phillips's career as a purser, for he was appointed to other ships, including the prestigious three-deck second rate *Albermarle*. In an account of officers in 1685, it was reported he was very well qualified and deserving preferment.[46]

1679

Exasperated by such absences, Beach wrote to the Navy Board on 27 January: "I have here enclosed sent your Honorables a list of the names of those officers of the ordinary that have neglected their day duties in one week that you may see what effect all your written and verbal orders, fair means and threatening have wrought among them. There is only the Purser of the *Anne* (who is newly married) that I gave leave to stay on shore some time ... and as for the Purser of the *Henrietta* he dare not appear on shore for fear of arrest that would be clapt upon him". The enclosed list shows that of sixteen warships, not one of them had a full complement of warrant officers for the week. *Lenox* had the best record, with only the purser having one day off. The list also reveals that by this time three more new third rates, *Anne*, *Hampton Court* and *Restoration* had been completed to join *Lenox* in ordinary at Chatham.[47]

Eager to keep up the pressure on the Navy Board to do something about the warrant officers, Beach sent another attendance record for the week following his first report. The results were much the same, with *Lenox*'s purser again having one day off. Even by the standards of the time, the warrant officers had abused their position for too long, ignoring repeated warnings and ineffective punishments. Beach's account further observed that few boatswains absented themselves, even though they had the same rights as the others; while the carpenters, pursers, gunners and cooks expected to have every second week at liberty to stay on shore days and nights "so that his Majesty can expect six months service from twelve".

Beach had established the seriousness of the problem of absenteeism. Before he took any action and suspended officers, he advised it would be "convenient" for the Navy Board to inform Samuel Pepys. It would then be up to Pepys to inform the king of Beach's intention to impose discipline on the absent officers.[48] Suspensions soon followed, among them George Sewell the cook of *Lenox* for being absent from his duties. After a couple

of months Beach wrote on 3 April that he should be replaced by another cook as he was "unfit for such an appointment as he thinks himself too good for it".[49] The next day the Navy Board passed on Sir Richard Beach's comments to the Admiralty and Sewell was suspended.[50] The cook's suspension was raised again in June when the Navy Board spoke to Beach.[51] After consideration, the incredibly tolerant Navy Board ordered the cook be reinstated. Officers' warrants, which had been signed by King Charles and Mr Pepys, were not lightly overturned in the Restoration navy.

While the new ships gathered, another third-rate ship similar to *Lenox*, the *Captain*, built by John Shish's brother Thomas at Woolwich, was commissioned to test the sailing qualities of the new third rates. Pepys read an account from her commander, Sir John Holmes, to the Admiralty Commission of the extraordinary proof of the ship *Captain*, to the great satisfaction of His Majesty and their Lordships. The *Captain* had been one of the third rates built with an upright stem as an experiment at the king's direction. Later, Sir Phineas Pett II, Master Shipwright at Chatham, would tell Pepys that they proved ill, and from the beginning he did not follow the king's directions to the extent the others did; indeed in the years that followed, when other ships were made to the same dimensions as those of 1677, the upright stem was less prominent.

Another of the new third rates, the *Elizabeth*, built by contract at Deptford by William Castle, was described by her captain John Berry thus: "Never was a better ship built, she outsails (I do swear) almost every frigate in the fleet, she rides easy and well at anchor, works well, bears a stout sail and does everything that we can desire, but I must inform you that the standing and running rigging are all too small".[52]

Below: Men in boats examining Lenox's *bows discover, to the consternation of Dockyard officers, including the Commissioner, the Master Shipwright, and his assistants, that the bows of* Lenox *were rotten only four years after she was built.*

Mr Pepys and the king however, had many more troubles to worry about than the sailing quality of the new ships or the abuses of the warrant officers. There had been fears for many years that Catholicism may be imposed upon the nation. That fear had increased as Louis XIV of France suppressed Protestantism in France and invaded the Protestant Netherlands. Some Protestant Frenchmen fled their country to seek asylum in England, bringing stories that only increased the tension. At the same time as Commissioner Beach wrote his reports about warrant officers, the dark political storm known as the Popish Plot broke over the king's administration. Trouble had been brewing since the previous August when a Christopher Kirkby requested an audience with the king and introduced him to Israel Tonge who disclosed the plot. A month later Titus Oates began telling fanciful tales. His colourful past included being a chaplain in the navy until he was dismissed for unnatural vice. The theme of his stories involved a papist plot to murder the king and replace English liberties and religion with a Catholic monarchy. His lies would probably have been contained but for the murder of a Protestant London magistrate, Sir Edmund Berry Godfrey. The alleged plot became inflamed and anyone with a Catholic connection became a suspect – including the King's brother, the Duke of York. Samuel Pepys became directly involved when one of his clerks, Samuel Atkins, was falsely arrested on suspicion of conspiring in the murder of Sir Edmund. While Atkins remained in Newgate Gaol, refusing to implicate either himself or Pepys, Titus Oates made further revelations that papists would seize the fleet at Chatham. Pepys responded by writing to Commissioner Beach on 19 November 1678 to keep a strict watch day and night "till this consternation under which all now justly lie be over which I pray God of Heaven send the whole government seeming at this day to remain in such a state of distraction and fear as no history I believe can parallel."[53]

Atkins was eventually tried but managed to prove that at the time Sir Edmund was murdered he had been aboard the yacht *Katherine* at Greenwich drinking with her commander, Captain Vittels, and two gentlewomen. One, Sarah Williams, returned home with him that night to share his bed. With such an alibi Atkins was acquitted. At the same time he revealed the sort of activity that went on aboard ships in harbour. Vittels soon became the First Boatswain of England, and in January 1679, the Master Attendant at Chatham.[54] He obtained his position on 7 March after serving as the boatswain of the *Royal Sovereign*.[55] Although competent, Vittels evidently had few gentlemanly manners, causing Pepys to write "to whom, with respect not only to his experience, but place, a great deal of regard is due, however the plainness of the man may render the delivery of his opinion less courtly".[56]

Atkins's acquittal was not the end of Pepys's problems. In the Commons it became difficult for him and the government to secure money for the maintenance of the navy. A general election was held which decimated the government supporters and forced Charles into forming a new administration consisting of many opposition members. A new Admiralty Commission was appointed full of Pepys's opponents; only one of them had voted for the building of the thirty new ships.[57] At the same time new false charges were brought against him in Parliament. It was claimed that during the Dutch Wars he had an interest in the *Hunter* privateer that had seized an English ship the *Catherine* of London. Even more damaging were allegations that he sold navy secrets to the French.

On 22 May 1679, Pepys and his alleged co-conspirator, Sir Anthony Deane, were committed to the Tower. While incarcerated Pepys wisely spent his time disproving the accusations against him and after six fearful weeks he was released without charge. Although his life was now safe, he was for the moment lost to the navy. The fate of the thirty new ships, which had been Pepys's greatest triumph, now lay in the hands of the new Admiralty Commission. Like their predecessors, however, they were more

politician than naval expert and decided policy rather than technical matters. Charles, astute as ever, undermined the new Commission and his political opponents by never calling Parliament again after March 1681. Unfortunately, of course, it was only Parliament who could vote money for the navy.

At Chatham the days were becoming longer and warmer as summer approached. The unseasoned planks on the sides and decks of the new ships were once again beginning to shrink, as they had the previous year. To carry out the essential maintenance work on caulking the opening seams some twenty caulkers and their materials were requested by Commissioner Beach.[58] A few of the ships may have been caulked but because of the lack of money most of them were left to the mercy of the elements. Although the officers at Chatham, and the new Admiralty Commission, were aware of the problems, they could do little without money. But for now, following the political upheaval money was something that would be in very short supply.

Towards the end of the year, the Navy Board heard that during the summer the masts and sides of the ships had not been scraped and payed with new stuff to protect them. In the coming winter months the shrunken, splitting seams would be open for the weather to penetrate and rot the plank. Much of the responsibility would fall on the newly appointed Master Attendant, Captain Richard Vittels, the same who had provided an alibi for Samuel Atkins. In a letter to the Navy Board he explained that it was not until 13 August that any tallow arrived and 20 September before any scrapers were in the stores. For want of simple tools such as scrapers a whole fleet would be exposed to ruin. By the time Vittels did receive the tallow and scrapers he was too busy securing the ships' moorings against the coming winter to pay enough attention to the protection of the ships' planking.[59]

Other problems peculiar to Chatham conspired to undermine the health of the moored new ships. Beach wrote on 6 May: "We are at more than ordinary trouble daily by reason of the shallowness of the river, every year it growing shallower and shallower, every spring (tide) most of the ships sewing [grounding][60] a foot, a foot and a half or more and not room in some places for the long keeled new third rates to wind up [swing with the tide]. I fear in a few years this part of the river will admit but of few great ships to be moored in security except ahead and astern [anchors both ends instead of in the bows only] which will be a great charge for so many anchors and cables".[61] Alarmed at the news, the Navy Board instructed the officers at Chatham to measure how much each ship "sewed" at the lowest of the spring tides. They replied on the 22nd, the same day the unfortunate Pepys was sent to the Tower.[62]

A footnote to the letter, which was enclosed with the table below, added that it was not known to where the ships could be removed to prevent grounding, nor what advantage would be gained by doing so. The damage done to the "white stuff" protection was not considered.

Grounding of the moored ships during the lowest of the spring tides:

From Rochester Bridge

Mooring	Ship	Grounding in inches
5th	*Montague*	12
6th	*St George*	14
7th	*Lenox**	18
8th	*Henrietta*	24
9th	*Restoration**	24
10th	*Triumph*	24
11th	*Victory*	14
12th	*Hope**	13

From Chatham Key from the 5th Mooring downwards

Mooring	Ship	Grounding in inches
1st	*Princess*	12 when swung with stern to eastward
4th	*Hampton Court**	12
5th	*St Michael*	6
6th	*Anne**	16
13th	*Prince*	26
14th	*Katherine*	18
16th	*Unicorn*	24

* new third rate ships

The fears created by the Popish Plot led to a review of the safety of the ships in ordinary, both from "treachery and an attempt by an enemy". Many people vividly remembered the Dutch raid on Chatham in 1667 and feared it could happen again. It was proposed on 18 June to move the *Triumph*, *St George*, *Unicorn* and *Rainbow*, all very old ships in need of repair,[63] to a position below the chain to guard the anchorage. The plan was approved and they were taken downriver to new moorings by 12 July,[64] armed with twelve guns and 100 men each. In addition to the existing chain across the Medway just below Upnor Castle, a boom was to be laid further downriver, made of masts and cables. Twenty pinnaces were also provided to patrol the river and gun batteries were placed at convenient points of land.[65] To make best use of the additional defences, Commissioner Beach made a plan dated 5 December showing precisely where they should be placed. In making it, he apologised to the Navy Board: "I hope your honourables will not expect an exquisite draft of each place with their true distances, my sight much failing me". Beach put a lot of thought into his drawing that agreed with the proposals made in June. Two of the four old ships were anchored fore and aft with their broadsides pointed towards the chain, while the other two ships covered it with their chase guns. The new boom would consist of two cables seized together with a chain at each end near the shore to prevent it being cut. Old pieces of masts were to be attached by chains to the boom with stream anchors to keep it taut in case any ships attempted to break through on a flood tide.[66] Part of the reason for moving the old ships was to make space for the new ships, which were now beginning to arrive in numbers.[67]

1680

During 1680, as more new ships arrived to join those already at Chatham, Richard Beach was suspended on corruption charges. It was decided he should exchange places with Commissioner Kempthorne at Portsmouth but before this could happen Kempthorne died. Beach nevertheless completed his move to Portsmouth and Sir John Godwin filled his old position at Chatham.[68]

On 24 September Godwin carried out another survey of the ships' moorings. It shows that because the old ships had gone, *Lenox* was moved upriver to the third mooring down from Rochester Bridge; at low tide she grounded by 18in abaft on a soft bank in the middle of the Lyme Kiln Channel. This bank, or middle ground, had formed seven or eight years before and had ruined one mooring and occasioned six ships, including *Lenox*, to take ground without damage. If the bank was removed there would be another good berth created and ships would not ground at all.[69] In spite of Godwin's reassurance, the Navy Board remained concerned about the situation and pursued the matter by consulting the "knowing men" of Chatham. They were generally foremen and assistants of the older generation, each with many years of experience behind them. Their opinion was that the bank could easily be removed but it mattered little, as it

was almost impossible for the ships to suffer injury, the ground being so very soft.[70] The "knowing men" were confident in their assessment, as they had inspected the keels and bottoms of the ships when they entered dry dock. The Navy Board informed the Admiralty of the situation and added that as the water rose considerably, the ships floated before they suffered from strain.[71]

To reduce grounding, however, the ships were lightly ballasted, leaving them riding high out of the water. But this practice caused problems

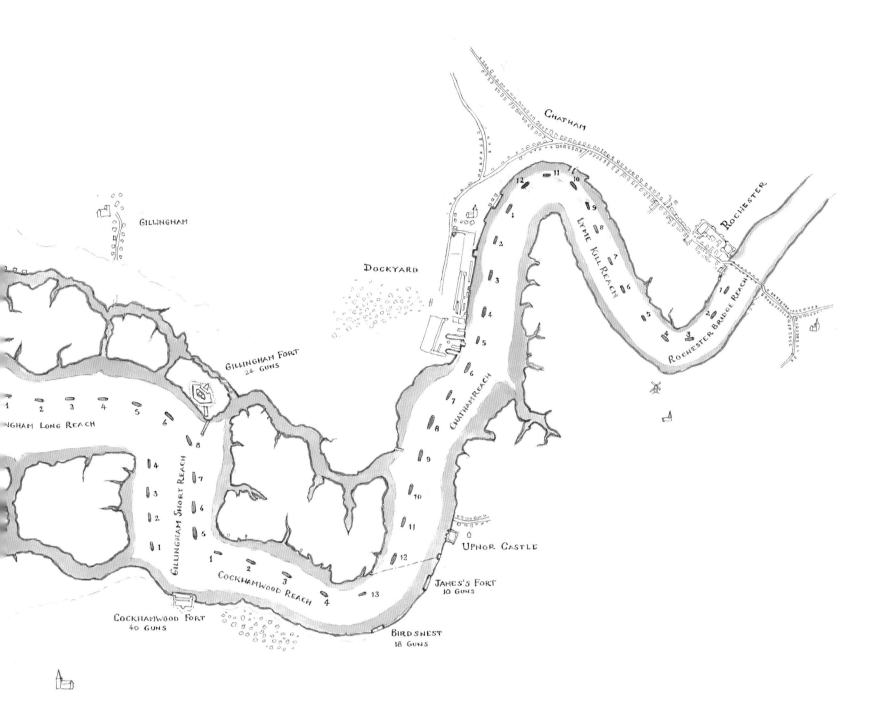

Above: *The anchorage at Chatham, based on a map of 1688 by Captain Greenvile Collins and anchorage descriptions in the National Archives. The principal moorings on the Medway are shown from Rochester Bridge Reach down to Gillingham Reach.*

of its own. The lack of buoyancy support at the bow and stern put considerable strain on the longitudinal strength of a ship. The keel could, with time, bend down at the ends, a condition known as "hogging". Another problem was the wetted area along the waterline. Normally when a ship rode at its moorings the wetted area remained in the same place as the ship rose and fell with the tides. But for *Lenox*, 18in at the stern was continually wetted and dried as she grounded with the tides. Furthermore, because the ships were lightly ballasted and high out of the water, more

of the shallow angled run fore and aft would be exposed. Eighteen inches in vertical height of the run aft translated into a band about 3ft wide. These are the best conditions to cause rot in oak that otherwise remains sound if kept either wet or dry.

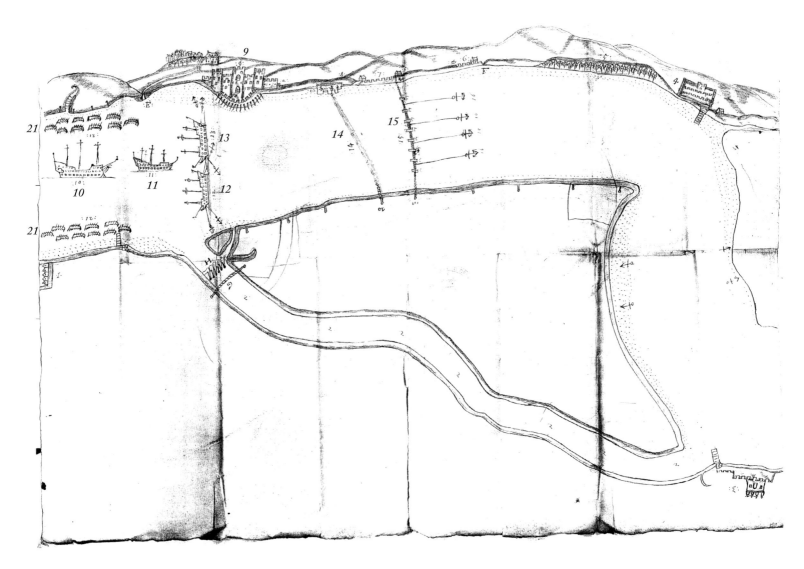

Draught of water of the new third-rate ships built to date and riding at Chatham

Above: *On 5 December 1678, the commissioner at Chatham, Sir Richard Beach, drew this plan for the defence of the dockyard. No. 10 is the* Triumph, *no. 11 the* St George, *no. 12 the* Unicorn, *no. 13 the* Rainbow, *no. 14 the chain, no. 15 the new boom, no. 9 Upnor Castle and no. 21 the guard boats. (NA ADM 106/330 f496)*

Ship	Afore		Aft	
	Ft	In	Ft	In
Berwick	14	2	15	2
Pendennis	13	2	15	0
Lenox	12	6	15	2
Hampton Court	14	0	15	3
Stirling Castle	13	9	15	8
Grafton	14	2	15	2
Hope	13	7	15	8
Fully loaded and ballasted	17	0	18	0

The Admiralty wisely decided something should be done about the anchorage and during December a model of a horse-powered "Engine" was presented to them. It was designed to tow the great ships up and down the Thames and Medway and also to act as a dredger. A contract was agreed for it to be built at a cost of £760 – horses, harnesses, anchors and cables excepted.[72] It was to be powered by fourteen able horses. Its builder, Mr Newsom, was reputedly the only person in England that understood the

working of it.[73] The craft was built but it is no surprise that in February 1683 the investors were obliged to petition for a trial of its ability, as they had not yet been paid. Eventually, when Commissioner Godwin at Chatham was asked about the performance of the tow engine, he replied in July 1684 "that it has for some time past been employed and taken up a considerable quantity of ooze but when the net comes to the hard ground, or gravel, it slides over it".[74] For many years a horse boat, perhaps the same vessel, was in constant use and often referred to.

Lenox was moored at a number of locations during the following years. In 1683 she had been moved from the third to the eighth mooring from Rochester Bridge and from 1687 onwards she was moored at the second mooring in Gillingham Short Reach, ballasted to draw 15ft 8in at the stem and 16ft at the stern.[75] The ballast was brought during 1687 from Greenhithe in two batches by hoy and totalled 70½ tons.[76]

Apart from grounding, the senior shipwrights at Chatham perceived another worrying trend that would colour their future judgement. They reported the *Charles* and *London* had been repaired using East Country timber from the Baltic that soon started to rot. Not wishing to use the same timber on the new first rate they were building, which was to become the *Britannia*, they desired "that no East Country plank may be wrought upon her, except the middle deck where it will lye dry and receive air".[7]

1681

During January, caulkers belatedly made their way out in boats to the *Lenox*, *Hampton Court* and *Restoration* to work on the shrunken seams between the wales and down to the water's edge. What they found alarmed them such that they immediately reported to the Dockyard officers. The officers, including the Commissioner, the Master Shipwright and his assistants, went out to the ships to examine them for themselves. They were shocked at what they saw and wrote to the Navy Board with their findings: "Upon a strict view thereof do find the particular works underwritten so very rotten (being all east country stuff) as that they must be shifted, part whereof cannot well be performed without a dry dock". As regards *Lenox*, they found the following plank rotten; "on the larboard [port] bows between the foremost harpins [main wales] one plank 33 feet long; under the lower harpin, one plank of 28 feet long with one hood [foremost plank that fitted into the rabbet of the stem] of 11 feet long; on the starboard bows under the lower harpin, three hoods and between the lower harpins, one plank of 33 feet and one of 30 feet long. Starboard buttock, nine planks and larboard buttock five planks all rotten".[78]

It must have been a great shock to find that the new ships were in such a deplorable state. *Lenox*, the oldest of them, had been launched only two and a half years before.

During the coming summer Commissioner Godwin did his best to procure stores for caulking. On 24 April he wrote to the Navy Board saying he needed 30 tons of blacking to preserve the new ships.[79] None was forthcoming due to the shortage of money. A month later he was writing again: "Great want of blacking to preserve them this hot weather".[80] He also appealed for caulkers to be sent from Deptford to help caulk the ships in the harbour during the hot weather. There had been scarcely any rain since Christmas, causing, as John Evelyn recorded, "the greatest drought ever known".[81] By the end of May, only fourteen ships had been caulked.[82] As the hot, dry weather continued into June yet more requests for blacking were sent. It seems the very best Godwin was able to do for the new ships was have them caulked and blacked. Under the circumstances he was unable to get money for the docking and replacement of rotten planks and no work was carried out on *Lenox*.

The shortage of money was felt elsewhere. The wages due to the shipwrights and other workmen had fallen fifteen months behind. On 11 July these men wrote a petition to the Navy Board stating that "they are reduced to great poverty ... all their creditors do earnestly press ... are constrained to pay six pence in the shilling extortion".[83] The situation did not change and nearly a year later they were still fifteen months wages in arrears.[84] By January 1688 this had reduced to six months, although a petition at that time stated "your poor petitioners have been exposed to a merciless extortion of their creditors and forced to pay unreasonably for forbearance and some of them lately compelled to quit the yard and list themselves soldiers to prevent their being dragged to a gaol".[85]

On 29 July 1687, the new purser of *Lenox*, Moses Turner, was reportedly "grievously afflicted with an ague and fever which weakened and disordered him so much that he is not able to perform his own duty on board without hazard of his life". He was excused from duty for two months provided he could be replaced in the meantime with a deputy.[86] Turner was still absent by 14 October and as a result was suspended for a time, and "fitting persons appointed to perform his duty".[87] The delays in paying the warrant officers aboard the ships understandably resulted in increased rates of absenteeism.

Matters reached something of a climax when three members of the Navy Board – Sir John Tippetts, Sir Richard Haddock and Sir John

Narbrough – took an opportunity to visit the ships themselves on 12 April. They had been downriver viewing Stanlett Creek (Stanly Creek on the Grenville Collins map of 1688 and Stangate Creek today), a little above Gillingham, for new mooring places.[88] They returned at eleven o' clock that night and took the opportunity to visit many of the ships at their moorings. On board they found many serious faults with the officers who were supposed to be keeping watch. The warrant officers made matters worse by offering lame excuses for their neglect of duty, excuses that the Navy Board officers found unbelievable. The officer of the watch aboard *Lenox* should have been Ambrose Fellows the carpenter, but he was not found on duty, "pretending to be indisposed by the toothache in his cabin, leaving his servant to watch who was likewise negligent".[89] Aboard the *Grafton*, the gunner at the end of his watch ordered the carpenter's servant to call the carpenter to relieve him. But the carpenter remained below pretending to be ill in his cabin, while the boys had all left to haul in an eel pot. All the negligent officers told similar stories. Only the carpenter of the *Henrietta* "owns his neglect of duty and begs pardon". Seventeen of the absentees were immediately suspended, among them Ambrose Fellows of *Lenox*. So serious did the Navy Board consider their offences to be that they wrote to Richard Vittels, the Master Attendant, ordering that "one diligent able seaman" from the *Royal Sovereign* or *Prince* take the place of the suspended officers. At the same time new orders were issued to make sure the officers kept better watch. They were to strike their ship's bell every half hour for which a half-hour glass was to be kept in every ship. They were also to keep watch on the quarterdeck or poop and not to leave their station until a succeeding officer had relieved them.[90]

The matter was brought before the Admiralty Commission shortly afterwards on Saturday 23 April. A report by the Navy Board[91] was laid before them together with a petition from the suspended officers. In it they acknowledged their "great crime ... and having humbly laid themselves at their Lordships' feet for mercy and pardon" they promised "they will be double diligent and extra watchful night and day to charge their Duties with all care and faithfulness".[92] King Charles himself was informed of the situation, and in accordance with his benevolent pleasure the suspensions were lifted. But as a warning to them and all others in the navy they lost their wages and victuals for the period of suspension and forfeited one month's wages to the Poor Chest at Chatham. To add humiliation to their punishment, notices were put up at the gates of the dockyards warning that further offences would be punished by the severity of the law.[93] The Navy Board officers must have been disappointed with the King's intervention. As often happened during Charles's reign, the punishments were not nearly harsh enough to deter offenders and were very lenient compared with later times.

1682

The reason Ambrose Fellows was often absent from *Lenox* was that his family lived at Deptford. After King Charles lifted his suspension, Fellows found a solution to his problem. In February salvation appeared in the shape of Obadiah Hancock, Carpenter of the *Foresight*, who had a similar problem. His family lived at Chatham but his ship was laid up at Deptford. Fellows and Hancock agreed to make a joint request to the Admiralty Commission to temporarily exchange ships while they were laid up in ordinary. The Admiralty Commission agreed, content no doubt that at least these two warrant officers were more likely to remain aboard their ships.[94] Fellows took his servant John Jones to Deptford with him, while Hancock brought his servant John Russell aboard *Lenox*.[95] Shortly afterwards, the purser of *Lenox*, Moses Turner, who had replaced Joseph Phillips at the end of 1678 also proved susceptible to absenteeism and was suspended for one month.[96]

During August orders were sent to Chatham for ships that had been "off the ground" (afloat) for more than three years to be docked for graving. Work began on the new third rate *Restoration* that had been moored in Lyme Kiln Reach close to *Lenox*. After cleaning they started to caulk her seams but some of the planks proved so rotten they simply crumbled away. Several planks between wind and water, especially in the bows and buttocks, were found to be in a similar condition. Robert Lee, the experienced fifty-year-old Master Shipwright supervising the operation, wrote that it was an absolute necessity the planking be removed and new put in its place. Unfortunately he could not find 4in plank in the yard for repairs, even though it was the most common plank used in ship construction. Ignoring his own good advice and in order to launch *Restoration* at the next tide he had the caulkers secure lead and tin plates over the worst parts, then grave her. His action trapped the moisture, allowing decay to spread.

Later the same night *Restoration* was launched with more harm than good being done to her decaying timbers. All that could be said of the work is that the plates stopped water pouring through the damaged seams and actually sinking her.[97] It was also "contrary to the known rules and practice of the Navy"[98] to leave ships unfit for service at sea. She was left to slowly rot until November 1685 when Isaac Betts, her builder, visited her to view the work that had already been done and assess what would still have to be done to her. After an understandable lament, he listed the surveys that had been taken on *Restoration* and the estimates to repair her. In 1682, when she was five years old, the cost was estimated at £420; in April 1684 it was £520; in July 1684 it had jumped to £902; on 13 May 1685 it was £1532, and the last survey on 10 October 1685 estimated the cost of repairs to be £2969.

Next ship in the dock after *Restoration* was the old third rate *Montague* of 1654. She was also found to have defects, and Lee kept her in dock for two weeks during the neap tide for repairs and was able to put her "into a good condition for sea service". Following her was the *Hope*, another new third rate that had the same problems as *Restoration* and received the same shoddy treatment. Lead plates were nailed over the rotting planks before she was returned to her moorings.

After the *Hope* came the *Breda*, yet another new ship, "being very leaky". Lee hoped to replace her rotten planks if new should arrive in time. It can be said in Lee's defence that although he wrote monthly demands for 4in plank, ultimately it was his own decision to treat the ships so badly and then simply inform the Navy Board what he had done. He also pointed out that if the fleet should ever be ordered to sea there would be too much repair work to be done without a new dry dock.[99]

Later the same month, other new ships of the 1677 programme were docked for graving. Their condition promoted Commissioner Godwin to write to the Navy Board: "Upon bringing into dock the ships who want graving it is found that several of the new third rates are very rotten in their plank and some of their [frame] timbers, and at this time the *Anne*, now in dock, has seven planks each side of her buttocks perfectly rotten, also a large piece of thick stuff in midships the same. The *Lenox* and *Hampton Court* [not yet docked] are found worse. The Master Shipwright demands for making good their works [are] about 14 loads of compass oak timber which may be bought in these parts but not without ready money which I pray you to consider".[100] While Godwin was in the mood for complaint he also mentioned that he understood Deptford and Woolwich yards were paid quarterly, which made his workmen extremely discouraged that they were not being equally treated – especially as they had been promised to be the first paid when money was available.

Two days after his first report of *Anne*'s condition Godwin wrote again, adding: "*Anne* in dock proves more rotten in her plank than I could then give you an account of".[101] The work performed on her lasted a week and on 8 September she was taken out of the dock and another ship docked in her place.[102] News of the condition of the *Anne* reached Samuel Pepys, now out of office, who made a record of the event in his papers.[103]

Godwin's reports were not ignored by the Admiralty Commission, for shortly afterward the Treasury ordered £200 be made available for the repair of ships. Godwin directed that the money would go toward the repair of the *Hampton Court* and other works as desired by the Admiralty Commission.[104] Considering the state of the ships, £200 was a very small amount and would have little impact. While the *Hampton Court* was in dock, on the 29 September, work almost ground to a complete halt because of a shortage of treenails. It was the intention to dock and repair *Lenox* after the *Hampton Court* was launched, but only if sufficient compass timber could be obtained for the work.[105] Unfortunately for *Lenox*, three Algerian prizes (the *Half Moon*, *Two Lions* and *Red Lion*) suddenly arrived and these were docked and repaired instead.[106]

1683

Since arriving at Chatham, the new ships had been plagued by the crippling state of finances and the willingness of the Chatham dockyard shipwrights to carry out inadequate repairs. The Navy Board wrote to the Admiralty that "the workmen are reduced by their wages being so long in arrears to the utter ruin of themselves and families, together with the want of almost all sort of materials for carrying on the works now in hand."[107] During May 1675, shipwrights were engaged on repairs to the first rate *Royal Soveriegn* that were not expected to be finished before January 1684. Only fifty-five shipwrights were employed on the repairs of all the other ships and even these numbers had to be fully justified by Robert Lee, the Master Shipwright, and Daniel Furzer his assistant.[108]

To help with the work, five caulkers were sent down from Deptford to Chatham. The wives left behind at Deptford, aware that it would be a considerable time before their husbands would be paid, wrote a petition to the Navy Board requesting they be returned "as our children are for want of relief ready to perish".[109]

The Chatham officers produced a list showing the order in which the ships should be docked and repaired. *Lenox* was fifteenth on the list. Unless things changed, even the most optimistic estimate would not see *Lenox* repaired for some years. The list also reveals that only six of the seventeen new third-rate ships lying at Chatham had been repaired or graved since they were built. The remaining three of the twenty new third rate-ships were laid up at Portsmouth. *Lenox* was one of those unfortunate ships that not been graved since the year of her launch. Winter was strangely late in arriving, with hardly any frost before Christmas; then suddenly, on 23 December; the Thames started to freeze.[110]

1684

By 9 January the ice on the Thames was thick enough to bear temporary booths in which meat was roasted and shops were set up selling all manner of wares. Coaches and horses were able to travel across from one side to the other. The cold continued throughout the month and the shops on the ice became so numerous that they were formalised into streets. Coaches plied their trade from Westminster to the Temple, there was sliding, skating, bull-baiting, horse and coach races, puppet plays, cooks, tippling and other lewd practices, as Londoners did their best to enjoy the harsh winter. Away from the fun on the Thames men and cattle perished, fowl, fish, birds and many deer in parks died. The air in London became so thick with sea-coal smoke that a man could hardly see across the street, and the fog filled the lungs such that it was scarcely possible to breathe. The frost penetrated trees so they split as if struck by lightning. The seas became locked up with ice; no vessels could stir. On 5 February it began to thaw but quickly re-froze, and even as late as 4 April there was hardly any sign of spring.[111] Considerable damage was done to ships' timbers.

Suddenly, and quite unexpectedly, circumstances changed. Opposition to Charles ceased and in May he revoked the Admiralty Commission of 1679. He once again assumed control with himself as Lord High

Above: Louise de Keroualle, Charles II's French mistress, whom he created Duchess of Portsmouth in 1673. **Right:** Charles Lennox, Louise's illegitimate son by Charles II, was born on 29 July 1672 and created 1st Duke of Richmond and of Lennox in 1675. He was aged 18 by the time this portrait was made, by Sir Godfrey Kneller, the leading portrait artist of the period. (Both images courtesy of the Trustees of the Goodwood Collection).

Below: A view of Greenwich with the Queen's House in the foreground and the remains of the Tudor Palace of Placentia nearer the river. The river is full of trade going to and from London and the church of St Nicholas at Deptford is visible in the background. By Jan Griffier the Elder (NMM BHC 1833).

Right: A flagon belonging to Jonas Shish, master shipwright, given to him as a gratuity by King Charles II for building the *London*, a 96-gun second rate, in 1670. It was made by Humphrey Stoaks, a goldsmith of London, at a cost of £22 4s. Stoaks also made a similar flagon to the value of £20 that the King presented to John Shish at the launching of *Lenox*. It was probably similar in style, but was recorded as being engraved with the Duke of York's arms with a stylised anchor through them, along with an accompanying inscription: "Att the launching of his Ma:ts ship Lenox ye 12th day of April 1678: Built at Deptford by John Shish his Ma:ts Master Shipwright there, Burthen 1096 Tunns, Menn 440 Gunns 70". (Courtesy of the Museum of London).

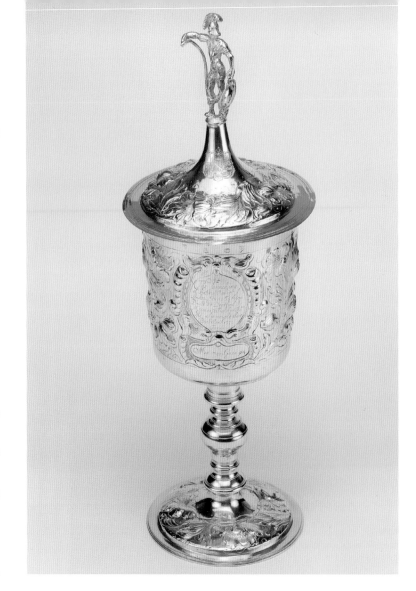

Far right: This masterpiece, by Willem van de Velde the Younger, has previously been identified as either *Hampton Court*, *Lenox* or *Captain*, all ships of the thirty ship building programme with 12 upper deck ports. The *Captain* has been subsequently ruled out because the hancing piece, intersecting the quarterdeck ports, is in the wrong place. Although there are a number of minor mistakes in the painting, it is likely to be *Lenox* as she was the first ship of the programme to be completed, and was thus the best documented and most commonly drawn third rate. Here, the ship is caught in a sudden gale and has let go her larboard fore course sheets and tacks to allow the mizzen to bring her head into the wind. (Birmingham Museums and Art Gallery).

Below: The launching of *Lenox* at Deptford on 12 April 1678, based on a Van de Velde sketch of the scene (NMM BHC 1109). Charles II, Louise de Keroualle, and their five year old son, Charles Lenox, are shown in the barge alongside. (Painting by the Author).

Above: The newly built *Lenox* passing Greenwich on 12 May 1678, on her maiden voyage from Deptford to Chatham. She was manned at this time by a mixed crew consisting of 93 seamen, watermen, shipwrights and dockyard workmen. *Lenox* would not return to Deptford until November 1699, when she was rebuilt. (Painting by the Author, after a work by Jan van Beecq).

Left: A view of the River Medway from Chatham toward Rochester Castle, where *Lenox* spent her early years laid up at the 7th or 8th mooring from Rochester Bridge. All the topmasts, rigging and yards are stored ashore or in the ships' holds. The barge in the foreground and flags on the ships seem to indicate the visit of an important dignitary, or possibly the King himself. (NMM BHC 0832).

Above: *Lenox* at the Battle of Beachy Head, fought between the Anglo-Dutch allies and the French on 29 June 1690. At a critical stage of the battle the Dutch squadron became detached and surrounded by the French. *Lenox* is shown in the foreground coming to their aid, followed by Admiral Torrington's flagship *Royal Sovereign* to the right and the *Stirling Castle* to the left. (Painting by the Author, collection of Randy B Mafit).

Below: The only English ship lost at the Battle of Beachy Head was the *Anne*, a 70-gun third rate, one of *Lenox*'s sister ships. She was dismasted during the battle, and then run aground and burnt at Pett Level near Hastings to prevent capture. Today, the lower part of her hull survives and is sometimes visible at low tide.
(Painting by the Author, courtesy of the Shipwreck Heritage Centre, Hastings).

Above: Captain William Kerr, by Michael Dahl, painted circa 1706. Kerr commanded *Lenox* in 1693. (NMM BHC 4146). **Right:** Captain John Munden, who commanded *Lenox* from early 1691 to early 1693. In the right background *Lenox* is shown engaged at the Battle of Barfleur, 1692. By Michael Dahl. (NMM BHC 2873).

Far right: A painting of *Lenox* by Willem Van de Velde the Younger, measuring 22½in by 28½in. *Lenox* is viewed from the starboard quarter with the wind on the larboard beam. The yards are braced round and all sails are set. Bonnets, the archaic sail laced to the bottom of the courses are set, although by the time *Lenox* went to sea in 1690 they were obsolete. This tends to date the painting to the early period of her career and could not have been painted from life. (Private collection).

Below: The burning of the French 112-gun ship *Soleil Royal* near Cherbourg, a scene witnessed by those aboard *Lenox*. During the night of the 2nd and 3rd of June, she was attacked by 17 ships, which she repelled until a fireship set her stern on fire, which soon reached the powder rooms. Painting by Adriaen van Diest. (NMM BHC 0338).

Left: *Lenox* at sea in a gale, viewed from the starboard bow. She is close hauled on the starboard tack. (Painting by the Author, after a work by Willem Van de Velde the Younger).

Below: *Lenox* in Plymouth Sound. On 17 June 1695 *Lenox* entered Plymouth Sound after escorting a small convoy consisting of the *Briget Galley*, a hired armed vessel, and 12 sail of small vessels from Spithead. She remained until the 27 June when she sailed out into the Soundings to join a small squadron of warships with orders to protect trade and frustrate the enemy. (Painting by the Author, on loan to the Peter LeFevre collection).

Admiral assisted by his brother, James, Duke of York. Samuel Pepys, who had just returned from a sea excursion to Tangiers, was recalled with the title of Secretary for the Affairs of the Admiralty of England. The terrible state of the navy and what to do about it confronted him. It was a daunting task, even for the energetic Pepys. After his return to office, the anonymity and neglect that the navy had suffered during the previous five years quickly disappeared, replaced by a mood of impending change. Once again it became worthwhile for men of influence to show an interest in naval affairs, and the fate of the new ships in particular. *Lenox* was by now in need of urgent attention. On 17 February, at the height of the winter of great frosts, it was intended to dock *Lenox*. The Navy Board was informed that her buttock planks were so rotten that they could not be caulked without them breaking up and that they would all have to be stripped off and replaced. Once the planks were removed and she was opened up it was expected that further considerable defects would be discovered.[112] Unfortunately, *Lenox* did not receive her necessary extensive repair, probably because of the extreme cold.

On 24 April more questions were asked about the practices at Chatham. Sir Phineas Pett II and Sir John Tippetts became interested in the new ships' decay and began an investigation. They wanted to know why some of the new ships had recently been caulked only down to the main wale and not all the way to the water's edge. The answer given by Robert Lee and Daniel Furzer was that this was usually done in the autumn but in the spring they were caulked only in the upper works. Even if this answer could be believed, it goes to show that rules appeared to be made up at Chatham to suit the yard's officers. Pett and Tippetts also wanted to know which of the new ships had not been dry-docked but had, according to the method prescribed by the Navy Board, been heeled, breamed and payed with black stuff. "Not any" came the reply; the danger of fire in the rotten powdery timber had been deemed too great.[113] This may well have been true, and illustrates the neglect the ships suffered when they were new and sound. *Lenox* was one of the ships that had never entered dock. In fact she received no attention at all except a minimal amount of scraping and paying of her sides. Almost as soon as Pepys entered his new post he ordered the Navy Board to make a detailed survey.[114] Dated 1 June, it described the condition of all His Majesty's ships[115] and revealed the steady decline of *Lenox* since the report of 1681:

> "The buttocks on each side to be wholly shifted as also the bows being decayed, as likewise several planks of six inch between the lower wales, with some pieces of spirketting on the gun deck containing about one hundred and eighty feet; and one hundred and eighty five feet of 4 inch plank there all decayed to be shifted, and likewise ten planks without board between the ports; on the upper and quarter deck, one hundred sixty one foot of Prussia deals to be shifted; an awning and house of easement wanting; the breadroom to plate, a crojack yard wanting, with other necessary works, caulking, graving, painting and repair of glass the sum of £846".

A breakdown of the cost was made to put the ship in full repair.[116] It was estimated that 46 workmen would be employed on her for 13 weeks.

	£ Cost
Timber, Plank, Board, Treenails	290
Masts, Baulks, Spars	22
Ironwork, Forge	43
Ironmonger	20
Pitch, Tarr, Rozin, Oil, Tallow, Brimstone	27
Copper, Lead, Leather	20
Sundry Stores	17
Total	439

Added to this sum were the additional costs for labour and finishing:

Workmanship	313
Carving, Glazing, Painting	94
Total	846

The Navy Board obtained approval for £24,250 for immediate repairs, although the Treasury was in a position to furnish only £12,000. It was some of this money that was at last spent on *Lenox*.[117] Work started about 12 July when *Lenox* was taken into the double dock.[118] She was the first of the thirty ships to have major repairs under the new administration,[119] because she was almost certainly one of the ships reported to be in actual danger of sinking.[120]

While *Lenox* was in dock, Samuel Pepys put the Navy Board to work preparing a new comprehensive report on the state of the ships. He visited Chatham himself on 12 August when he saw with his own eyes the deplorable condition of the fleet, recorded so graphically in his book *Memories Relating to the State of the Royal Navy* (1690). During his visit he would have seen repair work being performed on *Lenox* and the great, famous old first rate, the *Royal Sovereign*. On his return home to Derby House in London he wrote to Sir John Godwin, the Commissioner at Chatham, thanking him for "my late waiting on you at Chatham, and for the assistance wherein I now render you very many thanks, and particularly for that of your accompanying me in my melancholy visit to the new ships, and inspecting the deplorable state they are already in".[121]

In dock, the rotten planks in the buttocks and bows of *Lenox* were stripped away and Robert Lee ordered new English oak compass timber to replace them.[122] A contract for the timber was made with Sir Charles Bickerstaff for repairs to both *Lenox* and the *Royal Sovereign*. At the time, about 100 loads from a 300-load contract supply was waiting in the Chatham stores. This timber was suitable for wale pieces, plank between the wales and spirket rising, all relatively straight pieces. Unfortunately there were only three pieces of compass timber large enough to make buttock planks for *Lenox*. Lee sent the Purveyor to view the timber ordered but not yet delivered, to see if any was of sufficient compass to make buttock planks. None was, but the Purveyor did see Bickerstaff had other suitable compass timber and heard that some house carpenters at Maidstone had some small quantities of very good compass.[123] It was not a simple task to secure funding for more timber and Commissioner Godwin lamented the lack of money.[124] Even treenails ran out, and this further hampered the progress of repairs to *Lenox*.[125]

After the harsh winter the summer became excessively hot and dry as work continued. John Evelyn could not remember such a drought, when leaves dropped from the trees as if it were autumn.[126] At the end of the summer, work on *Lenox* was completed and she was launched on 15 October.[127] A record exists for the charge of workmanship for most of the time *Lenox* was being repaired. For the Michaelmas quarter, between 1 July and 30 September 1684, £407 9s 1d was paid in wages and during the Christmas quarter, between 1 October and 31 December 1684, the amount was £158 7s 4d.[128] The work was later listed as "a great repair";[129] perhaps the amount of work was far greater than the estimate and with it the realisation that repairing the whole fleet would be a much greater task than originally thought.

There was certainly a great deal of confusion about the ships needing repair and the costs. One report claimed the *Exeter*, *Kent*, *Restoration* and *Stirling Castle* were left untouched and in danger of sinking, while other ships in better condition were repaired. It also gave the estimated cost of £330 for repairing *Lenox* against the actual cost of £580, while another report thought the cost £492.[130] Shortly after *Lenox* was launched and many months later than Robert Lee's original estimate, the repairs to

Royal Sovereign were also completed. Van de Velde the Elder, perhaps accompanied by his son, came down from London to record the event.[131]

It was customary for the navy to bestow the gift of a piece of silver plate to the Master Shipwright when he built a new ship.[132] Lee was of the opinion that he deserved the honour for the repairs he had carried out to the *Royal Sovereign*. He appealed to King Charles himself, saying that old Jonas Shish had received one for rebuilding the *London* and John Shish for rebuilding the *Tiger*.[133] Treating Lee's claim seriously, a rigorous investigation was carried out. It concluded that the *London* was burned down almost to the wrongheads leaving only the bottom part of the hull. On top of these remains the ship was built up as new. As for the *Tiger*, which had been in the stern of the double dock at Deptford while *Lenox* was being built, she was taken completely to pieces with only part of her keel, stem and a few other timbers used in a wholly new ship. The investigation went on to say that the work on the *Royal Sovereign* constituted a "very considerable repair" but was not a rebuild to compare with the work done to the *London* or *Tiger*.[134] To prevent further confusion the Admiralty ordered, "We would have the work of rebuilding limited to in the description thereoff as to prevent it being applied to anything less than that."[135] The disappointment Robert Lee felt is not recorded.

Below: *The Fleet in ordinary at Chatham in 1685, by Willem Van de Velde the Elder, probably on the occasion of a royal visit. (British Museum, BM c75bcm 37-57)*

By the end of the year Pepys had prepared his report concerning the state of the navy. This was presented to King Charles on New Year's Day; it contained a general state of affairs and showed the deterioration of the thirty new ships. It also showed that the stores were exhausted while at the same time huge debts had been run up. Pepys also stated his opinion that the decay had been caused by neglect rather than poor-quality timber and rapid building.[136] The debate over blame would last for many years.

The first to come forward with an opinion was Sir Anthony Deane. He wrote a proposition for the Navy Board for preserving the new ships from decay, dated 23 March. It was written in a friendly, helpful tone. He addressed the two common beliefs for the reason of the ships' rapid decay: one; that the ships were built of foreign East Country plank; and two, that the ships had been built in too short a time; that is, before the timber they were built from had seasoned sufficiently. Addressing the first point he argued that the Dutch, Danes, Swedes, French and the East India Company built successful ships from East Country plank. The decay must therefore have been caused by the lack of adequate preservation applied to the plank after the ships were built. As to the second point, Deane reasoned that green timber would indeed dry out in the comparative warmth of a ship's hold, drawing in moisture from the outside which would then penetrate the dry wood and "putrefy it which then spreads like a leprosy". To prevent the penetration of moisture, he argued, the ships should have been heeled twice a year and the old "stuff" burnt off and new applied.

This had clearly not happened. By sealing the outside of the ship the timber would be allowed to season from the inside. The importance of keeping the holds aired was often mentioned at the time.[137] (See Appendix 11 for full transcript of Deane's proposition). Deane should have added that the outside protection was inadequate, and that water would migrate from the outside to the inside as much in seasoned timber as green.[138]

The arguments forwarded by Deane absolved himself and Pepys, who were responsible for building the new ships, and put the blame for decay on those who had taken care of the new vessels since. In this respect Deane's proposition may be a political as well as a technical work.

Less than a month later, and in response to Deane's proposition, another letter arrived at the Navy Board. It was written on the orders of the Navy Board by the Master Shipwright at Chatham, Robert Lee, and his assistant, Daniel Furzer, the men who were arguably the most responsible for looking after the new ships in ordinary. They put the cause of the problem squarely on the use of poor-quality over-mature East country trees: "We conceive past their growth and declining before their fall, or had lain long after, where by their nature all substance was abated and became spongy."[139] Over-mature oak trees are a potential problem, as a soft rot fungus could attack the heartwood, which is otherwise normally durable. The soft rot, sometimes called wet rot, makes the timber spongy when damp and powdery when dried.[140] It seems extremely unlikely that such timber was used during the building of *Lenox* because timber in much

better condition than that described by Lee and Furzer was rejected. More accurately, Lee and Furzer went on to say that there was no decay underwater but it appeared at the water's edge and for 5ft or 6ft above it. It first appeared in the buttocks and bows, where the plank sometimes appeared sound but after penetrating into it for between 1–1½in it was found to be "dirt rotten" through to the frame of the ship. Happily, test borings made on internal structural timbers found them to be sound and good. They recommended that the rotten East Country stuff should be removed and replaced with good English timber and plank.[141] (See Appendix 12 for full transcript).

Strangely enough, and in complete contradiction to what Lee said, the only planking on *Lenox* known for certain to be of English oak was that used on the buttocks and bows. It is the only planking for which John Shish specifically requested English oak at the time of her building. He wrote: "There will be wanting to carry on that work 40 loads of English Oake plank, to work round the buttocks and bows from the light mark up to the lower edge of the wale, which I desire your Honourables to be supplied with being in great want."[142] A careful search of the surviving records confirms that English oak was indeed used. A warrant was issued shortly afterward for Deptford to receive "Four inch plank English Oak for ye buttock of the new ship being of forty loads".[143] Three weeks later an imprest bill was made for the Purveyor of Timber at Deptford, Mr Lewsley, to pay £100 for "New ships 4 inch oake English for the new ships

de Charles   de Duches   de Winsor Castle   de Breda.   de Albemarle   de Royal
De Siarles (The Charles)   De Bretadie (Britannia) the Albemarle   The Neptune

buttocks".[144] Two receipts were made to Mr John Lee for 4in English oak plank totalling thirty-four loads 13ft and one to Mr John Brown for thirteen loads 14ft by the end of the year. All the English oak planks were 28ft long. For *Lenox*, it can be said with certainty that the first planking on her to rot was the English oak used on her buttocks and bows. The most likely conclusion to Lee's arguments is that they were intended to deflect any blame that may be attached to him for the rapid decay of the planking. In view of the evidence it also seems reasonable to suppose that good East Country oak was the equal of good English oak.

Further evidence of Lee's bad judgement came three years later when Sir Phineas Pett II, the most respected shipbuilder of his age and responsible for building many of the new ships, saw timber that Lee would have used when repairing them. He "observed the bad timber, very defective and not fit to be used for repairs to His Majesty's Navy". He added; "not withstanding the Master Shipwright, Storekeeper and Clerk of the Cheque are strictly enjoined by their instructions to inspect themselves both quantity and quality of stores received... and not to leave it to their instruments [underlings]."[145]

More shipbuilders' opinions were recorded in 1686 at a conference held by Samuel Pepys concerning the plank used for the new ships.[146] Among those present were the Thames-based private and navy shipbuilders who had built many of the new ships, including John Shish. Not unexpectedly they agreed with Sir Anthony Deane's view that the decay had nothing to do with the use of foreign timber or that the ships were built too quickly. They pointed out that three of the ships had taken some years in building yet suffered decay at the same rate. Regarding the use of East Country plank, it was noted that in the past some ships had been planked entirely with it and after being in service eight or nine years were better than the new ships after three. One of the new ships, *Northumberland* at Bristol, was built using only English plank, yet suffered as much as the rest.

The gathered shipbuilders estimated the shortage of English plank meant that of 120 loads of plank used only twenty were English and the rest foreign.[147] The Privy Council and the King agreed with the findings and declared the officers of the navy free to order foreign plank in the future. Pepys famously claimed that the new ships' holds were not cleaned or aired and therefore "suffered to heat and moulder, till I have with my own hands gathered toadstools growing in the most considerable of them, as big as my fists."[148] He further noted that they were not ballasted deep enough in the water. Gun-ports could not be opened for airing as port ropes were missing. Scuppers, designed to take away rainwater from the gun decks, were missing. Instead of replacing plank at the first sign of decay it was patched over with board and canvas, a fact Robert Lee admitted in a letter to the Navy Board.[149] Finally, that they had not been graved for years. Although Pepys's testimony should be treated with caution, it is clear from what we have seen of earlier reports that in this instance he can be taken at his word.

Many of the precautions for preventing decay were known to the navy officers. Their knowledge had been passed down from their predecessors and was based on years of shipbuilding experience. In spite of generally sound practices, fungal decay did not start to be understood until 1803, when it was first noted that dry rot was caused by a plant. No one in the seventeenth century knew that an unusual vegetable composed of fine branching, hair-like filaments that lived off damp timber caused the rot. The filaments actively penetrate wood and although individually too small to be seen with the naked eye they may be compacted and woven together in visible sheets. Alternatively the decay may be entirely within the timber. The fruit body of toadstool appearance is seen less frequently and only after considerable decay has already occurred.[150] Fungi derive their nourishment from organic material and not, as in green plants, by the action of sunlight on carbon dioxide and water.[151] They can therefore live in the dark, and grow generally twice as fast at 30 degrees Celsius as at 20 degrees Celsius.[152] These are the conditions that would have been found in the holds of the new ships. Fungi reproduce by single or multi-cell spores that germinate when in contact with moist wood.[153] Once the fungal attack has reached an advanced stage in late summer the fruiting body, or sporophore (which Pepys called "toadstools") germinates and rapidly spreads its spores. As the fungus takes nourishment, its enzymes break down the structure of the wood.[154]

Ideal conditions for growth occurs where timber has moisture content of between about 24 and 50 per cent. However, being aerobic, most fungi cannot live in total immersion.[155] At sea, the lower reaches of leaky ships – the floor timbers, keelson and heels of the lower futtocks – were washed in bilge water and it was the experience of every shipwright that these areas were never affected by dry rot, even when almost every other part was found in a state of decay.[156] *Lenox* and most of the other new ships had not been to sea, leaving the moisture content below 50 per cent and allowing fungi to grow, as witnessed by Pepys. It is difficult to create a much better environment for many types of fungi to flourish.

On the outside of the ships, in the area immediately above the waterline of the lightly ballasted new ships, conditions were excellent for fungi growth, especially on warm summer days. Here, large exposed areas of untreated, regularly moistened, unseasoned oak plank suited its growth perfectly.

There are many types of fungus that specialise to suit particular conditions and materials. The specie that Pepys found in the hold may well have been *Phellinus Megaloporus*. It has a characteristic large sporophore that fits his description. It is particularly fond of oak and flourishes in very moist conditions away from light. Its optimum temperatures are 27 degrees Celsius and it attacks heartwood very rapidly, perhaps more than any other fungus.[157] Another variety that most likely attacked the new ships' planking was *Polyporus Sulphureus*, the most damaging fungi found in ships. It is one of the few fungus found in standing trees that also affects felled timber. It is most common in large structural timbers and remains as long as the timber is moist.[158] Much of the new ships' timbers certainly remained moist from the moment they were cut down. Oak seasons very slowly, making it very unlikely that the moisture content would have dropped below 20 per cent to prevent fungus growth.

The timber subject to the greatest exposure to moisture was East Country plank. Pepys mentions that the best foreign plank came from Bohemia and was distinguishable by its black colour, rendered by its long immersion in water during its passage by river from the forests to the sea.[159] After its thorough soaking in fresh water, the timber was sawn into planks that were then normally stored in ventilated stacks before shipping. By keeping the plank surfaces dry, the fungus would not be able to grow.[160] When the plank was loaded into ships for transportation it was in all probability stacked in solid piles to save space. The moisture within it would then rise to saturate the surface once again. On arrival in England, when the timber was unloaded the damp surface of the plank would be exposed to the air, creating good conditions for fungal growth.[161] Although by nature foreign plank was as durable as English, the long journey from its source may well have left some of it deeply flawed. Certainly Robert Lee at Chatham was convinced it was a liability even if he did not know the reason why. In his experience he had probably seen how quickly rot could affect some East Country plank.

Sir Richard Beach, the Commissioner at Portsmouth, wrote: "Besides the thirty new ships having been built in haste by timber and plank not seasoned, how defective those ships have proved underwater in a few years."[162] Beach ignored the fact that it is very desirable to build a ship from good-quality green timber that can be easily worked by hand tools. Ideally it should then have been allowed to season on the ways for about three years with some of the planking left off under a covered roof. Such an option was unknown during the seventeenth century and none of the building slips or docks were covered. Nor, indeed, could thirty ships have been built in the time allotted by Parliament if they had lain idle in the dockyards for three years. The timescale also ruled out the possibility of allowing the green timber to season before being used on the ships.

It was known during the eighteenth century that a ship's frame, newly cut and exposed to the weather then covered with plank, would retain moisture and "is the first cause of putrification and will end in corrupting the whole to its original dust".[163] This was not a surprise, for rot was known to act faster in fresh water than salt. Proof that leaving a ship on the stocks without covering her was prejudicial to her integrity was demonstrated by the fourth rate *Sedgmore* in 1686. She was building at Chatham and left on the stocks for two years. She was not left to season; there was simply not enough money to complete her. Upon inspection by the Shipwrights Hall it was found that the keel and six strakes above would require replacing if she was not completed and launched that summer.[164] *Sedgmore* was eventually launched on 3 May 1687, an event that was attended by King James.[165]

In the light of what we have seen, *Lenox* and the other new ships were in a perilous state at the time of their launch. They should have received frequent attention in the area just above the waterline to prevent the timbers being continually saturated, especially in the first few years after their launch. This basic observation and maintenance could and should have been carried out by the Chatham dockyard officers in spite of the financial restrictions. A small amount of money spent in these early years would have saved a fortune later.

Unfortunately not enough was done to press the case for preserving the ships in which Robert Lee was the first link in a chain of indifference. Equally at fault were the warrant officers aboard the ships. A lack of ventilation is perhaps the most significant cause of deep-seated decay.[166] During the next century it was pointed out that a free communication of air through every part of the ship could be obtained by such drastic action

as removing the spirketting above the gun deck and the orlop clamps below so that air could reach the beam ends.[167]

Another reason the ships decayed so quickly must surely have been the extreme weather conditions of summer and winter during the 1680s. In view of the lack of a true understanding concerning rot, it is possible that nobody found himself carrying the lasting blame for the new ships' decay. Many accusations were bandied about, culminating in a 1690 Parliamentary Commission to investigate Pepys's repair of the navy. Ironically Robert Lee, for all the damage he had been responsible for, pointed a finger of accusation at Pepys. One of Lee's allegations was that Sir Anthony Deane returned his critical report concerning East country plank to be amended. The investigation rumbled on until November 1692, when it finally concluded that the ship repairs done by the commissioners were fully and well performed.

As if the decaying ships had not caused enough problems at Chatham another controversy arose concerning the ships' moorings. Long ago, as far back as anyone there could remember (and some of the old men could remember a long way back), warships had been moored with two cables.[168] Each cable was paid out across the river away from the ship toward the opposite shores.[169] In this position, it was unlikely that the anchors at either side of the line of ships could damage their bottoms as the river levels dropped. The ships could turn with the direction of the wind and tide but remain more or less in the same place.

The size of cables depended on the size of ship. The *Royal Sovereign*, the largest ship in the fleet, was moored with cables of between 17in and 21in, but at sea carried bower cables of 21in or 22in circumference, while the *Prince*, whose bower cables were 20in or 21in, was moored with cables

Below: Lenox *moored with 2 anchors and 2 cables. Two bridles lead from the hawse to the swivel and eye.*

Right: *The swivel invented by Captain Badily in 1660 that allowed ships to swing at their moorings without their cables crossing.*

between 16in and 18in. Generally the soft bottom of the Medway did not chafe the cables near the anchors and cause damage. But where there was hard ground the cables were "truncked and armed". During the winter months, when the weather was severe and the river flowed faster, an additional cable, the stream cable, was laid out upriver to the westward. The moorings were satisfactory although the cables must have sometimes crossed and fouled as the ship swung round with the tides, causing additional strain and wear to the cables.

The earliest recollections of this mooring practice came from the Assistant Master Attendant, Simon Dunning, who could remember as far back as 1639 and never heard of any miscarriages.[170] It was normal during the summer for worn-out cables to be replaced and new moorings laid out instead.[171]

At the time of the Restoration in 1660, the then Master Attendant introduced an idea of one Captain Badily that prevented cables crossing. This was the swivel; it was attached to two bridles that were the same size as the cables. The bridles led through the hawseholes and met the swivel under the ship's bows. At the outboard end of the swivel was an eye to which the cables were bent and at the end of the cables were the anchors. Now, as the ship turned, the swivel also turned, preventing the cables becoming crossed.[172]

Another development at about the same time was an increase to four mooring cables. As if this was not security enough, the sizes of the cables was allowed to increase to the size of their bower cables. They were laid out in the same manner from the bows as had previously been the practice, but now there were two cables each side of the ship.[173] The increase was related to efforts for making good use of worn cables from ships coming in from the sea.

As with many other practices there were exceptions to the rules. In August 1679, new ships arrived to be laid up and more swivels were required. Captain Richard Vittels, the Master Attendant, wrote to the Navy Board for a warrant to have new ones made.[174] Three months after Vittels requested them, the new third rate *Kent* arrived at Chatham but none could be found for her mooring. As a temporary measure, Vittels moored her with only one cable and anchor.[175] This successful solution encouraged him to innovate according to his own judgement.

Mooring became an issue during 1681 when the Admiralty Commission wrote to the Navy Board and instructed them to consult with Trinity House. They pointed out that ships were now moored with more cables than in the past, adding to the costs to the navy. The two bodies were to form a joint opinion about how ships should be moored and report their findings back to the Admiralty.[176] After consideration they recommended staying with the conservative, more expensive approach, reporting that ships "may not with safety be moored with fewer cables than they now ride by, and also that it is necessary that each of the said ships should be supplied with two new cables every year."[177]

There the matter may have rested, but the idea of saving mooring costs put forward by the Admiralty Commission was taken up by Captain Vittels. He was assisted by the Clerk of the Survey at Chatham, Edward Homewood, a man Pepys described as "industrious as well as of knowledge in all matters of this kind, I have long ago well observed and think him inferior to no officer in the Navy."[178] During the summer of 1683 the cable sizes of about twenty ships were reduced; *Lenox* was not affected but many of her sisters were given cables of 13½in and 14in instead of the accepted bower size of 18in.

During September, the Surveyor of the Navy, Sir John Tippetts, discovered what Vittels was up to. He had "at His Majesty's yard at Chatham presumed upon his own head to lessen the sizes of cables which have usually [been] allowed for the mooring of His Majesty's ships in that harbour to such a degree that we judged the King's ships so moored not to ride in safety", thereupon the entire Navy Board descended upon Chatham to investigate. When they confronted Vittels he defiantly told them he "judged it to be a piece of very good husbandry, a considerable charge of the moorings of the ships being thereby likely saved to the King." The infuriated Navy Board officers submitted to the Admiralty their "determination what reproof is fitting to be given to this officer who by his ignorant [sic] and presumption in lessening the sizes of moorings hath hazarded the safety of His Majesty's Ships."[179]

So that Vittels should be in no doubt about his error and future conduct the Admiralty sent him written orders based on the Navy Board report. The ships were to be moored with bower cables and "The third rates to have two bridles of the same size of their own moorings new at Michaelmas, and that every cable to have two sufficient seizing at the ring of the anchor and three at the horseshoes, and that the rings of the anchors be well puddined."[180] Rings of anchors were puddined by parcelling with tarred canvas and then seizing with twice laid stuff to prevent galling.

To make sure the moorings were returned to their former sizes, the Navy Board and four elder brethren of Trinity House went down to Chatham to report on the situation.[181] The brethren as well as the Navy Board took a more conservative view and concluded that ships of *Lenox*'s size should have four bower cables for mooring.[182] They recorded *Lenox* as riding at the eighth mooring down from Rochester Bridge, moored with a 17in upper and an old 20in lower cable on the northern side and a new 18in upper and a old 17½in lower cable on the southern side.

It might be supposed that Vittels was fortunate to retain his position and happy to carry out his orders. But with the return of Pepys to the Admiralty the matter took another turn. Pepys thought the Trinity House report "foolish" and supposed two old naval captains, Haddock and Narbrough, both members of the Navy Board, responsible for it.[183] Pepys was sympathetic toward Vittels's cost-cutting approach. To resolve the matter in his favour he brought the matter up with King Charles and the Duke of York, telling them the matter was of such importance it should not remain so long unresolved. With royal approval he wrote to Commissioner Godwin at Chatham on 23 August 1684 requesting reports from the officers at Chatham concerning mooring practice.[184] In the resulting report, signed by Vittels, Homewood and many boatswains, calculations were presented that showed costs could be halved by reducing the number of mooring cables to two and replacing them less frequently. They advocated a return to the successful mooring practice of many years before. The proposed savings amounted to nearly £100 a year for a third-rate ship. As there were more than forty ships at Chatham, the savings were enough to build a third rate ship every three years.

Godwin, with considerable understatement, commented on the difference between Trinity House and the Master Attendant at Chatham when he sent the report to the Navy Board. With the prospect of huge savings and support from the highest levels there could only be one outcome and in April 1684 Pepys declared Vittels had "the success of it".[185] In 1687, however, Phineas Pett II took up the challenge once again and a series of letters between him and the Navy Board ensued.[186] A testimony was made to Pepys pointing out that Vittels had saved £19,000 since 1683 and that he should continue supervising the moorings.[187] In July 1689, just after Pepys retired, Sir John Tippetts raised the matter yet again and asked for a survey of the moorings, as he "could never yet obtain to my full satisfaction".[188] However, by then the days of the fleet lying idly at their moorings were coming to an end.

The moorings did not escape the attentions of the warrant officers. It was discovered that the boatswain's mate of the *Rupert* had shortened the mooring bridles and carried parts of them off to his house. When questioned, he said that he was only following the orders of the boatswain, who in turn denied any knowledge of the offence.

Mooring Positions September-October 1683

From Rochester Bridge Down

Mooring	Ship	Rate
1	*Mary*	3
2	*Stirling Castle*	3*
3	*Restoration*	3*
4	*Newcastle*	4
5	*Resolution*	3
6	*Defiance*	3
7	*Hampton Court*	3*
8	*Lenox*	3*
9	*Bonaventure*	4
10	*Berwick*	3*
11	*Half Moon*	3
12	*Hope*	3*

Chatham Reach from the Key downward

Mooring	Ship	Rate
1	*Duchess*	2*
2	*Pendennis*	3*
3	*St Michael*	1
4	*London*	1
5	[*Hulk*]	
6	*St George*	2
7	*Triumph*	2
8	*Britannia*	1*
9	*Royal Prince*	1
10	*Windsor Castle*	2*
11	*St Andrew*	2
12	*Sandwich*	2*
13	*Unicorn*	2

Cockham Wood Reach Fort upward

Mooring	Ship	Rate
1	*Portland*	4
2	*Cambridge*	3
3	*Monmouth*	3
4	*Two Lions*	4

Gillingham Short Reach

Mooring	Ship	Rate
1	*Burford*	3*
2	*Kent*	3*
3	*Breda*	3*
4	*Essex*	3*
5	*Duke*	2*
6	*Royal Katherine*	2
7	*Neptune*	2*
8	*Albermarle*	2*

Gillingham Long Reach South Shore

Mooring	Ship	Rate
1	*Exeter*	3*
2	*Victory*	2
3	*Suffolk*	3*
4	*Elizabeth*	3*
5	*Captain*	3*
6	*Royal Oake*	3

* New ships of the 1677 programme.

1685

One would have thought Vittels was content with the trouble he had already created for himself. But further trouble was brewing with Mr Trevor, the Purser of the *Suffolk* and kinsman of Sir John Trevor, a Member of Parliament, Speaker and Master of the Rolls. The purser found it difficult to take orders, especially from someone like Vittels. It seems to have started when Vittels, in his boat, called for helping hands but Trevor, lying in his cabin, would not at first answer and when he did it was to refuse help. The Boatswain of *Elizabeth* had a similar experience when he called for hands, only to be told by Trevor that none were aboard. However, as the boatswain rowed away he saw them all laughing at him. On 1 September 1686, Trevor fell out with the gunner of his own ship, the *Suffolk*, for not passing on a warning from Vittels. Evidently Vittels threatened to tell the Commissioner that Trevor kept those in his ship from working. Trevor responded, saying Vittels could "kiss his arse" and could do nothing because he was "but an ignorant, bawling fellow and was so looked on by the Commissioner", while another warrant officer said he called Vittels the "son of a whore".

At 8 o' clock in the morning of 29 September, Vittels was moving some of the big ships and required all the hands available. He was out in his boat when he saw Trevor coming down the side of *Britannia* to be rowed ashore by two boy servants. "Sirrah, where are you going?" asked Vittels. An argument ensued in which Vittels threatened the two boys with being whipped on board the hulk if they didn't stop rowing and Trevor likewise threatening them, saying he would "break their parts" if they did not pull away. Vittels's boat caught up and he proceeded to beat the purser with a boat hook ten or twelve times about the head and body until his weapon broke. Trevor staggered in the stern sheets and fell, but Vittels hit him twice more with the broken staff until he fell a second time. The beaten boatswain was then towed back to *Britannia* but was later seen rowing himself ashore.

The injured purser consulted a doctor who found him dangerously weak with great disorder to the left side of his head, side and arm and was so weak he could not be bled. Trevor underwent "great tortures and miseries", had fainting fits and was not able to get out of bed for many days. The purser prepared a legal case and made a petition saying he spent £100 seeking relief, including a trip to Bath. Trevor insisted on £100 from Vittels as compensation. The matter was not resolved until 4 January 1688 when they both came before Lord Dartmouth and Pepys where it was agreed Vittels should pay Trevor £50 in five payments.[190]

During 1689 the Navy Board asked the Commissioner at Chatham for a testament regarding Vittels. "Captain Vittels has in divers capacities served ever since the Restoration of the Monarchy, and has been beyond all question a very stout seaman and a good officer and if it should please God to restore him again to the use of his limbs and health, he will be very fit to be continued, especially, if he can be prevailed with to desist merchandizing in this river for wood, coals, deals etc: a practice lately taken up by him, wherein though possibly there may be no real evil, yet in my poor judgement the appearance is too much."[191]

Pepys's far-ranging interest in the navy kept the officers at Chatham busy. During January there was a cruel frost that made the Thames freeze[192] and Pepys worried about the poor people of Chatham. He wrote to Commissioner Godwin reminding him that £150 was set aside to pay them for the unpicking of ocum, for "the poor people cannot subsist until payment be made them". An account from the Clerk of the Cheque, Jeremy Gregory, was demanded for the charge of workmen engaged in different works.[193]

Perhaps King Charles had never been more secure on the throne than at the beginning of the New Year. The intense political rivalries that marked his early years subsided and James, Duke of York, was able once again to take a leading role in the navy. This happy state of affairs did not last long, however, for Charles died during February before he had time to read the Pepys report on the navy. His brother succeeded him on the throne as King James II; in spite of being openly Catholic he was generally popular, especially in the navy, in which he had commanded and fought bravely during the Dutch Wars.

James immediately took an interest in Pepys's report. Perhaps to avoid partisan politics he planned to have the existing navy officers carry out the repairs. But the repair work dragged on and James ordered Sir Phineas Pett II, for the moment Comptroller of the Storekeepers Accounts, Sir John Tippetts and many Master Shipwrights, including John and Thomas Shish, to go down to Chatham to survey the present condition of the ships and advise which should be repaired first.[194] After his survey on 12 May Sir Phineas wrote "nothing of what I formerly proposed for their preservation hath been yet done … many of them having received prejudice by riding too light and found very much decayed for want of due care". He went on to list his proposals for applying preservatives, airing and putting ashore their gun carriages and rigging.[195] (see Appendix 13 for full transcript of Pett's proposals).

On the very same day, a reply to Pett's proposal was written and signed by all the Thames and Chatham shipbuilders. They added some of their own wisdom to Pett's ideas and a vital new one: that "in our humble opinions there is no better way to preserve his Majesty's Shipps of the thirty from future decay (of this nature) than with all possible speed to shift all such timber or plank as is now decayed on them. Judging the defective stuff infects the sound and (by continuance together) will daily spread its effects to the great damage of their frames if not preserved."[196] (see Appendix 14 for full transcript of the reply).

Two weeks later the shipwrights agreed how the new ships were to be treated while in harbour.[197] During the summer, the acting carpenter of Lenox, Obadiah Hancock, was helping the Chatham caulkers repair ships in the Middle Reach.[198]

Towards the end of the year such authoritative bodies as Shipwrights Hall carried out estimates of costs for the repairs.[199] Adding urgency to their work, the Navy Board noted the four worst affected ships, the Exeter, Kent, Restoration and Stirling Castle were in actual danger of sinking and the estimate to repair them had risen from £13,271 in May 1684 to £20,150 in June 1685.[200] From the reports, surveys of defects and estimates of repairs, Pepys made ready his proposals to present to King James.

1686

The king, in company with his Lord Treasurer, received Mr Pepys on New Year's day. Pepys treated them to a discourse in which he outlined his plans for the recovery of the navy. When he finished he was commanded to put his proposition in writing. The plan was simple enough: a Special Commission, with discipline enforced by Pepys, and £400,000 a year until the end of 1688 to repair the ships and complete their stores.[201] The money needed was available from a new Parliament willing to support the king. For the scheme to succeed in an industry that relied on manpower, it was vital that officials in control followed and would be motivated by Pepys. The trouble with the existing Navy Board was that not all of its members were of this persuasion. Pepys therefore introduced four new members whom he could rely upon: Sir Anthony Deane, Sir John Berry, Will Hewer and Balthasar St Michel. The members of the old Navy Board were retained but their duties altered to accommodate the new members.[202] The warrant for the new Commission for managing the affairs of the navy was signed for one year on 23 March 1686.[203]

King James visited Chatham in April and mounted a stage alongside one of the new ships and with his own hands felt the rotten condition of her timbers and treenails. Afterwards the King, the Duke of Grafton, Lord Dartmouth, Rear Admiral Herbert and Samuel Pepys among others went to Edward Gregory's house for a debate.[204]

At the time Sir Anthony Deane was a successful private shipbuilder. In the past Deane had suffered for his close friendship with Pepys and, being of a pessimistic nature, or perhaps blessed with foresight, he was not particularly enthusiastic about joining his friend's new venture. He was hauled before King James and lectured by the Lord Treasurer before being persuaded to participate. As part of the changes Sir Phineas Pett II became Commissioner at Chatham. Robert Lee, in spite of his difference of opinion with Pepys and Deane concerning the reasons for the decay of the thirty ships, remained at his post. In fact during 1688 he was described by the then Commissioner at Chatham as an able workman, diligent, "not only fitt to be continued in his present business, but highly deserving your further favor [sic]".[205] In general only officers at the Admiralty and Navy Board found themselves vulnerable to dismissal during political upheavals.

The new ships were to be docked and repaired in turn according to their condition. Orders were issued for associated work to be done while the repairs were being undertaken. The tin or lead lining in the breadroom, fitted to preserve the contents from rats,[206] was to be left off or removed, as condensation formed and rotted the timber beneath very quickly. This had not been realised at the time of Lenox's repair. New masts were provided to replace old ones found to be defective. The flooring of the great cabins and roundhouses was to be made but not fitted in order to stop the warrant officers lodging in them and using them for eating and drinking.[207] The new orders stated that officers " usually lodge in the great cabins and roundhouses of the said ships and make them their eating and drinking rooms to the great prejudice of the said cabins, their painting being thereby stained and spoiled, the floors worn out, and the joiners' works and other work much damaged."[208] The anchor lining, skids and standards were also to be carefully stored until the ship went to sea. Any ship which had not been fitted with a storeroom, breadroom or powder rooms, was not to have them fitted until they were needed, as the open deck space would aid the airing and drying of the ship.

As work started, the Commission members visited the ships to see the dreadful condition that had befallen them. They found most of them, especially the new ones, in a dangerous condition with a great number of boards and patches nailed on their sides within and without near the water's edge.[209] They also noted that ships were too light for want of ballast.[210]

Attention was not only paid to the repair of ships but to their maintenance also. Now that some of them, including Lenox, had been restored to seagoing condition, it was recognised that they must not be neglected and allowed to rot again. The Commission issued instructions for their preservation, constantly reminded by the foul weather that continued throughout June and July.[211] In fact new instructions were sent nearly every month during the summer, each with progressively refined details.

By July the Commission had formed a joint consensus with the Master Shipwright, his Assistants and the Master Carpenters of the ships. The Master Attendant, Richard Vittels, was ordered to scrape and tar the masts and black with tar the mastheads and trestle trees. The warrant officers now had to take a much greater responsibility for their ships. Vittels was to ensure they watered the sides and upper works of the ships in hot weather and make sure they opened the gun ports in fair weather and shut them again when it was wet. He was also ordered to ballast the ships deeper in the water to keep them from hogging. Finally he had to take care that the boatswains followed their instructions.

Robert Lee was instructed to take care that no patches were nailed on any ship. The carpenters were to shift and make good any planks or deals that they or the caulkers found defective. They were also to tar the bulkheads and keep the ships well tarred, especially in the summer when it softened and soaked into the plank. In addition, the carpenters were now also responsible for the upper masts and yards which were to be stored on pieces of wood between decks.[212]

The instructions seem to have been obeyed. During the following months payments were made for Greenhithe ballast to make the ships float deeper. *Lenox* received two deliveries, one of 355 tons and another of 50 tons.[213] Along with all the other ships, she was scraped, tarred and blacked.[214]

1687

The warrant officers also came under the gaze of the new Pepys regime. During January of 1687 it was noticed that many small arms kept aboard were "frequently lost and spoiled by carelessness and neglect of the gunners". To encourage the gunners to be more diligent they ordered that future losses were to be deducted from their pay.[215]

During November, *Lenox* was found to be without a carpenter. Ambrose Fellows, who was the warrant officer for *Lenox*, was still at Deptford on an exchange ship, the *Nonsuch*, so that he could be near his family. However, during 1687 and 1688 he left Deptford to spend long periods aboard *Lenox*. The *Nonsuch*'s carpenter who should have been aboard *Lenox* was also often at Deptford supervising repairs to his ship. Fellows also exchanged positions with John Marks, carpenter of the *Portland*, at Deptford in April 1688.[216] The situation was not properly resolved until November 1688 when Ambrose Fellows again joined *Lenox*.[217]

The warrant officers soon communicated the hardships imposed upon them to the Navy Board. They complained that they had no time to spend ashore with their families. The Navy Board agreed that they should have some relief and advised that "as an encouragement to the said officers, and no inconvenience to His Majesty's service, to give leave to one officer at a time of each ship (the cook excepted as formerly) to lodge ashore every week by turns."[218]

Regular wages and discipline imposed much-needed order on the ships and dockyard officers. Life ashore appeared convivial. Edward Gregory, who would become the Commissioner, often dined with Sir John Godwin, Sir Phineas Pett, Robert Lee and even Lord Dartmouth and Richard Vittels. Revealing another side of seventeenth-century life, he spent fully half his days in religious devotion, typically in abstinence, humility, solitude and repentance.[219]

As well as repairing the fleet and ensuring its ships were maintained, Pepys's Commission also set about making up deficiencies in stores. Each ship was supposed to have six months' carpenter's and boatswain's stores, but since the ships were built these had been severely depleted. A survey was carried out by Richard Vittels to establish what stores still remained. *Lenox* represents a typical example. It is interesting to compare her stores listed by Vittels[220] with those bought and paid for when she was constructed at Deptford. Among the items gone missing were 600 hammocks, one of her boats (followed by another during April 1689),[221] eight brass shivers, eighteen flags, bilbows and many other items. Her twenty-four original sails costing £489 had all gone, although twelve other smaller sails had been issued to her at the time of the survey. None of the expensive courses or topsails were left.[222] It may never be known what happened to *Lenox*'s stores; perhaps they had been stolen or allowed to rot or used on other ships. During January, the Navy Board requested the Admiralty to lay the matter of the stores of the thirty ships before the King. It was proposed that as the sails and sea stores are liable to decay they should be "employed to the best advantage".[223]

The attendance of the warrant officers under the restored Pepys regime seems to have marginally improved, as regular musters were taken.[224] One officer who refused to mend his ways, however, was Moses Turner, the purser of *Lenox*, who was an unreformed offender. He was first suspended in October 1680 for not returning to *Lenox* after two months' leave.[225] He was suspended again for absenteeism in 1682. In September 1683 he was given two weeks' leave but remained absent after it expired.[226] In June 1685 he was suspended yet again for absenteeism at night.[227] A muster was taken of the warrant officers during February 1686 and again the Purser was absent – this time supposedly due to sickness. The muster reveals the age of the officers. Richard Glinn, the boatswain was then fifty-two years old and Gabriel Walters, the gunner, thirty-five. The oldest member was Ian Lambert, the cook, who was sixty-three. Their servants by contrast were mere boys aged between sixteen and eighteen years old. Age was no barrier to employment; the average age of the *Berwick*'s officers was over fifty-eight years. The oldest was the gunner, who was seventy-four.[228]

Another time that the warrant officers fell foul of the new order came when Sir John Godwin, the Commissioner, checked they were on duty by hailing them in the evening to see if he would get a reply. The gunner of the *St David* and the boatswain of the *Dover* did not answer. The excuse they offered described "the many abuses offered them, both by day and night from such boats, wherries and hackney barges as pass up and down the Thames hailing the King's ships to no other end but to take occasion of giving foul, base and provoking language."[229]

Work continued on the repair of the ships and by April only six of the new third rates were not fit in all respects for sea service with eight months' boatswain's and carpenter's stores.[230] *Lenox* was not included, probably because she had not yet received her stores. During the repairs to the new ships some attention was paid to the decorative carved works. The surviving records relate mainly to Portsmouth but the problems experienced there must have been much the same at Chatham. Early in 1686 it had been found that a great part of the carvings on the ships were broken off and lost and what remained was almost all rotten. The carpenters of some of the third rates had taken off all the carved port work worth saving. Three of the new ships, *Ossory*, *Expedition* and *Northumberland* had their carved work of the ports removed and placed in storage.[231] Soon after hearing what was happening to the carvings, the Admiralty Commission issued orders confirming that carved work of the ports be taken off, marked to identify from whence it came then laid up in a convenient place aboard ship.[232] It was also reported that when ships went to sea and stayed out any length of time, half of the carvings, especially rails and hancing pieces, were broken off and lost before they came home. Orders were issued stating no port decorations were to be made except for ships with a *Royal* prefix. But old habits died hard, and the shipwrights spiked on a tier of port carvings when they repaired the old fourth rate *Jersey* in 1697. The act of defiance brought the shipwrights an official warning.[233]

During November further orders were issued that carved work was to be set up away from the side of the ship it was nailed upon so that air could pass under so it "may prevent any moisture from the weather to lodge and gather filth and corrupt those parts".[234] Later that same year at Chatham, however, sometime between July and December, carved work was performed on *Lenox* for the sum of £7 2s.[235] This work was then painted for £10 2s 8d.[236] At Chatham in 1688 the carved work of some ships was taken off to give the caulkers access to the seams.[237] During 1690, when ships were being made ready for war, a warrant was issued that stated no superfluous work be done by painters, carvers or joiners. The Master Shipwright at Portsmouth, William Stigant, doubted that he should repair one of the galleries of the *Rupert* that had broken away. Other ships had brackets, taffrails and other carved work missing.[238]

The carved work that did remain to adorn the ships did not glitter in golden glory as portrayed by the paintings of the Van de Veldes. In fact only the lion at the head and the king's arms in the stern were gilt.[239] All the rest were painted, probably in yellow ochre. Even this poor substitute for gold was defaced and blackened by smoke and flames when the filth and debris were burnt off their hulls before caulking and graving, a process known as breeming.[240] The paintings of the Van de Veldes represent an idealised vision of the baroque carvings, which in reality were sadly quite different. By 1687 the ships were much more utilitarian in appearance while their carvings were described by some as superfluous.

1688

King James had only been on the throne for three years but in that time he had managed to alienate people of all political persuasions. Only French Catholics were happy with his policies. In the navy a Catholic, Sir Roger Strickland, was appointed Vice Admiral and he nearly started a mutiny when he attempted to have a Mass read aboard ship. Not that the sailors were known for being particularly religious. Sir Richard Haddock told Pepys that Hugh Peters had long ago remarked "that there might be atheists in the Navy but never any Papists".[241] Of greater concern to the general populace was the increase in the size of the army and the appointment of Catholic officers in it. Many of the new army recruits were Irish Catholics who were camped threateningly near London on Hounslow Heath.

James pressed for the abolition of religious discrimination, particularly the Test Act against Catholics. Although seemingly reasonable, the fear was that James, aided by his army and France, would turn discrimination against Catholics into discrimination against Protestants. Horrors from the past were not forgotten and resentment increased. James completely ignored his countrymen's concerns, however, and pressed ahead with his policies. In May he ordered a Declaration of Indulgence toward Catholics be read from Anglican pulpits. Most clergymen refused to do so, including the Chaplain of the Ordinary at Chatham. The tension heightened on 18 June when seven bishops sent a petition to James suggesting that the declaration be withdrawn. Only two days later, with incredibly bad timing, the Queen gave birth to a healthy son, thus ensuring a Catholic succession.

Before the month was out, the bishops were tried in Westminster Hall for publishing a libel. To the joy of the populace, amid great celebrations, they were acquitted. James had suddenly lost his authority and the respect of the people. Eyes turned towards the next in line to the throne, his Protestant daughter Mary and her husband, William of Orange. Immediately after the acquittal of the bishops, a dissatisfied Rear Admiral, Arthur Herbert, on behalf of many leading political and aristocratic figures secretly carried an invitation to William to invade and save the liberties and religion of England. William was a man with little to lose. Louis XIV was making preparations to invade Holland with a force that the Dutch would find difficult to repulse.

In April work started at Chatham to bring the fleet into fighting condition. Apart from minor repairs, shot lockers and powder rooms were fitted. By early August *Lenox* had some of her top-hinged port lids replaced by twenty-four side-hung, half-port lids. The side-hinged half-port lids were used where a port was in close proximity to a shroud and replaced the normal top-hinged ports that interfered with them when opened. The ports involved were two in the wake of the fore channels, three in the wake of the main channels and one in the wake of the mizzen channel. Additionally, bucklers, fill-in pieces that fitted tightly around the cables as they passed through the hawsehole, were provided.[242] At least some of the work was carried out in dock for she was launched at the end of August.[243] Her masts also required work and her fore topmast was still in hand during September.[244]

One of the ships being readied for sea was the *Tiger*, the fourth-rate ship John Shish rebuilt at the same time he was building *Lenox*. Her captain, Matthew Tennant, had been a commander since 1678 and had joined *Tiger* in 1687.[245] Needing more men for his ship, Tennant sent his boatswain and gunner to press riggers from the yard. They rounded up fourteen men and when they showed their printed protection forms with the seal of the Navy Board, the boatswain told them he "cared not a fart

for Pett or his protection either". Acting Commissioner Pett confronted Tennant but he could not persuade the captain to release the men. The incident sent the yard into turmoil, and eighty labourers and riggers absconded, refusing to return until the pressed men were released.[246]

James was slow to react to William's preparations to fit out his fleet, as he did not believe that it would put to sea so late in the year. Finally, during September, James realised that he must act and fit out his own fleet in order to resist invasion. James went to Chatham himself to check progress. On 12 October, as events neared their momentous climax, Pepys's Special Commission officially finished its work on the repair of the fleet and was dissolved.

As well as being the first ship to have repairs carried out since the return of Pepys in 1684, *Lenox* was also the last in 1688. In the four years' interval since she was last worked on, some further decay must have occurred. A warrant from the Pepys Commission was sent to the Master Shipwright, Robert Lee, on 28 May asking him to report the true state of repair of fifteen of the ships in ordinary, one of which was *Lenox*. He was to get the present condition from the carpenters of the ships concerned as well as details of work already performed.[247]

During October, many newly repaired third-rate ships were brought into service at Chatham to join James's fleet. They were first docked for a day each to be graved before being rigged and manned. On 31 October, the *Suffolk* was graved and launched "and *Lenox* brought in her room".[248] *Lenox* was in need of more attention than her sisters, however, and her repair work lasted until 15 November when she was completed and launched. During the time she was in dock, Ambrose Fellows the carpenter returned to join her from Deptford.[249] Shortly afterwards, on the 20th, William Johnson was promoted out of the *Oxford* to be her new gunner.[250] *Lenox* was succeeded in the dock by the *Charles Galley*, which required the fire of breeming to kill the worm in her timbers.[251]

The cost of repairing *Lenox* as estimated by the Special Commission was listed as £354.[252] This is the lowest estimate for any of the new third-rate ships, repairs to which cost on average £1060. The calculation for *Lenox* was clearly not great enough to have included the eight months of work done to her in 1684, when she was in a worse state than any other new ship. After her repair, she was regarded as being in such good condition that the Commission could leave her until last before giving her any further attention. The sixteen days she spent in dock during November were almost certainly when the work estimated as costing £354 was carried out. In any event, the actual cost of repairing *Lenox* rose from £354 to between £699[253] and £797.[254] The reason the cost of the work doubled may be explained by the long delay between when the estimate was made and when the work was carried out.

The Pepys Special Commission had achieved its objectives and left England with a fleet that could once again contest the seas. In less than three years it had repaired 173 ships of all types, including fifty-nine of third-rate size and larger. This stands in dramatic contrast to the period immediately prior to the establishment of the commission, when *Lenox* and the *Royal Sovereign* were docked for repair. *Lenox* alone occupied a dock for eight months as work was constantly delayed by shortages of money to buy materials.

Yet although the huge enterprise was generally successful, there were failures. Perhaps the most notable was the *Expedition* at Portsmouth. During 1689 it was found that "all the thick stuff of her wale pieces and between the wales being decayed, the upper part of her stem as low as the scarf rotten" and later; "what a condition she is in and every day we find her worse and worse".[255]

ORDNANCE

THE 1677 ESTABLISHMENT

The Admiralty Commission and Navy Board were responsible for the construction of the thirty new ships for the navy. The guns, carriages and gunners' stores, however, were the responsibility of another body, the Board of Ordnance. This body was also responsible for supplying guns to the army. Guns generally lasted much longer than the ships that carried them and as a result a large stock of old guns was available for new ships. Unfortunately many of these guns were undesirable and obsolete. (A notable example was the drake, a lightweight gun designed to use little gunpowder. It proved inadequate during the Dutch Wars, being short-ranged, prone to bursting and wearing out quickly. Being light in weight it also had the unpleasant and dangerous characteristic of recoiling violently when fired.[1])

Periodically the Admiralty addressed the problem of the armament and manning of the fleet. The outcome was known as an establishment. An establishment was simply a list of His Majesty's ships alongside which was listed the number and types of guns and the number of men allocated to them. The Navy Board officers carried out this task, as they should have known both the number of gun ports in each ship and each ship's ability to carry guns.

The Board of Ordnance, for its part, periodically carried out the huge task of surveying all its guns. The surveys were carried out with varying levels of detail and efficiency. Unfortunately the Navy and Ordnance Boards rarely seem to have co-ordinated their efforts and as a result it was found that the navy's establishment could not always be satisfied, as the guns available did not match the establishment's requirements.

Ships which served during the Dutch Wars were grossly over-gunned compared with later practice and often exceeded their nominal establishment. Typically, the *Fairfax*, a ship with an establishment allocation of sixty guns, was carrying seventy-two guns in 1672.[2] This was a result of the local nature of the conflict in which a greater proportion of the load-carrying capacity of the ships could be devoted to ordnance. The end of the Dutch Wars in 1674 gave the Admiralty Commission time to consider the future. Samuel Pepys, Secretary to the Admiralty Commission and well known for his desire for order and regulated bureaucracy, became the driving force behind proposals for the new shipbuilding programme and a new establishment.

The first draft was submitted by the Navy Board on 16 March 1674 and listed the numbers of men and the nature and numbers of guns for the fleet.[3] Not much progress was made, although the Lords of the Admiralty Commission pressed the officers of the navy during October 1676 to dispatch the establishment "so long since referred".[4] Finally, new impetus was added to the proposals for a new establishment when Parliament voted to provide money for the thirty new ships and their guns during February 1677. At the next meeting of the Admiralty Commission, on 22 March, King Charles directed that the number and weight of guns for a ship of each rate be established.[5] In preparation, the Ordnance Office drew up an estimate on 26 April. It proposed that the new third rates be given 24-pdrs on the main gun deck, 12-pdrs on the upper deck and demi-culverin (9-pdrs) on the forecastle and quarterdeck. The total weight of the guns would be 120 tons.[6]

The matter was discussed at another Admiralty Commission meeting on Saturday 8 May. It was agreed that the manning and gunning of the new ships be debated between the officers of the Navy and Ordnance Boards and some of the principal officers of the navy before their attendance at an Admiralty Commission meeting in a week's time.[7] During the week available at least two conferences were subsequently held between the officers. At the first, the officers of the Ordnance proposed the type and weight of guns for the new ships.[8]

For Third Rate Ships

Type of gun	Length (ft)	Number	Weight each (cwt)	Total (tons.cwt)
Demi-cannon	9.5	26	54	70.4
12-pdrs	9	26	32	41.12
Saker	7	14	16	11.4
3-pdrs	5	4	5	1
Total Weight				124 tons

Then, on Sunday 16 May they agreed with the proposal made a few days earlier and added the number of men required to work the guns.[9]

	Number of Guns	Number of Men
Demi-cannon, to each six men	26	156
12-pounders, to each four men	26	104
Sakers on the forecastle, to each three men	4	12
Sakers on the quarter deck, to each three men	10	30
3-pounders, to each two men	4	8
To carry		20
To fill and hand powder		10

After the week's deliberations the king, the Commissioners of the Admiralty, the Navy Board and several flag officers discussed the outcome at the Admiralty Commission meeting of 17 May. The establishment, both for men and guns, was generally agreed and ordered to be drawn up fair for signing by the Officers of the Navy Board and Officers of the Fleet.[10] The Officers of the Fleet took the opportunity to declare that they were the

1677 establishment for the twenty new third-rate ships each of 1000 tons and seventy iron guns in time of war at home.[14]

No	Quality of ordnance	Weight of each in cwt	Full weight in tons cwt	Length in feet	No. of men
26	demi-cannon (32-pdrs)	54	70.4	9½	
26	12-pdrs	32	41.12	9	
10	Sakers (5¼-pdrs)	16	8.0	7	
4	Sakers	16	3.4	7	
4	3-pdrs	5	1.0	5	
70 guns			Total 124 Tons		460

1677 establishment for the twenty new third-rate ships in war abroad and in peace abroad and at home[15]

No	Quality of ordnance	No. of men in war abroad	No. of men in peace abroad and at home
24	demi-cannon (32-pdrs)		
24	12-pdrs		
12	Sakers (5¼-pdrs)		
2	3-pdrs		
62 guns		380	300

only proper judges of the establishment and could have done it as well as the Navy Board and Ordnance Board officers. The Admiralty Commission debated the matter again on 14 July; by this time the keel of *Lenox* had already been laid and work started on her floor timbers.

At the meeting it was again resolved that a draft be fairly drawn up and signed by the Officers of the Navy Board, Officers of the Fleet and the Lords of the Admiralty and then carried to the Council table to receive its confirmation from the king.[11] The members of the Admiralty Commission and the Navy Board duly signed the establishment document.[12] Charles added his royal signature on 3 November 1677. The status given to the new establishment made it clear the intention was to stop admirals and dockyard commissioners interfering with ships' authorised establishment of guns.[13] Included in the establishment was the number of guns and men "in war, abroad and in peace abroad and at home".

For the third rates, 32-pdr demi-cannon would form the armament of the gun deck and 12-pdrs the upper. Ten of the fourteen sakers would be mounted on the quarterdeck with the remaining 4 on the forecastle. The 3-pdrs would be mounted on the poop.[16] The changes, and in particular the change from 24-pdr to demi-cannon, resulted in the weight of ordnance increasing from 120 to 124 tons.[17] The 12-pdr was a comparatively recent introduction to bridge the gap between the culverin (18-pdr) and the demi-culverin (9-pdr).[18] The establishment lists no specialist chase guns firing fore and aft.

The thirty new ships would require 2310 guns, which the Ordnance Office estimated would cost £96,634 12s 2d including carriages, breechings and tackles.[19] To limit costs, the guns were to be standard "rough iron" rather than the expensive nealed and turned guns recently invented by Prince Rupert.[20] If all the guns required were cast new, they would cost 16 per cent of the total programme. Many old guns were available, however. In total, contracts placed with Mary Brown and Thomas Westerne for rough iron guns between 10 September 1677 and 7 March 1678 amounted to £45,533.[21]

The time taken to produce the 1677 establishment must be one of the reasons the twenty new third rates were to end up with such different gun port arrangements. Of the eleven or twelve ships identified, practically none of them have the same arrangement. They range from between sixty broadside gun ports for *Lenox* to seventy-four for the *Grafton*.[22] However, a count of the number of gun ports does not equal the number of guns carried. While the number on the gun deck and upper deck were of significance for the big guns, those of the smaller sakers and 3-pdrs were not. Their ports were, for the most part, decorative carvings worked into the plank sheer and could be added at will. Alternatively, small guns could simply be fired over the rails, provided convenient top timbers were nearby so that ringbolts could be driven through them for ropes and gun tackles.

Unregulated alterations had often been made to ships during refits and would continue for many years. So great was work in unauthorised "contrivances, joiners' work, carvers and painters, cabins, bulkheads, scuttles, cutting down of ships and through their sides" that new regulations were issued in an attempt to halt it.[23] For ships such as the *Grafton*, which had more gun ports than guns in the upper works, it was possible to shift some of her light guns from side to side as the tactical situation demanded.

As we have seen, the drawings of *Lenox* by the Van de Veldes probably date from the time of her launching when she had provision for only sixty broadside gun ports. The drawings show no sign of guns or gun ports on the forecastle or the poop. This soon changed, for only three months later the newly built *Lenox* had sixteen eyebolts for guns fitted on the forecastle and poop.[24] Proof of guns being mounted in these positions is given in a report saying they were temporarily removed from her poop during a refit on 5 December 1691.[25]

It is surprising that so much time and effort was necessary to draw up the new establishment, since the size, weight and number of guns for the new ships are very similar to those listed in the establishment for third-rate ships of 1666. The armament remained as powerful, but the ships became larger. Interestingly, the third rate *Rupert* of 1666, of 832 tons, carried 32-pdrs on her gun deck and 18-pdrs on her upper deck, the same size guns as a common third rate of 1727 tons of Nelson's time. In simple terms, the reduction in the proportion of weight of guns to the size of the ship resulted in better sailing performance and more storage space.

In spite of the effort that went into it, the new 1677 establishment was soon in trouble. Before the year was out a dispute arose concerning the gunning of the *Rupert*, and the matter was brought before the Admiralty Commission. The Master of Ordnance observed that it was impossible that the establishment could be put strictly into execution "in that the Office of Ordnance cannot gun his Majesty's ships otherwise than as the natures and weights of the guns his Majesty is at present master of will admit".[26] In other words the guns listed in the establishment did not match those available in His Majesty's stores. The 1677 establishment, so long worked on, ended up merely as a guide. Little or no effort was made to correct its defects, probably because Pepys became a victim of the scurrilous plot of 1679 and was forced to resign his office.

THE ADDITION OF CULVERINS

Lenox was launched on 12 April 1678 and sailed to Chatham shortly afterwards, carrying a temporary armament of about twelve guns[27] mounted to provide security during her short passage. Upon reaching the fleet anchorage at Chatham the guns, upper masts and rigging were removed and she was laid up in ordinary. On 1 March 1678, the Ordnance Office made a contract for the supply of seventy gun carriages for *Lenox*[28]

Above: *(Top) The 1685 gun establishment. The gun-port arrangement of* Stirling Castle *built by John Shish. Note how the thirteenth gun port on the upper deck is crowded in at the bow. The establishment would have been arranged as follows, although in practice, the guns of* Stirling Castle *did not match this list.*

Key	No. of guns	Type of gun
1	22	*32lb demi-cannon, 9½ft long*
2	4	*18lb culverin, 11ft long*
3	26	*12lb twelve-pounder, 9ft long*
4	4	*5¼lb saker, 7ft long*
5	10	*5¼lb saker, 7ft long*
6	4	*3lb three-pounder, 5ft long*
	70 guns	

(Bottom) The gun arrangement of Lenox.

Key	No. of guns	Type of gun
1	22	*32lb demi-cannon, 9½ft long*
2	4	*18lb culverin, 10ft or 11ft long*
3	22	*9lb demi-culverin between 8½ft and 10ft long*
4	2	*9lb demi-culverin, 10ft long*
5	2	*9lb demi-culverin, 10ft long*
6	2	*9lb demi-culverin cutt, 6ft long*
7	10	*9lb demi-culverin cutt, 6ft long*
8	6	*3lb three-pounder, 6ft long*
	70 guns	

which arrived on 28 June.[29] They did not comply with the 1677 establishment, either. On the gun deck four culverin carriages replaced four demi-cannon carriages. There seems to be no record of the decision to make the change that took place some time between November 1677, when Charles signed the establishment, and March 1678 when the

contract for gun carriages was made. The culverins, which fired a shot weighing 18lbs, were intended to double as chase guns. All of the other new third rates were similarly equipped with culverin carriages as they arrived at Chatham to join *Lenox*.

Lenox and most of the other thirty new ships arrived at the fleet anchorage at Chatham as the threat of war with France receded. Although they would not be fighting the French for many years to come, their boatswain's, carpenter's and gunner's stores were brought aboard. *Lenox* received all her gun carriages, including those for twenty-six upper deck 12-pdr guns. Nobody seems to have noticed the ship had only twenty-four ports. It will be remembered that *Lenox* was the first of the new ships to be built and was too far advanced in her construction to comply with the specification of principal dimensions for the new ships. William Hunt, Master of the hoy *Elizabeth of Rainham*, a vessel often employed by the Ordnance Board, brought *Lenox's* stores of war to her during 1678.[30] Later in the year the ironwork was fitted to the carriages and they were mended.[31]

THE 1685 ESTABLISHMENT

Only the gun carriages and gunner's stores were brought aboard *Lenox*; the guns themselves existed only in the theoretical list of the establishment. As we have seen, in the aftermath of the Popish Plot of 1678, the navy fell into a neglected state that lasted for some six years. In 1684 the King's political problems eased and interest was again concentrated on naval affairs. A new Ordnance establishment was proposed at the beginning of 1685. In relation to the new third rates, the 1677 establishment was confirmed and included the four gun-deck culverins whose carriages already lay aboard the ships.[32] For half of the new third rates the establishment would remain unaltered, in fact and theory, until the end of the century. The serious faults with the practicality of the 1677 establishment, where little thought had been given to the guns actually available, were addressed.

Now they did not simply compile a list of guns considered ideal for a particular ship. An abstract list was first prepared showing the number of guns required for all calibres compared with those on board ship and those in store. From this information a deficiency or surplus of guns, as the case may be, was calculated. At least three abstract lists dating from the 1685–1688 period were made. Unfortunately they vary considerably in their conclusions but do agree that there was a surplus of between 1500 and 2000 guns of all calibres. They also show a serious deficiency in the number of 12-pdr fortified guns (fortified guns are the strongest of their type, able to fire a full charge of gunpowder).

The Catholic James II succeeded his brother Charles II in 1685 but by 1688 he was so unpopular there were plans to replace him with his Protestant daughter, Mary, and her husband William of Orange. *Lenox* was not among the many ships fitted out by James to oppose William; she remained at her moorings in ordinary at Chatham when William landed in England. On 15 March 1688, a book entitled *The State of all the Ordnance with their Carriages for the whole Navy Royal as it now stands* was prepared for James.[33] It listed all ships in the fleet and their gun establishment, together with a list of the guns actually belonging to them. It was in agreement with the 1685 establishment and showed that *Lenox* now had a set of guns and carriages belonging to her. The carriages were still stored on board but the guns remained in store at Woolwich Arsenal. The guns that were marked out and assigned to her were exactly those in accordance with the establishment.

Guns listed as belonging to Lenox on 15 March 1688

22	demi-cannon
4	culverins
26	12-pdrs
14	sakers
4	3-pdrs
70	**guns total**

James fled to France while William completed the "Glorious Revolution" of 1688. Even when England entered into war with France nothing was done to bring *Lenox* into service. In fact, although she was the first of the thirty ships to be built she was among the last to be fitted out for war. Finally, during July 1689, she was surveyed and work was started to make her ready. Unfortunately the necessities of war cut across the 1685 establishment. During the race to get the fleet to sea, some ships had gone to sea with "borrowed" guns[34] and *Lenox*'s guns numbered among those officially allocated to other ships.

PREPARATIONS FOR WAR AND THE SEARCH FOR GUNS
The problem of finding guns for *Lenox* and several other ships was addressed at the beginning of 1690. On 17 February, a new establishment was produced specifically for twelve ships at Chatham ordered to be fitted out for war.[35] The practicalities of fitting guns to gun ports were addressed and a list of guns available versus guns required was also drawn up. The new establishment listed seven three-deckers, including the *Royal Sovereign*, and five third rates, all from the thirty ship programme. Two of the third rates had their establishments changed: "The reason why the *Restoration* and *Lenox* are appointed demi-culverin [9-pdr] for the upper tier instead of 12-pdr is that there is not any 12-pdrs in their Majesty's stores to supply them, the like was lately done to the *Expedition* and *Northumberland* at Portsmouth." As if this were not bad enough, all fourteen sakers were missing as they had already been allocated to other ships.

The establishment continued the emphasis on guns that could be moved from the broadside ports to fore and aft chase ports. The gun-deck armament was unaltered but the upper deck establishment of demi-

The Chatham gun establishment for twelve ships of 17 February 1690 as relating to *Lenox*

(The Deck column has been added to the original for clarity.)

Guns	Length	Number	Deck
Demi-cannon	9 ½	22	
Culverin	11	4	Gundeck
Demi-culverin	9	24	
Demi-culverin	10	2	Upperdeck
Sakers	7	12	Forecastle and
Sakers	10	2	Quarterdeck
3-pdrs	5	4	Poop

70 Guns

culverins included two 10ft chase guns. It was probably the intention to mount them in the foremost broadside position from where they could be moved to fire forward, through two chase gun ports in the beakhead bulkhead. The blast damage likely to occur when firing from this position would probably restrict the practice to battle conditions only. It was certainly dangerous: in 1691 the surgeon's mate of the *Centurion* was killed as he sat on the head easing himself when a chase gun was fired in celebration of the Queen's birthday.[36] A sad and inglorious way to die if ever there was one. An account for gunning some of the other third rates of 1677 shows that they had already received a pair of long chase guns for the upper deck.[37]

Of *Lenox*'s fourteen unavailable sakers, two were 10ft long. Ten of the sakers were intended for the quarterdeck and four for the forecastle, where the two long guns would be mounted so they could be moved to fire forwards in the chase position. Although the gun establishment mentions *Lenox* by name, it is clear that whoever compiled it had not taken the trouble to actually view the ship, for she is still listed as having twenty-six upper deck guns to fit her twenty-four gun ports.

In March 1690, the Assistant Surveyor of the Ordnance, William Boulter, sent his clerk, Edward Silvester, to Woolwich Arsenal. Although there were no sakers, he allocated other suitable guns to the *Hope, Exeter, Kent, Lenox* and *Restoration*.[38]

While the Ordnance Board officers searched their stocks for sakers, *Lenox*'s carriages were brought into good order. Edward Silvester arranged for new bolts, forelock springs, nails and straps to be replaced or repaired.[39] *Lenox*'s guns were brought to her by hoy in batches of about sixteen guns between 2 and 10 April while she lay anchored in the Hope weathering continuous gales.[40] After twelve years of painfully protracted establishments and idleness at Chatham, *Lenox* at last had a set of guns of her own actually sitting on her decks. The guns now aboard are broadly those listed as "What is at Chatham and Woolwich toward the same" in the recent twelve-ship establishment. The vacancy left by the fourteen sakers, of which none could be found, was filled by 6ft-long demi-culverin cutts.

That *Lenox* had only twelve instead of thirteen ports on each side on the upper deck could no longer go unnoticed. It is easy to imagine the head-scratching as Captain Granville and his gunner wondered why there were two more demi-culverin than gun ports. They soon resolved the problem, for on 4 April, the Admiralty wrote to the Duke of Schomberg, the Master General of the Ordnance: "Colonel Granville, Captain of their Majesties' ship the *Lenox*, having acquainted us that there are 26 demi-culverin appointed for the upper deck of the said ship, on which deck she has but 24 ports, and therefore he desires that two of the said demi-culverin may be changed for two 3-pdrs, to be used on her quarterdeck. We have thought fitting to grant the same, and do desire your Grace will give orders for the exchange of the said two guns accordingly."[41]

The list set out below is probably the result of Edward Silvester's visit to Woolwich Arsenal to mark out guns during March 1690.

Guns for *Lenox*

What is at Chatham and Woolwich toward the same

Guns	Length in ft	Number	Deck
Demi-cannon	9½	20	
Demi-cannon	9	2	
Culverin	10	4	Gundeck
Demi-culverin	10	2	
Demi-culverin	9½	8	
Demi-culverin	9	16	Upperdeck
Sakers	10	0	Forecastle and
Sakers	7	0	Quarterdeck
3-pdrs	6	4	Poop

56 guns

Having second thoughts, Captain Granville changed the plan and kept all twenty-six fortified demi-culverin and instead sent two short demi-culverin cutts ashore in exchange for the two 3-pdrs. He then moved the extra pair of upper deck fortified demi-culverins, displaced by having no gun port, onto the forward position on the forecastle from where they could double as chase guns. The two extra 3-pdrs were probably mounted on the poop to join the four guns already there.

Only twenty-six of her seventy guns now complied with the original establishment of 1677. So, after all the meetings, lists of guns and establishments the guns of *Lenox* arrived, owing more to accident than design. Not only that but the ship's captain made the final decisions concerning her armament against all the principles of Pepys's establishment of 1677.

It may be thought the extra two fortified demi-culverin could be kept on the upper deck permanently mounted at the two forward-facing chase gun ports, but this was not possible. There was simply not enough space for guns to be mounted at the foremost broadside port and the forward chase port at the same time. All the other new third rates that had twenty-six upper deck ports carried twenty-six guns and had no extra guns to mount permanently in the forward chase ports. Two of *Lenox*'s Shish-built sisters, *Hampton Court* and *Captain*, which also had only twenty-four upper deck ports, carried only twenty-four upper deck guns.

We have seen that in 1690, *Lenox* was armed with the last remains of the stocks of guns left over after the rest of the fleet had been armed. It is not surprising, therefore, that *Lenox*'s guns conformed less to the establishment of 1685 than any of her sisters, twelve of which conformed almost exactly. They were, however, rather special, consisting for the most part of Prince Rupert's patent high quality nealed and turned guns.

The history of gun metallurgy went back a long way. In 1585 all the heavy guns in the fleet had been cast in brass.[42] Brass was malleable and soft and was not usually apt suddenly to split or burst. The metal was easy to cast but had a tendency to droop when heated during prolonged firing. Brass guns were also especially prone to wear and very expensive. During the seventeenth century, as iron-casting technology improved and it was learned how to produce safe guns without honeycomb (gas bubbles), the use of iron guns became more widespread. By 1677 practically all new guns were being made from iron, although many brass guns were still in use. A major improvement occurred in 1628 with the introduction of "fine metal" guns by the Browne family of gun founders. Their development culminated in 1673 with the fabled Prince Rupert's patent guns, also cast by the Brownes. They were annealed by a lengthy cooling process which

would relieve casting stress and make the iron softer and less brittle. The guns were finish-turned and bored to high precision and because of their inherent strength they could be made lighter.[43]

In June 1681 a report set out the advantages of Captain Browne's nealed and turned guns of Prince Rupert's invention over both brass and ordinary iron guns:[44]

1. Brass costs £130 per ton and nealed and turned iron only £40 per ton.
2. Brass guns crack and dance [violently recoil] when hot and strain the ship's side. Ordinary iron guns scale and moulder [rust] and break in harsh weather.
3. Brass and ordinary iron guns weigh 2–3 cwt more than nealed and turned guns. Nealed and turned guns are turned straight while the others are crooked and leaning to one side by their casting.
4. Ordinary iron guns outshoot brass by ¼ mile. Nealed and turned guns outshoot brass by ½ mile.
5. Ordinary iron guns in service fly into pieces to the danger of all near them. Nealed and turned only crack and are more beautiful than brass or ordinary iron guns.

The recent recovery of a Prince Rupert patent demi-cannon from the *Stirling Castle*, wrecked in the great storm of 1703, confirms at least some of these claims. The outside surface is machined smooth and large areas of the gun seem unaffected by 300 years spent underwater. The colour and preservation suggest the gun is made of an alloy containing iron. It is the same size and came from the same batch delivered in 1690 as some of *Lenox*'s guns.

Prince Rupert's guns were expensive and only three ships were intended to be exclusively equipped with them. The guns' association with Prince Rupert of the Rhine and his patronage probably had something to do with their cost. Pepys later wrote, "Colonel Legg [Lord Dartmouth

Below: *Hoisting a gun aboard a ship using her tackles. Print from William Sutherland,* The Shipbuilder's Assistant, *1711, p114. (Author's collection)*

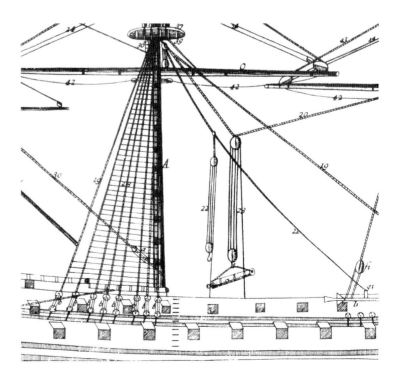

and Master General of Ordnance] gives me an account of the new invented guns not being more useful than the old [fine metal]. Observe the three partners in the profits thereof to be the Earl of Shaftsbury, the then Lord Chancellor, the Prince [Prince Rupert], and the then Master of Ordnance [Sir Thomas Chicheley]; that the King pays £60 a ton and has paid £5200 [for a set of ship's guns], whereas the founder [Browne] has offered them to Legg at £30."[45]

It is reasonable to suppose that part of the cost of the guns was the result of political rather than technical developments. At any rate Prince Rupert's patent guns were dropped in favour of ordinary "rough iron" guns that were not machined externally but left in their rough cast finish. Unfortunately, Prince Rupert's partnership had already produced guns in anticipation of future orders and left creditors with huge debts.[46]

In any case Lenox was especially fortunate to have such a high-quality set of guns. It is possible to speculate that her first commander, Colonel Granville, second son of the Earl of Bath and a gentleman captain of some influence, was responsible for them. In reality, however, it was down to chance. The establishment for the thirty ships called for ordinary rough iron, not nealed and turned guns. It was down to luck that the rough iron guns for the third-rate ships were all issued by the time Lenox came to be allocated hers.

WAR SERVICE

In 1690 James, supported by the French, threatened England with invasion and Lenox was ready just in time to join the fleet and oppose the French fleet at the Battle of Beachy Head. On 5 June, just before the battle, her armament was increased when she received two mortar pieces and their ammunition consisting of 100 grenade shells. The size of these weapons or where they were mounted is unknown. A likely site, with the clearest field of fire, would be the forecastle. There is a list of brass mortars showing they were produced in sizes from between 4½in to 18in in diameter. A common size seems to have been 7in, probably about the maximum that could be fired from Lenox's forecastle. They were removed from Lenox immediately after the battle and not used again.[47]

Three of Lenox's guns were badly damaged during the action and needed replacing. Although it cannot be known for sure, they probably started to crack during the act of continuous firing. They do not appear to have burst, for Lenox suffered no casualties. The guns concerned were a demi-culverin from the upper deck and two demi-cannon from the lower. They were replaced on 4 August 1690.[48]

During 1691 Lenox's guns were again mentioned in an establishment of several ships.[49] The establishment included all the surviving ships of the thirty ship programme except the second rate Vanguard. By this time three of the third rates had been lost with the destruction of the Anne and Breda witnessed by those aboard Lenox. The establishment was more than a theoretical ideal; now the fleet was at sea the establishment was a factual list of the guns aboard ships. The lengths of the guns were included so that a comparison may be made with the twelve ship Chatham establishment of 1690. It shows that Lenox's guns, especially her demi-culverins, were subject to a number of significant changes. The 1691 establishment of Lenox's guns is again confirmed in "A list of Their Majesties' whole Navy Royal together with an account of the natures and numbers of ordnance belonging to each of the said ships".[50] Unfortunately the list, which dates from around the end of 1693 or the beginning of 1694, adds nothing to what is already known. It merely lists the number and type of gun aboard ships.

A full complement of guns was not always carried. Establishment lists of guns often show Lenox and similar ships allocated with sixty-two guns for peace and abroad. The naval victory of the English and their Dutch allies in 1692 resulted in an increase in commerce raiding by the French, and Lenox was often employed in the cruising role against them. As no fleet

Establishment of 1691 as relating to Lenox

Note: The deck column has been added to illustrate the most likely disposition of each type of gun. This column assumes Captain Granville replaced two 6ft long demi-culverin cutts from the forecastle with two 3-pdrs as discussed previously. *N & T means Nealed & Turned, referring to the patented 'Rupertino' guns.

Guns	Length (ft)	Number	Type	Deck
Demi-cannon	9 ½	22	N & T	Gun deck
Culverin	11	2		
Culverin	10	2		
Demi-culverin	10	2		
Demi-culverin	9 ½	1	N & T	Upper deck
Demi-culverin	9	16		
Demi-culverin	8 ½	5		
Demi-culverin Cutt	6	10	N & T	Quarter deck
Demi-culverin Cutt	6	2	N & T	Forecastle
Demi-culverin	10	2	N & T	Forecastle
3-pdrs	6	6	N & T	Poop
Total		70 guns		

action was expected, some gun power was often sacrificed in favour of sailing performance. For example, by April 1695 Lenox carried only sixty guns.[51] Later, in March 1696, Ordnance officers boarded her to survey her guns and stores and found she was carrying sixty-four guns[52] with six fortified demi-culverin ashore. The gun deck was unaffected by the change, but how the guns on the upper decks were rearranged remains problematic and is difficult to interpret.

THE GREAT ORDNANCE SURVEY

The survey of about 1696[53] adds considerable information to that already known about the guns of Lenox and the rest of the navy. It was the greatest contemporary survey taken of naval guns. Guns belonging to individual ships are all listed together in numerical order. Every gun that was surveyed had its own survey number cut into it. Today, surviving guns from the period can be identified by their survey number, which can be checked against the survey record to see which ship the gun belonged to.

The weight of gun, condition and some important dimensions were recorded in the survey. These included the length, diameter of the trunnions and the diameter of the barrel at the trunnions – important information for the supplier of gun carriages. The condition of the gun was also noted; it was checked to see if it needed a new lining to the touchhole, or vent as it was known, a procedure known as venting. It was also checked for cracks near the muzzle; if found they were normally stopped by cutting off the end of the muzzle. The survey necessarily took a long time to complete. The date each ship was surveyed is not recorded although the location was. The survey seems to have started before November 1694, as this was when the James Galley was wrecked, and she is included. As the survey numbers run in sequence (continued on p155)

The 1696 gun survey and additional information relating to *Lenox*

Gun	Number Cut	Weight C Q L*	Length in ft	Diameter of Trunnion in inches	Diameter at Trunnion in inches	1)Taper bore 2)Wants Venting 3)To be cut 1 2 3	Founder	Bill book	Acceptance date
	6462	48 1 23	9½	7	16	0 0 0	RT	WO51/29	11-03-84
	6463	48 2 24	9½	6¾	17	0 0 0			
	6464	50 0 24	9½	7	17	0 0 0			
	6465	49 1 11	9½	6¾	16¾	0 0 0	RT	WO51/29	11-03-84
	6466	52 0 27	9½	6¾	18	0 0 0	JBr	WO51/16	25-08-74
	6467	49 2 12	9½	6¾	17	0 0 0			
	6468	48 3 04	9½	6¾	16½	0 0 0	MBr	WO51/29	10-05-78
	6469	49 0 12	9½	6¾	16	0 0 0			
Demi-cannon	6470	50 3 27	9½	6¾	17	0 0 0	JBr	WO51/16	25-08-74
	6471	48 3 26	9½	6¾	16½	0 0 0			
	6472	48 2 20	9½	6¾	16¾	0 0 0	RT	WO51/29	11-03-84
	6473	51 1 06	9½	6¾	17½	0 0 0	JBr	WO51/16	25-08-74
	6474	53 2 12	9½	7	17¾	0 0 1	TW or MBr	WO50/13 WO50/13	04-07-79 11-07-79
	6475	48 3 10	9½	6¾	16¾	0 0 0			
	6476	47 1 21	9½	6½	16½	0 0 0	JBa	WO51/41	15-06-90
	6477	51 3 23	9½	6½	17½	0 0 1	JBr	WO51/16	25-08-74
	6478	50 2 12	9½	6½	17½	0 0 0	MBr	WO51/20	10-05-78
	6479	50 1 11	9½	6¾	17	0 0 0	RT	WO51/29	11-03-84
	6480	52 3 06	9½	7	17¾	0 0 0	JBr	WO51/16	23-01-75
	6481	52 1 27	9½	7	17½	0 0 0	MBr	WO50/13	07-09-80
	6482	53 2 26	9½	7	18	0 0 0	JBr	WO51/16	23-01-75
	6483	51 2 27	9½	7	17½	0 0 0	MBr	WO50/13	28-07-80
	6484	43 0 12	10	5½	16	0 1 0			
Culverin	6485	44 2 26	10	6	16	0 0 0	TW	WO50/13	10-05-79
	6486	42 2 18	10	6	15½	0 0 0	TW	WO50/13	10-05-79
	6487	45 1 26	10	6	16½	0 0 0	GB	WO51/9	17-10-68
	6488	26 1 11	8½	4¾	13½	0 0 0	GBr	WO51/9	23-09-67
	6489	25 2 00	8½	4¾	13½	0 0 0			
	6490	28 2 26	9	5	13¾	0 0 0			
	6491	31 2 04	9½	5	14	0 0 0			
	6492	31 0 06	9	5	14	0 0 0			
	6493	33 1 25	10	5	14	0 0 0	JBr	WO51/14	5-10-72
	6494	36 2 05	10	5	14½	0 1 0			
	6495	30 2 11	10	4¾	13	0 0 0	TW	WO51/13	04-09-71
	6496	33 3 00	10	4¾	14	0 0 0		WO51/8 or 9	
	6497	30 1 13	10	4½	13	0 0 0	TW	WO51/13	04-09-71
	6498	28 3 26	9½	4	12½	0 0 0	JBa	WO51/41	13-06-90
	6499	31 2 18	9½	5	14	0 0 0	JBr	WO51/14	21-06-72
	6500	26 1 04	9	5	13¼	0 1 0	GBr	WO51/8	19-07-66
	6501	30 1 18	9	5	14	0 0 0	JBr	WO51/14	21-06-72
	6502	26 2 21	8½	5	13½	0 0 0	GBr	WO51/9	20-01-67
	6503	26 2 17	9	5	13	0 1 0			
	6504	24 1 14	9	5	12½	0 1 0	GBr	WO51/8	19-07-66
	6505	25 0 19	9	5	12¾	0 0 0	GBr	WO51/8	19-07-66
Demi-culverin	6506	27 0 10	8½	4¾	13½	0 0 0	GBr	WO51/9	20-01-67
	6507	29 3 18	9	5	14	0 0 0	JBr	WO51/14	21-06-72
	6508	27 2 06	8½	5	13¾	0 0 0	GBr	WO51/8	19-07-66
	6509	24 3 22	9	5	12½	0 0 0	GBr	WO51/8	19-07-66
	6510	31 1 04	9	5	14	0 0 0	JBr	WO51/14	21-06-72
	6511	30 2 04	9	5	14	0 0 0			
	6512	26 8 00	9	4½	12½	0 1 0	Dutch		
	6513	26 1 21	9	4½	13½	0 0 0	GBr	WO51/8	19-07-66

Gun	Number Cut	Weight C Q L*	Length in ft	Diameter of Trunnion in inches	Diameter at Trunnion in inches	1)Taper bore 2)Wants Venting 3)To be cut	Founder	Bill book	Acceptance date
	6514	13 0 13	6	4½	11	0 0 0	JBr	WO51/18	24-07-76
	6515	13 2 10	6	4½	11	0 0 0	JBr	WO51/18	24-07-76
	6516	13 1 00	6	4½	10¾	0 0 0	JBr	WO51/18	24-07-76
	6517	12 1 14	6	4½	10¾	0 0 0	JBr	WO51/18	24-07-76
Demi	6518	13 2 10	6	4½	11	0 0 0	JBr	WO51/18	24-07-76
culverin Cutt	6519	12 3 14	6	4½	11	0 0 0	JBr	WO51/18	24-07-76
	6520	13 0 00	6	4½	10¾	0 0 0	JBr	WO51/18	24-07-76
	6521	13 1 10	6	4½	11	0 0 0			
	6522	13 0 07	6	4½	10½	0 0 0	JBr	WO51/18	24-07-76
	6523	13 2 26	6	4½	11	0 0 0	JBr	WO51/18	24-07-76
	6524	13 1 14	6	4½	11	0 0 0	JBr	WO51/18	24-07-76
	6525	13 0 26	6	4½	11	0 0 0	JBr	WO51/18	24-07-76
	6526	7 3 20	6	3½	8½	0 0 0	MBr	WO51/20	01-04-78
3-pdr	6527	7 2 05	6	3½	8½	0 0 0	JBr	WO51/20	13-09-77
	6528	7 1 09	6	3½	8	0 0 0	MBr	WO51/20	01-04-78
	6529	7 2 26	6	3½	8½	0 0 0			
	6530	7 3 14	6	3½	8¼	0 0 0	MBr	WO51/20	01 04 78
	6531	7 3 07	6	3½	8¼	0 0 0	MBr	WO51/20	01-04-78

Key to gun founders/suppliers: RT=Richard Taylor, JBr=John Browne, GBr=George Browne, MBr=Mary Browne(Widow of JBr), TW=Thomas Westerne, JBa=John Baker
*C = Hundredweight, Q = Quarters, L = Pounds.

Below: *A Rupertino demi-cannon recovered from the wreck of the third rate* Stirling Castle. *It is the same size and type of gun with which* Lenox *was equipped. These demi-cannon would have formed the main battery of her gun deck. This particular example is now being conserved for the Ramsgate Maritime Museum.*

Right: Lenox *was equipped with four culverin chase guns situated on the gun deck. During her early years at sea two of the four guns were 11ft long, of the same type as this gun cast by George Brown and delivered in 1668. It is now conserved at the Canadian War Museum, Ottawa, Ontario. (Courtesy of the Canadian War Museum)*

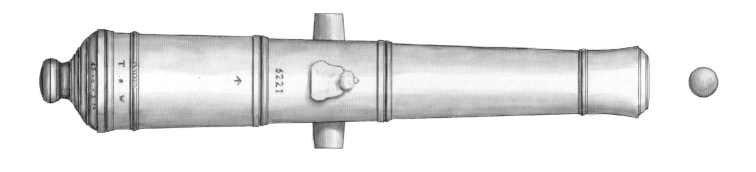

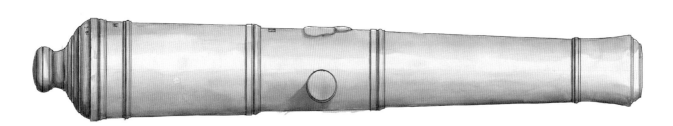

it seems that the ships were surveyed in the order that they are listed. *Lenox*'s guns are recorded as being surveyed at Chatham. *Lenox* visited Chatham in November 1697, at the end of the war with France. All the ships in the survey that precede her were there at the same time. *Lenox*'s guns were almost certainly surveyed at this time, at the end of her service career, just before she was rebuilt.

Strangely, the guns of the *Sovereign* that were surveyed at Chatham after those of *Lenox* are included, even though she had been accidentally burned to the waterline in January 1696.[54] The *Sovereign* was dearly loved by the navy and had survived since 1637. When a new ship bearing the same name was launched at Woolwich in 1701, she was recorded simply as a rebuild of the old ship, in effect denying her loss. In these circumstances it is not surprising that the salvaged guns of the ship were still listed as belonging to the *Sovereign*.

Guns were made in a great variety of designs and even those of the same design varied considerably in weight.[55] The weight in hundredweights, quarters and pounds was cut into the gun at its acceptance after proof testing. The broad arrow was added to confirm the gun belonged to the king. Only rarely do two guns, even of the same type, have identical weights. The length of the gun was measured from the back of the base ring, which is the part of the gun with the greatest diameter, to the end of the muzzle. A founder's mark may sometimes be found on the base ring or the trunnion.

The weight was cut into the gun on the first reinforce. A reinforce is one of the three sections into which a gun is divided. There are two reinforces and a chase which are separated by rings and ogees. The vent is in the first reinforce. The second reinforce contains the trunnions while the chase extends from the second reinforce to the muzzle end of the gun. Normally the broad arrow was cut near the forward end of the first reinforce. The 1696 survey number is generally cut into the second reinforce behind the trunnions. The rose and crown cipher, which typifies guns of this period, was cast onto the gun immediately over the trunnions. There

seems to be no standard pattern for the rose and crown, as the form changed often and different patterns were used at the same time.

Before acceptance, the gun was fired twice with a powder proof charge of four-fifths the weight of shot and checked for honeycomb – the small gas bubbles formed during the casting process. Iron guns are particularly susceptible to honeycomb and were checked by reflecting sunlight or inserting a candle into the bore to illuminate any defects. They were also checked for soundness by ringing with a hammer. Cracks were checked for by closing the mouth and the vent immediately after firing and seeing if any smoke escaped from defects.[56]

The 1696 survey relating to *Lenox* is reproduced on pp. 153–154 with additional information including, where known, the individual guns' dates of acceptance by the Ordnance Board and the gun founder's name.[57]

LENOX'S DEMI-CANNON

The 6in bore, 32-pdr demi-cannon had long established itself as the favoured heavy gun of the gun deck. It was not the largest sea service gun as the cannon-of-seven fired a shot 7in in diameter weighing 42lb. These huge guns were only mounted on the gun deck of the largest three-decker first-rate ships. Demi-cannon were mounted on Henry VIII's *Mary Rose* and would remain in service until the end of the age of sail.

To her good fortune *Lenox* was given a complete set of Prince Rupert's nealed and turned demi-cannon. They were all 9½ft long, the same length as her sisters' guns, but these weighed on average 3cwt less. These twenty-two guns fired two-thirds of her broadside by weight. They are not a set cast together but a collection of guns cast between 1674 and 1682. Among them are the first four demi-cannon made by Prince Rupert's partnership, numbered 6466, 6470, 6473 and 6477. Most of *Lenox*'s guns were supplied by the partnership but four were probably supplied by one of the partnership's creditors, Alderman Richard Taylor, after its financial collapse. Taylor obtained guns in lieu of money owed him. It is known that one gun was damaged at the Battle of Beachy Head. The gun that replaced it

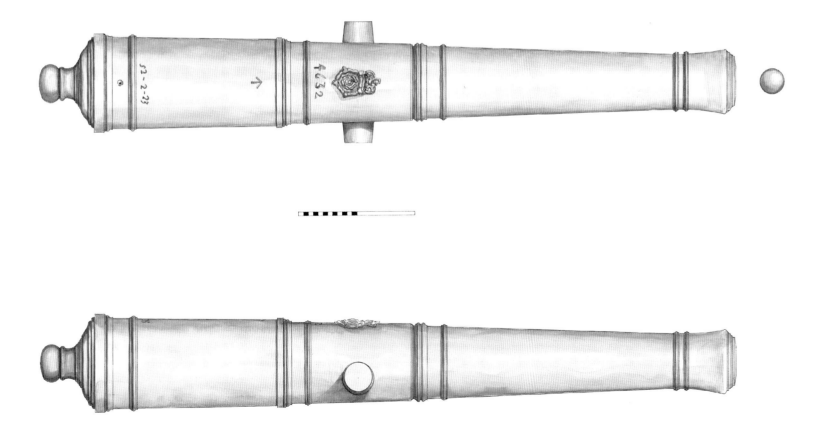

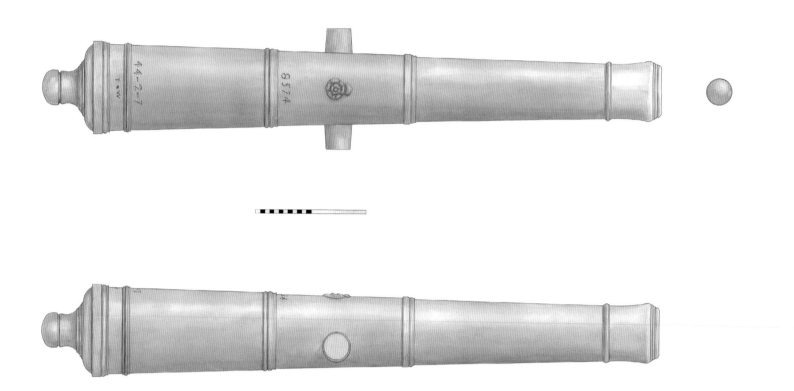

Above: A 10ft culverin similar to the four chase guns used on Lenox *late in her career. Two of* Lenox's *guns were also cast by Thomas Western, and are the same weight and date from the same time as this gun. In 1696 it was used in a battery on Drake's Island, Plymouth before becoming a bollard at the Tower of London. It is now being conserved at Fort Nelson, Portsmouth.*

is either 6476 or 6464. Both guns were supplied together in June 1690 by Thomas Westerne, another creditor, and taken aboard *Lenox* as replacement. By the time of the 1696 survey, two of the guns were cracking and needed their muzzles sawing off. Only *Stirling Castle* of the new ships was similarly equipped, although others did have a few guns of the same type.

THE CULVERIN CHASE GUNS

There is some contradictory evidence as to where the four 18-pdr culverins were positioned. Two of them may have been in the forward "luff" bow ports and two in the aft broadside ports on the gun deck. Alternatively, all four may have been positioned in the aft broadside ports. From this position they could easily be moved into the four aft-facing stern-ports. The four stern ports on the gun deck were somewhat smaller than the broadside gun ports and were normally left empty, as aft-facing guns would overlap the aft broadside guns. This area was known as the gun-room and to facilitate the movement of chase guns the deck was laid parallel to the sea, while the rest of the gun deck followed a slight curve. The levelling of the deck aft can be seen quite clearly in many Van de Velde drawings of *Lenox* and other ships.

The chase culverins are longer than the demi-cannon so the blast made during firing would be further away from the overhanging ship's counter. The blast would in any case be much less than from a demi-cannon. They would also be easier to move and in total weigh about a ton less than demi-cannon, an important consideration as the gunroom was structurally the weakest part of the gun deck. During the 1690s a debate on the disposition of gun deck ordnance for a new series of third rates

arose between the Admiralty, the officers of the Ordnance and the Navy Board. The Commissioners of the Admiralty proposed a set of 10ft-long demi-cannon for the gun deck together with four 11ft-long culverin chase guns. Even the demi-cannon were considered too long to be mounted in the bow ports, as the officers of the Ordnance and Navy Board noted in response: "The 10 foot demi-cannon will be too long – will also be found very difficult to be used between the bitts and the bow ports by reason of their extraordinary length".

Instead they proposed using 9½ft demi-cannon, the same size guns used in the thirty ships of 1677.[58] Clearly in this case the 11ft-long culverin chase guns were intended for the stern. However, it had been the practice some years before in 1666 to provide many ships with four culverin chase guns of which one pair may have been forward in the luff ports, since the ships had only two ports in the stern.[59]

In support of this, an account of 1689 for gunning some of the thirty ships proposed mounting four 11ft-long culverins on the gun deck as chase guns "fore and aft".[60] This suggests only two may have been mounted in the gunroom while the other two were mounted in the luff gun ports. Perhaps the experience of war in 1690 revealed the impracticality of mounting 10ft or 11ft-long guns in the bows; after 1690 all four culverins would be found in the gunroom.

The accuracy of a long culverin was better than that of a shorter demi-cannon. The muzzle velocity of a long-barrelled gun was not greater, but a long barrel resulted in the ball hitting the sides at ever shallower angles as it travelled towards the muzzle, making it fly more accurately after exiting the barrel. The obvious disadvantage of having culverins rather than demi-cannon was a reduction in firepower and the logistical problem of securing and maintaining ammunition and equipment for another type of gun. The 1691 establishment, which was written when the guns were actually on board ship, shows *Lenox* to have two 11ft guns and two 10ft guns. These guns were standard rough iron guns. By the time of the 1696 survey two guns had been exchanged so that all four guns were now

10ft long. One other ship with 10ft culverins was the *Stirling Castle*. This is surprising as the establishment called for 11ft long guns and the other 1677 third rates, almost without exception, had 11ft guns. The reason seems to be weight. The difference between a 10ft- and 11ft-long rough iron culverin is a considerable 7cwt. Altogether the guns on *Lenox*'s gun deck weighed 64.3 tons, 4.5 tons less than the average of her sisters.

THE FORTIFIED DEMI-CULVERINS

Lenox owned two distinct types of 9-pdr demi-culverins: her fortified guns mounted on her upper gun deck and a set of short demi-culverin cutts on her forecastle and quarterdeck. The fortified demi-culverins are an interesting set; most of them seem to be fine metal guns supplied by the Browne family. One of them is definitely not, for it is a Dutch gun, probably captured during the Dutch wars. It was probably a Dutch 8-pdr, equivalent to 8.72 English pounds, and could fire English demi-culverin shot.[61] The oldest of *Lenox*'s demi-culverins traced so far dates from 1666 and the newest from 1690. By 1696 there were five 8½ft, thirteen 9ft, three 9½ft and five 10ft-long guns, a veritable mixed bag. It is known that one gun was replaced after the Battle of Beachy Head in 1690.[62] The only gun belonging to *Lenox* that was delivered after the battle was number 6498, which was 9½ft long and weighed 28cwt 3qtr 26lb. It was a new Prince Rupert nealed and turned gun, for one of the same weight and length was delivered to the navy during 1690.

Between 1691 and the 1696 survey there were three other changes, probably made to replace guns damaged at the battles of Barfleur and La

Below: Lenox's largest demi-culverin from the upper deck. The battery number 12 engraved near the cipher indicates this was the foremost chase gun, which was moved to fire forward through the beakhead bulkhead gun port as Lenox *chased the French fleet after the Battle of Barfleur. It is now conserved at St Ann's Fort, The Garrison, Barbados. Measured by Charles Trollope and interpreted by Adrian Caruanna.*

Hogue. In order to agree with the 1696 survey, it seems that two 9ft guns were replaced by two 9½ft guns, and one 9ft gun by one of 10ft.

Two of the fortified guns had, as we have seen, almost certainly found their way up onto the forecastle above the upper deck to double as chase guns after Captain Granville's discovery that he had more guns than ports. Which pair of guns Granville moved may never be known for certain. It is nonetheless useful to examine the guns mounted on the forecastle of the other new third rates. Among their batteries of 7ft sakers, each had a pair of 10ft sakers weighing on average 29cwt that must have acted as fore chase guns. *Lenox* owned five 10ft demi-culverins. Two of the 10ft demi-culverins weighed only 30cwt each and would be ideally situated on the forecastle as chase guns. The heaviest pair of 10ft long demi-culverins would have been placed in the forward position on the upper gun deck where they could also double as chase guns.

Lenox carried a lighter load of guns on her upper deck than any of her sisters. The reason was of course that she had a lightweight set of guns and there were only twenty-four of them. The total weight of 34.4 tons was 10 tons lighter than the typical establishment of twenty-six 12-pdrs carried by *Elizabeth*.

The quality of *Lenox*'s demi-culverins is demonstrated by the fact some of them have survived to the present day. Remarkably, two have been identified. The first gun to turn up was found in Barbados. In the past many guns ended up there as additions to the fortifications, often as dues paid by visiting ships. After they became obsolete they were often disposed of by being buried in the sand. Fortunately for old guns, there was no iron founding industry in Barbados and they were not melted down for other uses. Today the guns periodically reveal themselves and are surveyed by the government of Barbados.

One of the guns surveyed was found to have the 1696 survey number 6494 and was a 10ft long demi-culverin. It has a distinctive reverse astragal in its second reinforce ring, just in front of the trunnions, identifying

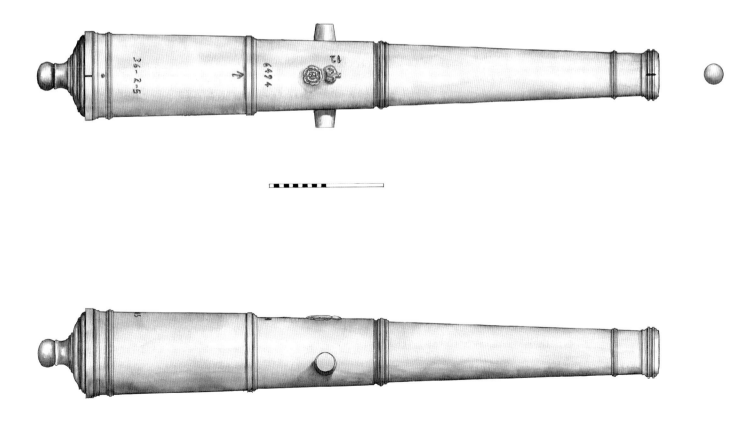

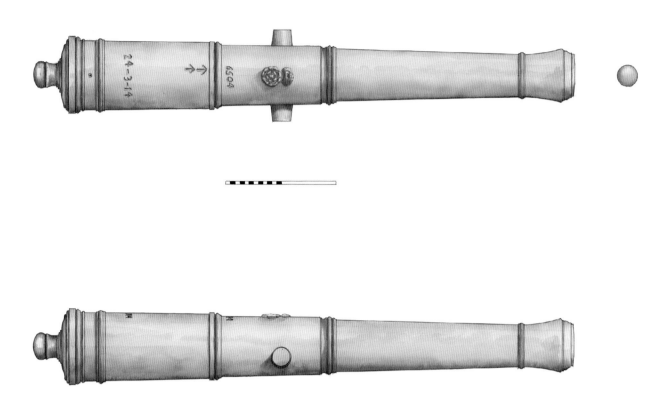

Above: *This surviving gun from* Lenox *was her smallest demi-culverin. It is now on permanent display at the Cobb Harbour wall at Lyme Regis.*

it as a gun made by the Brownes.[63] The 1696 survey shows that it was a gun that once belonged to *Lenox*. It is the heaviest demi-culverin she owned, by 3cwt, and therefore the most likely to be the foremost broadside gun on the upper-deck that could double as a fore chase gun – a fact confirmed by the number 12 cut into it near the trunnion, indicating it was the twelfth gun on one side. This is one of the guns referred to by the First Lieutenant, Edward Durley, in his journal entry for the battle of Barfleur: "We bore away after them, we firing our fore chase at them".[64] The gun was cast after 1660, for it bears the Charles II rose and crown cipher. It was well used; the 1696 survey says it required venting and today the gun can be seen to have had its vent re-bushed. What happened before the gun was issued to *Lenox* or what happened to it after she was decommissioned in 1697 is not known.

The gun is unusual in that it does not have a muzzle swell to reinforce the end of the gun; instead there is a series of rings of the type normally seen on land ordnance. The muzzle swell had replaced muzzle rings for sea service in Tudor times because the sharp-edged rings caused damage to the edge of gun ports. The gun was cast as a land-service piece but in the rush to arm *Lenox* in 1690 perhaps this was the best available piece that could act as a chase gun.

Equally remarkable was the discovery of a second fortified demi-culverin from *Lenox* that today can be seen at Lyme Regis.[65] A great deal is known about this gun for all its markings are readable. It is 9ft long and bears the 1696 survey number 6504. While the gun found in Barbados was the largest of its type on *Lenox*, 6504 is the smallest, being two-thirds the size of 6494. Distinctive features are the Brownes' reverse astragal in the second reinforce ring and the weight engraved into it. It was accepted after proof testing on 19 July 1666, the year of the Great Fire of London. It most surely would have fought in the Second and Third Dutch Wars.

After being in store it was issued to *Lenox* in 1690. It fought at Beachy Head, Barfleur and La Hogue. During inspection for the 1696 survey it was noted that that it needed venting, further evidence of its age and usage. After 1697 *Lenox* was taken out of service and the gun became anonymous for a time.

In 1740 John Scrope of Lyme Regis wrote a memorial pointing out that, as there has been such a long peace and the guns of the town had been so neglected, they were unserviceable and there was not a musket in the town that could be fired. Scrope requested a supply of ordnance and small arms be sent for the town's defences. The Privy Council, with the king himself in attendance, considered the matter on 27 October 1740. After consulting the Master General and the Principal Officers of the Board of Ordnance, it was agreed to send stores and ordnance worth £371 18s 9d to the custody of the Mayor of Lyme Regis. A condition was made requiring the town to make good the platforms for the guns and provide a place to keep the stores. They were also to send their unserviceable old ordnance to the stores at Woolwich.[66]

Iron ordnance duly arrived, consisting of six 9-pdrs complete with their oak standing carriages and iron trucks. One of them was *Lenox*'s old gun. Included in the delivery were carriages for two brass 9-pdrs, already at Lyme Regis, that had survived the decay suffered by the old iron guns. The stores included 300 rounds of shot, paper cartridges, copper powder measures, tampions and other small items.[67] The citizens of Lyme Regis appear to have looked after their "new" ordnance rather better than their old, for the guns remained at their defensive duties for many years. In 1783 there was concern in the Ordnance Board office that many old guns belonging to the king were proving inadequate through age, wear and obsolescent design. The Inspector of Artillery, Thomas Bloomfield, decided that all the old guns must pass a new proof test firing and inspection. He accordingly despatched the Assistant Inspector, Captain Fage and the Master Searcher, Sergeant Bell, with adequate hands around the coast to carry out his orders, a task which took many years.[68]

The great majority of guns failed the test and were scrapped, but *Lenox's* old gun passed and a second broad arrow was cut behind, 120 years after the original of 1666. The gun probably stood, proudly displaying its second broad arrow, until the 1840s when it finally became obsolete. Although its military role was over the good citizens of Lyme Regis found it another important duty. It was buried in the sand, muzzle up, as a foundation post for a boardwalk railway that led out alongside the west wall of the ancient and picturesque Cobb Harbour. Wooden posts supporting the structure were inserted into the muzzles. Long after the railway had disappeared the guns remained buried under the sand until exposed by a storm in 1974. A Lyme Regis resident, Mr Richard Fox, undertook the task of recovering them and finding them a new duty. They were mounted on gun carriages made of old railway sleepers and now serve as tourist attractions on the Cobb. The survey number 6504 can still be read on *Lenox's* old gun.

Their present duty would surely be a fitting end. Unfortunately, time spent in the sea is very damaging to old iron guns. Salt water enters the pores of the iron and causes decay. Without specialist conservation, the historic old guns' days are numbered.

THE DEMI-CULVERIN CUTTS

The 1677 establishment of fourteen sakers was carried by most of the new third rates. Nearly all of them had ten guns on the quarter deck and four on the forecastle. *Lenox* and *Stirling Castle* were the only 1677 ships not to have these guns; instead they carried twelve 6ft long demi-culverin cutts (9-pdrs), an armament often carried during the Dutch wars. The shortfall of two guns was made up by the two 10ft-long fortified demi-culverin chase guns displaced from the upper deck onto the forecastle.

Originally, as their name suggests, cutts were ordinary long guns that had their muzzle ends cut off. This was usually done to stop the spread of a crack. These guns were unbalanced when fired and became unpopular. However, the guns now delivered to *Lenox* were made as cutts and all came from one batch accepted by the Ordnance Board on 24 July 1676.[69] They were all Prince Rupert's patent nealed and turned guns cast short and would have been well balanced when fired. They were light pieces, weighing on average half that of fortified demi-culverins and 3cwt less than the sakers on board *Lenox's* sisters. In view of their light construction,

the demi-culverin cutts could only take half the gunpowder charge of the fortified guns on the deck below. If they were fired with a fortified demi-culverin cartridge from the deck below they would be very dangerous, causing them to recoil violently, and perhaps even to crack or burst.

LENOX'S 3-POUNDERS

Even *Lenox's* 3-pdrs did not conform to the normal establishment. Instead of four 5ft long rough iron guns, she had six of Prince Rupert's patent nealed and turned 6ft long guns. Four came from a batch delivered on 1 April 1678 and one from a batch delivered on 13 September 1677.[70] The guns each weighed 1cwt more than the 5ft-long guns supplied to her sisters.

It may seem surprising that heavier guns were chosen for her after so much care was taken to ensure all her other guns were as light as possible. They were probably chosen because they were Prince Rupert's patent guns, in common with the other guns aboard *Lenox.* Van de Velde drawings of *Lenox*[71] and two other Shish-built ships, *Hampton Court*[72] and *Captain,*[73] possibly show two circular rear-facing poop deck chase ports worked into the stern decorations for the 3-pdrs to fire aft.

Altogether, after adding up the weight of every individual gun aboard *Lenox*, it is found that her armament weighed almost exactly 112 tons and could fire a broadside of 568lbs. This compares with 125 tons for *Northumberland*, a typically establishment-armed ship, which could fire a broadside weighing 585lbs. In other words, for every ton of *Lenox's* guns she could fire 5.071lbs of shot and for every ton of *Northumberland's* guns 4.68lbs could be fired. *Lenox's* guns were thirteen tons lighter; this does not seem very much when it is considered she carried more than 300 tons of ballast. However, there is considerable difference in regard to the stability of a ship between the weight in the bottom of the hold compared with that some feet above the waterline.

GUN CARRIAGES

Gun carriages of the type used in the late seventeenth century with four trucks and a flat bed had been in use for a very long time. The *Mary Rose* had them in 1546 and so did the *Vasa* in 1628. They were necessary to absorb the reaction of firing. Guns could not simply be restrained to the sides of the ship, as the shock would soon pull the timbers apart. The guns were therefore allowed to recoil backward on carriages that absorbed the recoil energy until finally halted by thick ropes known as breeching ropes. The weight of the gun itself was an important factor; generally a heavy gun would recoil less than a light one. The friction caused by inefficient wheels, or "trucks" as they are properly known, also slowed the recoil. Another energy-absorbing device appears on the draught of a contemporary 70-gun ship at Wilton House, in which gunners are shown applying their linstocks in the act of firing (see page 66). The gun tackles, which were used to haul the gun forward to the gun port, are shown drawn up tight, holding the carriage to the side of the ship. The friction of the rope running through the sheaves would act as a considerable restraint to the recoil if this drawing can be relied upon.

Modern test firings have been conducted using a lightweight 1½pdr gun. Without restraint the gun performed a 360-degree backflip, reaching an altitude of about 5ft before coming to rest 15ft from its starting point. With the same charge and simulated gun tackles rigged, the gun recoiled gently for a distance of 4ft. These results do seem to indicate that the arrangement shown on the Wilton House draught may have been a practical method of minimising recoil, although admittedly tests have not been carried out using full size guns.[74]

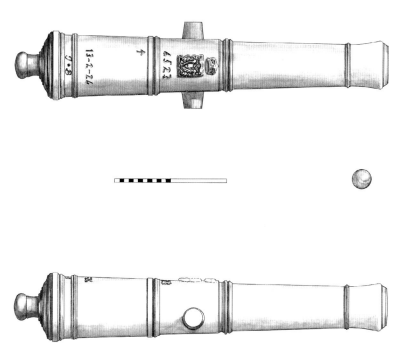

Left: *A reconstruction of one of* Lenox's Rupertino *demi-culverin cutts based on the dimensions given in the 1696 gun survey, an existing 6ft 6in Rupertino demi-culverin and a 6ft saker cutt also made by Brown.*

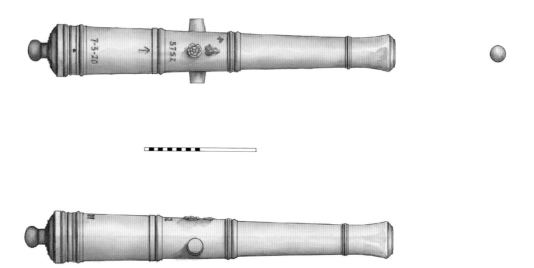

A sectional drawing by Edmund Dummer of the "body of an English Man of War"[75] illustrates a ship, possibly the *Britannia* of the 1677 programme. The carriages in the drawing show that "dead" trucks have replaced the rear trucks. Dead trucks are those that do not revolve but act as wheels with the brakes on. Dead trucks may well have replaced rear-revolving trucks because lightweight guns such as drakes and cutts had a tendency to be very "boisterous" when fired. The mechanical inefficiency of a carriage with rear dead trucks would have helped reduce the violent recoil of these types of guns.

Ship models of the period, although incredibly detailed, usually represent carriages as simple blocks of wood without trucks. Some contemporary model makers were more particular, however. The *Breda* model of 1692, a third rate of the same dimensions and armament as the 1677 ships, has very detailed carriages. Those on the main and upper decks have four trucks, while those on the quarterdeck, forecastle and poop have two trucks and two dead trucks of the type drawn by Dummer. A similarly finished model in Vienna, generally believed to be either the *Hope* or the *Elizabeth* of 1677, shows all the visible carriages with dead rear trucks. Only the guns on the gun deck would have had four trucks.

After the battles of Barfleur and La Hogue in 1692, many ships had extensive repairs made to their gun carriages. The lists of work mention three types of trucks: ordinary trucks, extraordinary trucks (presumably large trucks) and dead trucks. The quantities used are roughly in proportion to the trucks on the *Breda* model.[76] *Lenox*, as we have seen, had a lightweight set of guns and it is almost certain she had carriages with two trucks and two dead trucks for all her guns except those on the gun deck, as portrayed by the Vienna model.

While *Lenox* was in commission her gun carriages were periodically repaired. The ironwork was repaired at Portsmouth, late in 1691 during her winter refit, and any damaged carriages replaced. The Master Carpenter, Ambrose Stanyford, made the elm carriages and was paid £1 16s for a demi-cannon carriage, £1 7s 6d for a culverin carriage, £1 2s for a demi-culverin carriage, and 15s for a 3-pdr carriage.[77] The Master Smith, John Bryant, carried out the "binding" of the new carriages and the workmanship was certified by Francis Felton of the Ordnance Office.[78] During the war when *Lenox's* stores were removed before her winter refits, small vessels, usually hoys, were hired to transport them ashore. During June 1695, John Bryant again repaired *Lenox's* gun carriages when she visited Portsmouth.[79] The linchpins and forelocks always needed replacement, while iron crows (levers) required sharpening.

At other times *Lenox* underwent further repairs to her gun carriages. Henry Stanyford, probably a relation of the Ambrose Stanyford who had

Above: This 6ft 3-pdr nealed and turned gun, survey number 5752, served aboard the Cornwall *and is the same type and weight as* Lenox's *gun, number 6526. It was delivered between 12 and 20 January 1691. The gun has been preserved and mounted on a replica gun carriage and put on display at Fort Nelson Museum, Portsmouth.*

previously made gun carriages for *Lenox*, repaired the elm carriages and provided new ones where necessary.[80] John Bryant died but his wife Ann continued the business and a contract was made with her to bind new carriages and repair the old ones.[81] In August 1697 *Lenox* was refitted at Plymouth where Degory Cloake repaired the carriages at a cost of £5 19s 6d, while Thomas Basse repaired the ironwork for £2 19s 10d.

THE UPPER-DECK GUN CARRIAGE

In a very dark corner high up inside the thirteenth-century Curfew Tower at Windsor Castle, which has hardly altered since it was built, is an old gun. It was identified as a very large saker dating from around 1630[82] and appears to be the 9½ft long saker (5¼-pdr) on a ship's carriage mentioned in a list of ordnance at Windsor Castle dated 1679.[83] It has obviously been there for a very long time and it may have acted as a funeral minute gun, as the carriage is painted black. The gun has trunnions of 3½in diameter and the barrel is 12in diameter at the trunnion. It has no visible weight markings but must weigh about the same as one of *Lenox's* smaller fortified demi-cannon. The carriage has two trucks at the front and two dead trucks at the rear and is of the same type as those on the *Breda* and Vienna models. It is the only known gun carriage of the flat bed type to have survived from a non-archaeological environment and is precisely the type of carriage used on *Lenox*.

The bed is 4in thick and the cheeks 3½in thick. According to a list of 1684 a carriage 3½in thick is the size of a saker carriage "from the bottom to the trunnions". A demi-culverin carriage should be 4⅛in thick, indicating the carriage may be a little smaller than those used aboard *Lenox*.[84] It has ringbolts for the breech ropes to stop the recoil and eye bolts for the gun tackle to pull the gun back to the ship's side after firing. On the rear it has "tail rings", as they were known,[85] for the train tackle. The train tackle was used to pull the gun away from the ship's side when the ship was firing to leeward. There are no corresponding rings in the tower for the breech or tackle ropes, indicating the carriage was not made for its present location. The gun appears a little too large for the carriage and one of the cheeks has spread slightly outward by about 1in.

The carriage is made almost entirely of elm. There is a slot cut through from the underneath of the bed and into the stepped side-pieces, or cheeks, through which an oak wedge has been inserted and secured with

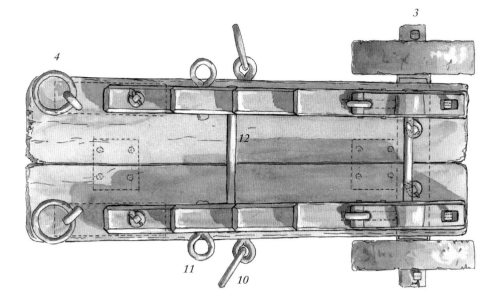

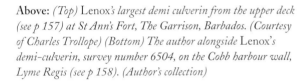

Above: *(Top) Lenox's largest demi culverin from the upper deck (see p 157) at St Ann's Fort, The Garrison, Barbados. (Courtesy of Charles Trollope) (Bottom) The author alongside Lenox's demi-culverin, survey number 6504, on the Cobb harbour wall, Lyme Regis (see p 158). (Author's collection)*

Left: *A large saker carriage in the Curfew Tower, Windsor Castle, dating from 1679 at the latest. It is the same type as was used on the upper deck of* Lenox. *The carriage shows various inconsistencies that wide manufacture tolerance allowed; one truck is wider than the other and is offset to one side. Thanks to the Dean and Chapter of St George's Chapel, Windsor Castle.*
(1) Cheek. (2) Bed. (3) Axletree. (4) Tail ring. (5) Dead truck.
(6) Truck. (7) Capsquare. (8) Capsquare pins. (9) Extree bolt.
(10) Ring bolts. (11) Guntackle eyebolts. (12) Side bolts.
(13) Lynchpin.

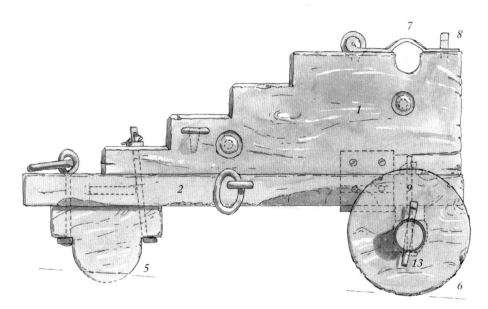

four treenails on each side. The grain of the wedge runs up and down to maximise the strength of the joint from the shock of firing. Similar wedges secure the two halves of the bed. The timber from which the bed was fashioned was not large enough and the outside of the curved tree trunk can be seen just behind the trucks. The carriage shows signs of considerable wear; the bottom of the dead trucks have been worn away by more than 1in and there are marks, probably made when hand spikes were used to move it sideways. There is 1in of wear between the truck and axle.

The wrought iron bolts that hold the carriage together are made specially to suit their duties. The axles are of ash[86] and properly known as axletrees. They required frequent replacing and are secured by iron "extree" bolts, bent over at 90 degrees for about 1in on the underside. There is a slot at the upper end through which a forelock is inserted. The forelocks are bent over to prevent them falling out but could still easily be removed and the axletree replaced. There is a 2in thick spacer plank between the axletree and the bed, possibly to raise the gun so that it fits

exactly in the middle of its gun port or to suit the camber of the deck. The through bolts, which are not normally removed, have 2½in or 3in diameter rivet rings or "roves" set into the cheeks over which the ends of the bolts have been riveted or "clenched". Large breech ringbolts are set blind into the bed of the carriage, which may be barbed or "ragged" for security. They are set so that the breech rope makes an angle of about 45 degrees as it goes over the button of the gun's cascabel. The tail ringbolts are not fitted with a forelock on the underside but clenched over a thick rove. This seems surprising because it would make the bolt, which also helps secure the dead truck, more difficult to remove. The eyebolts for the gun tackle are simply bent over on the inner face of the cheeks. The capsquares over the trunnions have been degraded by corrosion and are thinner than they once were. The eyebolt for the capsquare swivel and the capsquare pins are fitted blind into the carriage. All the detachable parts of the carriage are missing, including the capsquare forelock, which faced fore and aft, the quoin, stool and stoolbed.

The breeching rope for a demi-culverin was 5in in circumference (1⁵⁄₈in diameter) and 7 fathoms long. Four 10in-long pulleys, two each side, were used for the gun tackles. The tackle rope had a circumference of 3in (1in diameter) and was 13 fathoms long.[87] The train tackle was not listed, but the gun tackle was probably utilised by being unhooked and fitted to the rear of the carriage. The cost of a complete gun carriage of this type was 30s 6d, the same amount that a shipwright would earn in fifteen working days, while a demi-cannon carriage cost 40s 0d.[88]

GUN DECK GUN CARRIAGE

The demi-cannon gun carriage recovered from the wreck of the *Stirling Castle* is partly decayed, with large areas covered by concretion from its associated ironwork and the iron gun it supports.[89] In spite of this it can be seen to be strikingly similar in design principles to the saker carriage

Above: A demi-culverin and carriage rigged for use on Lenox's *upper deck. Heavy tackle prevented the gun from recoiling too violently when fired. The rear pair of dead trucks can also be seen clearly. These replaced rear-revolving trucks, acting as a braking mechanism to further help reduce the powerful recoil of these types of guns.*

in Windsor Castle. The only noticeable difference, apart from size, is that it has four trucks instead of a combination of two trucks and two dead trucks. In common with the Windsor Castle gun, the outside surface of the tree from which it was made is visible at the edges. There even appears to be some bark visible on the underside. The only parts not made from elm are the axletrees that have been identified as ash.[90] Because of the concretion the position of some of the iron furniture was impossible to determine with any accuracy. A list of 1684 gives the breeching rope for

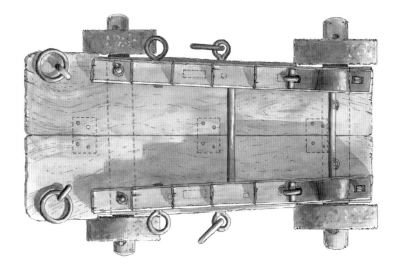

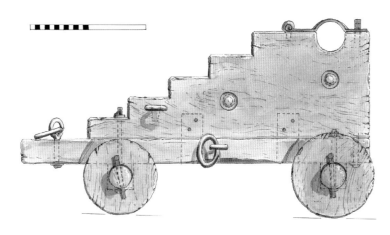

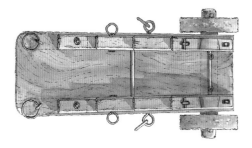

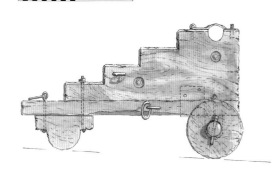

Left: *A demi-cannon gun carriage from the wreck of* Stirling Castle *lost in 1703. This is the type of gun carriage that was used on the gun deck of* Lenox. *The position of the ring bolts and side bolts is speculative, as they were covered in concretion at the time of viewing.*

Above right: *A reconstruction of a 3-pdr gun carriage based on models of the* Breda, *1692, the Vienna model of one of the thirty ships of 1677 and a small carriage from the wreck of the* Swan, *1653.*

a demi-cannon as having a circumference of 6in (2in diameter) and was 9 fathoms long. Four 12in long pulleys, two each side, were provided for the gun tackles. The tackle rope had a 3in circumference (1in diameter) and was 14 fathoms long. The carriage was recorded as being 6⅛in thick.[91]

COST OF ORDNANCE

In 1685 the value of the proposed rough iron for *Lenox* was put at £2322[92], and was calculated to cost £2153 in about 1690.[93] However, the guns she did receive were for the most part Prince Rupert's patent guns and would have been rather more expensive. The true cost is difficult to estimate, as Prince Rupert's patent guns varied from between £60 for early deliveries to a low of £18 a ton in 1685.[94] Altogether *Lenox*'s guns and gunner's stores might be valued at about £7000. As a comparison the cost of the hull rigging and stores of *Lenox* was £16,139 2s 11d.[95]

GUNNER'S STORES

There were four types of stores carried by warships: the victuals, boatswain's, carpenter's and gunner's. At the time of the building of *Lenox* a list was made at Deptford dockyard of her boatswain and carpenter's stores.[96] *Lenox* did not have any guns belonging to her at that time and therefore no gunner's stores were issued to her. The Ordnance Board periodically produced theoretical lists of stores ships should carry, and one was made out in 1677.[97] The Office of Ordnance issued another list dated 15 October 1695 entitled "An account of the allowance of Ordnance, Shot, Powder and other Gunners Stores to the War Ships of the Several Rates".[98] An actual Ordnance Storekeeper's record, written within a few months of the 1695 establishment being issued, closely matches this general establishment. This record is one of the few that has survived and

luckily it belongs to *Lenox*. It was made at the time she was laid up at Portsmouth after she had been out cruising, armed with only sixty-four guns.[99] It is valuable because it is a true record of the state of *Lenox*'s stores at that time. As one would expect, it closely follows the 1695 establishment but with alterations to suit *Lenox*'s unusual armament. The stores are listed in three columns. The "remain" is the quantity of gunner's stores that were actually on board the ship.[100] The "establishment" is the allocated quantity of stores *Lenox* should carry. The "supply" is the quantity of missing stores necessary to make up the establishment. In this case it probably means the quantity of stores used by *Lenox* since she last had a full establishment of stores.

Lenox's ordnance stores in early 1695[101]

Remain, supply and establishment deducted His Majesty's Ship *Lenox* By Order of the Board dated the 24th March 1695

	Remain	Supply	Establishment
Iron Ordnance			
Demi-cannon	22	–	22
Culverin	4	–	4
Demi-culverin	32	–	32 Remain includes 12 cutts
3 pound bullet	6	–	6
Ship Carriages for			
Demi-cannon	22	–	22
Culverin	4	–	4
Demi-culverin	32	–	32
3lb bullet	6	–	6

	Remain	Supply	Establishment
Round Shot for			
Demi-cannon	880		
Culverin	189		
Demi-culverin	2060	230	
3 pound bullet	329		
Falcon	415		
Falconet	2500		
Double headed hammered shot			
Demi-cannon	132		
Culverin	24		
Demi-culverin	228		
Tin Cases filled with Musquet shot			
Demi-culverin	200	51	Remain includes 51 unserviceable
Boxes for			
Tin Cases	20	5	Remain includes 5 unserviceable
Parchment Cartridges			
Demi-cannon	100	350	450
Culverin	-	80	80
Demi-culverin	150	900	1050
3 pdrs	-	160	160
Hand Granadoes	97		
Fuses for same	110		
Ladles and Sponges for			
Demi-cannon	4	6	
Culverin	2	2	
Demi-culverin	7	11	1 0
3 pound bullet	2	2	
Ladle Staves	25	5	
Cases of wood for Cartridges for			
Demi-cannon	33	2	Remain includes 2 unserviceable
Culverin	6	1	Remain includes 1 unserviceable
Demi-culverin	57	5	Remain includes 5 unserviceable
3 pound bullet	9		
Funnels of plate	4	3	Remain includes 3 unserviceable
Corn Powder barrels	254	34	Remain includes 7 unserviceable
Match (cwt)	4	3	
Three-quarter pikes	18	1	Remain includes 1 unserviceable
Short Pikes	18	2	Remain includes 2 unserviceable
Bills	8		
Hatchets	30		
Swords	33	7	
Hangers	39	3	Remain includes 2 unserviceable
Musquet Shot (cwt)	6	1	
Pistol shot lbs	87		
Sheet Lead (cwt)	1	1	
Aprons of Lead	70		
Crows of Iron	69	1	
Tackle hooks pairs	70	25	
Ladle hooks pairs	-	50	
Linch pins pairs	30	10	
Spikes	140	100	

	Remain	Supply	Establishment
Forelock keys pairs	50	30	
Sledges	2		
Melting Ladles great		2	
small		2	
Nails 40d	-	100	
30d	100	100	
20d	100	300	
10d	200	300	
6d	-	500	
4d	400	-	
3d	-	1000	
2d	-	3000	
Beds	30	12	Remain includes 12 unserviceable
Coins	170		
Trucks ordinary pairs	5		
extraordinary pairs	7		
Axletrees for			
Cannon and Demi-Cannon	4		
Culverin and Demi-Culverin	6		
Saker and Minion	2		
Tampions great and small	260		
great		200	
small		100	
Pulleys			
Great and small pairs	15		
Great pairs		30	
Small pairs		20	
Heads and Rammers			
Great and small pairs	40		
Small pairs		5	
Formers Great	6		
Small	2		
Budge Barrels			
Serviceable		3	
Unserviceable	3		
Tanned Hides	10		
Sheepskins	12	36	
Baskets	4	20	
Port Hooks	60		
Spare Hoops		2	
Barras Ells		75	
Paper Royal Rheams ½		6	
Fine Paper Rheams		1	½
Oil Gallons		4	
Tallow pounds	70		
(cwt)			¼
Starch (lb)		12	
Needles (dozens)		12	
Thread (lb)		12	

| | Serviceable | | Unserviceable |
	Remain	Supply	
Lanthorns Ordinary	5	4	4
Dark	2	-	-
Muscovia Lights Ordinary	2	1	1
Extraordinary	3	1	1
Wad Hooks	11	2	2
Hand Crow Levers	124	55	30
Rope Sponges	70	-	-
Powder Horns	90	-	-
Priming Irons	-	-	20
Linstocks	10	6	6
Marline lbs	30	-	10
Twine lbs	-	-	10
Wire	-	-	6
Hand Screws	2	-	-
Tarred Rope of			
6½in coils	1	-	-
5in coils	-	-	1
2½in coils	3	-	-
Breechings	110	-	-
Tackles	160	-	-
Port Tackle Coils	2	-	1
Junk (cwt)	30	-	10
Musquets Snaphance	79	-	1
Musquetoons	5	-	-
Musquet Rods	20	-	10
Blunderbusses	5	-	-
Pistols pairs	16	-	-
Bandaliers	30	-	-
Cartouch Boxes	78	-	2
Flints	1300	-	100

SMALL ARMS

There are a number of noteworthy changes between this list of 1696 and the earlier one of 1677. The 1677 list features ten matchlock muskets. The powder in these primitive muskets was ignited by a slow-match and was discontinued by 1696. The numbers of pistols decreased from forty-eight pairs to sixteen pairs while snaphance, or flintlock muskets as they were later known, increased from forty to eighty to compensate. The numbers of non-firing weapons, swords and pikes for example, also decreased. These changes seem to indicate that the ship was better equipped for longer-range engagements rather than close action and boarding.

THE GUNNER'S STOREROOM, POWDER ROOM AND FILLING ROOM

Minor structures, such as the powder and filling room, were deemed hardly worthy of mention by shipwrights in their correspondence, leaving us little evidence to interpret. They were almost certainly not fitted in *Lenox* when completed in 1678 as the *Hampton Court*, also built by John Shish, did not have her powder and filling rooms fitted until she was being made ready for war in July 1688. There were two powder rooms under the orlop deck, one forward and another aft and at least some ships had platforms in them to prevent dampness affecting the powder.

Apart from the main shot storage area near the mainmast, at least sixteen shot lockers, similar to cupboards, were disposed around the ship for shot.[102] For action, the shot was put inside a rope ring lying flat on the decks to lessen the danger should enemy fire hit them.[103]

From contemporary drawings it appears that up until about 1688, the filling of cartridges with powder was performed on the orlop deck in the gunner's storeroom. To reduce the risk from enemy shot, the filling room was sited below the waterline; the earliest drawing showing this is of the *Resolution* of 1708. To prevent powder escaping and to reduce the chance of sparks, the area was double-lined and plastered.[104] It was lit by screened candles and by 1691 contained at least two chests with canvas, paper and parchment cartridges.[105] In 1677 no ready-made cartridges are listed and the necessary formers, or moulds, were provided for them to be made. The

Above: *(Top) An officer's small sword, probably from the wreck of* Northumberland *lost in the 1703 storm, of which only knuckle bow, guard, and pommel survived. It may have belonged to Captain James Greenway, who had previously commanded* Lenox. *(Bottom) A seaman's lion-headed brass hilted hanger, forty-two of which were issued to* Lenox *in 1696. Although intricately decorated, it was cast to a common pattern and cheap to produce. The blade was relatively short, being about 26in long, making it ideal for fighting in confined spaces.*

a

b

c

d

cartridges were made from parchment, canvas or paper royal,[106] although ladles were also issued for direct gunpowder insertion. The wooden-framed former ensured accuracy of size and volume of the cartridge.[107] The cartridges were stored in chests (cases), which were kept in the gunner's storeroom in great quantity.

Above: *Small arms used aboard ship; (a) musket, c.1695 with dog lock and William III cipher; (b) blunderbuss, c.1695 with dog lock and William III cipher; (c) musketoon, c.1687. The brass barrel dates from the time of Charles II and is similar to an example found in the 1703 wreck of Stirling Castle; (d) a sea-service pistol with James II cipher, dating from c. 1687.*

GUNPOWDER

The quality of gunpowder was fundamental to the firepower of ships. By 1677 serpentine powder of earlier times had been replaced by stable, large-grained corn powder known as "pigeon's egg powder". By the late seventeenth century its ingredients as supplied to English ships consisted of 75 per cent saltpetre, with equal quantities of sulphur and charcoal making up the remainder. The most important ingredient was saltpetre. In the early days saltpetre was extracted from decayed farmyard waste, making it difficult to obtain and at the same time encouraging gunpowder manufacturers to keep the saltpetre content as low as possible. Since 1626 the East India Company had been supplying saltpetre from the Patna area of Bengal; the supply was so important that the Company's Charter contained a condition concerning it.[108] The quantity and quality of saltpetre made available by the East India Company was responsible for the good quality of English gunpowder. The French gunpowder industry did not have such good access to Bengal and, as a result, still had to rely on literally scratching around in farmyards. It was arguably this difference in the means of gunpowder production that resulted in the legendary superiority of English gunnery over that of the French.

The contracts for saltpetre were huge. Between November 1688 and June 1691, £98,000 worth was bought.[109] One contract alone, made with the East India Company in April 1691, was worth £65,000.[110] The sum of £65,000 would be enough to build four ships the size of *Lenox*. The saltpetre was made into gunpowder by separate contractors in England whose contracts invariably stated "to be made with the King's saltpetre".

Weight of powder and shot in lbs, c.1692[111]

Gun	Weight of powder	Weight of round shot	Weight of double-headed shot
Demi-cannon	15	32	34
Culverin	10	18	21
Demi-culverin	7	9	13
3-pdrs	2	3	5

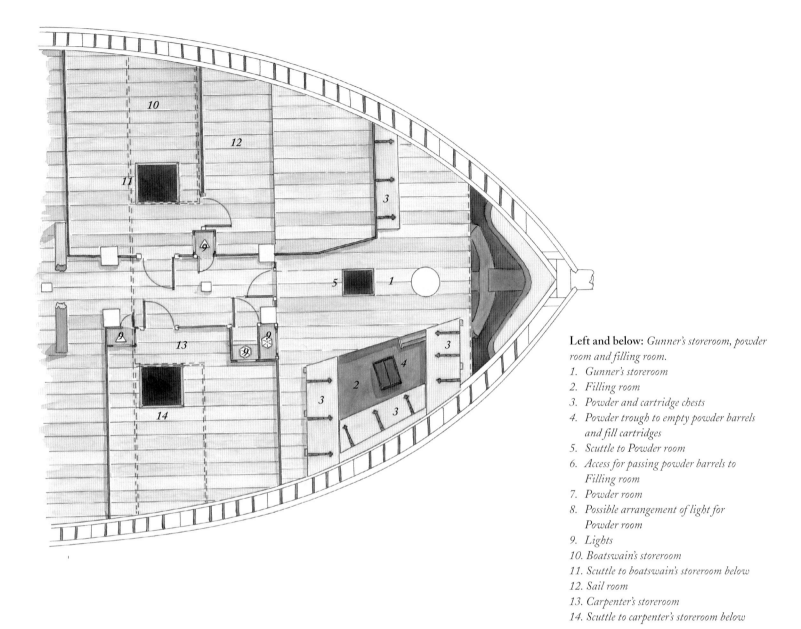

Left and below: *Gunner's storeroom, powder room and filling room.*

1. *Gunner's storeroom*
2. *Filling room*
3. *Powder and cartridge chests*
4. *Powder trough to empty powder barrels and fill cartridges*
5. *Scuttle to Powder room*
6. *Access for passing powder barrels to Filling room*
7. *Powder room*
8. *Possible arrangement of light for Powder room*
9. *Lights*
10. *Boatswain's storeroom*
11. *Scuttle to boatswain's storeroom below*
12. *Sail room*
13. *Carpenter's storeroom*
14. *Scuttle to carpenter's storeroom below*

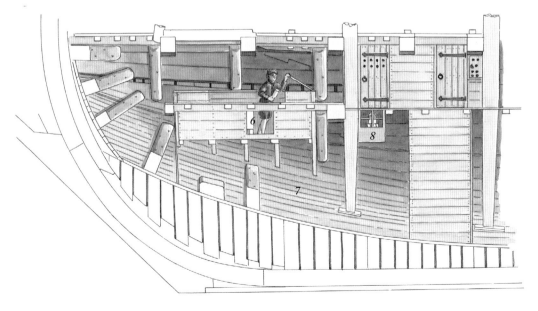

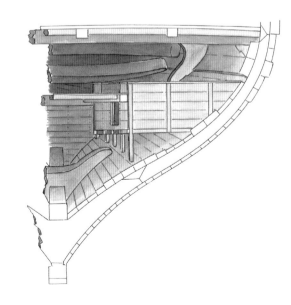

Section Two Plans

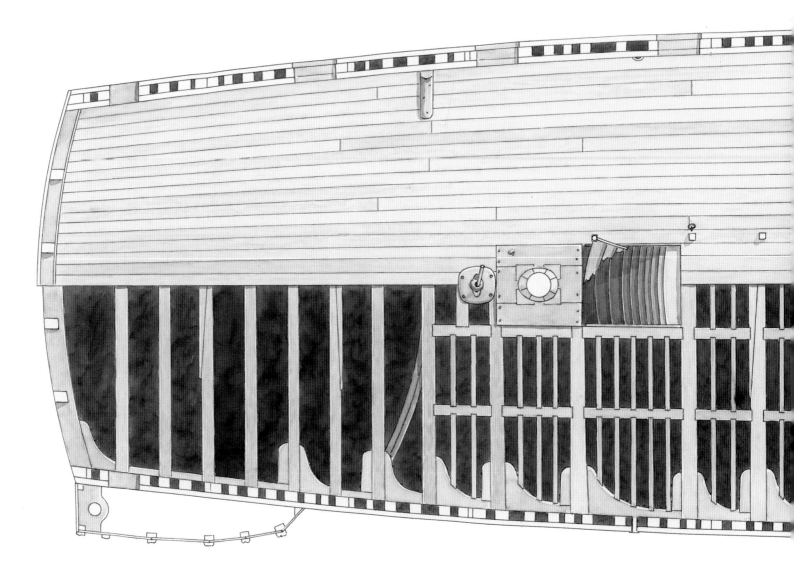

Feet

UPPER DECK

Above: *Upper deck (1:72). The upper deck, where the 24 demi-culverin were situated.*

BROADSIDE VIEW

Above: Broadside View (1:72).

Inset: Planking. The stern is shown on the left and the bow on the right. To prevent [pla]nks becoming too broad at the stern, extra stealer planks are inserted. Similarly, the [nu]mber of planks at the bow are reduced by joggle planks, to prevent them becoming [too] narrow.

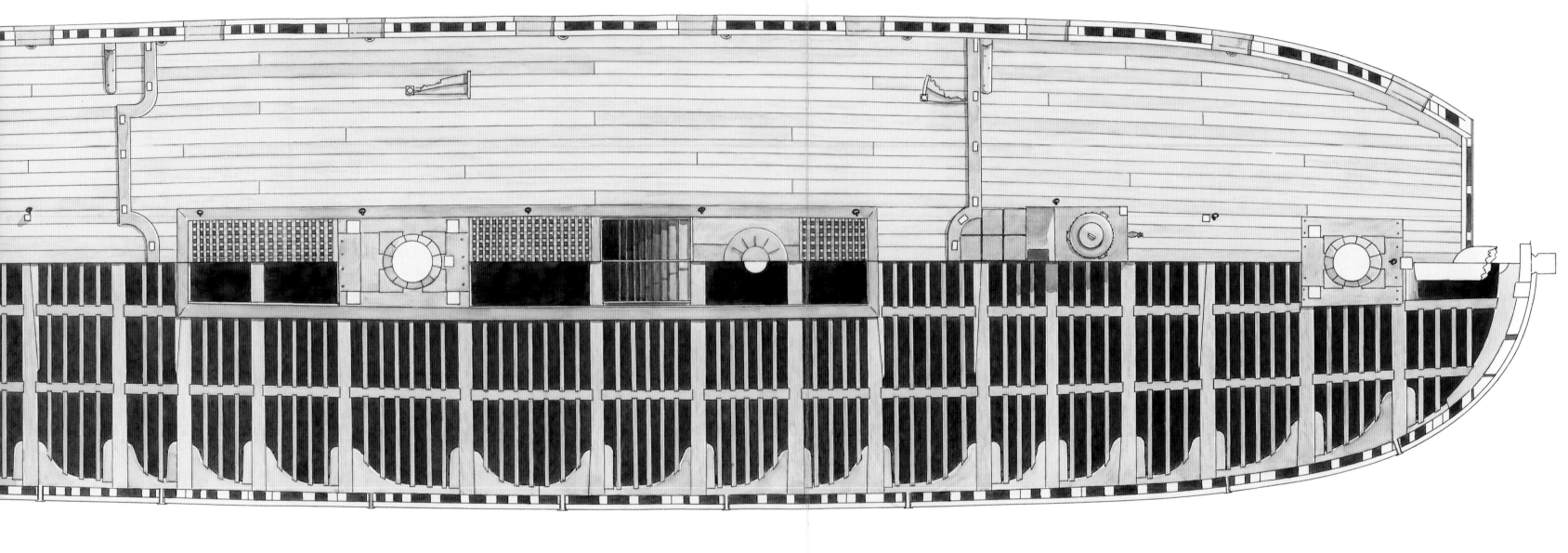
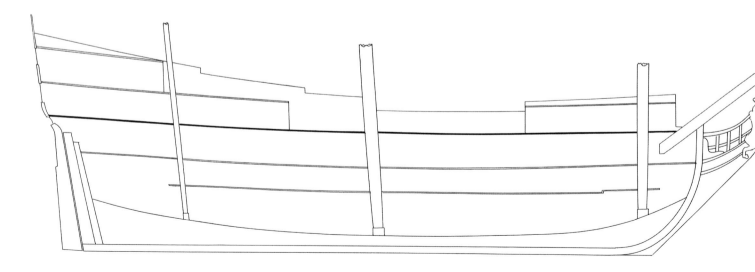

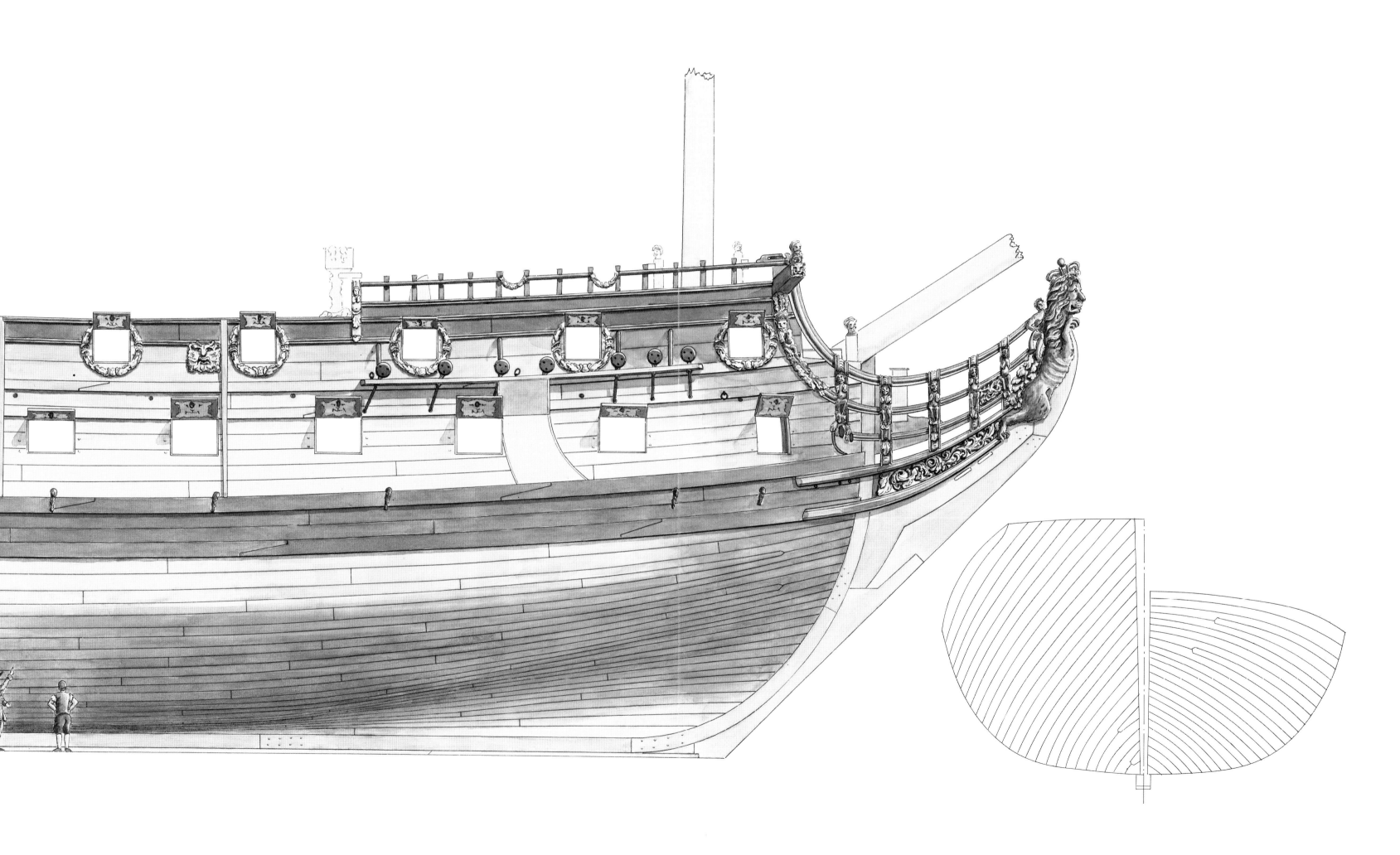

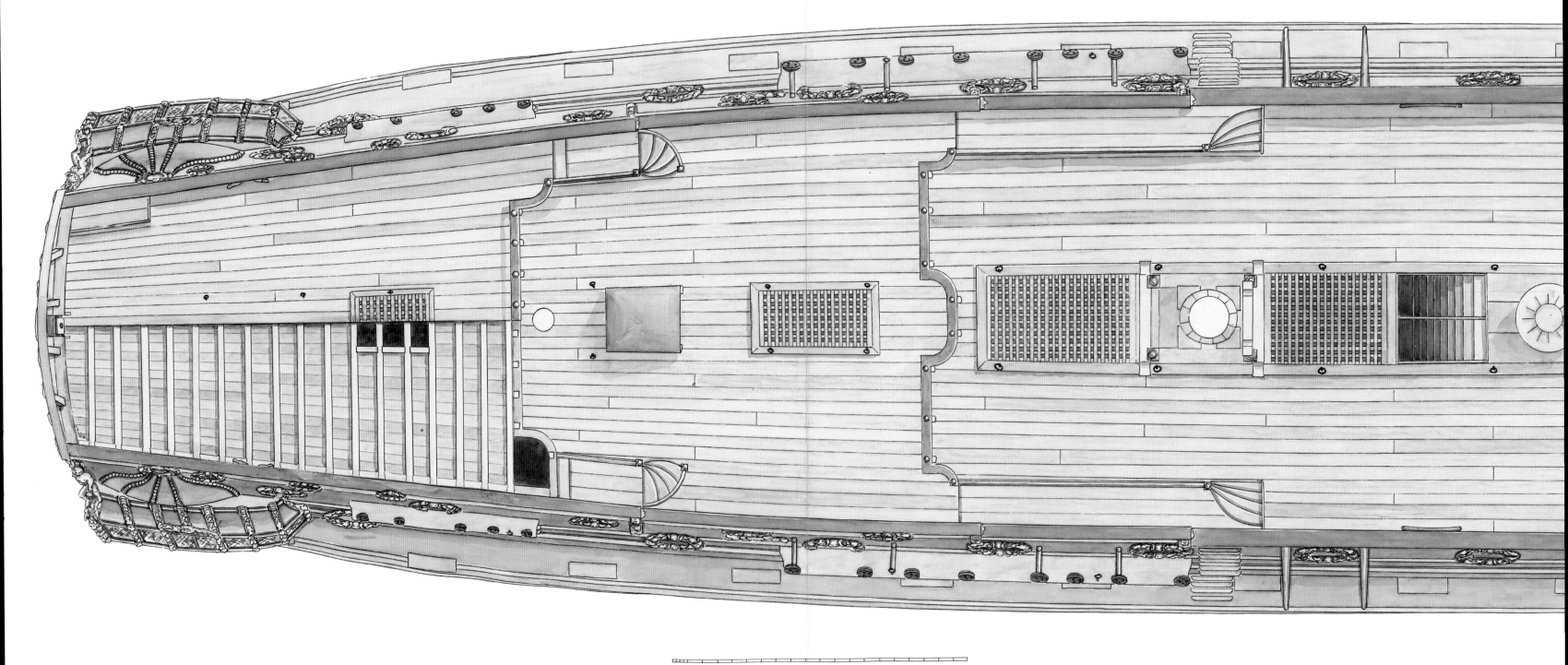

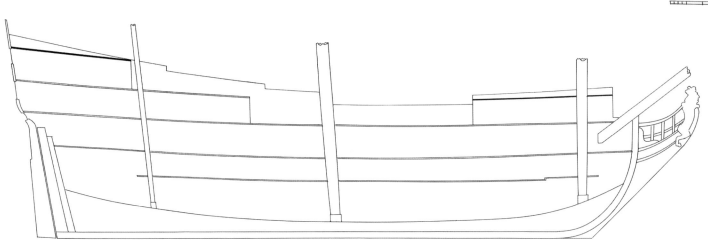

TOPSIDE PLAN

Above and right: *Topside Plan (1:72), less head – inset right.*

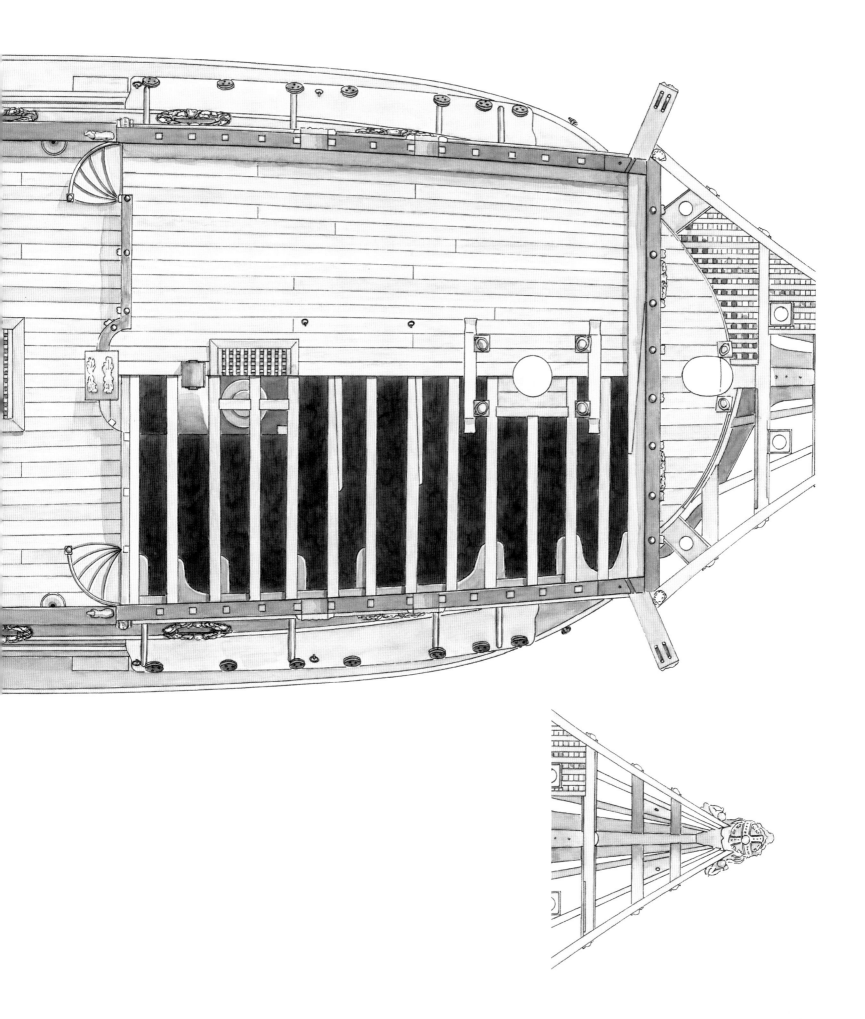

TAMPIONS

Tampions were inserted into the bores of guns to keep them dry. *Lenox*'s ordnance stores show that they were supplied in two sizes, great and small, and at least four were provided for each gun. The great tampions were undoubtedly for the demi-cannon and the small tampions for the demi-culverin. They were made of oak, turned flat on both sides to a thickness of 1⅛in with the outside diameter tapered by ¼in to ensure a good fit in the bore.[112] Once in place the tampion projected about ¼in beyond the gun and was covered with tallow to make sure it was watertight.[113] Tallow took on a dirty white appearance when wet and this can be seen on many Van de Velde paintings of the period.

Left: *Types of shot (from top left)*
(a) Lead apron, used for sealing the gun vent. (b) Hand grenade. The fuse consists of a hollow beech plug filled with slow burning powder. (c) Brass ring gauge used for measuring shot diameter, from the wreck of Stirling Castle, *1703. (d) Round shot. Forty rounds per gun were allowed for demi-cannon on the gun deck and sixty rounds for the smaller demi-culverin on the decks above. (e) Tin case filled with lead shot, later known as canister shot. (f) Base-and-burr, later known as grape shot. (g) Double-headed hammered shot of the type found aboard the wreck of* Stirling Castle, *1703.*

SHOT

There were several types of shot that could be fired by a ship's guns. In 1696 only three types of shot are listed: round shot, double-headed shot and "tin cases filled with musket shot", later known as canister shot. The most numerous type of shot was the cast iron round shot, which generally attained the greatest range and best penetration on impact. A third rate the size of *Lenox* usually carried about 24 tons of round shot, enough ammunition for about forty shots per gun. After round shot the next most popular ammunition was double-headed shot, of which 2½ tons were carried. It was usually forged to its distinctive shape and was more destructive than round shot against both the rigging and hull if fired at close range. It was the only other type of ammunition important enough to be mentioned in *Lenox*'s log books when "round shot and double headed shot" were taken aboard or unloaded during winter refits. In 1696 small quantities of case shot or "tin cases filled with musket balls" were also provided for her demi-culverins. This type of shot was ideal for clearing enemy decks at close quarters.

The 1677 list includes two additional types of shot, "base-and-burr" and iron bar. Base-and-burr was a bagged form of multiple shot that later evolved into grapeshot. The 1696 list does not include them, but does list 2500 falconet shot of 1¼lb each, and 415 rounds of falcon shot of 2¼lb. This is of note because no guns of falcon or falconet size were ever mentioned in any list of guns for *Lenox* or any of her sisters. They were small and light guns, usually reserved for the defence of smaller vessels. As such, the falconet shot was almost certainly sewn up into canvas bags aboard ship to make the base-and-burr for the demi-culverin, while the falcon shot was similarly made into base-and-burr for the demi-cannon.

The total value of the ammunition was about £275. At least 240 barrels of corned gunpowder, each weighing 100lb, were necessary to fire the ammunition and this was valued at about £720.[114]

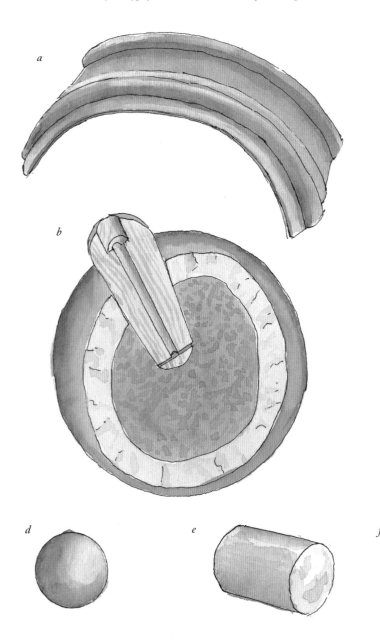

SERVICE HISTORY

THE SERVICE LIFE OF *LENOX*

Every movement made by *Lenox* during her service life, up until she was rebuilt at the end of the seventeenth century, is recorded. The primary sources for this information are her log books. It was the responsibility of the captain, the first lieutenant and the master each to keep a book, or journal as it was called at the time. It appears that officers kept a journal that was filled in day-by-day and remained their personal copy. At the end of a period of service a fair copy was then made which had to be submitted to the Admiralty. Richard Taylor, the master of *Lenox*, writing on the last page of his journal, "Parts of June, July and August were so fouled with filth – that I could not copy it", gives proof of this practice. However, the copy in the National Archives, including the passages fouled with filth, is clean. An example of a journal being delivered to the Admiralty is made by Captain John Guy who wrote to the Navy Board, "I have likewise delivered to him my journal for the *Crown* to be forwarded to the Right Honourable, the Lords of the Admiralty".

A survey of the preserved log books for the early part of the war reveals only 44 per cent survive,[1] which is not surprising after such a long period of time. Happily almost 90 per cent of the *Lenox* log books are still extant, although sometimes the ink has faded almost to a point of invisibility. Sometimes heavy cribbing has resulted in near identical works between some of the officers.

In the case of one log, written at the time of the Battle of Beachy Head, the vital days before, during and after the battle are missing. For the Mediterranean period of *Lenox*'s history the quality of entry is poor with few interesting details, while other periods are full of lively incidents. The logs often illustrate the responsibilities and natural concerns of their authors. For example, during a particularly violent storm the captain mentions his conferences with officers, while the lieutenant and the master give details of damage suffered by the ship.

The log books reflect individual style in typically seventeenth-century phonetic spelling. The masters' logs in particular are sometimes very imaginative and contain their owner's spoken accent, rendered into an approximate spelling – for example, 'Borfutt' for 'Burford' and 'fare' for 'fair'.

Unfortunately the log books contain little information about the orders that are being followed, or what the object of actions might be. The main source for linking log-book entries with historical events is Josiah Burchett's *Memoirs of Transactions at Sea during the War with France* (1703) and his later work, *A Complete History of the most Remarkable Transactions at Sea* (1720). Burchett was Secretary to the Admiralty, a position once held by Samuel Pepys. He sailed with the fleet as Secretary to Admiral Russell. His account and the *Lenox* log books fit together with precision and only rarely do minor discrepancies occur.

Another source of information is *Lenox*'s pay books, all of which survive and reveal details of deaths, desertions and so on. Incorrect information from the log books has not been corrected; in any case, errors are rare, the worst occurrence perhaps being a day's contradiction between journals, a result of the normal method of recording entries between noon the day before and noon the day of entry. Specifically, as in all naval combats before or since there is a tendency to exaggerate the size and power of the enemy ships destroyed.

As far as possible the men of *Lenox* have been left to tell their own story as they saw it; the author's interpretation has been kept to a minimum. I have not selected highlights of *Lenox*'s career but recorded all events and voyages as they happened. The normal problems and practices of keeping a third-rate ship of war at sea are therefore revealed in some detail. Major events such as battles and storms are described in much more detail than the mundane. The intention is to view events from the perspective of those on board this particular ship. The terminology of the time has been retained for accuracy and flavour where there can be no misunderstanding of the meaning, often using direct quotations of the originals. Where log books and other official records such as "disposition of ships"[2] and pay books[3] have been used, or where Burchett has been consulted, no references have been inserted.

NOTE ON LOG BOOKS

The surviving log books show several anomalies and also demonstrate some poor archiving, with lieutenants' logs stored among the masters' and captains' logs. The second lieutenant's log for 1690, not normally archived, is preserved among the captains' logs, and surely the captain's logs for 1695, 1696 and 1697 would be better preserved with others of the 1690s, rather than with those of the eighteenth century. Perhaps more worrying is finding Captain Myngs' log[4] signed on the last page in his own hand overlapping by a period of five months another log titled "Journal kept by Christopher Myngs".[5] It is likely that the former is a lieutenant's log that Myngs signed and which has therefore been attributed to him ever since. But these are relatively slight anomalies, and the survival and preservation of such complete records from a period of more than 300 years ago is quite remarkable.

All the dates are expressed as in the original texts according to the Julian calendar (that is, eleven days behind modern usage). The New Year is reckoned here to start on 1 January rather than 25 March, to enable easier understanding.

1689 – THE WAR OF THE LEAGUE OF AUGSBURG

The Glorious Revolution of 1688 was completed by-mid February 1689 when Mary, daughter of James II, arrived from Holland with William of Orange to jointly take the throne. England was thus brought into the struggle between the expansionist Louis XIV and the League of Augsburg, a combination of European states ranged against France. Louis

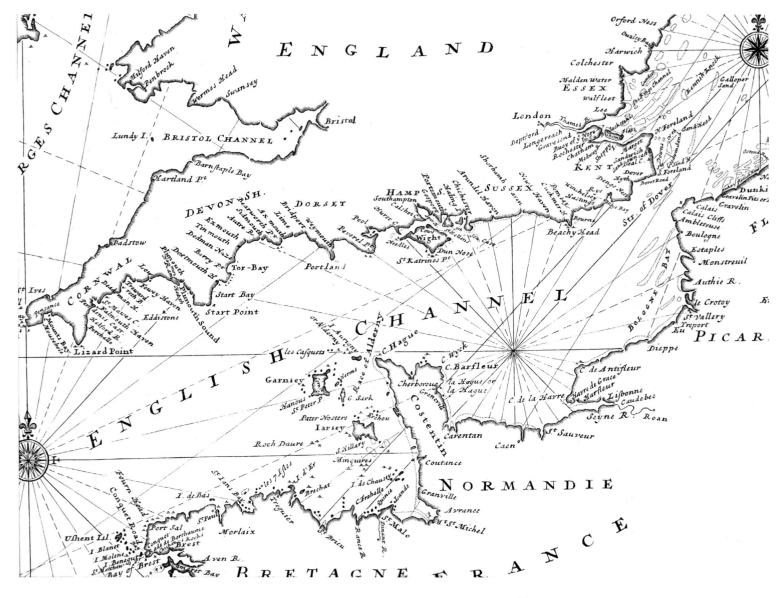

Above: *A chart of the Channel by H Moll for* Transactions at Sea, *Josiah Burchett, 1720. (Author's collection)*

was determined to restore James to the English throne and granted him a fleet and soldiers with which to land in Ireland as a preliminary to regaining his kingdom. The most notable naval action was the inconclusive battle of Bantry Bay on 1 May 1689, off the southern coast of Ireland.

Lenox remained at her moorings at Chatham during these early days while the French and Anglo-Dutch allies fitted out their fleets for the forthcoming summer campaign. Edward Gregory, the newly installed Commissioner, mentions *Lenox* in a letter to the Navy Board dated 23 April 1689. He noticed that *Lenox* and many other ships were not equipped with either longboat, pinnace or jolly boat and requested directions for building new ones.[6] A month later he wrote again, saying that in his opinion, as soon as the *Burford*, *Anne* and *Captain* had completed their fitting out and been launched, the *Lenox*, *Hope* and *Stirling Castle* should be taken in hand on the next spring tide and taken into dock.[7] The six ships were all third rates of the 1677 programme. At about the same time William Beckford, a naval contractor, supplied seamen's clothes for the ships. He listed *Lenox* as having 36 cotton suits, 288 blue shirts, 140 blue drawers, 48 blue jackets, 144 woollen hose, 72 shoes, 36 laces, and 18 yards of canvas, worth a total value of £120 15s.[8] The clothes or 'slops' as they were known were for seamen who entered a ship in rags and wished to

purchase new clothes. They were sold to the men by the purser for a commission of one shilling in the pound from the contractor, in this case William Beckford. A deduction for their value was then made from the seaman's wages when he was paid off. It is interesting to note that even in the seventeenth century blue would have been the predominating colour of the crew of *Lenox*'s "uniform".

A list was made, dated October 1699 and titled "Cost to Seamen of Slops"[9]

	£	s	d
Broad cloth coats		19	0
Kersey gowns		19	0
Waistcoats		6	8
Breeches, per pair		5	8½
Yarn stockings, per pair		2	10
Blue shirts		3	4
Wax leather shoes, per pair		4	3
Blue suits, viz. waistcoat & drawers		4	9
Striped suits lined with canvas		10	11¼
Viz. waistcoat & breeches			
Tobacco, per pound		1	7

Above: *An engraving of Arthur Herbert, 1st Earl of Torrington (1647–1716), the commander of the Anglo-Dutch fleet at the Battle of Beachy Head. (Author's collection)*

Lenox continued to lie in ordinary while most of the fleet was brought into commission. A month later on 2 July she was surveyed and listed in Admiralty records "as can be made ready for sea in two months her hull repaired, rigged and caulked if proving to be no worse than seen".[10] She was docked in the afternoon of 9 August but her condition seems to have been worse than expected, for Gregory reported on the 24th that large repairs were necessary. These were completed and by the end of the year she was in good condition.[11]

1690 – THE BATTLE OF BEACHY HEAD

Lenox and her sisters *Captain, Anne, Stirling Castle* and *Restoration* had all been docked, cleaned and launched early in the new year.[12] Then early in the morning of 10 February, the twenty-five year-old John Granville, second son of the Earl of Bath, arrived at Chatham and boarded *Lenox*. He held a commission as captain that had been signed four days before.[13] John Granville was descended from the youngest son of Rollo, first Duke of Normandy through Richard de Granville, Lord of Gloucester and Glamorgan, who lived in the reign of William the Conqueror. A later ancestor was Richard Grenville, Vice Admiral of England, who died gallantly defending his ship *Revenge* in 1591. John Granville had previously been Captain of the *Bristol*[14] and was a close friend of the Admiral of the Combined Fleet, Arthur Herbert, the Earl of Torrington. Other commissions followed: a boatswain on the 8th, the chaplain on the 10th

and a lieutenant, Edward Durley, on the 14th.[15] There were two lieutenants appointed to *Lenox*, Charles Brittish becoming first lieutenant and Edward Durley the second lieutenant. Captain Granville immediately wrote to the Navy Board informing them he was to command *Lenox* and requesting provisions be sent, as he would not be able to get men without them.[16]

The Earl of Torrington was the most senior English naval officer to accompany William of Orange to England during the Glorious Revolution. He was forty-three years of age and was renowned for being licentious and given to using colourful language. His face was marked with the scars of battle and he had a damaged eye after a fight with an Algerine, all of which supposedly made him somewhat repulsive in appearance. But these features were more than compensated for by his reputation as a fighting man, in which capacity he commanded considerable respect and admiration from most of those under his command, especially Granville.

It was clear to Captain Granville that, in spite of having many volunteer men join him, he would have to use the press gang if he was ever to acquire the 460 men needed to man his ship. In 1677 the normal method of pressing seamen started with press warrants issued by the Admiralty. Press Masters then rounded up the seamen "with as little grievance to His Majesty's subjects and the pressed men" as possible. A printed form in duplicate was filled in giving the names, ages and conditions of the pressed men. They were then handed over to a Conductor who escorted them to agents at a port. Here the agents delivered the pressed men to the captain or master of the ship they were to enter.

Sometime during this process it became a common fraud to exchange good men for bad. To prevent this happening, the duplicate list of pressed men was given to the ship's officer who checked and attested it to make sure the men now aboard his ship were those who were originally pressed. At the time, pressed men received 6d a day while waiting and 8d a day when travelling.[17]

According to Granville's wishes, the Admiralty granted him two printed press warrants dated 8 February.[18] These enabled him legally to impress men into his ship. A press warrant gave power to impress able seamen within the county of Kent, Blackstakes and in the Medway. The captain gave each imprest man a shilling and travelling expenses, or conduct money as it was known, at the rate of one penny per mile from where they were imprest. The money paid out by the captain would be repaid by the commissioners of the navy upon presentation of an account signed by the chief officer of the ship. The warrant also gave a warning against corruption and orders not to impress men from outward-bound merchant vessels. On the 13th, Granville sent Lieutenant Durley with a letter to the Navy Board requesting 200 printed tickets, in effect "IOUs" that could be given to pressed men instead of ready money. He further pointed out that if he didn't press those men available, others would.[19] There were many sad partings between couples; a typical ballad of the time illuminates their feelings.

What follows is an extract from 'The Dover Lovers', a pleasant dialogue between Jonathan, a courageous seaman, and Susan his mistress, a little before his departure.[20]

Seaman
Honor calls, I must away,
Yet Love bids me for to stay;
Why should I then fear a Storm,
When thy Beauty hath a Charm?
The Gods will guard me I don't fear,
For the sake of thee, my Dear;
When the Battle once is done,
To thy arms, to thy arms, my Love I'll come.

Maid

Oh! The thoughts of dreadful Fight,
Does disturb me Day and Night;
If you are resolv'd to go,
Let me put on Man's Cloaths too,
And your Servant I will be,
To fight against the Enemy,
And when the Foes about you swarm,
I will guard, I will guard you from all harm.

Seaman

Thanks a thousand times, my Dear,
For thy Love and for thy Care,
Better thousands of us kill'd,
Than one drop of thy Blood spill'd;
Take this Ring and Jewel too,
In token that I will be true;

Then they in each others arms
Hug'd and kist, hug'd and kist a thousand charms.

Gun & allocation	Number of men	Position of guns
26 demi-cannon six men each	156	Main gun deck
26 12-pdrs four men each	104	Upper gun deck
Men to govern the guns	6	
12 Sakers three men each	36	Quarterdeck
2 Sakers three men each	6	Forecastle
2 3-pdrs two men each	4	Poop deck
To carry powder	20	
To fill and hand powder	10	
To hand down wounded men	8	
Surgeon and crew	5	
Carpenter and crew	4	
Purser and his crew	5	
Boats: three	16	
To the tops if no flag	5	
For the helm	2	
To con the ship	2	
Small shot	35	
To the sails and rigging	36	
Total	460	

On joining *Lenox*, men were allocated to either the starboard watch, whose chief was the master, or the larboard watch under the mate. Each watch, when on duty, would be responsible for all the normal duties in working the ship, leaving the other watch to rest. Each watch lasted four hours and was divided into eight half-hour glasses. When there were sudden weather changes the call would be made for all hands on deck. Men would be allocated particular tasks in working the ship according to their ability. Care was taken that comrades who messed together were split into different watches so that the best use was made of the limited deck space. On *Lenox* each mess between guns would consist of about eight men, four on duty and four off. At the start of their last watch of the day, the chaplain would assemble them for evening prayers.[21] The frequency of the services, however, did not mean that seamen were particularly religious; in fact their reputation was quite the opposite.

In order to get her ready with sufficient men and provisions, *Lenox* had allotted to her a ketch, a small two-masted square-rigged vessel, and a smack, a smaller single-masted gaff-rigged craft to help speed the work. The ketch was the *Prosperity* of 120 tons commanded by Robert Emmerson and the smack the *Mary* of 30 tons commanded by George Watson. Other third-rate ships in the fleet were allocated tenders similar to those of *Lenox*.[22]

On 15 February Granville and the captains of the *Restoration, Stirling Castle, Lion* and *Defiance* received orders to fall down to Blackstakes, wind and weather permitting, to take in guns, stores and provisions.[23] It would be some time before the orders could be carried out, for there were only eight men aboard cleaning her on the 20th, and by the 25th still only fifty-four were aboard.[24] A few days later the supply smack brought some of her gunner's stores from Woolwich Arsenal, while a lighter brought down shingle ballast that was taken down into the hold. By the end of the month her foremast shrouds were up and some sails taken in. At the end of February Captain Granville received new orders to repair to the Hope to take in guns, stores and provisions and remain there until further orders. During the month two men's lives were lost, probably from accidents while setting up the rigging.

The following table (see top of next column) shows how men were allocated to a 68-gun third rate.[25] The guns listed bear a loose relationship to those aboard *Lenox*. In addition to the seamen, up to about 120 marines may have also been allocated from either of the two Marine regiments.

At the beginning of March there were still only 115 names entered on *Lenox*'s pay book. Granville had problems in persuading volunteer seamen to join the fleet and subsequently wrote a letter to the Navy Board:

"At our arrival here we find we have been disappointed of several men whom we sent down from London hither, who not being paid their conduct money [travelling expenses] at their ship coming, have chose rather to return to London than to go aboard our ships. We therefore make it our request to you that you will be pleased to make especial care for the encouragement of seamen in this time of need that they may have their expectations answered, and we not obliged to promise what we cannot perform."[26]

Granville's letter, written in his own hand, was signed by himself and three other well-known captains: George Churchill of the second rate *Windsor Castle*, William Botham of the *Restoration*, and Anthony Hastings of the *Stirling Castle*. The men were all close friends, and would follow Torrington in his flagship the *Royal Sovereign* in the line of battle.

Lenox was short of a master, pinnace and jolly boat, and Granville added a request for them at the bottom of his letter. The competition for seamen caused problems for Granville when a volunteer seamen, William Ailbury, to whom he had granted a ticket for eight days' leave, was pressed by the *Breda* in open breach of the rules. Granville immediately complained to the Admiralty who sent an order to Captain Tennant of the *Breda,* telling him to discharge the seaman forthwith "and to take care that neither yourself or those employed by you do for the future impress any men belonging to other ships of their Majesties".[27] Captain Tennant was not too concerned by his rebuke, for he continued to press anyone he could lay his hands on despite regular letters of complaint. As we have seen, he had made a reputation for himself in 1688 when, as captain of the *Tiger,* he carried off the riggers of Chatham dockyard to the annoyance of everyone.[28] Eventually he went too far by pressing the boatswain, gunner and carpenter of the *Colchester*. He was then ordered to write to the Admiralty to explain himself.

The frustration that Granville openly felt while manning his ship was surely turned to anger when Captain Fairborne of the *Warspite* took sixty men out of *Lenox*'s ketch and pressed them aboard his ship. Granville again wrote to the Admiralty to complain before the matter was resolved.[29]

Work continued on getting *Lenox* ready. Six cables and her anchors were stowed, the sails bent to the yards, and many tuns of beer taken down

Above: *Admiral of the Fleet Edward Russell (1653–1727), 1st Earl of Orford and a member of the Privy Council. (Author's collection)*

Above: *Admiral Sir George Rooke (1650–1709). He fought at Beachy Head and in May 1692 served under Russell at the Battle of Barfleur. (Author's collection)*

into the hold. On 6 March Edward Gregory wrote to the Admiralty informing them that the *Stirling Castle, Lenox, Restoration* and *Lion* had pilots on board but were prevented from sailing by contrary winds.[30] The weather in fact turned bleak for the next few weeks as it either rained or snowed while blowing incessant gales, forcing *Lenox* to lower her recently raised yards and topmasts. Striking topmasts and yards was a routine operation often carried out on a daily basis during bad weather. It finally cleared up on the 25th of the month and *Lenox* was warped the short distance down to Upnor Castle. The task took two days, as a new NNE gale sprang up to hinder progress.

Warping was normally undertaken by dropping an anchor from a boat in a position ahead of its ship. By winding in the cable with the capstan the ship was towed towards the anchor. In ports and harbours permanent posts were provided to which the cable could be attached, thereby avoiding the necessity of using an anchor.

Two days later *Lenox* set sail from the fleet anchorage, under way for the first time since she first arrived twelve years earlier. She sailed together with the *Stirling Castle, Grafton, Lion* and *Captain* to Blackstakes, where they anchored to join the *Sandwich* and *Katherine* before moving on to anchor, first at the Buoy of the Nore, the anchorage where the River Medway branches off from the River Thames, and then in the Hope. It was a month since she had received orders to proceed there; on arrival the ships

joined others who had gathered to complete their readiness, among them the *Defiance* and *Restoration*.

Lenox anchored in seven-and-a-half fathoms by her best bower cast off the port bow. As the fleet gathered *Lenox* received more stores and men, but as fourteen tuns of beer were brought aboard, a man fell into the hold and was seriously injured. The next day he died and was taken ashore for burial, becoming the third man to die aboard the ship. It continued to blow fresh from the NE and, as a precaution against *Lenox* dragging her best bower anchor; the sheet cable was spliced and bent to the sheet anchor for ready use. This was the largest safety anchor and was stowed on *Lenox*'s starboard channel. A week later in the continuing gales it was put over the side, followed by the small bower anchor a few days later. The small bower was carried on the opposite side to the best bower; it was taken out to the eastward to give a good spread. The only anchors left aboard *Lenox* were her spare anchor and two of much smaller size, the stream and kedge.

Lenox's guns, shot and double-headed shot were now brought to her in a hoy. On 2 April, sixteen guns for the upper deck were brought aboard followed by sixteen gun-deck, 32-pdr demi-cannon the next day. On the 5th, ten more gun-deck and six upper-deck guns followed. The arming was completed on 10 April when the rest of the guns arrived. In the continuing gales, the *Pembroke* was grounded on the north shore as she came out of Long Reach, but got off safely the next day. The guns of *Lenox* were mounted on their carriages just in time for Coronation Day when they were fired twenty-one times. The crew then drank the health of the King and Queen in further celebration of the accession of William and Mary. The small bower anchor was then weighed and moved further to the east for better purchase. To raise the anchor *Lenox*'s crew used the jeer capstan, which was evidently not strong enough as it became broken beyond repair. A sister ship of *Lenox*, the *Suffolk*, had recently broken her jeer capstan in a similar manner, causing reproving letters from Edward Gregory. Captain Granville was now obliged to write to him requesting a new capstan for *Lenox*.[31]

While *Lenox* was being readied for sea, her supply smack sailed down to the Buoy of the Nore to meet a fleet of colliers in the hope of pressing some of their men. *Lenox*'s guns soon fired again, for two days later a Dutch squadron arrived to join the fleet and "a Dutch man of war came up and saluted us with five guns, we returned with three". Dutch warships were never recognised or identified by *Lenox*'s officers, although they seem to have had no trouble identifying English ships at any distance.

Another accident occurred when a man from the *Stirling Castle* fell into the sea between *Lenox* and the *Stirling Castle* and was drowned. The rest of April was spent taking aboard provisions and men. *Lenox*'s ketch, the *Prosperity*, searched for men to press into service from the northward, while the smack *Mary* went to London. The "horse boat" arrived from Chatham with the new jeer capstan and spare mizzen topmast. The horse boat was powered by horses driving paddle wheels and was often referred to by Gregory.

With the coming of May, *Lenox* was still short of men and her ketch and smack continued to make press expeditions. A contingent of seamen for the fleet arrived from Scotland aboard the *Assurance*. On 12 May the Admiralty ordered forty of them to be sent to *Lenox*, and a few days later they arrived. Then the Comptroller of the Navy, Sir Richard Haddock,

Below: A plan of the Battle of Beachy Head, made for the Earl of Torrington. It shows the opposing French and Anglo-Dutch fleets at the start of the action. (Author's collection)

and some Navy Board officers arrived in the yacht *Charlotte* and came aboard the same day to pay them.[32]

On the 17th a sudden gust of wind broke the crojack yard. *Lenox* probably had a backed mizzen topsail set to hold her head to the wind when the accident happened. Captain Granville wrote to request another to replace it.[33]

The long delays in getting ships ready were causing alarm, and fears were expressed by Edward Russell, Queen Mary's naval adviser, that the French would be in the Channel with an invasion army before the fleet could sail to oppose them. Torrington himself made numerous protests as the time dragged by and even threatened to resign his commission.

Lenox received notice to sail together with the squadron at the Nore and join the main fleet gathering in the Downs, the protected anchorage between the Kent coast and the Goodwin Sands. But Granville wrote on 16 May: "This cometh to acquaint you that I am ready to sail for the Downs and have received your orders for the purpose, but have not any pilot here aboard. I do therefore desire you will be pleased to dispatch one down as soon as possible as no opportunity of sailing may be lost."[34] A pilot soon arrived and on the 21st Lieutenant Brittish went aboard the flagship to receive final orders for sailing. Next day, at 4.30 am, *Lenox* weighed anchor together with the three-deckers *Duchess* (with Rear Admiral Rooke, his blue flag at the mizzen topmast head), the *Royal Sovereign*, *St Andrew*, *Sandwich*, *Albermarle*, *Windsor Castle*, *Katherine*, the third rates *Grafton*, *Anne*, *Captain*, *Elizabeth*, *Defiance*, *Lion*, the fourth rate *Assurance* and six fireships.

The next morning abreast the Middle Ground the mighty *Sovereign*, originally built in 1637 as *Sovereign of the Seas* and still the largest ship in the navy, carried her spritsail topmast by the board with the knee of her bowsprit and at 6 pm off North Foreland split her main topsail. At 7.30 pm the ships anchored for the night but weighed again at 5 am the next morning during a fresh gale that swept them into the Downs. As the ships entered the anchorage, gales split the mainsails of the *Sovereign* and *St Andrew* and carried away *Lenox*'s fore topmast a little above the cap, breaking it in three pieces. The wreckage of the rigging was taken down and the foresail set in place of the fore topsail. It seems probable that the spars

A Prospect of the late Engagement at Sea between the English and the French Fleets on Monday the thirtieth of June 1690, the wind from NNE to the ENE with an exact List of y English Dutch & French Ships Engraven by the Earle of Torrington's Order & approbation

and sails of the ships, which had been so long in storage, had lost some of their strength and flexibility.

Lenox and the warships that accompanied her anchored off Deal Castle in eight fathoms to join other units of the gathering fleet. There was *Coronation* flying the red flag of Rear Admiral Delaval at the mizzen topmast head; the third rates *Berwick*, *Suffolk*, *Edgar*, *Hope*, *Plymouth*, *Expedition*, *Hampton Court* and *Breda*; the fourth rates *Nonsuch*, *Foresight*, *Pembroke*, and *Garland*; four fireships and a Dutch fleet of sixteen men-of-war. Meanwhile, *Lenox*'s carpenter, Ambrose Fellows, and his crew were busy making the spare main topmast into a new fore topmast. He found it defective and had to cut away some of the heel until it was sound. Unfortunately it ended up 3ft too short but for want of better it was got up and rigged within two days of the original being broken. It was not desirable to fight or risk weather damage without a spare topmast and Captain Granville wrote to the Navy Board asking for a spare topmast to be sent down to him.[35]

As May came to a close, *Lenox* and the fleet took every opportunity to fire their guns in salutes as flag officers joined the fleet. There is no evidence of formal gun drill and firing salutes were almost certainly used for the purpose. On 29 May she fired thirteen guns as Vice Admiral Sir John Ashby hoisted his blue flag and another fifteen for King Charles II's birthday. Another seventeen followed this as the Admiral of the fleet, Lord Torrington, returned from London in the yacht *Fubbs*, boarded the *Sovereign* and hoisted his flag at the main topmast head. "His Lordship returned his thanks with thirty-one guns to the whole fleet".

With the prospect of battle in the near future, *Lenox*'s smack and ketch continued to go on press expeditions, while some of her officers went ashore to take the Sacrament in Deal church. During the first week of June, as the final ships joined the fleet, the *Hopewell* fireship that had arrived from Shoreham was accidentally burned when a grenade shell blew up aboard her.[36] She was commanded by Thomas Warren and carried forty-five men and eight guns.[37] Several of her men were injured, one man lost a leg and another was badly burned. She was towed ashore onto Deal beach where she burned for an hour before the flames reached ten barrels of gunpowder and blew up.

The new fore topmast that Captain Granville had requested arrived at 8 pm that night. It was immediately got up to replace the short one recently made by the carpenter. The next morning it was rigged, the fore topgallant mast above it also got up and the sails bent to the yards.

Lenox's armament was increased on 5 June when two mortar pieces and their ammunition consisting of 100 grenade shells were brought aboard. Captain Granville with some other captains of the fleet then attended a council of war on the flagship, after which he went ashore to Deal Castle to receive sacrament as his officers had done before him.

Lenox was still seriously undermanned; of those men aboard about half were volunteers while the rest were pressed. With the likelihood of a pitched battle looming large, many of the pressed men were not about to stay on board if they could help it. With what must have been considerable ingenuity, many of them managed to desert ship and have an "R" for "run" recorded against their name in the pay book. This was on the statute as a capital offence, but in practice the only punishment was loss of pay, resulting in some of the money they would have received being donated to the Chatham Chest, the seamen's benevolent fund. Ten men deserted on the 9th followed by six more a few days later. Altogether fifty-six men had managed to desert. Losing men at this rate *Lenox* would never have reached her full complement had the loss not been made up by the transfer of 112 marine soldiers into her. This was the highest number of marines *Lenox* was to have. Although not seamen, they were more disciplined and less likely to run.

On 11 June the *Warspite* joined the fleet and the Admiral summoned a council of war by flying a red flag in his mizzen shrouds. Next morning

Torrington hauled home his fore topsail sheets, then made a signal by loosing his main topsail and firing a gun for the fleet to weigh anchor. The whole fleet, English and Dutch, set sails and steered south through the Downs and towards the Channel. The immediate intention was to join some third rates at Portsmouth before combining with two detached squadrons, one under the command of VA Henry Killigrew, which had been to the Mediterranean escorting a merchant convoy, and another under the command of Cloudesley Shovell in the Irish Sea. It was vital if the allied fleet were to be superior to the French that the detached squadrons join the main fleet before the French reached the Channel. The French under Tourville, however, hoping to destroy William's communications with Ireland where he was then campaigning, were first to put their fleet to sea and frustrated the intended rendezvous of the detached English squadrons.

On 15 June the Anglo-Dutch fleet reached St Helens, the anchorage off the east coast of the Isle of Wight, where *Lenox* anchored in seven fathoms of water. Captain Granville wished to be closer to Torrington, however, so the anchor was raised again and *Lenox* was moved to within a cable's length of the admiral's flagship. Granville was to keep *Lenox* close to the flagship throughout the forthcoming campaign as part of the centre or Red squadron.

On the 20th *Lenox* was heeled so that some of her bottom could be cleaned of weed, then, two days later, "This evening a Dutch privateer came in here and gave us an account of the French fleet being off the backward side of the Isle of Wight." That night the fleet unmoored, *Lenox* hauling in her cable so that she rode short on half a cable's length. The next day at 11 am, Torrington hauled home his topsail sheets and fired a gun as a signal for all the fleet to weigh anchor.

Off Dunnose, the *Portsmouth Galley* joined and brought confirmation of the news, Torrington anchored and dispatched the *Deptford* and yacht *Saudadoes* as scouts. The next morning at 5 am, the fleet weighed anchor in a fresh wind coming from the NE by E. During the day three Dutch men-of-war and the *Lion* joined them. At 8 pm they anchored again, four leagues off the west end of the Isle of Wight with the French in sight of *Lenox* for the first time.

Dawn of the 25th was accompanied by gales and fog but in "the early afternoon Lord Torrington made the signal for the fleet to fall into line of battle, the van of our fleet edged down toward the French, but they edged away, this evening we came to an anchor". Torrington, now realising he was outnumbered by about seventy-six ships to fifty, took the decision to retire slowly up the Channel and keep his fleet defensively to the eastward in the hope of delaying battle until he could be reinforced, fully aware that defeat could herald the invasion of England. For the next three days during continual fresh gales, Torrington maintained his position between four and six leagues off Beachy Head, anchoring twice a day according to the tides but keeping the French in sight.

On 29 June, however, this defensive strategy was countermanded by orders from Queen Mary, drafted by Edward Russell and the Secretary of State, Lord Nottingham, requiring him to give battle. They believed Torrington exaggerated the size of the French fleet and did not believe it more powerful than the allied fleet. There was also an underlying fear that many in the navy were sympathetic to King James for his long association with it. Nevertheless, according to their orders, *Lenox* and the fleet weighed anchor at 3 am the next morning with the wind from the NNE. The French were observed on the starboard tack heading north and, as seen from *Lenox*, about three leagues to leeward.

Lenox was stationed in the centre of the fleet close to Torrington's flagship the *Royal Sovereign,* which, at eight o'clock hoisted the red or "bloody" flag to engage the enemy. The allied fleet heading north followed the French course and fell into line of battle. Both fleets now sailed on the starboard tack with the allies bearing down toward the French fleet, helped by the ebb tide. The fleets converged and the allies attempted to match the length of the French line, but because of their numerical inferiority gradually became overextended. *Lenox* opened fire; the deafening roar of her first broadsides would have shaken her fabric and, combined with the billowing smoke from the guns' vents and smoke blowing in through the gun ports, would have seemed like another altogether infernal world to the crew. The sheer physical effort of swabbing, reloading, hauling on the train tackles, aiming the gun with crows, firing and keeping away from the great guns rumbling across the deck was exhausting, especially when accompanied by a constant fear of serious injury or death. It was necessary to keep the line the same length as the French in order to prevent them from turning round the head of the allied line and firing their broadsides into the bows of the leading ships and surrounding them. The Dutch, furthest north, led the van, and a gap opened up between them and the English. Between 9 and 10 o'clock in the morning the leading ship, the *Wapen Van Utrecht* of sixty-four guns, appeared to close too early and opened fire on the French van about nine ships away from the head of their line. The leading unengaged French ships, about eight in number, then tacked round the outnumbered Dutch to encircle them and engage them on both sides, causing considerable damage as they fought at close quarters. *Lenox* recorded that "when we engaged Beachy Head bore NNW six leagues".

Not wanting the French to fill the gap that had opened up between the Dutch and the English centre Red squadron, and double back on him, Torrington controversially kept well to the windward of the enemy. By doing so he became more detached from the Dutch and unable to give them support. At about 11 o'clock Torrington controversially luffed even further away and steered south "so that our rear might not be overprest with odds". He appeared to offer support to the English Blue squadron in the rear, rather than the Dutch in the van.

The bulk of the English centre, which included *Lenox*, was not fully committed to battle by Torrington and seldom came closer than twice gunshot range, although Sir Ralph Delaval, commander of the Blue squadron, closed to within musket shot of the French rear opposing him. The wind then died away and a calm set in until mid-afternoon. The Dutch, fighting alone and isolated, were very much disabled and the French took one of their ships of about eighty guns. Torrington, with *Lenox* close by, was about a mile to windward of the stricken Dutch and out of effective range. At about four o'clock, with the wind strengthening and the flood tide finished he was at last able to relieve the distress of his allies and ordered them to come to an anchor. Then, with his own ship and several others, including *Lenox*, he drove between them and the enemy.

Having at last come to the aid of his allies, Torrington attempted to withdraw from the battle and save as many damaged ships as possible. Boats passed hawsers to damaged Dutch ships and *Lenox* managed to pass one to the stricken *Wapen Van Utrecht*, during which she lost her longboat. The *Wapen Van Utrecht* was then taken in tow by *Lenox*, and anchored with the rest of the fleet in thirty-seven fathoms of water as the tide ebbed at about five o'clock in the afternoon. The *Wapen Van Utrecht* was a famous old ship. Built by the Admiralty of Amsterdam in 1666, she was Cornelius Tromp's flagship on the last day of the Four Days' Battle and fought in all four of the great battles of the Third Dutch War.

The Dutch came close to destruction and their casualties well illustrate their plight. They suffered the loss of about 2000 men compared to 350 English and 350 French. Of twenty-two Dutch vessels, seventeen were dismasted, damaged, run aground or later destroyed by fire. The allies were in a serious state with the French in a position to deliver a decisive blow. But, as so often happened, fate turned and now favoured the allies. The French failed to take notice of the ebb and allowed it to take them away from the allies. Later that evening at 10 o'clock the flood tide set in again and the allied fleet weighed and retired slowly up the Channel.

At about 2 o'clock the next morning, 1 July, *Lenox* weighed anchor again to continue her progress up Channel, pursued by the French. It was impossible to keep towing the *Wapen Van Utrecht*, so to prevent her being captured by the French her surviving crew were taken aboard *Lenox* before she was cast loose and burned. The French also burned the Dutch ships they had taken. Torrington held a council of war during which it was decided that if the fleet were to be preserved it must retire to the eastward rather than continue the action. His plan was aided by the French, who wasted time organising a formal line of battle rather than committing themselves to a general chase, as they should have done after their first day's success.

Most of the ships in the fleet including *Lenox* had only been at sea for a short time, with hastily assembled crews who had not had time to become friends or comrades. Many men had deserted and their numbers had been made up with soldiers. It is perhaps not surprising that the inexperienced fleet had not performed as efficiently as they were later to do. In the circumstances, therefore, although defeated, Torrington did well to ensure his fleet remained a fighting force.

With the fall of night, *Lenox*'s ketch came alongside and took off all the Dutchmen, many of whom were sick or badly wounded. The opposing fleets remained off Beachy Head for another day, both working the tides but without making further contact. The wind then freshened from ENE and *Lenox* could only just carry reefed topsails. The French pressed the allied fleet further eastwards until they were off Dungeness. On the morning of the 3rd, *Lenox* observed a Dutch man-of-war run ashore to the east of Hastings. Later, during the afternoon, three more were seen going into Hastings Bay. Four French warships stood in in an attempt to take them, forcing the allies to set them on fire to prevent their capture. On the 4th in thick weather yet another ship was observed to the west on the shore and burning. Finally at 6 pm in the evening of the 5th the *Anne*, one of the third-rate ships of the thirty ship programme, already dismasted and taking in water, was run ashore and also set ablaze, a scene witnessed by *Lenox* as she lay anchored with the fleet to the west of Dover.

The whole of England feared an invasion but the French were unable to follow up their victory. They were short of provisions and gunpowder and their crews were now falling sick, due to the poor quality of their provisions. All Tourville accomplished as he sailed back down the Channel was the pointless burning of the fishing village of Teignmouth. Tourville's failure to follow up his success after the battle resulted in his dismissal from service as soon as he reached Brest.

With the departure of the French, the half-beaten allies retired towards the Thames. *Lenox* lost no men in the battle, though Nicholas Randall lost his right leg above the knee and was lucky to recover and receive a pension.[38] Many men also fell sick. After passing the Gunfleet, she sailed west to the Buoy of the Nore. On 9 July Torrington hoisted a union flag in his mizzen shrouds and fired a gun to signal final council of war. The next day he struck his flag and went ashore to Chatham; eight sick and wounded men followed him from *Lenox*.

Queen Mary, seriously embarrassed by the losses suffered by her Dutch allies, blamed Torrington for the defeat. He was taken and thrown in the Tower the moment he reached shore. Reports into the conduct of the action were ordered and Captain Granville of *Lenox* was "amongst several persons of quality sent ashore at Sheerness to examine the matter upon oath". Granville defended Torrington against allegations that he had not done his best to damage the enemy, stating that "half an hour after the signal for battle the Dutch engaged. Soon after the Earl of Torrington came to cannon shot. Between eleven and twelve we were at half cannon shot, then the French towed off to clear a Dutch fireship." After Granville's evidence, and the evidence of other naval officers who supported their admiral in his conduct of the battle, Torrington was ordered to be court-martialled.

As soon as the ships anchored at the Buoy of the Nore, work was urgently started to repair the battle damage so that the fleet would be able to sail again to counter any further French action. Shipwrights and dockyard officials from nearby Sheerness surveyed the damage; those ships needing new lower masts were sent into the Swale to be replaced. The fourth rate *Bonaventure* needed all her masts renewing, the *Hampton Court* had a serious leak near the stern where a harping plank was shattered, but perhaps the most badly damaged ship was the fourth rate *Woolwich*. Her mainmast was shattered, the stern post and the false post next to it were shot through at the water edge, while at the head, the gripe underwater was shot away, making her very leaky. She required docking and was sent up the Thames to Woolwich for repairs. The damage to *Lenox* was probably as slight as any other ship in that part of the fleet which saw the least action. All she required was a new longboat.[39]

The rest of July was spent getting the fleet ready. All the empty casks aboard *Lenox*, stored in tiers directly on top of the ballast, were taken up on deck to create space in the hold so that it could be cleaned, probably in an attempt to reduce the sickness that had broken out. The ketch was sent to London with the empty casks and twenty-nine sick men. It was not soon enough for one soldier who died the morning he was due to go ashore. While in London the ketch attempted to press men to replace those taken ill. *Lenox*, having lost her longboat during the battle, borrowed the longboat of the *Windsor Castle* to bring fresh water aboard before a new boat was supplied from Sheerness dockyard. The master gunner of *Lenox*, William Johnson, voluntarily surrendered his warrant and his son, George Johnson, took over the position, the warrant being dated 24 July.[40]

Lenox's gunner's stores were surveyed and the gunner sent aboard a store ship to obtain replacements for those used in the battle. On the night of 18 July the ship was pumped dry during which, Captain Granville recorded, one of the two lieutenants fell into the well. Whether it was Brittish or Durley we do not know, for understandably neither mentions the incident in their journals. Granville went on, describing "five of our people going in to save him. The stench of the bilge water took their breath from them, that when we got them up they were in a manner dead." But their troubles were not over, for the well-intentioned Granville had thoughtfully "sent for the doctor of the *Stirling Castle*, he cupped them and bloodied them, so that we hoped they would all recover."

Granville reasoned it was the stench of bilge water that rendered his men unconscious but it was in all probability methane gas generated by decaying organic matter; it must have been present in some volume to have had such an effect. The incident perhaps helps to explain why so many men were sick and dying, for their food and drink was stored in the hold in close proximity to the unhealthy well. A week after a batch of sick men was sent ashore, the longboat took another twenty-three sick to the hospital ship to recover.

In order to prepare *Lenox* for sea, Captain Granville was ordered by Sir John Ashby to prepare an account of the provisions she required to last her men three months. The list prepared by the purser went as follows:[41]

Bread. Nineteen thousand three hundred and twenty pounds.
Beer. One hundred and ten tuns.
Beef. Two thousand seven hundred and sixty two pecks.
Pork. Two thousand seven hundred and sixty pecks.
Peas. Eighty six bushels.
Butter. One thousand and thirty four pounds.
Cheese. Two thousand and seventy pounds.
Oatmeal. One hundred and twenty nine bushels.

For beef and pork this represents about 25 tons of each.

To improve the ship's sailing qualities she was heeled and her bottom tallowed. On the 25th, at 4 am while gales were blowing, the Admiral loosed his main topsails and fired a gun signalling for the fleet to unmoor. (A ship is moored when anchors are laid each side of the ship so that she does not swing with the tides, very important when space is restricted by other ships or nearby shores.) *Lenox* weighed anchor and moved the short distance to Blackstakes where she anchored again in twelve fathoms of water.

A review of the state of the fleet by Edmund Dummer, dated the same day, revealed that *Lenox* required a new knee for her bowsprit. The damage was probably caused by the weather and was nothing to do with the battle.[42] This is the first indication of a weakness in *Lenox*'s bows, a problem that may have been caused by her rather long head which may have increased the leverage of the power of the sea against its side.

Meanwhile, men continued falling sick; on the 29th two men died at noon, and at 8 pm that evening twenty-two more sick were taken to Chatham in the pinnace. The next day the two mortar pieces and their grenade shell ammunition were taken away in a pink. During the recent battle three of *Lenox*'s guns were badly damaged during firing – a demi-culverin (9-pdr) from the upper deck and two demi-cannon (32-pdrs) from the gun deck. They were replaced on 4 August and *Lenox* was ready once more to sail.

The fall of Torrington resulted in the fleet being put under the joint command of Sir Richard Haddock, Sir John Ashby, and Henry Killigrew. Killigrew's squadron had returned from the Mediterranean and arrived in the Channel only days before the battle and was now safely at anchor in Plymouth. The main fleet was to proceed to Spithead and transport John Churchill, the Earl of Marlborough, and 5000 soldiers to Ireland. They were to follow up the victory gained by King William at the Battle of the Boyne, fought one day after the Battle of Beachy Head. *Lenox* was now transferred from the Red squadron to the Blue and took down her red pennant to hoist a blue one in its place. On the morning of the 6th, Sir Ralph Delaval, Vice Admiral of the Blue, loosed his main topsails and fired a gun for his squadron to unmoor. *Lenox* hove short on one anchor and weighed, and the fleet set sail with the Blue squadron leading the Red. At 3 pm they reached the buoy of the Gunfleet where they anchored for the night.

It was decided to detach the Blue squadron with orders to go on a reconnaissance into the Channel, then to return a week later to meet the main fleet in the Downs. Accordingly the next morning the Blue squadron set sail alone. Ahead of *Lenox* was the *Breda* with men on her yards reefing the topsails as the wind freshened. Suddenly two of them lost their hold on the yard and fell 120ft into the sea. Immediately *Lenox*'s boat went to their rescue and saved the first man they reached; unfortunately the second man sank to the depths before they could reach him. Shortly afterwards at noon, yet another man died aboard *Lenox* from the sickness. The fleet anchored off South Foreland during the first night and at 6 am the following morning *Lenox* set her fore shrouds and backstays before setting sail again with the wind blowing a gale from the west.

With the prospect of meeting some of the victorious enemy, Sir Ralph Delaval signalled his squadron to fall into line abreast by hoisting a jack flag on his mizzen peak and a blue flag on his mizzen topmast head. He then sent the fast sailing *Charles Galley* and *Bonaventure* out scouting. *Lenox* saw a Danish ship and sent her boat to it and was informed that they had seen two French men-of-war to the west the day before. Off Dover the *Charles Galley* and *Bonaventure* returned, reporting that they had not seen the enemy but had taken an English pink found trading with the enemy laden with tar bound for France. *Lenox* took the opportunity to send some casks into Dover for water as the squadron plied to windward. They continued until the 13th and as the wind then increased *Lenox* could only sail under her topsails. The squadron continued plying to the westward uneventfully for another day before returning east to the Downs where *Lenox* anchored about two miles off Deal on the 17th.

Fifteen men fell sick during the cruise and were sent ashore in the longboat. Suspecting that much of the beer had turned sour in the hold, twenty butts were brought up on deck for examination. To avoid fraud it was practice for the master from another ship to come aboard to inspect alleged bad victuals. An independent master duly arrived and he agreed that eighteen butts of beer were bad; the beer was then poured across the deck and hosed through the scuppers into the sea.

Four days after arriving in the Downs, the Blues were joined by the main body of the fleet, including the Red squadron with Sir John Ashby in the *Sovereign*. Sir Richard Haddock joined him. While in the Downs Captain Granville wrote to the Navy Board pleading for press money, as he was using every means to obtain men to replace those sent ashore sick.[43] The money would be used to pay for the press gang, which was usually sent ashore under the command of a lieutenant operating from the ship's ketch or tender. The press gang did not normally consist of members of the crew but private hands paid for by the navy.

Granville did not have to wait long for an opportunity for, during the afternoon of the 24th, 200 merchantmen arrived, having just been escorted up the Channel by Killigrew's squadron. The longboat and pinnace were quickly manned and sent to take some of the unfortunate seamen. Setting sail again a day later the fleet was joined by five English and five Dutch warships of Killigrew's squadron near Dover. The combined fleet arrived at Spithead on 28 August. The sickness that had struck *Lenox* during July after the Battle of Beachy Head continued through August with the death of another seven men.

The fleet took on board provisions to last until 26 October, or about two months, for an expedition to Ireland. *Lenox*'s purser made out an account for her supplies, which were delivered by a hoy; unfortunately the beer and 15 tons of provisions proved to be bad and had to be rejected. More supplies arrived a week later when thirty-three cheeses, two tubs of butter and sixty bags of biscuit were delivered. Space was also made for three companies of soldiers and their officers who came aboard on the 30th to be transported to Ireland. Lord Colchester and other gentlemen visited *Lenox* on 7 September.

A few days later King William arrived at Portsmouth. "We fired 21 guns for joy at His Majesty's arrival safely in England." During the second week of September the three Admirals in joint command decided that with the approach of winter the large three-deck first- and second-rate ships, which were not so seaworthy as the smaller third rates, would be sent back to Chatham and laid up until the spring. The flag officers then transferred their flags into the third-rate ships, *Kent*, *Burford*, and *Hampton Court*. The fleet then weighed anchor and set off down-Channel, bound for Ireland and aided by a fresh NE gale.

Lenox's supply ketch brought forty-four men aboard just as the fleet sailed. The ketch was soon in trouble for during the afternoon, in hazy weather, while lying waiting for *Lenox*'s longboat, the *Elizabeth* ran aboard her and nearly sank her, stoving in all her upper deck and leaving her leaking badly. The *Elizabeth* was also damaged by the collision, losing her spritsail topmast and the knee of her bowsprit. The ketch managed to reach Plymouth for repairs while the fleet steered west toward Land's End. By 19 September they were off the Scilly Isles, where there would be ocean swells and little use for the longboat that *Lenox* was still towing. It was therefore hoisted into the waist to avoid damage or loss.

After an uneventful voyage the fleet arrived off Cork on 21 September. Cork was still held by the Irish fighting for the cause of King James. On arrival the flood tide was found to be over and without its help the pilots advised against venturing in. The three admirals therefore held a council of war and a plan of attack was made for the next day. Morning proved calm and at eight o'clock the ships weighed anchor. With the help of their boats towing and the flood tide they entered the harbour. The *Kent* led as flagship with the union flag at her masthead, and was followed by the *Grafton* with *Lenox* lying sixth in the line. The fleet entered along the west side, and at 11 am the *Kent* engaged the first fortification protecting the harbour, from where several guns fired from a small platform. As *Lenox* passed she "fired very briskly beating it down about them so they ran from it". Several boatloads of men were sent ashore to secure the place. In it they found nine guns that had been dismounted and the carriages thrown into the sea. The king's colours were then hoisted at the highest place. Some Irish without colours continued fighting from some nearby houses. The landing force attacked and drove them from the houses before setting the buildings on fire.

The fleet continued its progress into Cork harbour without further incident, and by noon *Lenox* was safely at anchor in nine fathoms of water. Later she sent her boat to take Haulbowline Island, a small island with a castle on it, which guarded the entrance to the Passage at the mouth of the Lee River on which Cork stands. "The Irish ran from it and our men took it and hoisted the King's colours on the castle. In it there were about 11 guns and a small platform below which had four guns in it." The harbour was now secure and at 6 o'clock that evening *Lenox* was safely moored.

The reduction of Cork now began. At midnight the signal was made for the soldiers in *Lenox* and most of those from the other ships to take to the ketches, yawls and longboats. They were to be taken up the Passage, where large men-of-war could not enter, escorted by the *Owners Love* and *Salamander* fireships and landed six miles upstream from Cork. The next night several hundred seamen, gunners, and ships' carpenters were sent ashore from the fleet to set up cannon for battering the town. *Lenox*'s contribution consisted of her second lieutenant, Edward Durley, a carpenter's mate, a gunner's mate and fifteen seamen. They were given two days' provisions to take ashore with them. The following day some of the smaller ships of the fleet, the *Deptford* and the *Charles Galley* supported by several fireships, went close to the town to add their firepower. The bombardment continued until the 29th when the mortars caused explosions believed to have killed many of the defenders. At 3 o'clock in the afternoon Cork surrendered to the army and by 8 o'clock the next morning the victors marched into the town. The crew of *Lenox* took the opportunity to heel the ship and tallow her bottom.

With their main task completed, King William ordered the fleet to return to the Downs, leaving the third rates *Lenox*, *Breda*, *Grafton*, *Warspite*, one fifth rate, one sixth rate, the *Owners Love* and *Salamander* fireships behind, and a tender for each. As the fleet departed, the *Burford* ran aboard the *Defiance* causing her to cut her cable and briefly run aground. The remaining squadron was placed under the command of the Duke of Grafton, perhaps the most capable of King Charles II's illegitimate sons. He flew his flag in the *Grafton*, a third rate of the 1677 programme named after him. On 4 October the *Charles Galley* came in with a prize, while the *Breda* weighed anchor and moored to a position about 300ft from *Lenox*. Then, suddenly, the gloss of victory dissolved. On the 8th news was received of the death of the Duke of Grafton, killed by a musket ball as he lead some seamen ashore to work the cannon. Two days later all the ships fired their guns in respect to the Duke, with the *Grafton* leading. *Lenox*'s longboat returned from the Passage and Captain Tennant of the *Breda*, who had earlier in the year illegally pressed one of *Lenox*'s seamen, inherited the position of Commodore of the squadron and hoisted his broad pennant. Early in the morning of the 12th, the body of the Duke of Grafton was taken aboard the *Grafton*.

Shortly afterwards at 11 o'clock an even bigger disaster occurred when the *Breda*, moored so close to *Lenox*, suddenly blew up and was engulfed in fire. The explosion blew off *Lenox*'s topsails and yards, and flying debris destroyed the bulkhead of the great cabin. *Lenox*'s master, Richard Taylor, marvelled that *Lenox* was not seriously harmed: "it pleased god we had no damage". At the end of his journal for the period, he wrote, "Parts of my journal for June, July and August were so fouled with filth that came

Above: *The entrance to Kinsale from Charles Fort. The opposite shore was protected by James Fort. (Author's collection)*

in at my cabin when the *Breda* blew up I could not copy it". Only about seven men from a crew of more than 400 survived the destruction of the *Breda*. They were rescued and taken aboard *Lenox* to be entered on her books. Captain Matthew Tennant was found alive in the water and also taken aboard, but he had been mortally wounded and died a few hours later. *Lenox* lowered her colours in respect for the dead.

It is always assumed that explosions of this nature were caused by an accident in the powder room where many tons of gunpowder were stored. Although we will never know, it is possible the cause of the initial explosion was the detonation of methane gas in the confined area of the well. Captain Granville had already described how his men were rendered unconscious when they entered *Lenox*'s well. The problem was addressed by orders to throw away the upper ballast and clean the rest by heaving it ashore for airing when the ships were docked.[44] There were other precedents for this type of disaster and there were would be more to come; the *Exeter* was to blow up in similar fashion the following year.

At 10 o'clock the next morning, the body of Captain Tennant, accompanied by all the boats in the fleet, was taken to Long Island, a small island in Cork harbour, where he was buried. To honour him, three volleys of small shot were fired over his grave. *Lenox* fired twenty guns at a minute's interval. After the ceremonies, the remaining commanders conferred to decide who should be the third officer to hoist the broad commodore's pennant in the space of a week. The post fell to the eldest captain as of right, which happened to be Captain Crofts of the *Charles Galley*. In spite of these disasters, however, the military situation proceeded

satisfactorily and news was received that Kinsale had surrendered upon condition the defeated enemy could "march out with bag and baggage". Kinsale was soon made into a base for the fleet and would be a valuable asset for the rest of the war. On 18 October, *Lenox* made ready to return to England. Expecting bad weather, the gun ports were closed and caulked against the entrance of seawater. Then the fore topsail was loosed and the sheets hauled home, the best bower anchor was weighed and the ship hove short on the small bower. But the underwater wreckage of the sails and rigging from the tomb of the *Breda* fouled the nunn buoy and its rope (a buoy seized to the anchor to mark its position). After a great amount of difficulty it was decided they could not be untangled and were cut away.

Death was certainly in the air; on this day the sickness aboard *Lenox* reached its climax and claimed three more victims. At 12.30 in the afternoon, the Earl of Marlborough and Lord Colchester boarded *Lenox* to join her for the return voyage. Next day at midday, as final preparations were made to sail to England, twenty-five Irish officers were taken aboard as prisoners.

Lenox and most of the ships at Cork finally left and reached the open sea on the 20th. With a good wind from ENE they sailed south towards the Channel on the port tack, with all the reefs out of the topsails. *Lenox* was soon off Land's End but, sailing as close to the wind as she could, was obliged to continue on her course and head away from the shore, taking in two reefs from the fore topsail and one reef from the main topsail as the wind strengthened. At 4 o'clock in the morning of the 21st she tacked round to head N by W. The wind continued blowing from the east, at times increasing to gale force and obliging the crew to take in and let out the reefs of the topsails, and sometimes handing them, or taking them up completely. At the same time they tacked away from and then toward the shore. As the days passed and little or no progress was made the conditions on board worsened. To lessen the drift to leeward the longboat, which was being towed, was cut away and lost. Finally, on the 26th, as the Lizard bore north at a distance of six leagues the wind at last veered round to blow from WNW, and by the end of the next day *Lenox* was at anchor in the Downs. Captain Granville immediately wrote to the Admiralty informing them of his arrival and requesting time ashore as he himself had now fallen sick.

On 28 October, at 2 pm, Lord Colchester went ashore and *Lenox* fired a fifteen-gun salute. He was followed an hour later by the Earl of Marlborough who received a twenty-one-gun salute. A letter arrived from the Admiralty giving the captain permission to return to town to recover his health and conveying orders for the first lieutenant to bring the ship into the Quinborough Swale using all possible care.[45] That evening the Commodore in the *Charles Galley* arrived and anchored close to *Lenox*. For *Lenox*'s men it had been a terrible time as the illness claimed thirteen lives during the month.

On the last day of October *Lenox* and *Grafton* weighed anchor to make

their way to the Buoy of the Nore. The voyage involved negotiating the dangerous sand banks protecting the entrance of the Thames. The crew experienced difficulty working the ship in the confined channels and to get her in better trim they filled the empty cask with seawater while the gunner moved all the shot from the quarters down into the after powder room. After safely negotiating her passage she anchored at the Nore, where the guns were fired in celebration of 5 November. Her Irish officer prisoners were then sent to London aboard a yacht. They were lucky to be aboard *Lenox*, for those aboard *Northumberland* complained of losing several of their possessions to officers and seamen of the ship. The Admiralty immediately ordered their belongings to be returned and regular reports concerning the prisoners to be sent to the Admiralty.[46] *Lenox* was now made ready to be refitted during the winter; her guns were removed until only two remained in the hold, while many sick men were sent ashore at Chatham. On 1 December a pilot came aboard and Captain Granville, who had rejoined his ship, left to attend a court-martial of officers concerning the loss of the *Breda* and *Dreadnought*, and the conduct of the Earl of Torrington at Beachy Head.

The proceedings began a week later on 8 December when twenty-two captains, Rear Admiral Rooke and Vice Admiral Sir Ralph Delaval, who acted as president, convened aboard the *Kent*. The first officer to be

Below: *A model of the* Grafton, *a 70-gun third rate, launched at Woolwich Dockyard in 1679. (United States Naval Academy Museum. Photographer: Dr. Richard Bond)*

court-martialled was Captain Robert Wilmott of the *Dreadnought,* whose ship foundered through her leakiness. The court found that he used all possible means to preserve his ship and was acquitted. The next case concerned the destruction of the *Breda* in Cork harbour at which John Granville would have been able to give a graphic description of the horror that took place. The surviving senior officer of the *Breda,* First Lieutenant Wentworth Paxton, was court-martialled as a matter of course. At the time of the explosion he was ashore, as had been Edward Durley, serving the guns. The court found that "he was no ways an accessory" and cleared and acquitted him.[47] The next case was to be the Earl of Torrington's but some important witnesses were not present and the case was adjourned for two days.

The composition of the court when it reconvened was much the same as it had been before but with the addition of five more captains. Nearly all, like Granville, were Torrington's friends. In his account of the battle, Torrington held that the Dutch had brought disaster upon themselves by attacking the enemy before his Red squadron was in a position to offer the necessary support. The Dutch officers attending were incensed by the attitude of Torrington and his supporters. The proceedings degenerated into personal attacks during which the Dutch Vice Admiral Schey lost his temper and grabbed one of the captains by the coat. Another witness, who began to testify against Torrington, was challenged to a duel by one of the presiding captains. John Granville became involved in the exchanges when he challenged Schey to a duel. Delaval later wrote that the Dutch gave their evidence with too much heat.[48]

Torrington was unanimously acquitted but King William supported his countrymen and Torrington was dismissed from his offices and never employed again. William must also have felt that Dutch naval officers would never trust Torrington if he remained in command. His removal was therefore necessary if the Anglo-Dutch alliance was to hold together. Not surprisingly records of his incredible court-martial did not find their way into the court-martial record book.

Edward Russell later remarked that "it seemed odd for the French Admiral to be dismissed for not destroying the English fleet and for the English Admiral to be dismissed for not allowing it to be destroyed."[49] In arguably the most convincing account of the battle, made before his trial to the House of Commons, Torrington defended his position, saying "Most men were in fear that the French would invade, but I was always of another opinion, for I always said that, whilst we had a fleet in being they would not make the attempt."[50]

At the end of the year's campaign, forty-two other officers suspected of too close an association with Torrington were also dismissed.[51] High on the list was Granville. His close association and aggressive support of Torrington at his trial resulted in his never being employed again in the navy. His departure was certainly a loss, for although an aristocrat he was not merely a gentleman captain. From his journal it is clear that he calculated his ship's position and took an interest in the well-being of his men. His letters and journal are well written and full of authority concerning naval matters. It can be said that he lost his position for loyally supporting his admiral. Granville appears again in 1692 when a committee of Parliament recommended that "a commission of the Admiralty be of such persons as are of known experience in maritime affairs". The House, however, amended the resolution and deleted the reference to persons of experience in maritime affairs. Pepys wrote of the affair, "And herein also is to be noted that, at the time of the House's so unvoting their Committee's advice, none are said to have spoke so boldly against the present Admiralty Commission as Captain Churchill and Captain Granville, who of all the commanders of the fleet seem least furnished for so doing, though the thing in itself be most reasonable".[52] Granville represented Cornwall in Parliament, and in 1703 was created Baron Granville of Potheridge in Devon. He died in December 1707.[53]

Meanwhile preparations continued for *Lenox*'s refit. The pilot took *Lenox* into Gillingham Road where she was moored at the lowermost mooring. She was then warped to the upper mooring and later to a mooring against the dock. Once secure the topmasts were unrigged, the running rigging taken down, and the anchors taken ashore. A week later she was warped to the mooring against Chatham Church where her ground tier of cask was taken out followed by the ballast. The sickness claimed ten more men during November and six more in December. Altogether forty-six men died in 1690; almost certainly the bad quality of victuals was to blame. The sickness affected the whole fleet, and indeed the French fleet as well. The Navy Board, however, absorbed the lesson of good supplies and never again would *Lenox* and the fleet suffer deaths on such a scale. It is perhaps worth reflecting that if the 440 men of *Lenox* had stayed ashore, probably about fifteen would have died anyway during a typical year of the late seventeenth century.

Another problem for the crew of *Lenox* concerned their pay. It had been the custom in previous wars to pay off crews when their ships came in for their winter refit. Now, with the Treasury short of money, they would not be paid until the beginning of the following summer's campaign. The Navy hoped it would have the benefit of making it more likely that the men would report for duty for the next summer campaign and be available to help in refitting their ships in the winter. The men were allowed to go ashore on leave in turn, but they deeply resented the change to the ancient custom and wanted the opportunity to take a few months' leave to see their families, recover their health and spend their money. Thirty-two men responded to the Admiralty action by deserting the ship during December and thereby forfeiting their pay.

1691 – THE LOSS OF *CORONATION*

Lenox and the rest of the Anglo-Dutch fleet prepared for their end-of-season refit and made ready for the coming summer under Edward Russell. Russell had been appointed Admiral of the Fleet and now replaced the three admirals. An aristocratic Whig, Russell was a heavy drinker like Torrington before him. He had gained a reputation for roguish behaviour in earlier days under the Pepys administration. He was now thirty-eight years old and had, like Torrington, accompanied William to England in 1688.

It was hoped to prevent the French squadrons at St Malo and Dunkirk joining their main fleet at Brest and thereby head off a body of French troops in nearly 200 ships expected to sail to Ireland in support of James II. Allied merchant fleets also required protection from French privateers and small warships.

At the turn of the new year, the last of *Lenox*'s ballast was removed aboard lighters, leaving her ready for docking. The ballast of *Lenox* was almost certainly cleaned following the illness aboard the ship and in line with recent orders. A practice, as applied to the *Anglesey* in 1694, was to send all the provisions ashore, throw out all the ballast, scrub and tar the inside then wash the decks with vinegar.[54] *Lenox,* with all her guns, upper masts, stores, and ballast removed, was ready to be taken into dry dock. On 6 January the *Kent* and *Grafton,* having completed their refits, were launched at 1 o'clock in the morning. The same day the three-decker *Neptune* was docked upon the ways and the *Hope* and *Lenox* taken into the double dock and the gates shut behind them.[55]

There was a severe shortage of dry docks and work was carried out round the clock using artificial light, seven days a week, in order to refit the fleet during the winter. For ships in good condition, all that would be required was cleaning and tallowing, a task normally completed in a day. The work on *Lenox,* however, lasted ten days – possibly her old deal sheathing that protected her bottom was stripped off and her planking repaired and recaulked before new sheathing was nailed in place. After the work was completed on 18 January, the double dock was filled with water so that *Lenox*, in the stern, could be floated out and the *Expedition* docked in her room. Unfortunately the spring tide failed and *Lenox* could not be refloated for another two days, to the consternation of the newly knighted Sir Edward Gregory.[56] While *Lenox* was in dock Sir Edward wrote, "but in good faith had not the King surprised me into an astonishment by a sudden and unexpected command to kneel".

Having finally been launched, *Lenox* was taken back to her moorings against Chatham Church. The first load of ballast was then taken down into the hold and by the end of the month her sails, cables and anchors were all aboard.

John Munden was commissioned as *Lenox*'s new commander on 2 January in place of John Granville. He was a kinsman of Sir John Munden, another naval commander, although not an immediate relative. He was first commissioned as second lieutenant in 1677 and in 1688 was commander of the sixth rate prize *Half Moon*.[57] Charles Brittish, the old first lieutenant, left to further his career and would himself reach the position of captain by mid-1692. The second lieutenant, Edward Durley, received a commission dated 28 January promoting him to first lieutenant; Francis Gregory was appointed second lieutenant and Robert Yelverton replaced Richard Taylor as master.

During January proposals were made by the Admiralty to change the conditions of sea officers and make them more professional. The number of personal servants allowed was reduced but their pay was doubled in compensation. The number of lieutenants for a third rate was increased from two to three.[58] The proposals were approved and George Knitch became *Lenox*'s third lieutenant. The increase in pay made the position of sea officers very attractive and is illustrated by Captain Munden's attempt to forward a favourite of his own, one Nathanial Smith, to the position of master, in spite of the fact that the Admiralty had already appointed Robert Yelverton to the position. Munden had Smith write a testimonial:

"The occasion of this bouldness is urged by my small encouragement that I have meet with in our Majesty's service. When commander of a good merchant ship unfortunately taken into Algiers I thought never might I have the occasion to trouble your Honours with this impertinency. But since my coming thence with no small ransome have never been out of the navy, which is nine years since in the quality of Chief Mate under the command of Captain Newell in the *Rupert*, *Garland* and *Crown*, and under Captain Clemens in the *Cambridge*, and since at the request of Captain Clemens to Sir Richard Haddock was by his letter sent up from Portsmouth in order to be his Master in the *Expedition*, but being disappointed had a warrant for the *Berkeley Castle* where through gods [sic] assistance have carefully defrayed that charge. My request to the Honourable Board is that I may have some care taken of me and that I may not be forced to go volunteer, only with which I will not find bread for a family thats too large…"

Munden sent Smith to the Navy Board with this testimonial and a covering letter: "This bearer Nathanial Smith is of my Acquaintance and at this time out of employment. I desire that you will please to make him Master of my command the *Lenox* for I hear that the late Master Taylor is not in a condition to proceed the voyage".[59] Munden's letter is dated 6 February and was written aboard *Lenox*; he had clearly chosen to ignore his appointed master, for Yelverton had been aboard *Lenox* since at least 2 February recording daily events in his journal. Even more surprising is another letter Munden sent to the Navy Board dated 11 February: "The master you have allotted me [Yelverton] has appeared to me once and showed me his warrant, but where he is, or when he will attend his duty I know not, neither do I know him so have naught to say against him, I hearing his absence is caused by sickness, if it be possible that he may be provided elsewhere I pray that Mr Smith whom I know may be made Master and I shall acknowledge it as an urgent favour…". Munden's

intrigues proved to no avail – Yelverton remained master, hopefully unaware of the plots made behind his back, while Nathanial Smith, by his own account a very well qualified seaman, entered his name as a volunteer on *Lenox*'s books.[60]

Other seamen were not as willing to join *Lenox* as Smith; one of the first acts by Munden was to apply for press warrants.[61] Soon after, the press gang brought twelve men in the smack while the crew were busy bending cables to the anchors. Many press expeditions were made by her ketch the *Amity* and her long boat. The *Amity* would have been a hired vessel used to supply *Lenox* with provisions in home waters; normally one smack or ketch was provided for the use of two third-rate ships the size of *Lenox*. Early each year, when there were provisions and men to be supplied quickly, both a ketch and a smack were sometimes provided, as they had been the year before and were again now. The press gangs proved to be over-enthusiastic and Captain Munden began to receive letters of complaint from the Admiralty. He was ordered to discharge James Canon, "a coachman to John Buggin Esq.", and later "Abraham La Voste a shopkeeper and no seaman who you imprest to serve aboard your ship, you are forthwith to discharge him…"[62]

Munden was not the only captain to make the odd mistake, although no one reached the heights of the late Matthew Tennant.

Lenox was then ordered to Blackstakes to complete her readiness.[63] On 11 February *Lenox* was unmoored and warped down to moorings against the dock, supervised by the Master Attendant of Chatham. A few days later she was again unmoored and warped or "transported" lower down to Upnor Castle. It had snowed for days and must have been very cold and bleak for the new men coming aboard. The ketch came alongside with fourteen volunteers and pressed men while it continued to snow. The ketch only stayed until the evening, before taking aboard three weeks' provisions and sailing out to Deal with the captain's clerk in command of the press gang looking for merchantmen. They returned a week later with ten more men to find that *Lenox* had been moved to Gillingham. Yelverton the Master explained: "This noon small drifting rain and at half flood we carried out our stream anchor and three hawsers and transported our ship from Upnor Castle to Gillingham."

Captain Munden was having problems with the state of *Lenox*'s pay books, as they were becoming rather complicated. He blamed Granville in letters to the Navy Board, saying that the previous captain had not brought the books up to date and that several men made "run" in the books had in fact been discharged by ticket at London.[64] A day later on the 25th the Master Attendant came aboard again to supervise the warping of *Lenox* to Blackstakes, helped by boats from the *Neptune* and *St Andrew*. By the evening they had reached Short Reach were they anchored for the night. The next day the wind veered to the south and *Lenox* was able to set sail together with the *Hope* and reach Blackstakes by six in the evening. She joined *St Michael*, *Ossory*, *Grafton*, *Exeter* and *Eagle*. The guns were now brought aboard by lighter while other warships arrived from Chatham to join them, including the three-deckers *Royal Sovereign*, *Duke*, *Albermarle*, *Coronation*, and the third rates *Burford*, *Expedition*, *Stirling Castle* and *Restoration*.

The first week of March was spent taking in sea provisions, including 260 barrels of gunpowder. On one of the press raids the *Amity* suffered a mishap when she broke her yard as she turned to windward. The Carpenter of *Lenox* made another out of a main studding sail boom.

On 7 March at 6 am *Lenox* started to unmoor and by eleven had weighed anchor and set sail, together with the *Eagle* bound for the Nore where they anchored at five that afternoon. They joined a squadron of thirteen third rates under Sir Ralph Delaval in the *Essex* flying his blue flag at the fore topmast head. The next day Captain Munden had the frustration of seeing a squadron of six of the third rates under Delaval despatched to prevent French ships sailing from Dunkirk and joining their main fleet in Brest. *Lenox* would probably have been a member of the squadron had she the men. To redeem matters Munden sent the *Amity* to London for more men, while the ketch brought thirteen more from Deal. Captain Munden wrote again to the Navy Board concerning the pay books. The men "that are made run by Sir Edward Gregory have had leave from Colonel Granville, and as they were returning to their ship was taken up [pressed] by one tender or other which caused their so long absence. I therefore pray that you will order their 'R' to be taken off." For the rest of the year the only letters Munden wrote concerned the state of the pay and muster books.

Mid-March came and *Lenox* still lay at anchor at the Buoy of the Nore with the gathering fleet. The opportunity was taken to scrape her blocks and masts and black her yards. The sides of the ship were scraped clean, then payed with rozin, while the wales were tarred black. New sails were bent to the yards and the shrouds set up. She was still short of men and when a merchant fleet of twenty-seven sail of Hull merchant ships was sighted in a fog she tried to close in by warping. Evening came and in the gloom she failed to find them. Most of the officers managed to spend some time ashore. George Churchill arrived and became Commodore; he was still in the second rate *Windsor Castle*, the ship in which he served the previous year.

As the ship transformed into a fighting man-of-war the monotony was broken with incidents typical of the seventeenth-century navy. Robert Yelverton had another problem and described in his log how he fell out with Lieutenant Durley. As Durley came on deck at 8 pm one night, he found no master's mates on duty and, in the ensuing argument, Durley not only berated the master but "confined me to two hours with two sentinels at my door and ordered them if I came out to cut me down." Yelverton attempted to complain to Captain Munden but the captain, perhaps wisely, would not see him and the matter was dropped. On another occasion, an unfortunate whore was caught onboard and ducked three times from the main yard, no doubt to the amusement of the crew, before being returned to shore. What she did to deserve this treatment is not recorded. She probably came aboard as a "wife" but was spotted by an officer in a few too many different beds. Nathaniel Smith the volunteer also managed to spend some time ashore, going to London in *Lenox*'s jolly boat.

On 11 April all the ships fired fifteen guns each for Coronation Day, followed two days later by fifteen more as King William passed upriver from Holland in the *Mary* yacht. More royalty arrived in the form of Prince George of Denmark, who visited the fleet and was given three cheers. More ships arrived to join the fleet as preparations were made to sail to the Downs. A pilot came aboard *Lenox* to take her there.

On 19 April *Lenox* hoisted her red pennant to the mizzen topmast head as a private or "common" ship belonging to the Rear Admiral of the Red, George Rooke's, division. The men aboard *Lenox* were now due to be paid for their previous year's service. On the 21st the Commissioner, Sir Edward Gregory, and the Clerk of the Cheque from Chatham arrived to pay the new volunteer seamen their conduct money. Commissioner Sir Richard Haddock followed a few days later and paid the men their wages up until 1 October 1690. A few days later, with a much happier crew, *Lenox* loosed her fore topsail and sailed eastwards through the Gunfleet and arrived off North Foreland on 28 April. She was accompanied by eight third-rates and two fireships.

Lenox prepared to anchor in the Downs but Sir Ralph Delaval, Rear Admiral of the Blue, was there with the squadron that had left the Nore at the beginning of March. He had orders to blockade the French fleet in Dunkirk. Delaval ordered *Lenox* and her squadron to fall down as low as the ebb tide would take them. They managed to reach South Foreland before anchoring as the tide turned. The next morning, the combined squadrons weighed and sailed south and formed into a line of battle as they arrived off Dover.

The composition of the English fleet at the time was recorded by Yelverton as follows:

Kent*	*Dolphin* (fireship)
Dreadnought	*Expedition**
Swiftsure	*Northumberland**
Cambridge	*Centurion*
James Galley	*Stirling Castle**
York	*Captain**
Berwick*	*Spye* (fireship)
Resolution	*Wolfe* (fireship)
Monk	*Plymouth*
Monmouth	*Suffolk**
Happy Return	*St Albans*
Hampton Court*	*Lenox**
Hope*	*Speedwell* (fireship)
Defiance	*Grafton**
Woolwich Hound (fireship)	*Warspite*
Essex*	*Edgar*

* 70-gun third rates of the 1677 programme, the largest ships in this fleet.

The weather was good enough for the crew's hammocks to be put into the fore and aft netting for airing. The wind then died away completely and the fleet remained at anchor for a whole day, before returning to carry them to Dunkirk. The *Centurion* and a fireship were ordered by Delaval to close in and observe the harbour and they reported that about thirty ships were lying there with more in Flemish roads protected by guns. Next day, 3 May, Sir Ralph Delaval and several captains armed their boats and went to take a view of the ships in Gravelines, returning to *Lenox* as a gale sprang up at about midnight. *Lenox*'s yawl, which was now being towed behind *Lenox*, went missing when it suddenly broke loose and was lost. With morning some men were sent up to the topmast head to try and spot it. It was sighted and the pinnace with a spare mast and sail was sent to help; unfortunately the strongly running tide drove the pinnace under the yawl and sank her.

The weather turned dirty on 7 May as they weighed anchor and allowed the ebb to take them down Channel before lying still for a day during a gale. The squadron then sheltered off Deal Castle in the Downs on 9 May, where *Lenox*'s ketch went in for water. As a precaution against the bad weather, the spare anchor was taken up from the hold and a stock put on it ready for use. *Lenox* remained off South Foreland until the 12th, when she joined the *Hampton Court*, *Hope*, *Warspite*, *York*, and two fireships as part of a squadron under Captain Churchill, acting as Commodore in the *Kent* with orders to return to cruise off the French coast. They sailed back across Channel, arriving at a position two leagues off Gravelines the next day. On the 18th, at 9 o'clock in the morning some unidentified ships were sighted and *Lenox* and the *York* gave chase. At about 12 noon they made them out to be two French privateers, one of twenty guns and the other of ten. *Lenox* reported that "they outsailed us, we followed as far as Calais where they anchored inside the land so that we could not damnify them with our guns, there being but three fathoms upon the sand at high water". At 10 pm that night they gave up and sailed back to Gravelines and anchored in the middle of their squadron at four in the morning. On 19 May, *Lenox*'s yawl had some success when she chased and plundered a boat. The following afternoon the *Hampton Court* and *Hope* sailed to cruise to the eastward. *Lenox* herself stayed off Gravelines one day more before returning to Dover on the 21st, where her boats had to tow her into the road as there was so little wind. Her ketch brought her thirty tuns of water and took all her empty casks back to the harbour. *Lenox* set sail again at 3 pm the same day, and by the following dawn was once more anchored near the Commodore in eighteen fathoms of water off Gravelines. The same day the *Hampton Court* and *Hope* returned from their cruise. Over the next two days, raids were made by the ships' boats and several French fishing boats were seized. The squadron then slowly made its way down Channel and was off Calais on the 25th. Having completed its task, the squadron left the French coast and sailed north into the Downs to join the fleet. *Lenox* discharged her pilot and the ship was heeled and her bottom scrubbed while the sides were payed with rozin and tallow.

Lenox and the Delaval squadron had been successful in preventing most of the French ships in Dunkirk sailing and joining their main fleet at Brest. The English fleet, now combined with the Dutch under Russell, set sail on 9 June, heading down Channel in line of battle hoping to meet the French fleet from Brest. The French fleet had orders to avoid battle but destroy or capture the convoys that were the lifeblood of English and Dutch trade.

At 3 am Russell loosed his fore and main topsails and fired a gun for the whole fleet to unmoor. Two hours later they were ready to sail in their squadrons. The Dutch weighed first to form the van or White squadron, as they had the previous year. The English Red and Blue squadrons followed them. At 3pm the fleet anchored in fifteen fathoms off South Foreland as an adverse wind from the SW slowed the fleet's progress. Two days later they had still only reached as far as Beachy Head. The weather turned dirty and *Lenox* and the rest of the fleet sailed under double-reefed topsails. They anchored with each flood tide and by the 13th were off Dunnose where John Foley, a seaman of the boatswain's storeroom, died at 3 o'clock in the morning. Following normal practice he was buried at sea two hours later. Continuing westward the fleet was off the Start by the 17th. The constant anchoring with *Lenox*'s best bower caused the outer end of its cable to wear badly as it dragged on the sea floor. The outer end was cut off and a spare cable spliced to the anchor. The fleet called into Torbay for relief from the adverse weather and from there the *Lenox*'s ketch was sent into Plymouth for water.

The weather at last turned fair and the fleet set sail again on 23 June; Munden and the other ships' captains were ordered to go aboard Russell's flagship, *Britannia*, where they learned that a small ship from Copenhagen had sailed into Plymouth and reported sighting the French fleet off Ushant just six days earlier. The English fleet accordingly fell into line of battle and stretched across towards Ushant hoping to meet the enemy and at the same time provide protection for an incoming Smyrna merchant convoy bearing valuable cargoes from Turkey. For the last week of June they lay eight leagues off Ushant waiting for news.

During the afternoon of 3 July the three-decker *St Michael* caught fire, causing panic among her men. Fire was dreaded among sailors, as ships were highly combustible environments and contained many tons of gunpowder. Moreover it is believed that not many of the men could swim and while some of the crew fought the flames, others jumped into the sea. Those who stayed aboard soon put out the fire, but five men and two boys who did jump were drowned before the ship's boats could rescue them. The next day the fast-sailing *James Galley* came into the fleet with an account that the French fleet had been seen near Baltimore off the southern coast of Ireland on the 23rd of the previous month. Hoping to meet them the fleet turned northwards during an electrical storm that lasted two days. As the fleet neared Ireland, Yelverton in his log drew attention to the typical dangers faced by ordinary seaman, mentioning that "a man fell from the main yard into the longboat and there hung on the tackles and bruised himself but not killed him". After a three-day voyage the fleet arrived off Cape Clear but the French were nowhere to be found.

Learning that the Smyrna fleet had arrived safely at Kinsale, Russell took the fleet there to provide an escort for the final stage of its voyage. On 10 July they were offshore by Old Head and soon saw some of the Smyrna fleet coming out of Kinsale. It began to blow hard and some of the merchantmen were forced back in. *Lenox* then witnessed the *Adventure* come into the fleet firing her guns and with her top gallants flying in the traditional signal of enemy in sight. She reported seeing the enemy to the southward in latitude 48 degrees and that some of the French fleet chased her as far north as 49.30 degrees, indicating that the majority of the French fleet was still in the latitude of Brest. Russell consulted with his flag officers and decided to stay with the Smyrna ships. A gale blew up from the WSW and *Lenox* double-reefed her topsails and stood off the coast. Several ships split their topsails, the *Victory* broke her main topsail yard in the slings and two fireships carried away their main topmasts. On the 12th the wind eased and veered to the west, allowing the Smyrna ships left in Kinsale to depart. Russell was now able to sail with the combined naval and merchant convoy to safety past the Scillies, before returning once again to his cruising station off Ushant.

Lenox reduced her men's rations from six to four men's allowance to conserve her provisions. On 20 July Russell, in hopeful anticipation, formed a line of battle and ordered Sir Cloudesley Shovell with a ship from each squadron to approach Brest to see if the French fleet was there. The next day Shovell returned with four French barks he had taken captive and told Russell the French fleet lay forty leagues WSW from Ushant. Russell immediately set sail with the fleet to meet them and a few days later the intelligence was confirmed when the *Coronation*'s boat captured a small bark loaded with fresh vegetables. The French sailors informed their captors that their orders were to sail on a further twenty leagues if they did not find their own fleet. Soon after, ten more sail were seen and several ships gave chase, managing to capture four of them before nightfall. All the captures contained fresh green provisions for the French fleet. The actions of the French supply vessels seemed to indicate that when they first saw the allied fleet they mistook it for the French. The conclusion drawn was that the French fleet had been in the area but had now withdrawn and wished to avoid battle.

At dawn on 29 July *Lenox* found herself to windward of the fleet with her main course and mizzen set as the wind blew a hard gale, creating very great seas. To regain her position she bore down towards the flagship and in "bearing down the men at the helm let fly the whipstaff out of their hands". The 12ft ash pole that pivoted in a rowle held by a retaining scuttle flew sideways to its extremity and jammed. Out of control, *Lenox* was brought by the lee and could easily have been dismasted but "blessed be God we received no damage". The incident illustrates the difficulty of containing the tremendous forces exerted on the whipstaff through the rudder in a rough sea from astern. It was liable to snatch the whipstaff left or right out of the hands, even though a number of men were holding it. The problem was not solved until the introduction of the steering wheel early in the eighteenth century. In search of the enemy, the fleet had ventured far out into the Atlantic but the French, wishing to avoid battle, managed to lose them in the vastness of the ocean.

As the summer wore on, men began to show signs of scurvy. It was also apparent to Russell that, as he could not force an action against the enemy, he would have to consider returning to a home port. On 5 August, again off Ushant, several ships came into the fleet with news that the French fleet had slipped past the allies and entered Brest. There was now no reason to keep the fleet at sea and Russell immediately ordered it to bear away for Torbay.

The fleet was off the Lizard by the 6th when the *Saudadoes* came in from westward and reported seeing forty-four Frenchmen to the west. The fleet tacked round and next day found the ships, which to their great disappointment proved to be the *Foresight*, *Deptford* and some victuallers. Turning once more, the fleet resumed its voyage towards Torbay. Many of *Lenox*'s men were now falling sick from scurvy and Captain Munden borrowed a small prize to carry thirty-three of *Lenox*'s sick men to Plymouth where a large furnished house with outhouses had been purchased for use as the first ever Royal Naval Hospital. Three men died aboard *Lenox* during June and two more in July; before that only one man had died since the refit at the beginning of the year. The deaths were to be expected after being so long at sea, but the sickness that had killed so many during the previous year did not return.

The final frustration of the long cruise occurred off the Start when it became so foggy that one end of the ship could not be seen from the other. *Lenox* lay by for two days before entering Torbay on 11 August and her ketch was able to replenish her with thirteen tuns of much needed fresh water. At the same time forty more sick men were sent ashore to recover. In calm, hazy conditions *Lenox* was heeled over for the larboard side to be scrubbed followed by the starboard side the next day. After cleaning, the mainmast was stayed and provisions taken on board. The ketch came from Torbay with twelve tuns of water and some pork, butter and peas; in the afternoon they had a survey of 577 loaves of bread but it was condemned. The bread was replaced and thirty-four tuns of sea beer and water were brought aboard for the crew's refreshment. The sick men who had been ashore for more than two weeks were now recovered and returned.

Even though it was growing late in the year for a fleet to be at sea, Russell decided to send a small squadron to make a reconnaissance off the French coast. On 23 August, Captain Munden of the *Lenox* was called aboard the admiral's flagship and made commodore of a small squadron, consisting of *Montague*, *Foresight*, *Greyhound* and the *Fubbs* yacht, and given orders to cruise off the coast of France between Seven Islands and the Isle of Bass for two days. The squadron set sail immediately after taking on board a pilot and by the end of the day was off Start Point heading in a southerly direction. On the 24th they chased two ships that turned out to be Dutch "capers" or privateers preying on French merchantmen near the Isle of St Pauls off the French coast. Two more sail were seen to the west and were chased by *Lenox* and *Montague*; the English hoisted their colours and the quarry hoisted French colours. Unfortunately *Lenox* and *Montague* could not get near enough to open fire and finally tacked away to the northward.

The next day the squadron met a Danish flyboat and *Lenox* "spoke" with her. She reported seeing the English fleet leave Torbay heading westward. Having finished their reconnaissance, *Lenox* and her small squadron sailed north as far as the Lizard and then eastwards. They met the *Chatham* and discovered from her that the Danish flyboat had misinformed them, for the English fleet had not sailed and was still in Torbay. While *Lenox* was off the French coast, Russell had received orders from the Admiralty, prepared by Queen Mary, to proceed to sea. In spite of the time of year, he was to lie in a position he thought most likely to meet the French fleet should they make any attempt upon England. He was also to provide security for homeward-bound trade, the ships in Ireland and to intercept any support the French fleet may try to give the Irish.

Lenox continuing sailing eastwards and rejoined the fleet as it was setting sail on 28 August. Next day they anchored off Start Point in a calm and by the beginning of September the fleet was again off Ushant. Gales forced *Lenox* and the fleet to bring down their topgallant yards and masts and lie by with their heads to the northward. Another man was lost, Robert Ratclife of the carpenter's crew, who died by foolishly swallowing a couple of small shot and "could get them neither up nor down".

Soon after, the worst fears of a fleet at sea late in the year were realised. At noon on 2 September it blew very hard from the SW for two or three hours, and then went about to the SSE. Admiral Russell put out the signal for his Red squadron to bear away for the safety of Torbay. The violence of the wind caused damage to the rigging; first the strop of *Lenox*'s larboard topsail sheet block gave way. Then, during the evening, more damage was caused when the starboard main topsail sheet gave way at the knot. *Lenox* soon found her flagship and the whole of the fleet to

windward of her, probably as a result of the damage. For the coming night, with the storm increasing in violence coming from the SSE and S by E, the vulnerable topsails were taken in and lowered to the tops, the main tack was taken on board, the lee clew garnets hauled up and a backed sail made for the mizzen to keep the head as near to the wind as possible. With her sails set in this way *Lenox* managed to steer away as close to the wind as possible between ENE and NE by E for the whole night.

The Reverend Richard Allyn described the conditions on board the fourth-rate *Centurion* during a storm in 1692:

> "But though the wind was so boisterous, yet the running of shot, chests and loose things about the ship made almost as great noise as that. We had about 16 or 17 butts and pipes of wine in the steerage, all gave way together, and the head of one of them broke out. We shipped several great seas over our quarter, as well as the waist. Sometimes for nigh the space of a minute the ship would seem to be all under water; and again, sometimes would seem fairly to settle on one side. The chests etc, swimmed between decks; and we had several foot of water in the hold. In short the weather was so bad, that the whole ship's company declared they thought they had never seen the like, and that it was impossible for it to be worse. Notwithstanding all this, our ports were neither caulked nor lined, the want of doing which, was supposed to occasion the loss of the *Coronation*. During this dreadful season I quietly kept my bed, though very wet by reason of the water that came into my scuttle. The behaviour of our pugs at this time was not a little remarkable: some few of them pray, but more of them cursed and swore louder than the wind and weather. I cannot forebear writing one instance of this nature, and that is in the story which was told me in the next morning of George the caulker, and old Robin Anderson. Poor George being very apprehensive of his being a sinner, and now in great danger of his life, fell down upon his marrow-bones, and began to pray, 'Lord have mercy upon me… Christ have mercy' etc., etc. to the Lord's Prayer: All the while old Robin was near him, and between every sentence cried out; 'Ah you lubberly dog! Ah you coward! Zounds, thou hast not got the heart of a flea'. Poor George thus disturbed at his devotion, would look over his shoulder, and at the end of every petition would make answer to old Robin, with 'God Damn you, you old dog, can't you let a body pray at quiet for you Ha! A plague rot you, let me alone, can't ye'. Thus the one kept praying and cursing, and together railing for half an hour, when a great log of wood by the rolling of the ship, tumbled upon George's legs, and bruised him a little; which George taking up into his hands, and thinking it had been thrown at him by old Robin, let fly at the old Fellow, together with a whole broadside of oaths and curses; and so they fell to boxing. I mention this only to show the incorrigible senselessness of such tarpaulin wretches in the greatest extremity of danger."[65]

Through the storm and thick weather, land started to appear at seven or eight o'clock the next morning, while off to windward lay the rest of the fleet. The land, bearing N by E, four or five leagues, was first thought to be Start Point, the intended landfall for reaching Torbay. They soon realised that it was in fact Rame Head, the rocky peninsula that protects Plymouth Sound. It was a much more difficult and dangerous place to enter. The admiral accepted the danger and made a signal for the fleet to bear up to Plymouth, just as the wind eased in violence. *Lenox* set her mainsails and topsails with two reefs but the lull in the storm was temporary. The extra sails soon proved too much for the main topmast and it split just above the cap. The topmast could no longer carry sail so the topsails were taken in and the mainsail set with the sheets on board. Yelverton was afraid the main yard would give way as it buckled under the strain. The storm of wind drove *Lenox* to within two miles to windward of Rame Head, while most of the fleet passed her and bore away for

Plymouth. There was now no safety margin left for *Lenox* and any further damage would result in her smashing into the massive rock of Rame Head. It continued to blow a violent gale, whipping up a great sea, leaving the master doubtful if she could weather the Rame. Sailing as close to the wind as they could, *Lenox* was "conned" round the headland. The helmsmen, who could not see out from inside the ship, usually steered with the aid of a compass. He now received shouted instructions from the deck above from where Rame Head, towering above, could only too easily be seen. *Lenox* just cleared the rocks and entered Plymouth Sound safely at about 11 o'clock. She hauled up her main sail, and bore away for the Hamoaze under her foresail and spritsail. Her troubles were not yet over, however, for ahead the *Warspite*, *Exeter* and *Foresight* had run ashore. As *Lenox* came within Sir Francis Island, she endeavoured to bear up to the ebb tide but it took her on the starboard bow and forced her ashore abreast of Edgecombe House.

To prevent her being driven further ashore, *Lenox*'s sails were laid to the mast and the anchor dropped so that she lay in the wash of the shore. The *Harwich*, *Royal Oak*, *Northumberland*, *Dreadnought*, *Stirling Castle*, *Elizabeth*, *Hope* and *Edgar*, coming in after *Lenox*, now had to avoid her. Ships ran aboard each other, breaking heads, quarter galleries and masts and causing a great deal of damage. During the afternoon *Lenox* got herself clear from among the crowd of ships and moored with her best bower anchor to the NE. The stream anchor was laid right ahead and four hawsers taken ashore and fixed to trees for extra security. The following morning the split topmast was unrigged, taken down and Ambrose Fellows and his crew fitted a replacement. During the afternoon it began to blow hard again but a pilot came aboard and they managed to take *Lenox* into the Hamoaze and moor her. Yelverton gives credit for the handling of *Lenox* during the dangerous manoeuvres to Nathaniel Smith, the volunteer member of the crew.

The storm took a heavy toll on the fleet. *Lenox*, however, was lucky; she lost only one man and a small bit of her taffrail as another ship hit her stern. The *Harwich*, which ran ashore just astern of *Lenox*, fell on her side, bulged and was wrecked. Two other ships, the *Royal Oak* and the *Northumberland*, whose rudder was half torn off, also ran aground, but after their upper tier of guns and provisions were taken out to lighten them they were got off. The *Exeter* and *Foresight* suffered badly damaged hulls as they ran ashore and lay on their sides.

The worst incident by far occurred west of Rame Head. The second-rate three-deck *Coronation*, one of the thirty new ships, lost all her masts at about the same time *Lenox* split her main topmast. The only thing left standing was her ensign staff. In desperation she anchored but the crew could not hold her and she was lost with all her company, except for about twenty souls preserved in the longboat and one from her wreck.

On 5 September, *Lenox* was kedged into Plymouth Sound. The storm had dispersed the fleet widely and in order to regroup the ships at Plymouth set sail bound for Spithead the next day, leaving the disabled ships to return later. They arrived on 9 September, when the seaworthy remainder of the fleet was reunited. Further work was carried out on *Lenox* to repair the storm damage. The main yard was found to be sprung and a new one and a spare topmast taken on board. The new yard was rigged, the bowsprit gammoned and the shrouds reset. The ship was then heeled each way and the sides scrubbed. Beer, water, bread, beef, pork and peas were stowed in the hold. The "Great Ships", that is, the first- and second-rate three-deckers, were not to be risked any further in the winter seas. Instead they received orders to sail to Chatham to be laid up for the winter.

It was feared that the French would try to reinforce Limerick, still holding out for King James and under attack by the Anglo-Dutch allied army. Russell received orders from the Admiralty to send thirty ships or as many that were fit for sea to stop a French convoy escorted by thirty

men-of-war under the command of Château-Renault.[66] A squadron of third rates, including *Lenox*, was ordered to sea. To make up the number of men to replace those volunteers who had left, she received eleven men from the *Essex*, twenty from the *Restoration* and ten from the *Stirling Castle*. She set sail on 19 September under Delaval, flying his flag in the *Berwick,* with orders to cruise off Cape Clear.[67] Severe gales continued to blow and *Lenox* lowered her yards to the deck to lay portlast. The squadron could make no progress in the bad weather and returned to St Helens. To improve her seaworthiness, two lower-deck demi-cannon were struck down into the hold. The squadron set sail again but were once more driven back. This time, to further lower the centre of gravity three longboat loads of ballast were taken into the hold. Again the squadron loosed the fore topsails and hove short, before setting sail with a NW wind. The ships reached a position off Dunnose where the wind increased to a gale, which veered round to the WSW, sending the squadron back into St Helens for the third time.

Finally, on 9 October, the gales came round from the northward and the squadron managed to get out of St Helens and into the Channel. The wind veered to come from the WSW, so in order to make headway they were obliged to tack against the wind. The strain soon told on the ships: *Lenox* sprung her newly fitted main topmast, while the *Burford* broke her fore yard forcing her back to St Helens. The severe weather continued, forcing Delaval to take his squadron into shelter at Torbay where he held a consultation aboard his flagship. It was agreed to set sail and continue towards Ireland but off Start Point a fresh westerly gale forced them back once more into Torbay. *Lenox* used the delay to take down her sprung main topmast and put up a new one. Fresh gales soon obliged her to strike her yards and topmasts again, however. The weather at last improved on 19 October and the ships weighed anchor.

The squadron, now consisting of twelve third rates and some Dutch vessels, duly set sail in far from ideal conditions. The ships were not only short of provisions but also in bad states of repair. The following morning sixteen sail were seen to the south and Delaval ordered the squadron into a line of battle, but as they closed the ships disappointingly turned out to be Dutch merchantmen. The weather continued to improve and off Land's End the topgallant masts and yards were set up. The *Plymouth* and *Adventure* departed with the squadron bound for Virginia, the *Centurion* left with some store ships bound for Kinsale and some Dutch merchantmen left to their trade in the east.

Lenox's crew was reduced from six to four men's allowance for all things to conserve rations. The full weekly ration for each man was 7lbs of biscuit, 4lbs of beef, 2lbs of pork, 2 pints of peas, $1\frac{1}{8}$ cod 2ft long, 6oz of butter and 12oz of cheese. Substitutes were sometimes made to the ration depending on the availability of victuals. Although more than adequate in quantity, the diet was severely deficient in vitamin C, and was the cause of illness and disease such as scurvy. Drink allowance was also generous – a gallon of weak beer a day was provided as well as water. When men's rations were reduced, the value of the savings was added to the men's pay.

The gales soon returned, and as the ships struggled to work westwards, the weather continued to deteriorate, causing a great sea. The sound of many guns could be heard being fired from ships in distress. To the horror of those aboard *Lenox*, the *Mary* ketch sank before their eyes. During a lull in the storms, the *Chester* brought in a captured small vessel, from which it was learned that fears of the French returning to Ireland were true and Monsieur Château-Renault with a squadron had gone to Ireland to relieve Limerick. Sir Ralph signalled his captains to him and it was agreed that they should continue to make their best way to Ireland. He sent his decision to the Dutch in his squadron but received no reply. The gales renewed their fury and the master of *Lenox*'s ketch was washed overboard and lost.

Provisions aboard the Dutch ships were now very low, and all the crews, *Lenox* recorded, were grumbling for want of provisions. On 5 November the ketch tendering *Lenox* lost her mast by reason of an old spring it had received the previous year. *Lenox* tacked and took the men out of her for fear they should be lost. The ketch was abandoned and left to founder.

Sir Ralph Delaval was now at the limit of his men's and ships' endurance and consulted his captains. This time they all agreed they should not continue towards Ireland but return to Portsmouth. He sent word of his decision to the Dutch who sent their thanks and added they would "positively not sail for Ireland in any case". A westerly wind helped the passage home but its violence that night was worse than anything they had yet encountered, with the men aboard *Lenox* fearing they would all be lost. Dawn revealed an empty ocean, but soon the Lizard and other ships of the squadron were in sight. By the evening of 8 October *Lenox* was safely at anchor in St Helens where the other members of the squadron joined her, the last being *Burford* on the 11th. The heroic efforts of Delaval's squadron had been in vain. Limerick fell to William on 13 October, while news that the French had set sail to reinforce it proved untrue.

To the dismay of the men, orders were then received to complete victualling for forty-two more days at sea. Two days later, before a revolt among the men could take place, new orders were received by Sir Ralph from the Admiralty that, notwithstanding his former orders, the *Lenox* and *Hampton Court* would remain at Spithead in order to be refitted at Portsmouth.[68] The news must have been very welcome aboard the two ships. Sir Ralph then struck his flag and was saluted by many guns from the squadron. On the 22nd the Dutch and English ships to be refitted at Chatham sailed to the east and *Lenox* fired a total of thirty guns in salutes as they left. Captain Munden also left for London after receiving orders from Parliament.

By the beginning of December, *Lenox* was in harbour being made ready to dock. On the 2nd some of the lower tier of guns and anchors were taken out. They were followed by the rest of the lower tier and 110 barrels of gunpowder the next day. Two more days were spent bringing the upper tiers of guns and some empty casks out. The 5th was spent taking the quarterdeck and poop-deck guns out. The crew then made a start to "rummage in the hold" and take out the casks. The gun carriages were the next out, followed by the cables. By the middle of the month the first of the ballast was taken ashore in a hoy. The shot was then taken ashore and finally the topmasts were unrigged and taken down.

During 1691 only nine men died aboard *Lenox*, in spite of the hazards presented by the many storms imposed upon the men, particularly those aloft in the rigging. The worst period was in June and July when *Lenox* was constantly at sea with the fleet. During the year sixty-five men deserted the ship, fourteen alone during December while she lay at Portsmouth, the majority not returning from leave and forfeiting their pay. A muster was taken on 26 December and of 419 men borne on her pay book 311 were listed as absent on leave.[69]

The dates of men being enlisted and discharged conform to little pattern. They came and went in an irregular manner that tested the pay clerks' ability to correctly calculate their wages. So many men left during December that Captain Munden was granted a new press warrant.

1692 – THE BATTLE OF BARFLEUR AND LA HOGUE

The failure in Ireland left James II with no alternative plan but a direct invasion of England. In early spring, he returned from Ireland to muster his French and Irish army at Le Havre and La Hogue. Before he could embark in his transport ships he would have to wait for the French fleet to join him at Brest. James was encouraged by widespread dissatisfaction with William's government and its lack of success. He hoped to assemble his invasion force, cross the Channel and land in Torbay before the combined English-Dutch fleet was ready tp put to sea. Yet for James' highly

ambitious plan to succeed, speed was essential. Meanwhile, work on *Lenox*'s winter refit continued. On 9 January the *Resolution* was lashed aboard her and the remainder of *Lenox*'s ballast, except for about 30 tons, hove out into her. Only 121 of her 418 men were available to carry out the work, as 297 were absent on leave.[70]

While *Lenox* waited to dock, her new second lieutenant, Thomas Williams (who had replaced Francis Gregory earlier in the year) was sent to Bristol to impress men. While there he received orders from the Admiralty to board the *Joseph* under the command of Captain Brook. He found some volunteers and pressed men and was ready to sail to the Thames where they were to be dispersed into the navy.[71]

On the 25th, as snow lay on the ground, Captain Munden sent the Navy Board the monthly muster books and a note stating; "if anything be wanting I pray your advice and will endeavour to mend it for the future", adding "this day we endeavour to haul the *Lenox* on the ways to clean her".[72]

As planned, *Lenox* was docked and the same day one side of her was breamed, the marine growth being burned off with reeds or straw.[73] The other side was done the following day. To complete the process she was graved with either white or black stuff and probably tallowed. The dock was then flooded and she was hauled off the ways to make room for the next ship. She had been out of the water a little more than twenty-four hours and work was immediately started to make her ready for sea. The topmasts were set up and ballast brought into the hold; by 8 February she had 338 tons aboard.

January and February was spent taking in ballast, guns and provisions, re-rigging and pressing men for whom two more press warrants had been issued on 26 January. The guns were cleaned or "scaled", as it was known, probably by firing sand propelled by a small charge. The external surface would have been cleaned and treated on land while the ship was being refitted. Blacking was applied to protect the yards and the heads of the masts. A blue flag was hoisted at the mizzen topmast head on behalf of Richard Carter, Rear Admiral of the Blue, whose flagship *Lenox* became while she lay in harbour. Some of *Lenox*'s equipment was worn out and replaced; she received a new fore yard, new best and small bower anchors, and spare cables.

On 11 February Captain Munden was able to report that *Lenox* was ready except for want of provisions but that they were expected to be complete in about a fortnight.[74] Two days later, all the ships fired nineteen guns for the King's Proclamation Day and by the 20th of the month the sails were brought to the yards. Munden had done well to man ship and get her ready for sea early. With a core of experienced seamen and having already spent two years at sea, *Lenox* was at the peak of efficiency.

Lenox and *Stirling Castle* received orders to sail with a convoy of merchant ships requiring escort into the Downs. *Lenox* was short of her allowed surgeon's supplies and needed an order from the Navy Board in London to obtain them. Captain Munden wrote a letter requesting the order be sent by return, as he was ready to set sail.[75] Rear Admiral Carter did not sail with the convoy and he left *Lenox* on the 25th after being aboard for three weeks. The following morning *Lenox* loosed her sails and by one o'clock she was at Spithead. Captain Munden moored his ship, then wrote to the Navy Board reporting he was all ready to leave, but only stayed as he was waiting for the *Stirling Castle* to join him.[76]

A few days later on 1 March, as a gale blew, the convoy consisting of *Lenox*, *Stirling Castle* and fourteen merchant ships made ready to unmoor. The next day they weighed anchor, set sail and steered east. After sighting the coast of France they hauled north toward the Downs, which they entered on the 3rd. They kept the lead going for safety and recorded depths of between six and twenty fathoms before anchoring. They found the Downs full of shipping, much of it conveying regiments of English soldiers bound to fight the French in Flanders.

On the 7th, as it snowed and blew fresh gales, Captain Munden sent his tender to Dover for water. He was not too impressed with the vessel and had seen something much more to his liking. The same day he wrote to the Navy Board:

"I have a tender taken up at Portsmouth a very sorry thing for I dare not send her anywhere to get men;- Here now offers to me the *Tartan* which has been out a privateering, not anything in the sea beats her with a sail, she has four cannon, six peteraroes, and is 65 tons. She cannot sail with less than eight or nine men, she has also 20 oars which upon occasion may be manned and do great service. She is also fit to be sent with any particular without fear. I have talked with the Captain of her, and for all their quality he desires no more for her service than you give for any ketch of her burthen. I pray that you please to let me have her for the present (and I'll leave off what I have) and when the fleet is got together she will serve two of us. I want men and can send her to the westward or eastward without fear, she'll stow 200 men, which is all from yours faithfully."[77]

Only one day later Munden received a reply, giving him permission to use the *Tartan* at a rate of £42 per month. He spoke to the *Tartan*'s master, William Humble, and the owners, John and Richard Tartan. They expressed hopes that the navy would pay £45 per month, "but would serve and stand by the award" already agreed. The *Tartan* prize was recently captured from the French by a packet boat and was of a burthen of 49 tons, not the 65 tons estimated by Captain Munden.[78] The four guns would have been two- or three-pdrs. Peteraroes were swivel guns with a tail-directing bar, and they fired a half-pound shot. They were probably muzzle-loading, but may have been breech-loading at this point.[79]

Lenox had still not received her allowed surgeon's supplies, which Munden had requested while *Lenox* lay at Portsmouth. He wrote again to the Navy Board informing them he would send his surgeon's mate to London to obtain them.[80] The gales continued for another week, during which time the yards and topmasts were lowered and got up again as the weather improved. Munden and Captain Waters of the *Stirling Castle* received orders dated the 11th to sail without any loss of time to the Buoy of the Nore where the fleet was gathering and to wait there for further orders.[81] The fore topsails were loosed and the sheets hove home but the gale increased once more, obliging the sail to be taken in and the yards and topmasts lowered.

On 16 March, as the weather moderated, a squadron under the command of Sir Ralph Delaval arrived from Cadiz to join the ships in the Downs and anchor. During his voyage the French fleet, following a policy of trying to attack detached squadrons, had nearly intercepted him. Nevertheless he arrived, whereupon *Lenox* greeted him with a nine-gun salute; then she, the *Stirling Castle* and the *Monk* weighed anchor bound for the Nore.

All the firing of salutes may have been good gunnery practice but Russell considered the benefits not worth the expense. He issued an order to the ships at the Buoy of the Nore and at Blackstakes that a great quantity of powder had been expended in saluting flags and even by private captains saluting each other, and that in future they should stop the practice and were not even to salute flags.[82] It must have been a very unpopular order. The following evening, the pilot aboard *Lenox* nearly ran her aground upon the shoals of Sunk Head but was saved by promptly letting go the anchors. After anchoring for the night the voyage was resumed in the morning. She arrived at the Nore and moored on the 20th while a hoy brought water and four pressed men from London. Unfortunately the men were on leave from the *Centurion*, whose captain had already taken the matter up with the Admiralty. The Admiralty reacted so quickly that Munden received the order to return them to their ship before the hoy had even arrived with the men.[83]

Munden found himself in trouble again when another Admiralty letter arrived ordering him to discharge John Frith, "who was neither a seaman or a waterman, but a house carpenter", and should not therefore have been pressed. Sir Richard Haddock came into the fleet to pay the men their wages. He and his clerks came aboard the *Lenox* on 29 March and paid the men according to the pay book,[84] covering the nine-month period from 1 October 1690 to 30 June 1691.

At the start of April marines were allocated to the ships. Admiral Russell ordered the marines of the Earl of Danby's regiment to be put on board ships of the Red Squadron, beginning with the third-rate ships, and particularly such of them as were in greatest need of men.[85] At this time *Lenox* was was well manned and as a result received only forty-five. Having had only ten men "run" since the turn of the year helped her situation. Unfortunately *Stirling Castle* was still undermanned and *Lenox* received orders to turn over eleven men to her in an attempt to balance the crews.

On 16 and 17 April many of the great three-deckers including the *Royal Sovereign* and *Britannia* arrived from Sheerness. Among them were all the ships that were to form Sir Ralph Deleval's Red division for the forthcoming campaign, of which *Lenox* was to be a member. On the 19th Sir Ralph boarded his flagship the *Royal Sovereign* and hoisted his flag. *Lenox* and the other ships in her division shifted their pennants to the fore topmast head. At 8 am the following morning *Lenox* weighed anchor and moved closer into her division till she was abreast her flagship before anchoring again.

Sir Ralph sent Captain Munden a letter ordering him to convoy some transport ships to Flanders, whereupon Munden immediately wrote to the Navy Board for a pilot.[86] He then received a hoy load of provisions and water. Gales struck before he could sail and the yards and topmasts were taken down. The weather cleared early on 25 April and *Lenox*, two yachts, a fireship and nine vessels full of soldiers weighed anchor at 4 o'clock in the morning. Twenty-four hours later, aided by a fresh NW wind, the coast of Holland appeared. The detachment crept eastwards, heaving the lead, recording at first twenty-five fathoms that gradually shoaled to fourteen; the lead brought up from the bottom "a coarse grey sand with black shells". By 6 am the convoy was close enough to safety and Captain Munden sent them away and stood off with only the fireship in company. That night it blew a gale and the two ships anchored; they weighed again at dawn and continued east. By noon they were anchored in the Gunfleet, where they found Sir Ralph Delaval flying his flag in a sister of *Lenox*, the *Berwick*. With him were seventeen third rates and five fireships.

While *Lenox* was off the coast of Holland, Delaval received orders from Russell to cruise along the coast of France as far as Cap La Hogue, seeking information on the whereabouts of the French fleet. He was then to return, calling in at the Isle of Wight, Dover and the Flats of the Foreland for new orders. In case Delaval should meet the French in superior numbers to his own, he was ordered not to engage but retreat towards the Flats of the Downs until he should meet the rest of the allied fleet under Russell. If he met with the other squadron out cruising under Admiral Carter, he was to join it to his own.[87]

Meanwhile, Tourville, reinstated as Admiral, was under strict orders and timetable to be at sea before the allies. He was to set sail from Brest with only thirty-seven warships and land James II and his army in England. Twenty more ships were left in harbour, as they were not yet fully manned and the Mediterranean squadron under d'Estrées had been delayed. Tourville's orders, ominously echoing Torrington's in 1690 before the Battle of Beachy Head, instructed him to engage the enemy, whatever the circumstances.

Lenox, as part of Delaval's squadron, made her way across the Thames estuary into the Downs where fifteen Dutch men-of-war were found lying at anchor. By 1 May they were off Dungeness and with the wind blowing from the NE they stood over towards the French coast. Delaval ensured the French fleet could not cut off his squadron by lying strategically between Beachy Head and the mouth of the River Seine, while individual ships carried out reconnaissance duties along the French coast. The cruisers did not find the French and Delaval, following his orders, returned north, anchoring at St Helens on the 8th. At 7 o'clock the next morning the squadron weighed anchor and stood out to sea to search for Carter's squadron. Almost immediately it was seen to the westward consisting of nineteen sail of warships, causing Delaval to lay by for three hours waiting for them to join him. The combined squadrons remained in the vicinity of St Helens to wait for Russell with the bulk of the main fleet, including the great three-deckers. In the meantime the *Advice*, *Ruby* and *Greyhound* brought in three French fishing boats as prizes.

Delaval's and Carter's squadron was joined by the main fleet of about fifty sail on the morning of 13 May and at noon the signal was made to moor. The following morning further reinforcements arrived when a Dutch squadron sailed in. *Lenox* took on board 15 tons of beer from Portsmouth. It was an impressive performance to put such a large fleet to sea so early in the year.

The combined English-Dutch fleet now consisted of more than eighty warships with the Dutch again forming the White van squadron. The fleet was about twice the size of the opposing French fleet under Tourville, even though that was reinforced off Plymouth by seven ships from Rochefort on 15 May. Tourville had hoped to meet with the transports containing the invasion army and escort them across the Channel before the English and Dutch were ready. England feared an imminent invasion, but the easterly winds had delayed the French long enough for the allied fleet to assemble.

There were fears at court that disaffected persons in the fleet might not fight against King James, their old commander. A paper was prepared and signed by all the officers in the fleet declaring their zeal and loyalty, which Russell then sent to Queen Mary.

Lenox formed part of the Red squadron under Delaval, who had transferred his flag from the third rate *Berwick* to the 100-gun first rate *Royal Sovereign*. The Red squadron consisted of three divisions, of which Delaval's was the leading. A list of the ships in it follows overleaf.

Below: *An engraving by the celebrated English engraver William Woollett (1735–1785) of the Battle at La Hogue, executed in 1781 after an oil painting by the Anglo-American artist Benjamin West (1738 - 1820). Admiral George Rooke is depicted; he can be seen in the left foreground, standing in a boat. His raised sword embodies the spirit of heroic command as the battle rages around him. The French flagship* Royal Sun *is also visible through a parting in the thick smoke. Although beached in the centre distance, the ship was actually sunk a few days before this encounter, and figures as a symbol for the French defeat. This is emphasised at the right-hand side of the work, where a bald Frenchman deserts his craft with its fleur-de-lis motif. Having lost his wig, he becomes an object of ridicule.(Author's collection)*

Ships comprising Vice Admiral Sir Ralph Delaval's Red division:

	Guns	Fireships
St Michael	98	*Extravagant*
Lenox	70	*Wolf*
Bonaventure	48	*Vulcan*
Royal Katherine	86	*Hound*
Royal Sovereign	100	
Captain	70	
Centurion	50	
Burford	70	

While the fleet lay in a calm at St Helens, *Lenox*'s boatswain was moved into the *Sandwich* and a warrant issued for a new man to be appointed in his room. It was normal practice for two boatswains from other ships to check the state of the stores left by a departing boatswain. Accordingly orders were given by Russell for the boatswains from the *Royal Katherine* and the *Royal Sovereign* to survey *Lenox*'s stores.[88]

A council of war was held and it was decided to sail at the first fair weather towards Cap La Hogue and Barfleur to seek out the French fleet. At dawn next morning on the 17th, there was a slight breeze from the NNE and *Lenox* unmoored with the rest of the fleet. The light airs prevented the fleet getting very far and at 9 am *Lenox* anchored again two miles SSE of Culver Cliffs in twelve fathoms of water as the tide turned. She weighed again at 2 pm and anchored again at 4 pm for the night.

The next day news came of the whereabouts of the French fleet and Russell made the signal to form the line of battle. The light winds coming from the WSW were hardly strong enough for the fleet to steer into position and as a result the line was somewhat ragged.

Next morning, 19 May, proved hazy with a moderate wind from the WSW, as the Allied fleet stood south on the starboard tack. They made

Above: *Loading the demi-culverin chase gun, which has been moved from the foremost broadside gun port on the upper deck.*

contact with an unidentified fleet heading NE, which was seen by those aboard *Lenox* at 4 am. Between 7 and 8 o'clock in the morning they identified them as the French fleet. Russell ordered the rear, Blue squadron of his fleet to tack round and head north in the same direction as the French. Tourville responded by bravely turning south, also on the starboard tack and bearing down on the allied line; by doing so he maintained the weather gauge and stayed to the windward of the allies. His action also detached Russell's Blue squadron and he would now only have to face the White, Dutch squadron, and the Red, English squadron for some considerable time.

Between 10 and 11 am *Lenox* cleared for action. The guns were loaded, all the temporary cabins between the gun ports were removed and the men nervously stood at their battle stations. The two fleets closed to short range, with the *Lenox* lying second to the *St Michael* at the head of their division. *Lenox* recorded her position at the time to be nine leagues south of Dunnose.

Both fleets opened fire at about noon and were soon engulfed in smoke and flame. At the beginning of the action *Lenox* witnessed the fireship *Extravagant* from her division accidentally set alight and uselessly burned. The weather turned calm and visibility deteriorated as gun smoke added to the haze. The men aboard *Lenox* once more fought their guns amid the noise and smoke of battle. Russell, worried by the unevenness of the line, issued an order to all his commanders who could see it "to use all possible means to tow your ship into line of battle during this calm and not to go out of line".[89] As battle proceeded ships in *Lenox*'s division astern of her began to suffer considerable damage as the French centre closed with them. The *Royal Katherine* suffered damage to her rigging and had twenty-two guns disabled, while underwater damage to the *Centurion* flooded her hold to a depth of 7ft. To her good fortune, *Lenox* seems not to have had an enemy ship opposite her, and as a result she remained unscathed. She was able to fire her guns and inflict damage without experiencing the horror of incoming cannon shot. The nearest French ships across from her were probably the *Henri* and the *Fort*, ships of about sixty guns.[90] The situation became confused amid the thick smoke and did not

change until between four and five o'clock in the afternoon, when the wind started to blow again and, changing direction, came from the east.

The French, despite having suffered more than the allies, maintained their discipline and attempted as quickly as possible to get away from the battering they were receiving. Helped by the fog and gun smoke they set all the sail that their damaged masts and rigging would carry and stood away westwards, helped by their boats towing ahead. *Lenox* gave chase and as she bore away after them she observed the Dutch squadron also sailing after the French. Admiral Russell's division was seen lying off to the east. *Lenox* stayed close enough to the enemy to continue firing with her fore chase guns. Ahead of the French, part of the Blue squadron managed to rejoin the action and some of the Red division was seen through the haze to leeward. The French fleet continued on its course and as evening wore on it fought its way through the ships in its path. *Lenox* continued firing until nine o'clock by which time it was too dark to see the enemy. Shortly afterward *Lenox*'s crew saw two ships blow up, "but could not discover who they were". In actual fact neither side lost any warships. The explosions, heard by other ships as well, must have been caused by abandoned and burning fireships.

During the night *Lenox* continued to chase in a NNW direction and in the dark and confusion became detached from her division. In all probability the damage sustained by some of the ships prevented them chasing as far as *Lenox* had done. Incredibly, in spite of fighting in a pitched battle for much of the day, no one aboard was killed, though Thomas Swallow dislocated his left knee.[91]

At 4 am the next morning, the 20th, it was hazy and nothing could be seen of the enemy. The wind then sprang up and veered to the westwards while *Lenox* steered on a northerly course in search of the French. At 8 am it was realised that they were heading in the wrong direction so they tacked to the south as other ships were seen to do. At about 10 am the haze cleared to reveal the French off to the SW; *Lenox* and other allied ships, no longer in any formation, resumed the chase and stood after the battered but not yet beaten French, most of whom retreated westwards towards Alderney. The wind became variable, blowing sometimes from the SW, then gusting strongly and dying away. At 5 pm, with the ebb tide starting to turn, *Lenox* anchored in 40 fathoms with Cap La Hogue bearing SW six leagues. The French were anchored just out of reach only four miles to windward of her.

At midnight, with the next ebb tide the chase continued. *Lenox* weighed anchor and by four o'clock in the morning of the 21st, the early dawn light revealed her among the most westerly ships with the Dutch and some of the Blue squadron. Alderney was observed bearing SW by S two leagues. The force of the next flood tide caused the ships of both fleets to anchor, *Lenox* eventually doing so at 7 am in forty-eight fathoms of water with Alderney now bearing S by SW five leagues and clear of the Casquets. *Lenox* and the ships near her were now in a position to meet the French if they attempted to pass to the west of Alderney. The only way the French could effect an escape would be to wait for the next ebb tide and negotiate the dangerous Alderney Race to the east of the island and make a run for St Malo. As the sea came up the Channel on the flood, about thirteen of the waiting French ships near La Hogue, many of which had cut away their heavy anchors earlier in the battle, found themselves unable to hold their position. They reluctantly cut away the cables to the dragging anchors and ran to the east before wind and tide. There was no harbour in which the damaged ships could shelter, and with a superior fleet waiting for them they were in a dangerous situation.

Russell and the bulk of the fleet, now consisting of about forty ships, also cut their cables to chase the French ships. *Lenox*, however, with undamaged sails and rigging, stayed forward at anchor near the twenty-one French ships that remained. With her were the *Bonaventure* that had sailed next to her in the line of battle and the *Rupert*, also of the Red

squadron. The other ships that remained were the *Victory*, *Expedition*, *Defiance*, *Edgar*, *Suffolk*, *Adventure* and the *Speedwell* fireship, all from the Blue squadron,[92] and twenty-six ships of the Dutch squadron. At noon came a new ebb, and *Lenox* weighed anchor again to continue plying forward in a fresh wind. At about 6 pm she witnessed the *Adventure* take a French fireship. An hour later with the next flood she anchored in forty-five fathoms, the Casquets now bearing SE by E three leagues and the French three leagues to windward. There was now only one escape route left for the remaining French ships.

At about midnight the anchored and desperate French made use of the next ebb tide to run the gauntlet of the Alderney Race, while the senior remaining allied officer, Ashby, Admiral of the Blue in the *Victory*, decided not to follow. It was a tactical mistake for the Blue squadron; the Dutch and *Lenox* could not sail round the Channel Islands and meet the French before they could get into St Malo.

The French gamble paid off and at 4 am the next morning of the 22nd, *Lenox* bore away eastwards to rejoin her squadron in their attack on the French ships that had fled to the east. At 11 o'clock that morning a great plume of smoke was seen near the shore to the east, although the men aboard *Lenox* had no means for the moment of knowing the reason. The wind became variable and *Lenox* worked the tides. At 5 pm, off Cherbourg, she came up to rejoin the squadron of Sir Ralph Delaval with a division of ships heading from the shore and learned the reason for the smoke. Richard Studvill, the master, who had replaced Robert Yelverton during the recent winter refit, wrote:

> "This morning Sir Ralph Delaval went on board the *Burford* [from the *Royal Sovereign*] and hoisted the union flag at the main topmast head. He stood in for Cherbourg with some fireships where he found three of the French three deck ships with their masts cut by the board that they might not be discovered, but two of them was [sic] set on fire by our fireships and the other Sir Ralph went on board of with his boat and brought off the wounded men that was in her, and set her afire. Two of them that was set on fire was the *Rising Sun* of one hundred and ten guns, the *Admirable* of one hundred guns the other of ninety guns."

Lieutenant Durley thought the ships destroyed were the *Sun*, *Admirable* and *Conqueror*, while Captain Munden gave the *Royal Sun* 100 guns, the *Admirable* 96 guns and the *Wonderfull* 100 guns. In fact they were the *Soleil Royal* of 104 guns, the *Admirable* of 90 guns and the *Triomphant* of 76 guns. At 8 pm *Lenox* anchored in Cherbourg Bay against the next tide, along with Delaval's division, the Blue squadron and the Dutch.

Lenox weighed anchor at 5 am on the 23rd and steered ESE with a light wind from the WNW. Three hours later the main fleet was seen to the SE and by noon Barfleur bore WNW five leagues. At 2.30 pm she joined the main body of the English fleet three leagues off La Hogue, where thirteen French warships were lying close to the shore, with many transports of King James's invasion fleet sheltering in the harbour. She was just in time to take part in the final act of the engagement that occurred later that night and the following morning.

At about 4 am on the 24th, the "General" (as Admiral Russell was generally known) made a signal for all the boats in the fleet to be manned and armed, and then to come alongside him. *Lenox* contributed forty men under the command of her indomitable first lieutenant, Edward Durley. Durley duly boarded the *Britannia* and received orders to obey the commands of George Rooke, Vice Admiral of the Blue, who had transferred his flag from the second rate *Neptune* to the 70-gun *Eagle*, a sister of *Lenox*. They were given orders to assault and set fire to the French ships that had warped themselves onto the rocky shore in the Bay of La Hogue. The boats were to be escorted in by some third and fourth rates and the remaining fireships. After a preliminary bombardment by the larger ships, the boats attacked as darkness fell, led by Rooke flying his flag from a boat. At 9 pm that night *Lenox*'s boats were involved with others in boarding and setting fire to the most easterly of the French ships. Soon after, four or five others were also attacked and set on fire. Using the light from the six burning ships, which Durley thought were all three-deckers, they retired on the ebb tide. Many of the boats returned to their parent ships but some, including *Lenox*'s boats, remained with the inshore squadron.

Early the next morning, at about 5 am, as the tide turned the attack was resumed on the second group of French warships further along the bay to the west. Some small frigates led the attack and battered down French defensive platforms near the shore on which guns had been mounted. Then the boats went in, Lieutenant Durley again commanding *Lenox*'s contingent. They enjoyed outstanding success, as five more 70-gun ships and two three-deckers were burned.

Encouraged by the attack and the modesty of the opposition, Rooke decided to enter La Hogue harbour itself and attack the transport ships with which King James had hoped to invade England. Two fireships, the *Half Moon* and the *Cadiz Merchant*, were towed in but could not get near enough to destroy their targets and were brought out again. Unfortunately they ran aground and to prevent them falling into French hands their

crews set them on fire where they lay. The boats had more success and burned between fifteen and twenty merchant transport ships. Lieutenant Durley captured a small vessel and decided not to set fire to it until he inspected its contents, a wise decision for he found it contained 130 barrels of gunpowder. He made the vessel his prize and took it out of the harbour with him. *Lenox*'s boats returned to their ship at 2 o'clock in the afternoon in triumph, their crews tired but elated and bringing accounts of their success without the loss of a single man. Captain Munden, reflecting on the dangerous attack into the harbour, wrote "they had guns mounted ashore and an army of men, but all signified nothing".

Lenox remained off La Hogue until the morning of the 25th when she made sail for home. A boat under a flag of truce came out of La Hogue as *Lenox* and most of the fleet plyed away from the shore; aboard it was Captain McDougall, a Jacobite Scots officer, and two French officers. They brought a packet for the admiral and news that King James himself was at La Hogue with his army when the boats attacked.

Lenox arrived in St Helens Roads on the 26th and moored. The next day she greeted the remaining ships of the victorious fleet arriving after her, among them the battered 100-gun *Prince* of 1670, now renamed *Royal William*, and six other sail. So extensive was her damage that the *Royal William* would be out of action and under repair for the rest of the year. Shortly after the 26th the weather turned stormy and *Lenox* was forced to lower her yards and strike her topmasts. On 3 June at about 3 pm the body

Below: *A model of the first rate* Britannia, *built by Phineas Pett II at Chatham in 1682. (United States Naval Academy Museum. Photographer: Dr. Richard Bond)*

of Rear Admiral Carter, killed during the battle aboard his flagship the *Duke*, was carried ashore for burial. His body was put into a barge with his flag flying at half-mast, attended by all the barges in the fleet. The *Duke* began to fire a salute and was followed by the *Britannia*, then all the flagships in order fired twenty-four guns while the ships in Carter's own division followed with twenty more. The same day *Lenox*'s main topmast, which had been damaged by a cannon ball during the battle, was brought down to the deck and repaired.

During the following week repairs were carried out to bring *Lenox* back into fighting order. The mainmast, also damaged by shot, was fished and woulded; *Lenox* was then scrubbed and a survey taken of the stores. Finally the sides were scraped clean.

Lenox had clearly been hit during the battle but not one man was listed in the pay books as slain – although two men did die during the campaign, one on the 20th and another on the 21st. While *Lenox* remained at St Helens most of the other ships in her squadron, which had all suffered more than her, went to Spithead for refits.

This was no time for the Protestant allies to rest on their laurels and the English government was eager to press home its advantage. If the remains of the French fleet could be destroyed, thousands of troops stationed in England to oppose an invasion could be released to fight elsewhere. It was also planned to descend on the coast of France with an army, thereby forcing King Louis to call back troops that were currently invading the Netherlands. As a first step in this plan, Russell intended to put the fleet to sea and try to destroy the French squadron that had escaped through the Alderney Race after the Battle of Barfleur and which now lay, safe but isolated, at St Malo.

On 12 June Russell gained intelligence from the master of a captured French snow (a type of brig) that some twenty-five ships were in St Malo harbour. Their worst battle damage had been repaired and the squadron was preparing to set sail for the main French port of Brest. Russell seems to have had no doubts about the reliability of the intelligence, for he temporarily went aboard the second rate *Neptune*, hoisted the standard at the main topmast head and made the signal for all the flags to come aboard. His strategy, he informed them, was to keep to the west of St Malo and intercept the French during their passage. Furthermore, orders were given that if he should miss them, then soldiers should be embarked on transport ships at Plymouth in preparation for a land assault. The next morning Russell loosed his main topsail and fired a gun as the signal for his division to unmoor.

Lenox, which had her topmasts and yards up, weighed anchor and set sail in company with the ships that were not undergoing lengthy repairs. The fleet sailed down Channel, anchoring off Dunnose on the 15th and Portland on the 16th. The following day, off Bolt Point, Russell sent written orders to Captain Munden of *Lenox*:

"You are hereby required to take under your command their Majesty's ships *St Albans*, *Dragon*, *Greyhound*, and the small vessell lately taken from the enemy at La Hogue, whose commanders are to obey your orders and to make the best of your way over to the French coast and to cruise off the Seven Islands; You are to endeavour to intercept some persons that may give intelligence of the enemies proceeding, by sending in the small frigates in the night, at some place near the aforesaid island, or otherwise as you shall find most feasible; You are not to stay above forty eight hours on this station, and then to return to the fleet, with all possible diligence, according to the rendezvous lately delivered to you dated on board the *Britannia* off Bolt Head the 17 June".[93]

The rendezvous in case of separation was, with a westerly wind, Torbay, and with an easterly, the eastward of Forne Head. The Seven Islands are to the west of St Malo and if the enemy ships there had already sailed, anyone from the local seafaring community would be sure to know. Even if they had not sailed, the locals might well have knowledge of their state of readiness. *Lenox* was probably chosen for the task because of her good sailing qualities; she was often chosen for similar duties. Not all of her sisters were so blessed, for Sir Edward Gregory commented on the unsuitability of some of her sister third rates to act as cruisers.

At 4 o'clock the same afternoon, as a gale blew up, *Lenox* and her small squadron took their departure from Bolt Head and steered south. At two o'clock next morning, the 18th, they brought to as they neared land. Dawn revealed the coast six leagues off, bearing S by E, taken to be one of the Seven Islands. They continued on their course until 9 am to be sure of their position, then tacked away north out of sight of land. That evening they returned and Munden hoisted a pennant at *Lenox*'s mizzen peak as a signal for his captains to come on board. Following his orders from "the General", Munden ordered the boats to be manned and armed for a landing ashore, to be escorted in by the nimble *Greyhound*, a 16-gun sixth rate, and the French prize, *La Volage*. Lieutenant Durley was again to command the boats as he had at La Hogue. Next morning Durley headed for a place he refers to as Pero Road, where a great flyboat was seen at anchor; but likewise the English boats were soon spotted and to avoid capture the flyboat cut her cables and ran ashore. Durley and his men pursued and boarded her, finding her to be of about 400 tons burthen and laded with wine, salt, tobacco and pitch – all in all, a very valuable prize. Unfortunately the tide was falling and there was no possibility of getting the ship off, leaving Durley with no other option but to set her ablaze. About noon, as she caught fire, eight other French vessels were seen coming from the east. Realising what was happening they ran in among the rocks with Durley after them. By now the alarm was raised and as the boats attempted to enter a narrow passage to follow the French vessels, people on the shore fired down on them with small shot. Durley could see that most of his men would be killed if he dared entered the passage. Proving he was not just a hot-headed fighting man he decided he could do no more, and headed out to sea to join the *Greyhound* and *La Volage*.

At 8 o'clock that night the boats' crews were back aboard their ships. The next morning the squadron was still off the Seven Islands; they had certainly caused loss and disruption to French trade but had gained no intelligence of the fleet in St Malo. At four in the afternoon a fog descended and their forty-eight hours expired. Munden ordered *Lenox* to tack and steer NNW. By 6 am the next morning they made out Bolt Head and a few hours later they saw the fleet off the Start, from which came the *Centurion* and *Bonaventure* to speak with the newcomers.

Back with the main fleet, Captain Munden went aboard the flagship to make his report to Admiral Russell. A council of war was held two days later, which took into account Munden's report and debated as to what action should be taken, having learned that Russell's contingency plan to land troops for an assault was thwarted by the government sending the troops to Portsmouth, instead of Plymouth as previously arranged. It was decided to lie about twenty leagues north of the Isle of Bass in a position to intercept the St Malo ships, should they decide to sail.

As *Lenox* and the fleet bore away, Russell's undertaking suffered yet another setback. A violent storm blew up from the NNW, as *Lenox* steered SSW under double-reefed topsails. *Berwick* and the new 80-gun third rate *Cornwall* had just joined the fleet, when the *Cornwall* immediately lost her main topmast over the side. The next day, the 25th, the gales increased in force and *Lenox* took in her topsails and set her three main courses, the normal storm canvas. The violence of the gale increased but with plenty of searoom the fleet was not in danger of being driven ashore. The sails were therefore taken in and the yards lowered to lie portlasted across the deck, allowing the ship to drift to leeward. The gales continued for another two days during which many ships suffered damage to their masts and rigging and *Lenox* lost all her topsails. Two Dutch ships were seen from *Lenox* to be in trouble, one with her bowsprit gone and the other with her bowsprit and foremast missing.

The fleet was swept out of the Channel twenty leagues past Ushant. After the storm abated, "the General" hoisted a striped flag at his main topmast head as the signal for the Red, Blue and Dutch squadrons to form into three lines of battle against the possibility of meeting the French.

At the beginning of July the fleet worked its way back east to anchor in Guernsey Roads, where *Lenox* sent her longboat ashore for water. The fleet lay in a dangerous anchorage with a lee shore nearby and poor grounding for its anchors. Russell therefore decided to take the main body of his fleet into the shelter of Torbay while a detachment under the command of Vice Admiral of the Blue, George Rooke, was ordered to observe the area near St Malo to ascertain whether an attack on the ships there was feasible. Rooke's orders were to proceed to the French coast off Cape Frehel, near St Malo, to see if it was possible to anchor a fleet there in any security from the weather and tides. He was also ordered to send small vessels near St Malo to observe the readiness of the ships sheltered there.[94] On 5 July Rooke transferred his flag into the third rate *Berwick*. His squadron consisted of:

*Northumberland**	*Berwick**
Defiance	*Warspite*
Monmouth	*Lenox**
*Hope**	*Cambridge*
Swiftsure	*Restoration**
Resolution	*Dreadnought*
Monk	*York*
Woolwich	*Adventure*
Despatch brigantine	*Griffin* fireship
Speedwell fireship	

* 70-gun third rates of the 1677 programme

The Dutch contributed twelve ships of between fifty and eighty-six guns, two frigates and three fireships. As the main fleet weighed anchor and sailed north, Rooke's detachment sailed SSW towards Cape Frehel. During the night of the 6th, the *Adventure* chased two sail close to the shore but lost sight them in a fog. For the next two days the squadron lay at anchor near St Malo and from *Lenox* it was possible to see at least nine of the elusive French warships safely anchored within. On the morning of 8 July, as *Lenox* still lay at anchor off Cape Frehel about three leagues west of St Malo, a French dogger pink was sighted near the shore, sailing up from the westward. *Lenox* manned two of her boats with a lieutenant in each, who together with several boats from other ships gave chase. The pink made for land as the boats closed in and ran ashore under a small castle. The castle had no guns and lay in ruins but a small army of men camped nearby ran down and fired at the boats. Undaunted, the attack continued and the first boats to reach the pink were those from the *Lenox*, whose men promptly boarded. After a short but bloody struggle most of the French crew managed to flee ashore and the pink was captured. Captain Partridge of the fireship *Griffin* was seriously injured by a shot to the head and seven or eight other men were killed, among them Samuel Jeffries of the *Lenox*. He was the first man from the ship to be listed in the pay book as slain. Captain Munden reported another man from his ship wounded. He was Thomas Humperbee; a shot had passed through his left arm, into his body and lodged in his right breast. Although the shot could not be removed, Humperbee recovered from his wound and received a pension.[95]

The pink was brought off into deep water at midday, and on being searched it was found she was armed with six guns and laden with salt bound for St Malo. The next day, after consulting his officers, taking soundings and completing his observations for a possible landing, Rooke ordered his detachment to weigh anchor, head northward and return to Torbay where the main fleet lay. As the ships passed the north end of Guernsey, *Lenox* recorded "The Commander of the fireship *Griffin*, shot when boarding the dogger pink died and was buried, his fireship fired ten guns. We followed with as many."

On 11 July, as the detachment lay off the north of Guernsey anchored in thirty-five fathoms, a gust of wind caught and drove *Lenox* so her best bower cable fouled the anchor cable of *Swiftsure*. In desperation to avoid collision with the *Swiftsure*, *Lenox* cut her own cable and then let go her small bower anchor. Before the anchor could bring her up *Lenox* swung round and drove into the *Restoration*'s stern, breaking her mizzen topsail yard. To avoid further damage *Restoration* cut her cable and drifted clear, while *Lenox* let go her sheet anchor. The lost best bower cable of *Lenox* was marked with the marker buoy and later taken in through the spare hawse of the *Swiftsure*.[96]

The homeward-bound ships became scattered as the Dutch and four or five English ships found they could not get about Guernsey before the tide was finished. *Lenox* and most of the English ships arrived off Berry Head the next day and were joined by the rest of Rooke's detachment a day later.

Lenox spent the next few days anchored two miles SE of Dartmouth Castle where her provisions were replenished. Russell was concerned about the condition of the hard-working third rates. He ordered the carpenters of the *Britannia*, *Royal Sovereign* and *Duke* to jointly board those in the Red squadron and together with their own carpenters, take a survey of the hulls and report back to him. He further ordered they be careful and exact and report what repairs would be necessary when the ships were brought into dock. *Lenox* was among the seventeen ships listed for surveying[97] but sadly the report for her seems not to have survived.

Rooke meanwhile reported to Russell that about thirty-five ships lay in St Malo but that the coast was very dangerous, with strong tides, and that "not one of the pilots would undertake to carry in any ship-of-war or fireship, to make any attempt on the French ships at St Malo, though I offered £100 encouragement to each man". After studying the gloomy report, Russell concluded that a land assault supported by the fleet would be very hazardous.

In a subsequent council of war, Russell and his officers concluded that although they were opposed to making a land assault, the isolated French ships would not winter at St Malo where there were few facilities and would soon set sail, probably for Brest but possibly for the Mediterranean. The best chance to attack them would be during their passage. Acting on this assumption, the fleet was divided into two squadrons, one under Russell and the other under Captain Neville, who hoisted a broad pennant as Commodore in the third rate *Kent*. Neville's squadron consisted of about twenty-five English and Dutch ships of which *Lenox* would form a part. Russell sent Neville written orders on 16 July to proceed without loss of time to a position about ten leagues north from the west end of the Isle of Bass, where he was to lie with his squadron in order to intercept the enemy if they sailed from St Malo to the westward. Ships were to be sent out to the north and south in search and on meeting any French Neville was to take or sink them. He was to continue on station until he saw the enemy or received new orders.[98]

At eight o'clock the same day *Lenox* and her squadron raised anchor. Two days later they had reached their station and the Seven Islands were once more in sight. They remained off the dangerous shore for two weeks in almost continuous gales, tacking back and forth. The French did not sail, but the nimble *Greyhound* managed to capture three French fishing boats, while the *York* and *Woolwich* captured a prize each. On the 28th a large number of sail was seen off to the east but it was not until the next day they made contact, only to discover it was a squadron under the command of Sir John Ashby, Admiral of the Blue. He had orders from Russell to replace Neville's squadron, which was to return home, but to take as many of Neville's ships that would make his own squadron up to a strength of thirty. He was also ordered to report back to Russell at Dartmouth as conveniently as possible.[99]

Ashby, still in the *Victory* in which he had he fought at Barfleur, had twenty-five ships and among the five he selected to remain with him was *Lenox*. On 30 July *Lenox* took in a month's provisions from a victualler and a month's gratis pay for all the crew awarded for their actions at Barfleur. Ashby then dispatched *Lenox* from the squadron to report back to Russell.

Lenox arrived off Dartmouth the next day, 1 August, and anchored in Dartmouth Range, two miles from the castle. Captain Munden immediately went ashore to report that no French warships had sailed; meanwhile his ship was heeled and a start was made at scrubbing her bottom. Two days before, Russell had convinced a General Council that it was too late in the year and too dangerous to make a landing on the French coast. At about the same time Queen Mary, growing impatient at the lack of action, issued fresh instructions urging that something be attempted against the ships at St Malo by her soldiers, now aboard their transport ships and ready to be escorted by the fleet. Russell held another council of war and remained steadfastly opposed to such an attack. The council resolved to send the fleet and transports to St Helens and await further orders.

The failure to act more positively undermined all the prestige that Russell had gained in battle earlier in the year. All the plans involving the army and transport ships were abandoned, but Anglo-Dutch naval power and the threat of landings upon the French coast had probably achieved the desired effect and caused thousands of French troops to be stationed in Brittany rather than on the borders of Holland.

As a result of the council of war, Munden was given an express letter to be delivered to Sir John Ashby. The contents were orders for him to forthwith return to St Helens with the English and Dutch under his command, leaving only the *Oxford* and *Adventure* to cruise off the Lizard.[100] *Lenox* immediately made all the sail she could and stood south. As she neared Ashby her bowsprit began to show signs of strain and to make sure nothing gave way the spritsail and spritsail topsail were taken in and yards taken down. The next day she turned westwards, found the squadron and made the signal to speak with Admiral Ashby.

Munden delivered the express in person to Ashby on the 4th, containing orders to leave for St Helens. Ashby, however, wished to remain longer in the belief that the ships in St Malo would soon set sail. He conveyed his request to Captain Munden who was immediately sent back to his ship with orders to return to Russell at Dartmouth. *Lenox* made sail and steered north in squalls of rain and accompanying gales. As soon as she arrived, Munden went aboard Russell's flagship to report, while *Lenox*'s crew heeled the ship again and scrubbed more of her bottom. Russell decided it was time to return to St Helens with the main fleet and end the year's campaign but agreed to Ashby's request to remain with his squadron off St Malo.

Lenox set sail once more to give Ashby his latest orders. The gales and squalls continued but by 5 o'clock the following morning Ashby's fleet was sighted to leeward. Munden went aboard the flagship, delivered the orders, returned to his ship and made sail for Dartmouth.

Munden reported again to Russell who gave orders for *Lenox* to sail out to the Isle of Bass and seek out four ships cruising there and give them orders to return. *Lenox* sailed out again steering SE during the night and by dawn of the 9th "espied two ships to windward and two to leeward"; after delivering Russell's orders *Lenox* stood away from the English coast

to join the main fleet now at St Helens. She arrived off the Isle of Wight in a thunderstorm on the 14th. The next day *Lenox* anchored and struck her yards and topmasts. Her log recorded that "here rides at anchor Admiral Russell with almost all the fleet, English and Dutch".

About this time *Lenox* lost the services of her tender, the *Tartan*. Since early June she had been mentioned at Admiralty board meetings with a view to taking her into the navy and a Mr Rainsford was appointed to survey her at Falmouth. The report seems to have gone awry for nothing happened, and as a result the Admiralty minuted a note to speak with the Navy Board about "Captain Munden's *Tartan*". On 11 July the question was raised again and as a result an order dated 2 August required the *Tartan* to be fitted out at Plymouth as an advice boat.

Captain Munden wrote to the Navy Board on routine matters. For delivery to the Board's offices he had given his muster books to a Mr Bully, whom he stated was often employed by their Lords of the Admiralty on business, and asked them to order some necessities for the surgeon.[101] A cable was very worn and Russell issued orders for the masters of the *Hampton Court*, *Burford* and *Eagle* to carry out a survey. They reported it so very much worn that the ship could not safely ride by it. As a result Russell issued a further order for the old cable to be delivered into the stores at Portsmouth forthwith[102] where it was probably unpicked and used as oakum for caulking ships' seams.

On the 19th, those aboard *Lenox* would also have been talking about John Pyke, the second lieutenant of the *Swiftsure*, who was court-martialled for cowardice in the last engagement during which, it was alleged, he forsook his station and was discovered hiding in the captain's storeroom of his ship, having been there for two hours. So serious a crime could carry the death sentence. President of the court was Admiral Russell and John Munden was one of the captains presiding.[103] Pyke was found guilty and subsequently carried in a pinnace with a halter round his neck and made to board all flagships where, to the beat of a drum, a declaration was read according to the sentence in due solemnity by the Provost Marshal. Then, "all the ships companies crying three times a coward", he was then sent ashore in disgrace. Humiliating though this must have been, seventeenth-century punishment was often more fitted to the offence than the strict discipline imposed later, when Pyke might easily have been hanged from a yard-arm.

With the approach of winter and the withdrawal of Ashby's ships from the French coast, some of the ships in St Malo set sail. Three 50-gun French ships, which had escaped after Barfleur, attacked a convoy of seventy Dutch merchantmen escorted by two warships off the Lizard at the end of August. The merchantmen were lucky and escaped but at the cost of both their Dutch escorts. Anticipating their arrival, Russell sent orders to Captain Munden dated 28 August to command a small squadron for their protection consisting of *Lenox*, *Hampton Court*, *York*, *Montague*, two fireships, four Dutch vessels and two victuallers. He was to proceed to Plymouth where the Dutch merchantmen were expected; there he was to signal to the masters of the Dutch ships and inform them that he would take them to St Helens.[104] *Lenox* and her squadron weighed the following morning and tacked down Channel on a light wind. By 1 September they were off the Start and next day met the undefended merchantmen as they approached Plymouth. All the ships entered the Sound where Munden, in accordance with his orders, spread a white ensign on the ensign staff and fired a gun for the masters of the merchantmen to come aboard *Lenox* for their orders. The weather remained calm but on 5 September Munden received another order from Russell instructing him to add the *Exeter* hulk and *Sunn* prize, already at Plymouth, to the convoy. He was further instructed to provide men from his squadron to man the two ships. The order came just as the Dutch merchantmen were setting sail bound for the east. Munden sent Lieutenant Williams and twenty men from each of the warships to man the two ships. The squadron was joined by the *Deptford* and nine further sail of merchantmen on the 9th and aided by a westerly gale, arrived safely at Spithead two days later. The Admiralty later granted Lieutenant Williams £10 for his care and pains in the operation.[105]

At Spithead, as gales blew up, *Lenox* joined the bulk of the fleet waiting to be refitted during the winter, and lowered her yards and topmasts. On 15 and 16 September a court-martial was held aboard the second rate *Albermarle*, presided over by Sir John Ashby, and Captain Munden was among twenty-seven captains who attended. Among matters dealt with was the captain of the *Oxford* for not forcing a Swedish man-of-war to strike his topsail in acknowledgement. He was discharged after witnesses stated that the Swede saluted by firing guns at the meeting and parting and did not fly his pennant. The purser of the *Windsor Castle* was discharged after being accused of selling brandy at excessive rates; but the most serious case involved seven men from the *Resolution* who mutinied after being turned over into the *Cornwall*. They were found guilty and the two ringleaders were sentenced to be hanged, while the rest were to receive 100 lashes. The harsh sentence caused a stir and Lieutenant Durley noted the verdicts in his journal. He mentioned the incident again two days later when the prisoners received their punishment and were whipped from ship to ship.

A Dutch privateer then arrived with news that twenty-two of the French warships in St Malo, which Russell had spent all summer hoping to bring to action, had set sail bound for Brest. A powerful squadron including *Lenox* was ordered out into the awful winter seas in the forlorn hope of meeting them. Sir John Ashby hoisted his flag aboard the third rate *Eagle* and made the signal to unmoor. The fleet of thirty-three English and Dutch weighed anchor at six in the morning of the 19th and steered down Channel aided by a NE gale. The squadron worked its way out to the Lizard by the 22nd, where it remained in continuing gales until the end of the month. The French, however, had made too much progress to be caught and when the wind changed to blow from the SW, the squadron made its way back up-Channel to shelter in Torbay. Ashby ordered five of *Lenox*'s sisters, *Suffolk*, *Hope*, *Expedition*, *Northumberland*, *Captain* and two fireships to cruise to the westward before weighing with the rest of his squadron to return to St Helens, arriving on 11 October.

Lenox anchored in Stokes Bay where a survey was made of some bad provisions aboard. The ships moved on to Spithead where *Lenox* lowered her yards and topmasts and received fresh provisions for her crew. On the 18th, twenty-one guns were fired in celebration of King Williams' safe return from the continent where he had been fighting the French. On 1 November it snowed, but some relief from the conditions occurred on the 4th when *Lenox* fired her guns twenty-one times in celebration of the king's birthday, followed by the same again the next day in commemoration of the anniversary of the Gunpowder Plot. Admiral of the Blue, Sir John Ashby, moved into *Lenox* and hoisted his flag at the main masthead. Many of the ships under his command were sent to Chatham for the winter and Sir John Ashby left for London on the 16th, taking took down his flag, which was replaced by a private ship pennant. A month's provisions were taken aboard and orders were received from Sir Francis Wheeler to cruise off the coast of France for four days. *Lenox* unmoored and rode short on her best bower anchor but, unusually for the time of year, the wind died away completely and it remained calm for two days. The planned cruise to France was cancelled, probably to the immense relief of the crew, and *Lenox* joined several other ships bound for Chatham for refitting.

Lenox was anchored in the Downs on 22 November when Captain Munden received orders to seize twenty ships of a Hamburg convoy and their man-of-war escort. They had been stopped by some English privateers on suspicion of carrying contraband goods bound for France. The Hamburg warship and convoy had fired at one of the privateers, ran it

down and put her men in irons. Hamburg was a member of the League of Augsburg and the convoy was therefore an ally of England trading with the enemy. On the 24th some of the merchantmen were delivered into the custody of the privateers who had originally seized them. Munden then received written orders from the Admiralty to join together with the *Expedition*, whose commander was to follow Munden's orders, and take the rest of the convoy and the warship to Spithead.[106] Captain Munden sent for *Expedition*'s lieutenant and told him to make a signal for the masters of the merchantmen to set sail. At the same time Lieutenant Durley went in *Lenox*'s smack to encourage the merchantmen to obey the signal. Durley found some of the merchant flyboats had three anchors down, as a period of pleasant weather was coming to an end with a gale rapidly developing. Despite his efforts, there was nothing Durley could do to persuade the masters to set sail. Captain Munden himself boarded one ship to hasten it in getting under sail but the merchantmen stubbornly remained at anchor.

The following day, 28 November, the gale increased to a hard gale and Munden scornfully wrote that most of the ships were riding with two or three anchors ahead while he was riding at a single. The merchantmen's decision not to sail was fully justified, however, for the gale increased in ferocity and several of them were observed flying distress signals and Munden was forced to lower *Lenox*'s yards and topmasts. During the last day of November the gale reached its climax, one flyboat overset while another was driven ashore a little to the north of Sandown Castle. *Lenox* was obliged to cast her other anchors and moor. Munden wrote to the Admiralty informing them of the condition of the merchantmen, adding that one had lost all her masts and many had lost anchors.

By 3 December the gale had eased off sufficiently for *Lenox* to get up her masts and yards. Communication was made with the merchantmen and Munden gave their masters leave to go ashore to obtain new anchors and cables. The Admiralty, realising that it was now impossible for the merchant ships to sail for Spithead, issued new orders for them to be taken into Quinborough Swale and men from *Lenox* and *Expedition* were to go aboard to help.[107]

Munden tried to send pilots to the merchantmen but the weather became so bad their boats could not approach. By the 5th the weather had eased to become fair and Munden signalled for all the masters to come aboard *Lenox*. The merchantmen were now in some sort of condition to sail and even the overset ship was righted. Captain Munden informed eleven of the thirteen masters that he now had pilots for their ships and would set sail before nightfall bound for the Swale. The masters themselves would not be allowed to go with their ships but sail all together in the last one. The arrangement made sure they could not cause problems during the voyage after which they would be returned to their own ships. Munden then sent Lieutenant Durley and thirty-two men from *Lenox* and twenty-four from *Expedition* aboard the ships as ordered by the Admiralty. The next day pilots were found for the remaining two ships and they followed the others toward the Swale. The implications of what had happened brought attention to the papers aboard the merchantmen as they would be used to decide the legal claims against them. The Admiralty ordered Munden to "send up the writings and papers belonging to them carefully sealed up distinctly and writ on to whom they belong to into the Judge of the High Court of the Admiralty".[108] Subsequently, at an Admiralty Board meeting attended by Mr Jacobson, agent for the City of Hamburg, it was resolved to put the merchant ships into the hands of the commission for prizes to be prosecuted according to law.[109] The only ship of the convoy now left in the Downs with *Lenox* and *Expedition* was the Hamburg man-of-war. The hard gales continued and the two merchant ships delayed while waiting for pilots lost all their masts. Other ships entered the Downs, including two Dutch men-of-war; one saluted *Lenox* and the second did so only after Munden sent a boat out to demand that he do so. On 16 December orders were received for both *Lenox* and *Expedition* to sail at the first opportunity of wind and weather to Spithead for refitting and to take along with them the Hamburg man-of-war which they were to deliver to the custody of the Prize Officer at Portsmouth.[110] The weather eased enough by the 18th for Munden to make the signal to *Expedition* and the Hamburg warship to unmoor. The wind blew helpfully from NNE but dropped to a calm two days later as the ships entered St Helens roads. Captain Munden must have been thankful to conclude his involvement with the Hamburg convoy and be rid of the unusual difficulties it had brought him.

At Spithead the gales soon returned and *Lenox*'s masts and yards were lowered, and a sudden gust of wind made the small bower anchor "come home" (give way): "I let go my sheet anchor under foot and brought both bowers to bear." The stormy weather continued and on the 30th a West Indiaman, the *Loyalty Of London*, under the command of Captain Priest bound for Barbados was observed being driven ashore in Stokes Bay. The people aboard were saved after her masts were cut by the board.

1693 – THE TURKEY CONVOY

After their defeat in battle, the French resorted to attacking English and Dutch merchant ships using small numbers of warships and privateers. The ravages of these attacks on trade caused concern and convoys had to be escorted through the Channel to safety by warships. Also requiring escort past the Scilly Isles and into safe waters was a small squadron under the command of Sir Francis Wheeler in the *Resolution*. His squadron was to be assembled at Spithead with orders to attack French possessions in the West Indies. It was ready to sail by early January 1693 and consisted of two third rates, six fourth rates, three fifth rates, one sixth rate, three fireships, a store ship, a hospital ship, a bomb vessel, and ships to transport 1500 soldiers.

The pressure on the naval yards to refit the fleet during the winter caused many problems, as there was only a limited number of dry docks available. It had originally been the intention to refit *Lenox* first at Chatham, then at Portsmouth, after she had been diverted there with the Hamburg man-of-war. Now she, along with the *Elizabeth* and *Expedition*, were to be made ready for sea with orders to be part of the escort squadron sailing with the West Indies ships.[111] The Admiralty wrote to the Commissioners of Victualling, ordering them to give an account of how long it would be before they could supply the ships with a month's provision each.[112] As part of the preparations the boatswain's and carpenter's stores were replenished, a spare anchor and cable stowed and a new topmast got up. Some men, probably volunteers, left the ship, and to make up the numbers forty men were turned over into *Lenox* from the *Plymouth*, which had recently arrived at Portsmouth. Captain Munden wrote to the Navy Board that his men, who had sailed with the Hamburg merchantmen into the Swale, had now rejoined him at Portsmouth but had not been paid conduct money for their journey. He also stated that some necessities were exhausted and pills and medicines required replenishing.[113] The Admiralty immediately ordered the men of *Lenox* and *Expedition* who were owed conduct money to be paid.[114]

Lenox spent the first week of January continually setting up and taking down topmasts and yards according to the weather. She then took on board final provisions of bread and water before unmooring and hauling home her topsail sheets to sail together with *Expedition* on the 7th. It took another two days, in the teeth of gales, to get out and join the small squadron sailing past the Needles where Sir Francis Wheeler flew his flag from the mizzen topmast head of *Resolution*. The ships then sailed down the Channel and were joined off the Lizard by seven other warships, including the *Dover* with a French privateer she had recently taken. On 13 January Sir Francis made the signal for *Lenox* and *Expedition* to leave as they had reached safe waters in the previously agreed position off the Scilly

Isles where "we gave Sir Francis three cheers and eleven guns. He returned us three cheers and nine guns. We got our larboard tacks on board and stood away NE by E."

The next day the weather deteriorated to a severe gale during which the main yard split in the slings, preventing *Lenox* from setting her main course. The main topsail yard was taken from the main topmast and slung in place of the main yard and the mizzen topsail yard set for the main topsail yard. The damaged main yard was temporarily repaired on deck by the carpenter's crew who fished it the best they could. They got it up again two days later and moved the yards back to their correct masts. It must have been a very unpleasant voyage for her crew, never really dry, in freezing conditions, and with the only heat coming from the cooking stove in the forecastle to keep a lucky few warm.

Lenox returned up-Channel and moored at Spithead on the 21st to wait for her turn in dock for her delayed refit. As she waited, a new main yard and spare topmast were taken on board. On 13 February Munden failed to notice the anniversary of the king's day of accession but was soon reminded by Sir John Ashby's smack, which came alongside and ordered him to fire nineteen guns. A few days later the king himself came to Portsmouth and every ship in the fleet fired twenty-one times. William left the shore to go aboard the *Royal Oak* and the royal standard was hoisted at her masthead. He then visited a Dutch warship and was rowed in his barge to view the fleet, to the accompaniment of more salutes.

On 18 February *Lenox* sailed into Portsmouth Harbour and moored at the fourth mooring. Ten days later, and covered in snow, she was taken to Jolly Head where she was again moored and made ready for docking. To lighten her, the topmasts and yards were taken down, and all her guns, ballast, and stores removed. Captain Munden had leave from the Lords of the Admiralty and left *Lenox* for London.[115] He would not return, but had distinguished himself in a highly successful period of *Lenox*'s career; he was promoted into the second rate *St Michael* and in 1695 made captain of the first rate *Victory*. He became a Rear Admiral and was knighted by King William on the *William and Mary* yacht in 1701. This would be the zenith of his fortunes, for after commanding an unsuccessful expedition, he was dismissed by Queen Anne and he lived in retirement until his death in 1718.[116]

Hearing of Munden's appointment, Lieutenant Durley arranged for his possessions to be sent round to the Nore. The actions of Lieutenant Durley had not gone unnoticed by the Admiralty either, for in the minutes of a board meeting at this time it was recorded, "Lieutenant Durley be remembered for promotion to a bomb vessel or brigantine".[117]

Lenox's new commander would be Captain William Kerr who received orders while at Plymouth dated 12 February to repair to London to take command of a third rate ship.[118] Kerr was originally appointed second lieutenant of the *Pendennis* in 1688 and on 14 May 1690 promoted commander of the fifty gun fourth rate *Deptford*. He and his men had made a name for themselves by being pursued and exchanging fire with many ships of the French fleet just before the Battle of Beachy Head. In November 1691 in company with the *Chester* he captured a large privateer of twenty-two guns that had caused a nuisance in the Channel for some time. He then captured the *Fortune* from Nantes in October 1692, another large French privateer carrying twenty-four guns, eight patararoes (guns firing stone shot) and 180 men.

The following month, in company with the *Portsmouth*, he took an even more powerful vessel called the *Hyacinth*.[119] Kerr must have made himself and his men rich with prize money and persuaded many men from his old ship to follow him to London. Once there he would be able to tell them where to proceed to their new ship. In London at the Admiralty Board meeting of 27 February, a minute records a visit by Kerr who delivered a list of fifty men who were coming up to town voluntarily to serve under his command. He was worried other ships would press them before they were able to join him in their new ship, and requested a written protection. The board resolved that if any of Kerr's men were pressed aboard other ships they would issue a discharge on request by Captain Kerr.[120] Such was the trust between Kerr and his men that he paid them conduct money of 25s each from his own pocket at their departure.[121] Many of the men made the long journey from Plymouth to London and then on to Portsmouth. They arrived in small parties, seventeen arriving aboard *Lenox* on 7 March.

At the Admiralty on the 9th, Kerr officially received his commission to command *Lenox*. He took the opportunity while at the Admiralty offices to have leave granted for eight days in London and to have two printed press warrants to help him man his ship.[122] At the end of March Kerr wrote to the Navy Board politely requesting the 25s conduct money he had paid to each of his men from the *Deptford* be reimbursed, while pointing out that Captain Laton had been reimbursed conduct money under similar circumstances when he was removed from the *St Albans* into the *Hampton Court*.[123] The Admiralty at first ordered conduct money for only twenty-five men be paid to Kerr but a week later issued new orders for him to be paid for all the men who joined him from the *Deptford*.[124]

Since her last refit in January 1692, twelve of *Lenox's* men had died, through various causes. Two had died at the time of the Battle of Barfleur and another had been slain when taking a French ship. There were no clusters of deaths, except perhaps during October when three died. Taking into account the considerable time the ship spent at sea with 500 men aboard, *Lenox* was a very safe place to be as regards both health and physical injury. During the same period thirty-two men deserted the ship.

Lenox finally entered dry dock on 13 March 1693. The dock gates were then closed behind her and her hull was cleaned but she was not to be refloated the next day, unlike the previous January. The ravages of a year's hard service required repair and she was not ready for launching until the 25th, when she was taken out of the dock and moored once again at Jolly Head.

The next day her new captain came aboard for the first time. The same three lieutenants who had sailed in her the previous year greeted him. Edward Durley remained as first lieutenant, while Thomas Williams was Second and George Knitch third. Captain Kerr wrote to the Navy Board acquainting them that *Lenox* was now under his command and that she was out of dock, "and being in great want of a tender for the impressing of seamen desire that you will give your orders to the Commissioner of this place that I may have a ketch for that service."

All the stores and equipment that had been taken out before her docking were now taken back on board. The masts and yards were hauled up and set, the ship was rigged and any decayed or worn ropes were replaced. Each day a lighter brought alongside a load of ballast averaging some 40 tons of shingle, which was taken down into the hold until there was a total of 337 tons. The hard work was eased by the fact that 300 of *Lenox's* crew were retained in pay from the previous year, serving as extra labour to help speed up the refit.

For the allies, the euphoria of victory after Barfleur and La Hogue had now given way to disappointment because the French fleet had eluded them at the end of the previous year. The issue became political as vast amounts of money had been expended in keeping the fleet at sea. The Whig Admiral of the Fleet, Russell, took most of the blame and was accused of dither and delay. He now fell into conflict with the Earl of Nottingham, the Tory Secretary of State, over his conduct of the campaign. Pepys, by then retired, wrote of the affair: "Remember also on this head of the imperfectness of our judgings in sea-matters, the irreconcilable difference between the resolutions of our two Houses of Parliament upon this very action, the Lords in favour of my Lord Nottingham and the Commons of Mr Russell, and their thinking it of so little moment to have the knowledge of the very truth therein".

William and Mary, however, supported their Secretary of State and Russell was obliged to resign. He was replaced by an unsatisfactory split command of three Admirals: Henry Killigrew, Sir Cloudesley Shovell and Sir Ralph Delaval, who had orders for the coming summer campaign to destroy the French fleet in Brest. It was agreed to assemble a fleet of seventy warships carrying five regiments of foot that would be landed to destroy the batteries of artillery protecting the harbour entrance before the warships entered. But there were other considerations; a Mediterranean convoy bound for Turkey consisting of hundreds of merchant ships required escort and it was agreed that it should sail with the fleet to Brest. A squadron under the command of Sir George Rooke would be detached to continue with it through the Mediterranean as protection against a possible attack by the French squadron from Toulon. If, on reaching Brest, it was found that the French fleet was no longer in harbour and therefore at sea, the whole fleet would continue with the convoy. The opportunity was also taken of adding smaller merchant convoys bound for Virginia and Bilbao to the fleet until they, too, were in safe waters.

On board *Lenox*, First Lieutenant Edward Durley received notice of promotion. He had joined *Lenox* as second lieutenant when she was first commissioned in 1690 and risen to first a year later. Admiral Rooke received orders dated 6 April 1693 for Lieutenant Durley to be sent to London to receive a commission to command a brigantine. On arrival an Admiralty commission was duly made for him to command the *Shark*, an 8-gun vessel of 58 tons.[125] A week later Durley began his journal aboard her.[126] He was later to achieve fame in his next vessel, the fireship *Charles*, which he blew up against the fortified Quince Rock causing its destruction during a fleet attack on St Malo. He survived the action and lived to be placed on the list of half-pay officers at the end of the war. The vacancy left by Durley resulted in promotion for *Lenox's* second and third lieutenants; Thomas Williams moved up to first and George Knitch second. Their commissions were dated 9 and 10 March 1694.[127] Her new third lieutenant was George Delaval, descended from a different branch of the ancient family to which Sir Ralph Delaval belonged.[128]

The Admiralty issued an order on 11 April to keep two or three ships cruising in the Channel for the security of merchant ships against French privateers.[129] As the fleet struggled to find enough men, the well-manned *Lenox* left Portsmouth Harbour to be part of the small cruising squadron, mooring in Spithead on 15 April. A week later she sailed out into St Helens where she was joined by two of her sisters, *Hampton Court* and *Northumberland*, together with the *Newcastle*, two fireships, and two Dutch vessels. On 28 April, after being delayed for two days by gales, they got up their yards and stood out to sea. By 3 May they were off Land's End where at 4.30 pm they made out sail to the SW, about four miles distant. They gave chase but by 8 pm could not come up to them and, judging them to be Dutch privateers, left off and stood northwards.

Two days later off the Scilly Isles they met the *York* and later the same day the *Deptford*, *Portsmouth*, and *James Galley*. After a few days of further cruising the squadron anchored off Deadman's Cove before sailing on to Berry Head. They then sailed back up the Channel on 11 May to rejoin the main fleet which was still being assembled by the three admirals for the attack on Brest, anchoring in six fathoms off St Helens Point. They joined the combined English and Dutch fleet. *Lenox* was assigned to the Red squadron which included the *St Michael*, commanded by John Munden, and the flagship *Britannia* with the three admirals aboard.[130]

During the final two weeks before the fleet sailed, a survey was made of the men aboard each ship. *Lenox*, whose complement was 460, bore 434 on her books, although only 406 were actually on board at the time. Compared with her sisters she was slightly better off than most.[131] A few days later soldiers stationed at Portsmouth were transferred into the warships and *Lenox* received her share to help make up her complement. On 14 May the longboat brought on board a company of soldiers belonging to Colonel Collyer's regiment. A day later Sir Richard Haddock, Comptroller of the Navy, and his clerks came aboard *Lenox* and started paying the crew, a task he finished the next day. Having earned a sizeable amount of pay and aware of the dangers posed by another summer campaign, Richard Studvill, the master, made his will, witnessed by the first and second lieutenants, leaving everything to his wife Francis.[132] In the chaotic final preparations of the fleet another company of soldiers belonging to Colonel Villiers was brought aboard but sent ashore again the same day.

By 30 May and after a council of war the fleet was ready and the signal was made to set sail. The Anglo-Dutch fleet, together with the Mediterranean convoy, weighed anchor and stood out to sea. The fleet made good progress and were off the Lizard by 2 June and thirty leagues WSW of Ushant by 4 June – the position off Brest where it had been agreed to leave the Mediterranean convoy and its escort under Rooke. The position of the French Brest and Toulon ships was still unknown, however, and the indecisive three admirals did not send out scouts to gain intelligence on the French ships as the fleet passed. Without knowledge of the whereabouts of the French, the three admirals sailed a further twenty leagues with the convoy and its escort before leaving and sailing northwards again.

Lenox took her station in the line of battle, formed in the hope of meeting the enemy. The French were not seen and after arriving once more off Brest the *Warspite* was sent in to gain information of the French fleet in harbour. To her crew's surprise it was empty, the only vessels observed being two or three small fishing boats. The assault on Brest was now pointless so the fleet continued the search by heading north towards the Scilly Isles, which were sighted from *Lenox* on 20 June. The fleet was now very short of provisions and the three admirals, suddenly aware of their primary duty of protecting England from possible invasion, abandoned the fruitless search and sailed east to Torbay where the fleet moored on 22 June. *Lenox* anchored in ten fathoms of water and for the next two weeks took in provisions – not salted cask meat, but fresh beef, pork and mutton, butter, bread, oatmeal, and many tuns of beer. During a calm the opportunity was also taken to heel and scrub the ship's bottom.

As the fleet replenished itself, news finally came in of the whereabouts of the Brest fleet. The Consul of Oporto reported alarming news that on 1 June the ships entered Lagos Bay, an ideal position from which to intercept the Mediterranean-bound convoy. Other reports came in that the Toulon ships had also sailed. The three admirals had made the mistake of assuming the Brest fleet would remain in northern waters when in fact they had sailed south. The English court was now very concerned for the fate of the Mediterranean convoy and Queen Mary ordered the fleet to put to sea again.

On hearing of the danger to the convoy, the three admirals, still at anchor in Torbay, held a council of flag officers and captains on 9 July. It was resolved to proceed to a position forty leagues SW of Ushant and then consider whether it was best to remain there or to sail to some other station in the hope of intercepting the enemy as they returned to Brest. Three days later *Lenox* and the fleet weighed anchor and after weathering Berry Head, set course down Channel.

As the fleet set sail orders arrived, dated 13 July, from the Admiralty to the three admirals, which instructed them to send Captain Kerr to attend the Admiralty Board and answer charges of ill-treatment towards the muster master of the Red squadron, one Robert Wilkins: "Specifically for not complying with what the muster master demanded of him pursuant to his instructions". The Admiralty were also to appoint another captain in Kerr's room.[133] The charge seems rather lame on the surface and it appears the real problem was never put to paper. Wilkins's duty was to periodically list or "muster" all the names of the men aboard each ship. These lists were then compared with ship's pay books with the object of revealing frauds which ingenious captains and pursers devised to increase

their personal wealth. The simplest form of this practice was to enter fictional men's names on the pay books and pocket their pay. Needless to say muster masters were not popular visitors aboard warships. The three admirals, traditionally siding with a sea officer, decided not to comply immediately with the orders and wrote that they were just setting sail as the orders were received and that Captain Kerr was very sick and could not be sent ashore.[134] To tell the Admiralty that Kerr could not be sent ashore because he was sick seems a very strange excuse, as everyone knew that sick men were precisely those who were sent ashore. Indeed, the Admiralty was not convinced by the letter and over the next two weeks other orders arrived stating that their orders be "forthwith complied with".

A day after the fleet sailed, at 9 am on the 14th, the sixth rate *Lark* came in from the Straights (as the Mediterranean was generally known at that time) and brought an account that our Straights fleet had fallen in with the whole French fleet. Sir George Rooke, whose squadron was escorting the Straits fleet, skilfully kept to windward of the French who failed to press home their attack, thereby allowing the bulk of the convoy to escape and eventually reach Cork. But trade had been disrupted and ninety-two merchant vessels, mostly Dutch, were lost.

The fleet and *Lenox* continued down Channel and arrived off the Scilly Isles on 20 July and formed a line of battle. They continued cruising in the vicinity but the autumn seas were causing damage to the weather-worn ships and many were sent home for repairs. It was believed within the fleet that the quality of the refits that had been made during the winter was poor and as a result the three admirals ordered a survey of each squadron. For the Red squadron the carpenters of the *Britannia*, *St Andrew*, *Ossory* and *Lenox* made written accounts concerning the hulls and lower masts. For *Lenox* they reported:

> "In the carpenters storeroom a beam broke off and worked down and likewise the hanging knee is broke and the foremost beam of the main hatchway is broke and the knee with the standard under it upon the orlop beams which occasions the ship working through very much weakness. In the pursers cabin we find a lodging knee broke and the false post disabled by rot. The rudder head very weak and ought to be strengthened with bolts and plates. All the masts and yards very good except the main mast which rings through much weakness, requires to be strengthened and caulked within and without which cannot be done without a dock."[135]

It must have been distressing for Ambrose Fellows the carpenter, as he would have remembered building the ship long ago in 1677. If *Lenox*'s report is alarming, many other of the third rates of 1677 were far worse; the *Essex* reached and stretched so much that she threw oakum out of her sides while the carlines, lying fore and aft between the beams, had to be cleated to stop them falling out. The entire head of the *Restoration* was so weak it nearly fell off when five men shook it. The *Hampton Court*, built alongside *Lenox*, worked so much that the bolts of the standards were drawn 2in up into the beams. As the three admirals observed to the Navy Board, "As we have so often observed to your Honours many of the ships with continual use for so many years together are grown weak and sickly and not fitting for winter service".[136]

At 9 o'clock on the night of 7 August, *Lenox*, sailing under her main and fore courses and with her topsails reefed in, was suddenly struck by a gust of wind that carried away her fore topmast. It took her two days before she managed to make a new one and get it up and rigged. Her damage and the continual orders for Captain Kerr to attend the Admiralty board resulted in *Lenox* being ordered back to Torbay where she arrived on 16 August. The next day, by chance, George Rooke arrived from Ireland after his recent action in support of the Mediterranean convoy and added her to his depleted squadron. *Lenox*'s departure from the main fleet did not result in her missing any encounter with the French fleet as it was now safely in the Mediterranean, where it intended to stay until it was safe to return north. If the events after Barfleur and La Hogue in 1692 had been disappointing, those of 1693 were considered disastrous. There was talk of treason by some in the Whig party, with accusations and suspicions that some of the fleet had been acting in the interests of King James. The Lords and Commons started their own inquiries and witnesses were examined about the disaster.

Two days after arriving at Torbay, Captain Kerr of *Lenox* left his ship to return to London to give his evidence concerning his treatment of Robert Wilkins. The Admiralty had prepared for Kerr's examination by ordering the muster master to attend them, bringing along his clerk as a witness. They also ordered Wilkins to "apply yourself to the Admirals of their Majesty's fleet for their orders to such persons also to come to town as you shall think most proper to give an account of the difference between Captain Kerr and yourself". Orders were also sent to the admirals to comply with Wilkins' requests concerning witnesses and to send the purser of *Lenox*, Mr Henders, to them as well.[137] The Lords of the Admiralty clearly took the matter very seriously and gave the muster master assistance in his case. *Lenox*'s purser was the first to be summoned into the Board of the Admiralty but loyally would not say a word against his captain. He was subsequently dismissed from his employment "for not giving the board a true account of what he knew concerning the difference between Captain Kerr and the Muster Master of the Red".

Captain Kerr attended the Board the next day. He was dismissed from his employment but mysteriously no mention of it appears in the minutes of the meeting.[138] It was a strange case that would warrant the dismissal of a captain but not be considered serious enough for a court-martial. Interestingly there was much talk of "disaffection" in the fleet at the time and Kerr's dismissal may have been partly political. At any rate this was not the end of Captain Kerr for, three years later, he turned up again as captain of the *Burlington*. The admirals replaced Kerr with Lord Archibald Hamilton, the seventh and youngest son of the Duke of Hamilton, to command the *Lenox* as temporary captain. He was at the time first lieutenant of the *Britannia*.

The normal season for fleets to be at sea was coming to a close and ships were in need of refitting and repairs. *Lenox* was ordered to Chatham for her refit. A storm on 25 August forced her to strike her yards and topmasts but the next day she was able to get them up again and she sailed for St Helens, arriving at the beginning of September. After mooring, the stormy weather again obliged her to strike her yards and topmasts. The next week King William sent orders to the three admirals to put ashore the four regiments of soldiers that were aboard the fleet.[139]

Lenox sent her soldiers, who had been with her all year, ashore in the longboat. On 17 September, Captain Lord Hamilton was ordered aboard the flagship *Britannia* and given command of the *Woolwich*, a fourth rate that was being made ready for sea as part of a small squadron for Channel service. To man the squadron, the Admiralty ordered men from the ships being refitted to be turned over into it, thereby making up each ship's complement to thirty above their normal complement for Channel service. On leaving *Lenox*, Lord Hamilton obtained an order for forty-five of *Lenox*'s remaining crew and all her marines, except commissioned officers, to be turned over into the *Woolwich*. A further thirty men were turned over into the *Vanguard*. On 26 July 1694, the turned over men from *Lenox* were still on board the *Woolwich* at the Buoy of the Nore under Hamilton, from where he wrote to the Navy Board office asking for the *Lenox*'s books to be sent down to him, so that the men might receive their pay.[140]

The practice of "turning over" men from ships coming in for refit into ships manning for service was resented by the men. After being at sea for many months most were in need of recuperation and longed to see their families and friends. Complications arose over the payment of wages as men's names were transferred to a new ship's pay books. In theory the only

men exempt from being "turned over" were volunteer seamen who had not been pressed into the navy and who also received two months' wages bounty during their first six months of service. Volunteers also had the right to leave a ship when it was refitted at the end of a season. To ease resentment on this occasion, the Admiralty further ordered that the turned-over men be paid the wages owed them up until the previous March, before the ships were to sail.

Lenox and eight of her sister third-rate ships were ordered to Blackstakes under the command of Rear Admiral David Mitchell to be made ready for their refits at Chatham.[141] Following the departure of Lord Hamilton to the *Woolwich*, her captain Christopher Myngs was given command of *Lenox*. He had served as lieutenant in several ships since 1684 and was appointed as commander of *Woolwich* in 1693.[142]

Following a calm, the squadron set sail on 21 September and anchored in the Downs that night. After arrival at Blackstakes, *Lenox* went on to Chatham where she arrived later that month. By the end of 1693 all her guns, powder and ballast were taken out and her topmasts and yards taken down. Captain Myngs wrote a number of letters to the Navy Board on routine matters. He still had a month's sea provisions on board and advised that they should be used up before he went onto petty warrant provisions.[143] Petty warrant provisions were provisions supplied directly from the victuallers on the authority of a petty warrant supplied by the Dockyard Clerk of the Cheque and only applied while a ship was in port.

While *Lenox* prepared for refitting, the fate of the three admirals was being decided. King William proclaimed those responsible for the loss of the convoy would be punished and ships' officers, including those from *Lenox*, were summoned to the House of Commons to give their evidence in preparation for the debate. The political tide had turned. The Whig party was now in the ascendancy and Killigrew and Delaval were both Tories, as was the Earl of Nottingham. In Parliament the motion that the three admirals were guilty of a high breach of trust was debated and voted upon but was narrowly defeated. Killigrew, Delaval and Shovell were dismissed from office, however, and Nottingham was forced to resign.

The fall of the Tories and the return to political power of the Whigs resulted in a change of naval command. Edward Russell, who had spent the last year ashore advancing the Whig cause, was the main benefactor of the political change. The day after the fall of his Tory opponent, Nottingham, Russell became Admiral of the Fleet, and later in April he was appointed senior Lord of the Admiralty. He was at the height of his power, underpinned by the appointment of his friend, Charles Talbot the Earl of Shrewsbury, to the position of Secretary of State. One of Russell's first acts was to send written orders to all flag officers and commanders, including Myngs, to attend him at Chatham on 11 December to discuss matters relating to the fleet.[144]

During the autumn, as the allied fleet returned to port, the French fleet sailed from the Mediterranean and returned to Brest. It was not a tactical move, but dictated by the lack of facilities for a major fleet in the Mediterranean. During the year *Lenox* had lost only seven men, three of them during September when the ship was barely at sea. The number of men deserting was also down from the previous years to only twenty-five. *Lenox* was entered in the Admiralty list to join the main fleet for the forthcoming year but Captain Myngs, having lost so many turned over into other ships, had only 353 men. He was granted press warrants and wrote to the Navy Board that one James Gother, master of a ketch, was available to act as tender for pressing men.[145]

1694 – THE MEDITERRANEAN

After a number of years of almost continual service, most of the ships in the English fleet required a major repair during their annual refit. The report prepared by the three admirals provided details of the work to be done. At Chatham, Sir Edward Gregory prepared a scheme whereby twenty-three warships, including *Lenox*, were allocated a month each in dock for repair. In the scheme he observed "that every ship we meddle with this year must be caulked under each bilge which will unavoidably take up an abundance of time, but is indispensably necessary to be done."[146] *Lenox*'s turn for docking came early in the new year and the defects listed in the report were made good. At the end of January she was taken out of dock and secured alongside a mast hulk to be fitted with a new mainmast to replace the one condemned in the report.

During the next four weeks *Lenox* was made ready for sea, her guns were brought aboard and she was loaded with 368 tons of ballast. The ship's boats were employed bringing men aboard to complete her crew including fifty men turned over from the *Edgar*. She received powder and provisions and a new 18in circumference cable was spliced to the best bower cable and an 18½in cable spliced to the sheet cable. Captain Myngs wrote to the Navy Board to certify that his surgeon, Robert Kirkton, had supplied *Lenox* with physical medicines and drugs for the whole year and requested bills be granted for supplying his chest.[147] It was not until the beginning of March before *Lenox* was ready to sail to the Buoy of the Nore.

The fleet orders for the coming summer campaign were in principle unchanged from the previous year: the destruction of the French fleet at Brest. It was again planned to land soldiers and reduce the land defences before an attack against the French warships and the dockyard facilities. The attacking squadron would be under the command of Lord Berkeley. Russell would support the attack and sail with enough provisions to enable him to pursue the French if they managed to sail before he arrived.

It was considered a possibility that the French fleet would again sail for the Mediterranean to support their army attacking Spain along its southern coast. It was therefore vital that the fleet sail as soon as possible if it were to catch the French fleet in harbour. Six thousand soldiers assigned for the Brest attack would be embarked at Portsmouth where the fleet would assemble. *Lenox* unmoored on 3 March but in the stormy weather made slow progress out of the Medway and was temporarily forced to strike her yards and topmasts. She eventually arrived and anchored at the Nore on the 23rd, where Sir Richard Haddock and his clerks came aboard to pay the crew for the period between 2 May 1692 and 3 September 1693.

Lenox was then ordered to join a squadron of ten third-rate ships and four fireships being assembled in the Downs under the command of Sir George Rooke. He had orders to cruise in the Soundings and meet homecoming Cadiz and Canary Isles merchant convoys and other ships due to arrive in March from the West Indies and Canaries. He was then to return to Spithead.[148] The squadron's orders were influenced by a Parliamentary decision which stipulated that warships must be provided for trade protection as a condition for voting money for the fleet.

On 25 March, a pilot came aboard *Lenox* for her passage down the Swale. She set sail in company with the *Suffolk*, *Monmouth*, *Expedition*, *Burford*, *Restoration*, *Captain* and *Kent*, and arrived in the Downs two days later. Sir George Rooke transferred his flag to the third rate *Grafton* from the second-rate three-decker *Albermarle*. On 2 April the signal was made

Below: *A chart of the Mediterranean by H Moll for* Transactions at Sea, *Josiah Burchett, 1720. (Author's collection)*

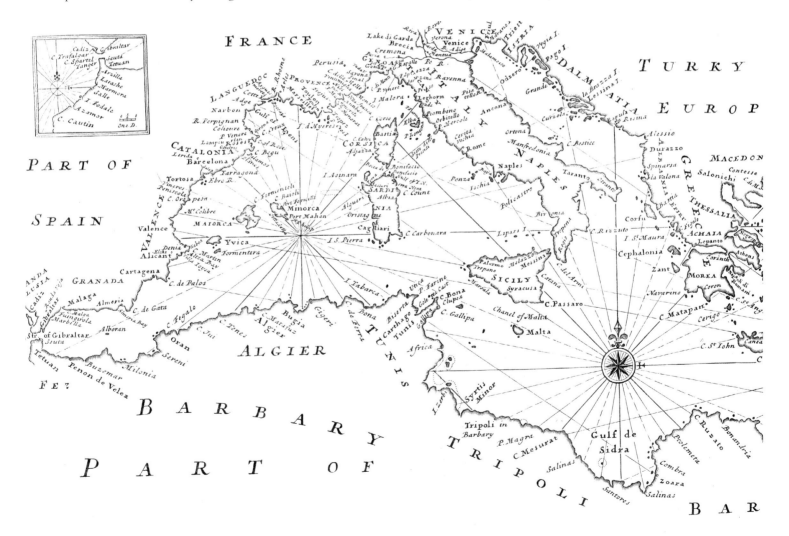

for unmooring and *Lenox* hauled short on her best bower anchor but a fresh gale blew up and the signal was made to remoor once again, and the yards and topmasts were struck. The continuing bad weather prevented the squadron from sailing and it was still at anchor in the Downs on the 9th when the Cadiz merchant ships safely arrived, followed by the Canaries fleet a few days later; as a result the cruise was cancelled.

The *Hampton Court*, which had lost her rudder in bad weather, and *Canterbury* joined from the coast of Holland. Later a court of inquiry into the damage suffered by the *Hampton Court* revealed that she ran aground on the Galloper sands after striking three or four times. Her fore topsails were hoisted and the helm "clapped a weather", which got her off with the loss some of the false keel and all her rudder. The pilot was blamed for the accident and discharged from his position.[149]

On 11 April, twenty-one guns were fired for the coronation of William and Mary. Two days later, Rear Admiral Mitchell, in the second rate *Duke*, together with several first- and second-rate ships, joined the gathering fleet and anchored nearby. *Lenox* was to be part of his squadron and accordingly moved her pennant from the main to mizzen topmast head. A few days later on the 18th, at 6 pm, during a calm, a French privateer with four guns and sixteen men came into the Downs about Northsand Head and stood to the south. Borrowing a longboat, George Rooke made the signal for all the ships' barges to chase. "We fired several guns and small shot and he was taken and brought into the fleet. Our Lieutenant brought aboard one of the Frenchmen at two in the morning". Later the same day the entire fleet of twenty-nine warships and four fireships set sail to join Russell and the main fleet at Spithead.

At sea the weather became unseasonably calm, and the fleet was obliged to anchor with each flood tide as it progressed slowly down Channel. The Admiralty ordered Rooke to send the two most suitable ships to cruise off the enemy's ports to gain intelligence.[150] On nearing Spithead on the 21st, the chosen ships, the *Lenox* and *Monmouth*, left the squadron, sailed west and next day met a convoy of fifty English ships. During the afternoon a gale sprang up and it became foggy. *Monmouth* fired a gun to attract *Lenox*'s attention and stood to the southward. *Lenox* followed but soon lost sight of her. The next day, continuing alone, *Lenox* met a ship flying Danish colours bound for Bordeaux; she was searched but found to be clean of contraband and was allowed to continue on her way. Continuing her cruise *Lenox* arrived off Cape Barfleur on the 24th before heading north again to rejoin the fleet. Two days later she was off the Isle of Wight and at 6 o'clock the following morning the body of a seaman was sent ashore for burial in Stokes Bay. Captain Myngs, meanwhile, went aboard the flagship to report the results of his cruise.

Russell was now impatient to sail together with the squadron destined for the Brest attack. Preparations were still not complete, however, and some of the soldiers had not yet arrived. Many of the ships' crews had not been paid for their last year's service and Russell, mindful of the hardships of the men's families – not to mention their growing anger – would not allow ships to sail until they had been paid. He therefore decided to sail into the Soundings with those ships that had been paid, to gather information of the French. He ordered Cloudesley Shovell to stay with the unpaid ships at Portsmouth, to take in provisions and receive the land forces for the Brest attack.

As part of the fleet of thirty-five English and Dutch ships that would sail with Russell, *Lenox* took stores on board including fifty tuns of beer. Due to the uncertain length of the voyage her crew was put on short allowance for its duration. On 3 May the fleet moved out into St Helens Roads and the next day, during a gale, sailed out into the Channel. By the 9th, as *Lenox* sailed under her fore course and main topsail, the fleet arrived at its appointed station fifteen leagues SSW of the Lizard. Russell then learned from the master of a Swedish ship who had been into Brest that the French fleet had sailed for the Mediterranean two weeks

previously. The disappointing news was not entirely unexpected, given the delays in sailing, but was slightly alleviated by the *Monmouth* and *Resolution* who reported destroying thirty-five French merchant ships and capturing two more during a cruise away from the fleet. Russell could do no more for the moment and therefore sailed back up Channel to join the ships being made ready at Portsmouth. His fleet moored in Torbay on the 19th and at St Helens on the 22nd, where Cloudesley Shovell joined it in the second rate *Neptune* in company with the ships from Spithead.

The plan to land soldiers at Brest was still to go ahead but Russell would now pursue the French fleet into the Mediterranean first. The combined fleet was now almost ready to sail and *Lenox* took on board more provisions. She was also transferred back into Rear Admiral of the Red Sir David Mitchell's division after her detachment. Meanwhile her master carried out a survey of provisions and rations in the fireship *Vulture* at the request of Admiral Russell on the 27th. Three days later the fleet weighed anchor but in the calm weather it only managed to sail as far as Portland by 2 June. The weather then freshened and by 6 June they were off the Lizard where Lord Berkeley, Admiral of the Blue, and Sir Cloudesley Shovell, Admiral of the Red, took part of the fleet southwards for the attack on Brest. Russell sailed on with the main body of the fleet, including *Lenox*, in search of the French fleet thought to be in the Mediterranean. The French in Brest had plenty of warning of the assault and made preparations to receive it; the attack went ahead and some 600 soldiers were pointlessly lost.

During the voyage southwards, the *Medway* captured a pink, the *Ferod*, laden with water. *Lenox* was ordered to tow the pink but on the morning of 17 June the wind suddenly backed to the westward and *Lenox* was taken aback, causing her to foul the pink against her hawse. *Lenox*'s cathead was damaged and the pink lost her bowsprit. Captain Myngs sent his carpenter aboard to make repairs, a task that took two days. On the 21st *Lenox* took on board provisions from a supply pink. A week later another slow-sailing pink was taken in tow but during the following night the hawser broke. *Lenox* remained under easy sail near the pink and took her in tow again the next morning. *Lenox* sprang a severe leak and began to make 4ft of water a watch, a problem that would not be satisfactorily resolved until she had returned to England. Myngs was sufficiently concerned to send a message to Russell to inform him of the situation.

At the end of June, after a voyage lasting twenty-eight days since leaving the Lizard, the fleet neared Cape Spartel on the North African coast. As a precaution against meeting the French, the fleet formed a line of battle sailing on the starboard tack. Russell informed the Spanish that he was off the Cape and they contributed to the fleet by sending a vice admiral and ten ships to join the English and Dutch allies. Russell was later to write dismissively of the state of the Spanish ships: "four might indeed (for want of better) have been admitted into the line of battle, but that the rest were of but little force, and were so rotten that they would hardly bear the firing of their own guns". Eight English and eight Dutch warships that had been waiting at Cadiz under Rear Admiral Neville, flying a union flag at his mizzen topmast head, reinforced the fleet. He kept the union flag flying for nearly three more weeks, before finally complying with regulations and taking it down to hoist a blue one instead.

The confederate fleet of sixty-three warships, about the same size as the French fleet, entered the Straits at the beginning of July. As they passed Gibraltar three men from *Lenox* were drowned in an accident when one of the ship's boats was overset.

Meanwhile the French fleet was blockading the Spanish coast near Barcelona in support of the French army. It responded to the news of the confederate fleet's entrance into the Mediterranean by continuing its policy of avoiding action, and retired toward Toulon. Russell continued sailing eastwards in the hope of meeting the enemy but was delayed by contrary winds. *Lenox* recorded that they were off Cape de Palos on

12 July when the wind died away, completely becalming the fleet for days. On the 16th the Spanish ships all fired their guns in celebration of St James's Day, their piety being rewarded the next day by a fresh wind. Progress remained slow, however, and it took nearly a week before the confederate fleet arrived in Altea Bay near Alicante, where *Lenox* took on board sixty-seven tuns of water. Continuing eastwards the fleet finally reached Barcelona at the end of July.

The fleet was now at the limit of its endurance and the French had retired from the scene. If Russell did not wish to risk the cumbersome three-deckers sailing home during the winter, they would have to leave soon. Provisions were running out and although the Spanish were incapable of replenishing so large a fleet, *Lenox* did receive six butts of wine. The Spanish were desperate for the allied fleet to remain in order to keep the French from descending on their shores and taking Barcelona. Their protests persuaded Russell to tarry for two more weeks before he finally made the decision to set sail for England. He first took the fleet eastwards until it was out of sight of land and then west, hoping to deceive French spies into thinking the fleet was to remain in the Mediterranean.

In order to improve her sailing qualities, *Lenox* put her spare anchor and two of her lower-deck chase guns into the hold. As the fleet began to sail west towards Gibraltar, the thoughts of many in the fleet must have turned to home and those left behind in England. One seaman aboard *Britannia* put his feelings on paper:

Extract from *The Faithful Mariner* (to the tune of *The Languishing Swain*) Writ by a Seaman on Board the *Britannia* in the Streights and directed to fair Isabel, his loyal Love, in the City of London.[151]

Fair Isabel of beauty bright,
To thee in love these lines I write,
Hoping thou art alive and well,
As I am now, as I am now, Fair Isabel

On board the brave *Britannia* bold,
I have the fortune to behold,
The sweet delightful banks of Spain,
While in the streights, while in the streights, We do remain.

Then dearest do not grieve nor mourn,
With patience wait my safe return;
And then we'll both united be,
In lasting bonds, in lasting bonds, Of Loyalty.

The fleet made steady progress; *Lenox* reported passing Cape St Martin on the 24th, Cartagena on the 28th and Cape de Gata on 1 September, sailing an average of thirty miles a day. Then, to the consternation of many in the fleet, new orders came in from King William instructing Russell to remain in the Mediterranean and to spend the winter at Cadiz. The implications were enormous; vast quantities of provisions and all the equipment and men necessary to refit the fleet would have to be transported from England to Spain. Russell held a council of war and it was decided to return to Alicante where there were at least some provisions. It was further resolved that the fleet would now be able to remain in the Mediterranean until October before leaving the Straits and entering the harbour at Cadiz.

After spending nearly a week at anchor while new plans were made, *Lenox* and the fleet set sail on 7 September to return to Alicante Roads, arriving on the 11th. While the fleet lay at anchor an outbreak of dysentery occurred. Among those struck down with the "bloody flux" were Russell and his secretary Josiah Burchett, who both retired ashore to recover. Despite his "distemper", Russell ordered a squadron of ten ships to

sail immediately to intercept any enemy ships that might try to pass to the westward. He also appointed Rear Admiral Neville as commander of a small squadron, including *Lenox*, to collect firewood for the ships' stoves. After two days' sailing the detached squadron lay off Formentera, a small island south of Ibiza where *Lenox* sent her longboat ashore to cut wood and bring it aboard. The cruise lasted another week, during which the wind blew a continual gale, driving the *Falmouth* aboard *Lenox* and carrying away one of her stern lanthorns. The squadron arrived back in Alicante Roads on 22 September. The *Vesuvius* fireship was then sent out for two days to collect firewood and eighty of *Lenox*'s men went aboard her to hasten the work.

An unhappy soul aboard *Lenox* at this time was Robert Wilkins, muster master of the fleet, who in the previous year was allegedly roughly treated aboard *Lenox* by Captain Kerr. Wilkins seems to have been overzealous in his duties of uncovering fraud, and had forgotten the damage he could do to his reputation by falling out with the officers of the fleet. *Lenox*'s pay book shows he had been marooned aboard her since the squadron detailed to attack Brest had left the main fleet in early June. While off Alicante he began a series of correspondence with the Navy Board that became increasingly desperate, as his growing unpopularity was matched by his revelations of abuses by ships' officers. The pressures of a friendless existence were eventually too much and against orders he returned to England with a homebound convoy. On arrival the Admiralty discharged him from his office. It seems ironic that he who was responsible for the dismissal of *Lenox*'s captain should bring about his own downfall by disobeying orders aboard the same ship.

Lenox remained only a few days at Alicante before she sailed with the fleet under a restored Russell for its winter base at Cadiz. In fair weather the fleet passed Cape de Gata on the 28th, Gibraltar on the 30th and was off Cape Spartel by 1 October. Once in the Atlantic, gales sprang up and the foul-bottomed ships made slow progress. On the 6th, off Cape Trafalgar, *Lenox* split her main topsail and sprang her fore topmast before finally dropping anchor in Cadiz on the 8th.

The naval administration efficiently arranged stores and equipment to be transported from England to Cadiz for the refitting of the fleet. The task was enormous, and considerable credit was given to those successfully involved. The work progressed smoothly but in order to retain a powerful fleet at readiness in case of a sudden appearance by the French, only a few ships at a time were sent within the Puntals, the inner harbour of Cadiz, to be refitted and cleaned. At this time the fleet was also very short of men, being nearly 3000 under strength.

Having been deprived of fresh provisions for such a long time in hot conditions, the men in the fleet began to fall ill. *Lenox* suffered the loss of about three men a month while in the Mediterranean and losses reached a peak during this month when seven men died. A week after arriving at Cadiz on 15 October George Knitch, the first lieutenant who had served aboard *Lenox* since 1691, died. Over the next few weeks others followed including Richard Studvill, the master, a veteran of three years' service aboard the ship. His journals, all of which survive, are *Lenox*'s best record for much of the period, including Barfleur and La Hogue. During the last two weeks of his journal he touchingly recorded the names of those who perished before him. The last entry, written in another hand on 7 November, marks his end: "This morning at eight died our Master in my arms". In accordance with his will, made out shortly before the voyage, his wife received his worldly goods. Francis Studvill and Elizabeth Knitch, the widows of the two officers, eventually received the wages owed to their dead husbands.

The crew's health slowly improved after the fleet arrived at Cadiz as a result of better supplies, which arrived from England. The tragic losses sustained at the end of 1690 were not repeated. Four men deserted while *Lenox* was moored at Cadiz.

The remainder of October passed while *Lenox* lay at anchor and she witnessed the comings and goings of ships sent out cruising. On the morning of the 17th a pair of young English folk came aboard to be married, presumably because an Anglican service was impossible in Catholic Spain. Although *Lenox* still had three months' supply of other provisions, the butter ran out and was replaced by an allowance of half a pint of oil for three men per day.

Winter set in at the beginning of November and the harbour was rocked with violent thunder and lightning storms, often causing *Lenox* to strike her yards and topmasts. On the 4th, "The whole fleet English and Dutch spread all the colours they could make in respect of this day, being the landing of King William in England. All the commanders celebrated by dining with Admiral Russell." The next day "At twelve thirty we fired 15 guns to express joy for the King's landing in England, being his birthday besides." The ship rolled noticeably more while she lay at anchor than when she was under way and steadied by her sails.

English beer was now in short supply and the crew were served two butts of wine as a substitute beverage, their opinion of which was not recorded. As the weather became worse the fleet was moved further into the bay. The great ships were moved on 13 November and *Lenox* was unmoored and berthed in a new position in six fathoms with the best bower anchor two days later.

Admiralty orders sent to Russell dated 12 November stated that two convoys would be sent to him carrying victuals and stores. The *Suffolk* and *Edgar* would escort the first convoy to Cadiz and the second would be shepherded by the *Lancaster* and *Ipswich*. Upon receiving the orders, four third-rate ships were to be sent home in room of the four ships sent out.[152] The condition of the hulls of many of the third-rate ships was giving cause for concern and Russell decided to send home those most in need of docking and repair. To make matters worse, *teredo navalis*, or ship's worm was present in the warm waters of the Mediterranean and some ships were badly affected. The hull of *Lenox* was causing particular concern and on 22 November Russell ordered three carpenters from the *Royal Sovereign* and *Britannia* to make a survey of her condition. They reported that in their opinion *Lenox* was unfit to remain in the Straits and she should have her bows caulked between wind and water before being sent home to England. Their report was accepted and *Lenox* was ordered to escort a convoy home. The rest of the month was spent fitting her for the voyage and making essential repairs. Ambrose Fellows and his carpenters brought two fishes on her weak bowsprit and the ship was brought 3ft deeper by the stern in order to expose her bow planking for repair in an attempt to stop her persistent leaks.

On 1 December *Lenox* made ready to sail to England together with the *Northumberland*, another of the thirty ships of 1677, under the command of Captain Lambert acting as commodore. The convoy also included two Dutch men of war and twenty-seven sail of English and Dutch merchantmen. The next morning they loosed the main topsails in the tops and hauled home the fore topsail sheets. The convoy spent the next few days endeavouring to leave Cadiz Bay during which time the admiral ordered the longboat of *Lenox* be sent to the *Edgar* and 100 bushels of peas sent to the *St Michael*. *Lenox* fired fifteen guns in salute as she finally departed. As the convoy sailed north two other merchantmen joined for their protection but one of them, a small slow-sailing merchantman, could not keep up and was taken in tow.

On 11 December a sail was seen to leeward. *Lenox* sent a boat to board her and took three casks of pork from her. Later two sail were seen and chased by *Lenox*; one proved to be from the convoy and the other a Portuguese flyboat with a cargo of salt bound for Cork. As the convoy reached further north the weather became worse and *Lenox* was forced to cast off her tow. Two ships even managed to sail through the convoy without being identified in the heavy seas. The convoy took twenty days to sail into the

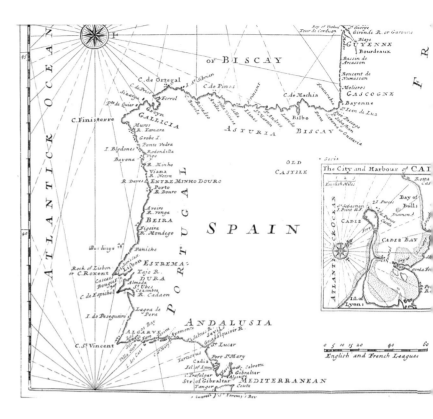

Above: *A chart of the coast of Spain by H Moll for* Transactions at Sea, *Josiah Burchett, 1720. (Author's collection)*

Soundings from Cadiz with the wind generally blowing from NE. John Swan, the new Master, calculated *Lenox* sailed 1180 miles during the voyage at an average of 58 miles a day. The best day's sailing gained 105 miles and the worst, just 16 miles. The convoy approached the Channel at the worst possible time of year.

1695 – STORMS

January greeted the convoy with a NE hail storm. *Lenox* at first lay a-try under bare poles before setting the mainsail with her head into the wind. The wind then came round from the opposite direction and *Lenox* wore and stood east. The next day, the 3rd, *Lenox* lay by between 8 am and noon; there followed a day's respite before the storm resumed with lightning from the NW, during which two of the foremast shrouds were broken. *Lenox* scudded away before the storm, heading ENE with only the foresail in the brails set.

The next day, the 6th, the storm increased in intensity to a great storm with lightning and rain from the SW. In such filthy and dangerous conditions the ships sought only to survive. *Lenox* continued scudding away under a goose-winged foresail – that is, with only the lee clew set while the weather side of the sail was lashed to the yard. In this manner only a fraction of the sail area was exposed to the wind but such was its fury the sail suddenly split and half of it was blown from the yard. The crew managed to bend the spare foresail to the yard and the vessel kept scudding before the wind. The seas became so mountainous from the stern that they pooped *Lenox* and broke in the great cabin window. With water flooding in from the stern it was essential the head be brought round to face the sea and wind. In what must have been a very dangerous manoeuvre the mizzen sail was set, which successfully brought the ship round with the head to the southward. During the fight for survival contact was lost with the rest of the convoy.

After a week the storm began to ease and *Lenox* bent her spritsail and set her foretopsail and headed ENE. She was then brought to and managed to establish her position at the entrance to the Channel. A sounding was made which recorded a depth of fifty-five fathoms "with yellow sand with things like needles and black specks". The next day land was seen which proved to be the Start. The wind moderated to a fresh gale, and three ships were seen which proved to be merchantmen from the scattered convoy. One of them, the *Merchants Delight*, under the command of Captain Cotterell, had lost her mizzen mast during the storm, so she was taken in tow by *Lenox*. The four ships continued up-Channel, passing Beachy Head on the 10th, and into the Downs the next day where they anchored in eight fathoms during a gale with snow falling.

Captain Myngs wrote to the Navy Board informing them that he and three merchant ships of the convoy had arrived but that he had been separated from Captain Lambert and the rest of the convoy by a storm.[153] *Lenox* was replenished by the hoy *Rebecca*, allowing the crew to be issued with whole allowances of fresh provisions. Myngs had Ambrose Fellows survey the damage to the ship and sent a report to the Navy Board: "We have sprung a leak in the bread room which hath damaged our bread and another a midships against one of our orlop beams and another under the step of our foremast which we have done what we can do but makes a great deal of water as much as we can do to keep her free with one pump and our ship damnified very much by reason of very bad weather that was met with coming home and our bowsprit sprung'.[154] That evening Lambert in the *Northumberland* arrived with the mast ship flyboat, *Firr Tree*, laden with masts he had retaken as a prize from the French. Lambert also wrote to the Navy Board and received orders by return instructing him, his prize and *Lenox* to proceed to Sheerness and then on to the Buoy of the Nore.[155] Over the next few days before they set sail, small groups of merchant ships from the convoy joined them to add their harrowing tales of the storm.

On 18 January a Dover pilot came aboard and the topmasts and yards were hoisted and the ships got under way, sailing as far as Sandown Castle before anchoring again. *Lenox*, *Northumberland* and her prize remained at anchor for the next four days in snowy weather with lowered topmasts and yards before making their way north to the entrance of the Kings Channel. They anchored near Margate Roads on the 24th, four leagues north of North Foreland on the 25th, and near a shipwreck off the Kentish Knock on the 26th. They then proceeded into the King's Channel where they anchored to the east of the Gunfleet buoy in thirteen fathoms of water. At this location Myngs wrote to the Navy Board to inform them of his arrival.[156] In response, the Admiralty ordered the ships *Northumberland* and *Lenox* to make their best way to Blackstakes, to take out their guns and powder, and then to proceed to Chatham for their refits. To prevent men deserting during the refit, they were to be put aboard the *Swiftsure* at the Nore until *Lenox* was ready to receive them again.[157]

The ships were further delayed by the weather before a pilot came aboard to carry the ships on to Blackstakes. They unmoored on 8 February and anchored again on arrival, when forty-eight guns were taken out of *Lenox*. The rest, except those in the hold, and all the gunpowder followed the next day.

The Commissioner at Chatham Dockyard, Edward Gregory, was becoming agitated by the delays in bringing the two ships to Chatham. He wrote on 14 February: "I every day look out for the *Lenox* or *Northumberland* or both. If they arrive not suddenly we shall be in danger of losing the approaching spring. We have now a hard hearted wind at the NW that will neither suffice the ships that we have to go down, nor those below to come up". He wrote a letter in similar vein two days later.[158] Yet another pilot came aboard to take *Lenox* on to Chatham but the severe weather delayed her again for five more days; she then got as far as Gillingham where she was moored against the Golden Horse Inn near the bridge. The cables and empty casks were then taken out and put into a lighter. Edward Gregory reported that *Lenox* had been transported (warped) to Gillingham and that he would work on her incessantly in order to dock her.[159]

After all they had suffered at the hands of the weather, forty of the crew were then turned over into the *Greenwich* followed by eighty more "suspected deserters" into the *Swiftsure* the next day. It is difficult to imagine how the men felt after such a long hard voyage to be sent to other ships without a chance to touch land. They could not have been in the best of health and at the very least tired and in need of recuperation. Their families and friends would have to wait to hear their stories, although they would have been allowed to visit them aboard ship. Captain Myngs, aware of the problems of keeping records of men's pay when they were transferred from ship to ship, wrote to the Navy Board enclosing their pay tickets so that the "poor men may be paid".[160] The problem of pay tickets and turning over into other ships could cause real resentment:

Extract from *The Sea-Martyrs*
The Seaman's Sad Lamentation for their Faithful Service, Bad Pay and Cruel Usage.[161]

> Good People, do but lend an Ear,
> And a sad Story you shall hear,
> A sadder you have never heard,
> Of due desert and base Reward,
> Which will English Subjects fright,
> For our New Government to fight.
>
> Their starving Families at home,
> Expected their slow Pay would come;
> But our proud Court meant no such thing,
> Not one Groat must they have till Spring;
> To starve all Summer would not do,
> They must still starve all Winter too.

The rest of *Lenox*'s stores, ballast and gun carriages were removed and she was made ready for docking. An attempt was made on 21 February to dock her on the ways but the tide was not high enough by just 6in. During the night Edward Gregory endeavoured to lighten *Lenox* further to lessen her draught and at the same time lower the ways some inches. His efforts were successful and at high water the next day she was safely taken into dock and repairs begun. Gregory wrote "We happily docked the *Lenox* and shall prepare vigorously to prepare her for the sea."[162]

Lenox's repairs and refit lasted one month before she was floated out of the dock and made ready for service. It is not known what work was carried out but apart from repair to the outside planking, additional reinforcing timbers, including extra standards and riders, may have strengthened the interior. The foremast was found to be rotten above the partners and it was removed. Four loads of ballast totalling 276 tons were then taken down into the hold. All her stores were taken back on board including anchors and new cables of 19½in circumference. A new foremast was then set up and the old, sprung bowsprit replaced with a new one. By 11 April, *Lenox* was ready to sail and a pilot was brought on board to carry the ship to Blackstakes. There, dockyard riggers were brought aboard to help in handling her. Having been delayed by a southerly wind for two days she was taken down and moored abreast the salthouses where she was to receive her guns, nineteen the first day, twenty-one the second and twenty guns on the third to make her complement of sixty. While the battle fleets of France, England and Holland were far away in the Mediterranean there was no need for her full complement of seventy guns. *Lenox* was to spend the summer escorting convoys to safety past the menace of French commerce raiders, and gunpower could be sacrificed for improved sailing qualities.

Lenox remained at Blackstakes taking aboard provisions and gunpowder and would have been ready for service had she not lost so many men turned over into other ships. Captain Myngs therefore obtained three press warrants. Orders were then received instructing Myngs to sail to the Downs but he was unable to do so for lack of men. Further orders arrived to turn over men from the *Berwick* and *York* into the *Lenox* and *Northumberland* to make up their complements.[163]

Lenox was ready to sail on 25 April and a pilot came aboard at 4 o'clock in the morning to take the ship to the Buoy of the Nore. She unmoored at noon and arrived at 3 pm where she anchored by her best bower. The next day Sir Richard Haddock came aboard and paid the ship's crew up to the previous 1 October. Those of the crew who could not be paid because they had been turned over into other ships would have to wait until they were discharged from their new ships, at which point they could claim their wages from the pay office in London. The measures to prevent men deserting ship had been successful, as only ten men had run since their arrival in England. Soon after being paid, however, thirteen managed to desert. Myngs requested a ketch to bring men in.[164] The supply smack went to London to collect some new sails and to press men but came back with only one man.

The plight of pressed men is often described but it is not the whole story, as many seamen volunteered for service but are seldom mentioned. On one occasion three of *Lenox*'s men, who had been on leave, wished to return to her but had been made "run" by the Clerk of the Cheque at Chatham Dockyard. It required a certificate and a letter from Captain Myngs to restore them onto the pay book of their ship.[165]

On 12 May Lord Berkeley arrived in the Nore. With the main fleet away in the Mediterranean he was to command the Channel fleet during the coming summer, with orders to bombard French coastal towns. Later, "at twelve midnight the Gracious King sailed by in the *William And Mary* yacht bound for Holland, we all fired guns as he passed". The next day the *Henry* ran ashore in a gale but was got off at the next high water without serious damage. Orders were then received requiring *Lenox* and other warships to sail to Plymouth.[166] She was still very short of men so the Admiralty ordered 164 men from other ships to be loaned to her. They were not to be entered onto *Lenox*'s books but kept on a separate list. By not entering the men on the pay book, there was no method for deducting their pay for any goods the men might receive. Captain Myngs wrote to the Navy Board concerning the matter: "Now these poor men being very destitute as well of clothes as bedding I desire your honours will be pleased to give directions in this matter whereby the men may be supplied and where to charge them when I so do."[167]

Lenox weighed anchor with the ebb tide to depart but Sir Cloudesley Shovell, also at the Nore, was aware of the unhappy manning arrangements and ordered her back again. He instructed *Lenox* to return the men she had borrowed from the *Queen* and *Neptune* and take 102 from the *London* instead. The delays resulted in *Lenox*'s orders for joining the fleet being cancelled and new ones issued ordering her to cruise in the Soundings. She set sail once more and arrived in the Downs at the end of May where in the evening the *Charles Galley* anchored and gave an account of a fight with a 50-gun Danish man-of-war that lasted all morning. The smaller, lightly built *Charles Galley* had lost seven men killed, thirty wounded and was very much disabled.

Orders arrived from the Admiralty instructing Captain Myngs to sail at the first opportunity for Spithead and remain there until Peregrine Osborne, Marquess of Carmarthen and Rear Admiral of the Blue, came aboard. Myngs was then to follow his orders.[168] On 1 June *Lenox* therefore set sail together with two small frigates from the Downs and a convoy of about twenty ships bound for Spithead. They arrived the next day where they found a West India merchant fleet at anchor. *Lenox* immediately sent her lugger to press some of their crews and took fifty

experienced seamen out of them which she exchanged for fifty landsmen she had received earlier. While *Lenox* waited for the Marquess of Carmarthen, her ballast was increased by 50 tons brought aboard by two hoys.

One man who had been pressed into *Lenox* made a complaint to the Admiralty which immediately wrote to Captain Myngs concerning the matter. "A petition has been presented by James Phillips late boatswains mate of the *Portland* saying he was in St Bartholomew's Hospital in London for cure of a lameness contracted in his ship and that going toward Plymouth in a wagon when returning to her he was impressed to serve in the ship under your command. You are ordered upon meeting the *Portland* to cause the said James Phillips to be returned to her giving him a ticket for his service".[169]

Lenox remained at Spithead for two more weeks before being joined by the Marquess of Carmarthen. He had orders from the Admiralty to take command of the ships that were cruising in the Soundings for the protection of trade and to annoy the enemy. His squadron, apart from *Lenox*, consisted of *Dreadnought, Kent, Stirling Castle, Litchfield, Hawk, Owners Love* and two fireships.[170] *Lenox* was to become his flagship, so the longboat was sent into the dock for flagstands and a top lantern; these and the rear admiral's blue flag were hoisted at the mizzen topmast head.

Lenox weighed anchor and plied into St Helens Road with a fine gale of wind from the ENE; there she anchored and was joined by a small convoy consisting of the *Bridget Galley*, a hired armed vessel, and twelve small vessels all bound westward. They were off the Start on the 16th and next day entered Plymouth Sound in light winds where *Lenox* anchored in seven fathoms with Drake's Island bearing NNW and Penlee Point SW. In these calm conditions the opportunity was taken to heel the ship over and scrub her bottom.

Lenox and the *Bridget Galley* remained in Plymouth Sound until the 27 June when the *Bridget Galley* unmoored and went out to escort a convoy of merchantmen bound for Torbay. Fresh orders arrived from the Admiralty for Carmarthen, informing him he should join the *Rupert* and *Foresight* at Plymouth and take them under his command into the Soundings and send in two other ships of his squadron already at Plymouth for cleaning and refitting. Two days later, the small squadron under the command of Carmarthen in *Lenox*, together with the small third rate *Rupert* and fourth rate *Foresight*, weighed anchor in a "pleasant gale" to go cruising to the westward. As they left Plymouth Sound the *Bridget Galley*, returning from Torbay, joined them. The small squadron steered WSW past the Lizard where three sail were spotted to the eastward. These were judged to be French privateers, so the English gave chase but the French crammed on sail and they were unable to close. The squadron remained in an area about ten leagues SW of the Scilly Isles for the first few days of July. On the 3rd, at 3 pm, another ship was seen sailing to the eastward and the *Foresight* and *Bridget Galley* were ordered to chase. During the following night they closed with the ship and fired single guns but were unable to stop her. At 5 o'clock the next morning they gave up and rejoined the convoy. Sailing further eastwards, the ships made Cape Clear at the southern tip of Ireland on the 6th before sailing eastwards along the coast to Kinsale where they anchored in five fathoms. The Admiralty arranged for the *Canterbury* storeship to be at Kinsale, thence to be escorted to Bantry Bay to be loaded with ship's knees. It was so valuable a cargo that it was to be escorted by two ships of Carmarthen's squadron.[171]

They were joined the next day by a huge incoming convoy from the Straits consisting of ninety English, Dutch, Spanish and Hamburg ships escorted by three warships, the *Mary*, a famous old commonwealth third rate, the *Berkeley Castle*, a new 48-gun ship and the new 80-gun *Devonshire*. *Lenox* and her squadron had orders to join the escort and sail with it until it was safely past the Scilly Isles, before returning to the Soundings. In the event most of the fleet was prevented from leaving Kinsale by contrary winds. Only the *Bridget Galley* and two merchantmen bound for

Bristol managed to leave. The next day at 10 o'clock in the evening there was an explosion in the gunroom of *Devonshire* caused by an accident with a powder horn. Forty men were wounded.

The fleet continued having difficulty leaving the Irish coast in the teeth of persistent gales and it was not until 15 June that *Lenox* was able to make the signal to sail. The convoy weighed anchor and all except the *Foresight* and twelve merchantmen reached open sea. By the following day they reached Cork Harbour where *Lenox* anchored in ten fathoms. *Foresight* and the merchantmen left behind arrived to rejoin the fleet over the next few days. The gales continued and it was not until the 21st that the fleet was able to set sail again. Two days later they were off Land's End where seventeen unidentified sail were seen far to the southward. Then a boat from the shore gave an account that the seventeen sail had been lying off for a week. Hearing this, Carmarthen called a council of war and it was agreed they must be French. It was decided to make the best way to Milford Haven and Carmarthen immediately sent a dispatch to the Admiralty informing them of the situation. The fleet changed course and headed north away from the perceived danger and toward the southwestern tip of Wales, which was reached the next day.

Some of the men were now becoming sick and thirty-five were sent ashore at Pembroke to pitch a tent. On 1 August the *Bridget Galley* returned from her detachment and rejoined the fleet. It was decided to wait for two weeks before the coast was clear of the French. After a week, orders were received from the Admiralty acknowledging Carmarthen's dispatch and adding "we have received advice from Commissioner St Lo [Plymouth Dockyard] and Captain Coal that the squadron which was supposed to be French ships is in fact the West Indies fleet and some of the ships of your [own] lost squadron, as you will be more particularly informed by copies of their letters which come enclosed."

Now that it was decided there was no danger posed by the French, half the ships stood out to sea from Milford on 7 August under easy sail to be joined by the rest next day. Two days later the fleet was once more off the Lizard and heading up Channel; as they passed Plymouth, Carmarthen received orders from the Admiralty explaining that as the ships in his squadron were dispersed among several stations and there was not sufficient number together for a flag officer, his squadron was to follow the orders of Captain Coal. He, personally, was to proceed to Plymouth with the *Lenox* and *Bridget Galley* and from there repair to London after leaving orders for the two ships to be cleaned.[172]

Lenox and *Bridget Galley* duly left the fleet and headed for the Sound. They anchored in Plymouth Sound in seven fathoms of water at noon. At 2 o'clock they weighed again, then ran into the Hamoaze and anchored in front of the docks in fourteen fathoms. They struck the yards and topmasts and got the hulk on board in order to clean the guns and deck.

Following Pepys's reforms, officers of the navy were generally competent and although they may not always have been honest by later standards, they usually showed remarkable loyalty to their brother officers. Carmarthen, however, proved the exception. He turned on Captain Coal for pointing out that the "French" ships had in fact been ships of his own squadron. Carmarthen pursued Coal until the end of the war, trying to have him court-martialled for not obeying an order. Later, when he himself failed to obey an order, he informed the Admiralty that he was wholly ignorant of it, to which the Admiralty politely pointed out that he should therefore read his orders. On another occasion he turned up at a board meeting of the Admiralty with a letter from the Secretary of State, ordering that a new ship that was nearly complete should be fitted out and furnished as Carmarthen's flagship according to his personal desire. The Board made it clear that such orders were invalid and were not to be carried out.[173] Carmarthen would not normally have been tolerated, but he happened to be the second son of Thomas Osborne, the 1st Duke of Leeds, who was one of William III's chief advisers.

For fast-sailing cruising ships, it was essential that the underwater surface of the hull was clean and free from marine growth. *Lenox* was to enter dock to be tallowed with the application of a mixture of animal fat and soap to give a slippery surface; this treatment unfortunately wore off after only a few months. In order to lighten the ship for entering dry dock, *Lenox* spent the next three days removing her ballast, guns and provisions. She was taken into dock on the morning of 16 August and by the afternoon she was cleaned and tallowed and floating alongside the hulk ready to take on her ballast again. As the ship made ready for sea, the Marquess of Carmarthen, in accordance with his orders, left and set out for London by land after arranging for his goods to be taken aboard the *Bridget Galley* for transportation by sea.

Captain Myngs, once more in full command of his ship, wrote to the Navy Board concerning his men. Many from the *Berwick* and *York* who had been turned over into *Lenox* had run away and a list of their names was submitted so that their pay from their previous ships could be stopped. Incredibly, seventy-three men had deserted *Lenox* since she left Blackstakes earlier in the summer. Myngs continued, "I have a wonderful sickly ships company. I left behind 20 men at Milford and am like to leave 40 more here."[174] Although many men fell sick, the death rate remained at about one per month. It is clear that the quality and morale of *Lenox*'s crew had deteriorated from the previous year, made up as it was of numerous turned-over and pressed men. *Lenox* spent the next week taking aboard her guns, powder and provisions. She then left the Hamoaze and anchored in Plymouth Sound in seven fathoms. New orders were received to escort five merchant vessels to a position of safety sixty leagues out to sea, and then to cruise in the Soundings. The merchantmen *Sarah and Mary*, *Sarah and Eleanor*, *Hopewell*, *Katherine* and *Elizabeth* were bound for Barbados and Jamaica with cargoes of merchandise and perishable provisions.[175]

On 24 August, twenty-one ships from Virginia, a convoy from Falmouth, and the old rebuilt commonwealth fourth rate *Crown* arrived in the Sound. Captain Myngs took the opportunity to press some men from

Below: *(a) A sailor's leather hat. (b) A brass uniform button. (c) A wooden screw-top container, probably for medical pills. (d) A glass union bottle. All drawings based on items recovered from the 1703 wreck of* Stirling Castle, *Ramsgate Maritime Museum.*

the merchant ships. As the month came to an end, he joined the number of men struck down with sickness; a week later he developed a fever and was forced to go ashore on 3 September. It became clear that he would not be well enough to go to sea and George St Lo, the Commissioner at Plymouth, wrote to the Admiralty explaining the situation. Acting swiftly, the Admiralty sent St Lo a letter with four blank commissions, one for the captain of the *Crown* to command *Lenox*, the rest for junior officers to move up, thereby filling the vacancies above them. *Lenox*'s log records "This day Captain Guy lately in command of the *Crown* had a commission to command the ship in Captain Myngs room". Lieutenant Hopson, who had joined *Lenox* at Cadiz, was promoted out of *Lenox* into the *Hawk* fireship. Captain Guy wrote to the Navy Board: "The Right Honourable, the Commissioners of the Admiralty hath been pleased to grant me a commission to command his Majesty's ship *Lenox* which I am now in which is very badly manned. We are ready for the sea with the first fair wind, the wind now at NW a hard gale".[176]

Captain Myngs later recovered from his fever and by November was given command of the *Woolwich*. He later became Superintendent at Portsmouth, retired in 1714 and died on 23 October 1725.[177]

Lenox remained riding in Plymouth Sound waiting for ships of the escort and convoy of five merchantmen to assemble, drying her sails as the weather turned fair. The escort, now under Captain Dilks, acting commodore in the *Rupert*, consisted of the *Lenox*, *Crown*, and *Owners Love*. He had orders to remain with the merchantmen until they had sailed into a position sixty leagues west of the Scilly Isles. They were to take care of the bearing and distance of land so that, should they meet incoming merchantmen, they would be able to give them an accurate account of their position.[178] The convoy unmoored on 9 September and rode short on a single anchor for the night. The following morning *Rupert* made the signal to weigh anchor and the convoy stood southward out of Plymouth Sound and steered west. A day later the Admiralty sent advice of intelligence that four men-of-war from Brest had set sail and that they would probably cruise together. Captain Dilks was to take particular care to keep his ships together to prevent an "inconvenience" if they might happen to separate.[179] The small squadron sailed before it received the intelligence and was unaware it was in some danger of encountering a squadron of French ships considerably more powerful than itself.

Two days later at night, *Lenox* observed the Scilly Light bearing W by N five leagues distant. The next day a strange sail was seen to leeward and *Rupert* made the signal for *Lenox* to chase, but by noon it became clear that *Lenox*, even with a recently tallowed bottom, could not catch the ship. Commodore Dilks therefore hoisted a white flag at his fore topmast head and fired a gun as signal for *Lenox* to rejoin and by 6 pm she was again in company with the convoy. Two days later another ship was seen off to windward and again *Lenox* set off in pursuit but night came before she was able to speak with her. The warships remained with the merchantmen until 17 September, by which time they were 100 leagues west of the Scilly Isles, a distance the commodore judged to be safe from French privateers. Then, during a squall at 5 o'clock the following morning, the escort turned east and by 8 o'clock the merchantmen were out of sight.

The next day three Portuguese ships were sighted. When hailed and boarded, they proved to be laden with wine and brandy bound for London and Holland. The following morning, the small squadron retook a little square-sterned London-owned vessel that had been captured the day before by two fifty-gun French privateers. The squadron then stood to the westward seeking them. They continued during the night on their course in variable winds. Then, at 2 o'clock in the morning they gave chase to the two French privateers which had taken the Barbados ship, and followed until 4 o'clock in the afternoon when the Commodore made the signal to give up the chase. There was so little wind that it would have been impossible to catch them.

They turned their heads to the eastward and continued their cruise uneventfully until the 26th when the wind blew hard from the SW by S. The ships sailed under their mainsails with their heads to the south. Towards the end of the month, the ships cruised in an area about thirty-five leagues south-west of the Scillies. Three ships were seen on the 27th and another on the 29th but each time lost sight of the chase in thick weather.

On the night of 1 October, as the warships neared England, *Lenox* heaved the lead and recorded fifty-eight fathoms with small round sand. Land's End was sighted next morning and by the 3rd the ships had entered Mounts Bay where the prize they had taken was sent in. The four warships departed the same evening in company together and headed toward Kinsale, arriving on the 9th. *Lenox* anchored in Kinsale Harbour in four and half fathoms with her best bower. The next week was spent taking in provisions and water. At the same time a new set of mizzen shrouds was set up and overhauled, which was followed by heeling of the ship and the scrubbing of its bottom. By the 18th a fine gale from the north gave *Lenox* the opportunity to continue her cruise; she unmoored at 4 am and was under sail by 11 am accompanied only by the *Crown*.

Lenox and *Crown* had orders to cruise off the Scilly Isles for the protection of English and Dutch trade until 20 October. Several sets of orders were issued to Captain Dilks during this time, typically informing him of expected merchantmen arrivals and directing which ships to send in for cleaning. They illustrate an impressive display of control and awareness by the Admiralty. The day after leaving Kinsale two distant ships turned and headed towards the two warships but as they neared and realised they were closing with a 70-gun warship, they bore away. *Lenox* and *Crown* crammed on sail in pursuit. *Lenox*, however, was again disappointed as the two unknown ships, almost certainly French privateers, sailed off and were hidden by the coming night. Morning luckily revealed the two ships once more and the chase was resumed. Again *Lenox* and *Crown* were outsailed by the lighter-built vessels and failed to close. Night came down once more and still the privateers could not lose their pursuers and the chase continued into the third day. Then luck intervened and the wind died away to a calm favouring the lighter ships. *Lenox* launched all her boats to tow but their efforts were in vain, as the privateers at last left their pursuers behind. Soon after this disappointment, a convoy of seven Swedish merchantmen coming from Lisbon was seen and spoken with. Then on the night of the 24th two strange sail were again seen and chased all night "but in the morning found them to be two Dutch men-of-war from Cadiz". *Lenox* and *Crown* had by chance met ships from the great Mediterranean fleet returning home, of which *Lenox* had once been a part the previous year. In the afternoon they met another, the *Deptford*, out scouting and received an account of the news from General Russell's fleet. *Lenox* had not missed any battles. The French fleet had managed to avoid action and the return of the combined English and Dutch fleet was to be the last time during the war a great fleet would be fitted out. That evening, at dusk, *Lenox* observed the fleet itself pass at distance to windward.

The next day, 26 October, during a moderate gale, three other strange sail were seen and chased; the first lieutenant's journal forlornly records "but they made three feet for our one". *Lenox* and *Crown* remained in their cruising station in fine gales until the end of the month without seeing any other vessels. They then made sail and steered towards Plymouth, arriving on 6 November to join several other warships, the largest of which was the *Stirling Castle*.

Captain Guy wrote to the Navy Board on the 8th informing them of his arrival and that he was to re-victual. He added that the *Crown* had sprung her fore topmast and the furnace was unserviceable.[180] The next day orders were sent to Captain Guy from London ordering him to complete ten weeks' provisions and cruise as long as provisions lasted between twenty and forty leagues SW and WSW of the Scilly Isles in the fairway of a considerable fleet of rich merchantmen from Barbados and New England, and escort them to safety. He was also to look out for the *Harwich* and take her under his command. After escorting the merchantmen to safety he was to continue cruising on his station.

Orders were also sent for the *Crown* to be refitted and cleaned.[181] Three days later on receipt of his orders Guy wrote another letter acknowledging them.[182] In spite of the distance between London and Plymouth it seems that letters took only about two days in transit.

At Plymouth it was decided *Lenox* and *Crown* should escort a convoy of merchantmen to Ireland as they sailed west. The convoy was not yet ready, however, and during the interval the two ships were cleaned and provisioned, *Crown* having ten days work carried out on her in the Hamoaze near the dock. On the 15 November *Stirling Castle* and five other warships sailed out to escort a convoy of forty merchantmen while Guy wrote that he had taken in thirty tuns of beer. It took some time for the ten weeks' provisions to be supplied, and a few days later Guy complained from Plymouth Sound that he was still short of most of his bread, beer, butter and cheese but would have it below as soon as it was dispatched from the shore.[183] On the 21st, two men were caught cutting cloth out of an ensign, presumably to make clothes. Captain Guy decided to make them "run the gauntlet" as a punishment. This probably required the offender running between two lines of seamen who would strike him with a single twisted and knotted cord as he passed. To make sure he didn't go too fast, the master-at-arms led with his sword pointing back at him.[184] Finally on the 23rd, at 2 o'clock in the afternoon, the wind came round from the NE to a position favourable for the convoy to sail; *Lenox* made the signal to unmoor. By 5 o'clock she rode short on her best bower anchor and at 10.30 the next morning *Lenox*, *Crown* and nineteen merchantmen got under way.

As expected for the time of year, hard gales blew up and in the heavy seas sight of *Crown* was temporarily lost off the Lizard. The convoy managed to stay together and by the 26th it had been safely shepherded into Cork and Kinsale. *Lenox* and *Crown* then steered west to their cruising station and as the month came to a close, two ships were chased during which contact was again lost with *Crown*. On 1 December *Lenox*, now alone, met a ketch from Bristol bound for the West Indies and next day gave chase to two French ships of forty or fifty guns, which were soon lost in the gales. On the 4th the weather worsened and in a great sea *Lenox* lay a-try under a reefed mainsail and mizzen with her head to windward until noon. The next day the main shrouds were set up as the wind eased to a fresh gale lasting two days. It then increased to a storm and *Lenox* was forced to lay a-try again under the mainsail and mizzen. In these conditions the fastenings and stressed timbers of the hard-worked ship began to fail and "The ship began to complain making a foot of water a watch."

The next day the wind eased and it was decided to see if the topsails could be used, but they soon split in the wind. Over the next few days the storms abated to hard gales during which time other ships were seen, one of them giving an account of George Rooke's arrival in Cadiz. Yet again the wind increased to a storm with a great sea from the NNE, and during the night the new main topsail blew away and "we hauled up the foresail and lay a-try under a mainsail and a reefed mizzen. At six o'clock in the morning the mainsail blew out of its bolt rope and most of it lost. We brought a foresail to the main yard". The ship made 1ft of water in a watch and leaks were found in the breadroom, gunner's and boatswain's stores. There was no respite; a day later the storm increased to a great storm from the NE and the weakening timbers of *Lenox* began to complain more and more, even though she lay with her head to windward.

A day later, on the 16th, they found the cutwater and knee of the head very loose. Equally alarming, the following morning they had to bail water out of the gunner's after storeroom, as the water would not run into the well amidships because the ship had hogged so much toward the stern.

Conditions aboard *Lenox* must have been appalling; freezing water would be dripping through the working deck planks, leaving nowhere and nobody dry. Added to that was the misery of short rations served cold, tiredness caused by constant pumping and fear that *Lenox* would founder. The hard storms continued for another two days before Captain Guy recorded, "I called a consultation with my officers to hear their advice whether it was safe to keep to the sea any longer or make for some harbour, the result was the latter, the ship being so very weak and leaky it was not safe to keep the sea any longer but make for Corunna if we could fetch it, if not some other port". By abandoning his orders, Captain Guy knew he would come under the scrutiny of the Admiralty, and that his formal journal entry involving his officers may lead to them all being court-martialled. A similar situation confronted the officers of the *Swallow*, who in 1692, found that after a storm the head of their ship became loose and the ship very leaky. They abandoned their orders and managed to reach the safety of Kinsale; the officers were all court-martialled but were acquitted. In fact, the court gave them some credit and approved of their actions.[185] As *Lenox* stood SSE under her lower sails, on the morning of 19 December land was seen to the eastward and it was determined they had passed Corunna. Cape Finisterre was then sighted bearing N by W seven leagues. Continuing on the same course, *Lenox* sailed on towards Vigo. Next day she anchored off Bayona in twenty-two fathoms, five miles to the west of Vigo. Ironically the wind now dropped and the next two days were spent moving the ship nearer the town where she would remain for the rest of the month while the crew endeavoured to repair her damage. *Lenox* warped to within a cable's length of Vigo and was moored with the best bower anchor. As soon as he could, Captain Guy wrote a letter to the Navy Office at Crutched Friars in London:

Right Honourable Sirs,

These are humbly to acquaint your Honours that by a storm of wind between the north and east I was forced out of my station, the storm continued for fourteen days which blew a new main sail to bits and strained the ship so much that she sprung several leaks which caused her to make thirteen inches of water in half an hour. The ships [sic] head and cutwater so very loose that I expected to lose it every minute. The knee of the head very loose. If the head had gone away I had lost my bowsprit and foremast and with her labouring so long hath loosened the upper part of the stem and also loosened several of her standards between decks which obliged me to stand in for land for the security of his Majesty's ship under my command and sailing in with Cape Finnestair put for Vigo where I arrived the 22nd instant where in the harbour she makes ten or eleven inches water in half an hour. I will use all my diligence to search for her leaks and am in hopes in one weeks [sic] time to secure her defects as well as I can, then make the best of my way to England or Ireland the first place I can fetch. The *Harwich* never joined me, the *Crown* lost company the 29th November about three o'clock in the morning with a very hard gale of wind and have not seen him since. In my driving the southward I have seen very few ships, but on the 17th instant saw a fleet of ships bound to the southward, but what they were I cannot tell. By the reports of English living here there is on this coast the *Diamond* and another such and several privateers which have taken several ships as they tell me. They likewise tell me the *Oxford* and some part of the Newfoundland fleet are arrived at Lisbon. I am forced to put the ships [sic] company to half allowance of bread, butter and cheese because I could not be supplied with my proportion of them at Plymouth, nor could they be got at Kinsale to be supplied there. The bread is very bad here it cannot be had under twenty seven [?] hundred. I have nought else to trouble your honours with at present, but that I will use all diligence to dispatch as soon as possible and away with the first opportunity of wind and weather I remain, your most humble servant
John Guy.

Casks were immediately sent ashore for water and the carpenters set to work on repairs to the cutwater and fastenings to the knee of the head. The loose bows were caulked and then, to ease the cutwater the magnificent lion figurehead, the first to be made to the new fashion wearing a crown (which had been so often drawn and painted by the Van de Veldes), was cut off. It was probably stored in the hold for use later when major repairs could be undertaken in a home port. Bread and cheese rations were halved and Captain Guy hired a Spanish shallop to fetch water and take out the stream anchor and cable to northward in order to prevent the ship swinging towards the shore. The shallop cost Captain Guy four pieces of eight; he spent another two at a blacksmith's shop to repair the chain plates, bolts and other ironwork.

Four guns were struck down into the hold, probably from forward standings, to ease the strain near the bows and to lower *Lenox*'s centre of gravity. She was then heeled to starboard for repairs to the larboard side. Two known leaks, a fresh one in her run aft and another abreast the chesstree under the wale were all stopped. Soon *Lenox*'s temporary repairs were complete and the last supplies of water were brought aboard as the crew made final preparations to sail. As they did so seven merchantmen escorted by the *Oxford* joined her.

1696 – REPAIR

On 1 January *Lenox* unmoored and hove in her cable to ride short on her best bower. The wind was unfavourable and she was obliged to wait until the 3rd before being able to depart. The weather seemed to have no pity for the battered ship and its crew, for immediately *Lenox* left Vigo Bay the wind rose to yet another great storm. Her sails split again as she fought a large sea from the south. Next day the problems increased when three main weather shrouds broke, causing the mainmast to split two thirds through at the level of the partner on the upper deck. To effect repairs the ship scudded under a foresail, the main yard was then lowered and three fishes put on the mast.

To add to the dangers, leaks again became a serious problem and 12in of water entered the ship per hour. The great storm continued into the next day but by the evening the main topmast was taken down in extremely difficult circumstances to ease the strain on the damaged mainmast. The storm at last eased to a hard gale and the main yard was got up again and the mainsail set. At 4 am on the 7th, the welcoming lights of Scilly were seen and next day, five days after leaving Vigo, *Lenox* found the sanctuary of Plymouth Sound. During the storms *Lenox* was fortunate to have lost only one man.

Her misfortunes were not yet quite over, however. A pilot came aboard at 7 am to take her into the Hamoaze. Anchor was weighed and the ship taken between Drake's Island and the mainland but she came aground starboard on the north side. She was got off again almost immediately and taken against the dock. The master builder's assistant, two carpenters and officers of the dockyard carried out a survey of the ship and the mainmast.

Lenox was in a very sorry condition; she was eighteen years old and had suffered seriously from the storms during her last cruise. Her fastenings had been worked severely, causing movement in her timbers: "complaining", as her journals have it. The loosened timbers resulted in the keel bowing downwards towards the ship's ends, a condition known as hogging. This is the reason that water was bailed out of the gunner's after storeroom, as it could not run uphill into the pump well amidships. The surveyors estimated that the repairs would take two months to complete and were too severe to be carried out at Plymouth.

Captain Guy wrote to the Navy Board describing his voyage home and the findings of the survey.[186] As a result, the Admiralty ordered *Lenox* to sail to Portsmouth.[187] In order to get there more temporary repairs were necessary. The damaged mainmast was newly fished and the main topmast

re-rigged. While the work was being carried out the *Crown*, which also survived the storm, came into the Hamoaze and was taken into dock.

On 20 January *Lenox* was ready to sail. At 8 o'clock in the morning both anchors were aboard, but as the wind was too calm to move the ship under sails she was warped out into Plymouth Sound. The wind then picked up and by 2 pm she was out at sea. The next day *Lenox* passed Dunnose at 4 pm and later that evening she anchored at St Helens. The following morning during a fresh gale she entered Spithead where she found seven or eight English warships, some Dutch, and a fleet of English and Dutch merchantmen. Captain Guy wrote informing the Navy Board that he would deliver the monthly books for October and November, the old general muster book and the ship's books for last year into the hands of the Commissioner at Portsmouth.[188] The most frequent subject of captains' letters concerned the delivery of books to the Navy Board.

The crew of *Lenox* would not be needed during the forthcoming repairs and the Admiralty ordered some of her crew to be turned over into other ships. During February 39 men went to the *Berwick*; 26 men to the *Northumberland*, with a promise to be paid before she sailed; 117 to the *Dreadnought*; 13 to the *Vulture* fireship; 61 to the *Southampton*; 8 to the *Resolution*, and 8 into the *Experiment*. Surplus butter and cheese was sent to the *Advice*.[189]

On 4 February, as *Lenox* waited to enter Portsmouth in a sharp wind, the small bower cable, being old and well worn, parted in the hawse and forced the release of the sheet anchor. Next morning they got hold of the end of the small bower cable and put on a line. A week later *Lenox* was run into Portsmouth Harbour and work started taking out her stores.

On the 25th there was a sudden alarm, with (false) rumours of an impeding French invasion. Intelligence was received that a French army of 30,000 soldiers would embark in transports at Calais and Dunkirk and cross the Channel before an English fleet was ready to oppose them. All the English and Dutch men-of-war in commission sailed to the eastward at the news that a French squadron was out. Both English and Dutch merchantmen crowded into Portsmouth Harbour and the Hampton river for their security.

On 5 March the Admiralty ordered the *Lenox* to be laid up in Portsmouth and her officers and remaining men (except the commissioned and warrant officers and their servants) to be turned over into other ships at Spithead. They were further directed to take care that they should be put aboard such ships of the third rate as were in greatest readiness for sea and that the inferior officers of the *Lenox* be provided for among these ships at Spithead and Portsmouth in the same stations as they served in *Lenox*.[190] On 7 March, after a few days of snow, *Lenox*'s master, John Swan, wrote concerning the few remaining men aboard: "This day we was all discharged from the *Lenox* except standing warrant officers." No plans were made to repair *Lenox* for the moment, although a warrant was sent to Commissioner Greenhill to go in hand with her repairs as soon as the pressing works of the port would allow.[191]

Lenox was removed from the fleet lists and her decks fell silent for the first time since 1690. Commissioner Greenhill's weekly reports to the Navy Board show that *Lenox* remained in harbour until 22 April when at 12 o'clock at night the newly graved *Dreadnought* was hauled out of dock to be succeeded in the early morning at high tide by *Lenox*. Details of the work carried out were not recorded, although it was probably very similar to the work carried out at the same time to another ship of 1677, the *Elizabeth*, which seems to have been in much the same condition as *Lenox*. Additional timbers were added to the inside of the hull, including two floor riders, fourteen futtock riders, two pairs of top riders and a new beam in the middle of the breadroom with double knees. The bows were strengthened by the addition of a pair of standards and three pairs of knees. The cheeks were taken off and refitted. A considerable leak was found and repaired under the wale in the hooding ends. Finally, she was newly caulked.[192] The repairs to *Lenox* took a month before she was hauled out of dock on 21 May. The repairs would not have returned the ship to anything like a good condition but would buy her some time before she had to be taken apart and rebuilt.

Lenox was taken to her mooring in Portsmouth Harbour and laid up for the summer, as there were not enough men available to man her. She had to wait a further six months, until 16 November, before the Admiralty found men for her. The third rate *Edgar* was to be refitted at Chatham and her captain, James Greenway, was appointed as *Lenox*'s new commander. *Edgar*'s men and their commissioned officers were to be turned over into the *Lenox*. Some of the men were to travel to Portsmouth in the *Cornwall* and the rest overland.[193] Greenway had been promoted to lieutenant in 1688 and made a name for himself by destroying the French 96-gun ship *Conquerant* at La Hogue with the *Wolf* fireship.[194] On 24 November Greenway wrote: "Yesterday myself and ships company were discharged from the *Edgar* at Chatham, and this day we entered for the *Lenox*. The chests and chattels of all our men being put aboard the *Cornwall* at the Nore and part of my men in order to be carried to Portsmouth to the *Lenox*. The other part of my men I gave tickets to travel by land and after having completed the pay books and all other accounts of the *Edgar* I made the best of my way to the *Lenox* and on the 6 December I came on board and found her with part of her ballast in and the topmasts rigged and in no other readiness for sea."

Captain Greenway was very popular, for shortly afterwards fifty of his old crew, who had journeyed by land, arrived to join him. Unfortunately he had no money to pay them and some returned to London. Captain Greenway must have been mortified at this, for more and more ships' companies were made up of pressed landsmen, men not bred to the sea.

There was certainly a great deal of comradeship between Greenway and his remaining men. He wrote to the Navy Board informing them of the arrangements he had made, also stating that he would be thankful if the master of the *Edgar*, who was on leave in the country, was granted a warrant for the *Lenox*.[195]

By the end of November there were 163 men from the *Edgar* aboard *Lenox*. The men from the *Edgar* would spin stories and ballads about their war experiences, and one that must have been near the top of the popularity list is related below:

Extract from *The Maiden Sailor*
A true relation of a young Damsel, who was pressed on board the *Edgar* man-of-war, being taken up in a seaman's habit; after being known she was discharged, and at her examination, she declared she would serve the King at sea, as long as her sweet-heart continued in Flanders.[196]

This Maiden she was press'd, Sir,
and so was many more,
And she, among the rest, Sir,
was brought down to the Nore,
Where ev'ry one did think they had
Prest a very pritty Colliers Lad;
But yet it prov'd not so,
When they the truth did know,
They search'd her well below,
and see how things did go,
and found her so and so,
And then (she) swore, the like was never known before.

On 1 December *Lenox* started to receive petty warrant provisions. Boatswain's and carpenter's stores were brought aboard and hoy-loads of ballast taken in. The *Cornwall*, which Captain Greenway was eagerly expecting, arrived with some more of his old crew but his pleasure at seeing

them was spoiled by being ordered to send fifteen of them straight to the *Queen*. The rest of the month was spent setting up rigging, gammoning the bowsprit, and re-coiling cables on the orlop deck. Stowing cables was a very skilled art; the coil should be started at the outside with the next coil inside it. The layer on top was then started on the inside and worked out. It was important that the bends were made as gently as possible to avoid damage to the fibres.

Orders were sent from the Admiralty to Matthew Aylmer, Vice Admiral of the Red, to use all possible diligence to get *Lenox* ready and out into Spithead.[197] Suddenly, however, Greenway's preparations hit a major problem. The Clerk of the Cheque to Portsmouth offered fifty of the men who had travelled overland conduct money in old clipped coinage. The Mint had recently introduced coinage with a milled edge that could not be clipped. The men wanted not just the face value but the full weight of silver coin. Captain Greenway explained to the Navy Board "the Clerk of the Cheque have none but old money which will not pass, the want of this money makes the men uneasy, and their clothes being on board the *Cornwall* which is not yet come here. I cannot keep them aboard to their duty."[198] A crisis in the money supply, partly brought on by the introduction of milled coinage, left the dockyards rapidly running out of supplies. On 20 December Captain Greenway wrote dejectedly that he had orders to be fitted for sea as soon as possible "but as yet have no provisions aboard of any kind and the victuallers agent have told me there is none in the stores".[199] To make up the number of men needed to man *Lenox*, three press warrants were issued to Captain Greenway. By the 23rd the gun carriages and some sea beer were aboard.

1697 – CONVOY DUTY
On 1 January *Lenox*'s men were sent to help get the *Newark* out of dock after her refit. While they were ashore they witnessed the launching of the new 90-gun *Barfleur* and no doubt joined in the customary celebrations. The money crisis put an end to any plans for *Lenox* going to sea in the near future, as the victuallers told Captain Greenway there were simply no provisions available. Greenway wrote to the Navy Board informing them of the situation and adding that no tender could be hired for his ship and could one therefore be sent round the coast from London. Despite the shortage of provisions he would be able make her ready with the 150 men he had.[200]

During the next few weeks *Lenox* slowly became a warship again. The decks were cleaned ready for receiving the guns, the anchors were brought aboard, the running rigging rove, the sails got in and bent to their yards, and powder was taken below. The guns were brought aboard mounted and scaled. But shortages continued: Captain Greenway ran out of paper and incredibly found that there was none at Portsmouth to be had except common muster-book paper. He wrote pleading that paper be sent down from London so that he could make up ship's books and tickets necessary for the business of his ship.[201] More problems arose when bedding, which belonged to the *Edgar*'s men, went missing when being transported to *Lenox* by the *Cornwall*. By 20 January *Lenox* was ready to be moored at Spithead and men were borrowed from the *Devonshire* and *St Michael* to help move her. For the rest of the winter *Lenox* remained at her moorings, short of men and provisions thanks to the money crisis. Her new lieutenants arrived: Churcher was first, Daniel Tilsley the new second and Wheeler the third. On 31 March there was brief excitement when the *Torbay*, which was then in dock, caught fire but it was soon brought under control and put out.

With the coming of April activity increased. Admiral Benbow, popular and capable but of lowly birth and coarse manner, departed with a squadron to cruise off the Scilly Isles and news was received that a French squadron of four warships had attacked a convoy bound for the West Indies. It was decided to get *Lenox* ready for sea and orders arrived from the

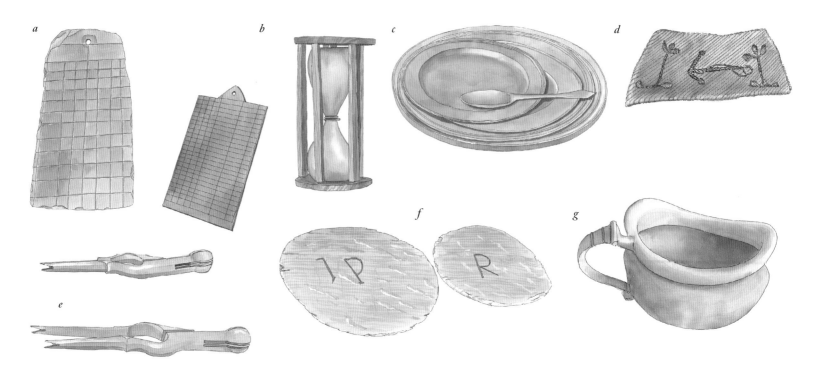

a   b   c   d

f   g

e

Above: *(a) A crude and fine navigational slate, used by the watch to record details such as course, wind direction, wind strength and speed for calculating the ship's position. (b) A time glass, consisting of two separate glass flasks held in place by turned wooden ends and spindles. In 1677 Lenox was issued with 1 four-hour watch glass, 1 two-hour half-watch glass, 18 one-hour glasses and 6 one-minute glasses. (c) A set of pewter plates for officers. The quality emphasises the difference in status compared to seamen. However, both officers and men enjoyed generous allowances of food and drink. (d) Part of clothing, probably from an officer's coat, made of woven wool on a leather stiffener depicting a fouled anchor and floral decoration. (e) Navigational dividers. (f) Wooden platters for seamen, inscribed with the owner's initials. (g) A small pewter chamber pot, probably used by an officer for relief during the night. All drawings based on items recovered from the 1703 wreck of* Stirling Castle, *Ramsgate Maritime Museum.*

Admiralty for her to take in twelve months' sea stores for a foreign voyage.[202] She was examined on 7 Aril and found to be "very weak now and will require rebuilding suddenly".[203] She received 100 men from the *Association*, brought aboard by a yacht on the 8th, followed a week later by forty-four men from the *Lancaster*. On 13 May *Lenox* received orders to take in a month's food provisions in preparation for sailing; the next day a tender brought aboard twenty-two pressed men, thirty-three more from the *Association*, and finally, on the 20th, fifty from the *Cornwall*. So concerned were the Admiralty about the attitude of the men that they sent orders to Admiral Mitchell, the senior officer at Portsmouth, to "tell the captains to call together the men and assure them their wages will be paid punctually as soon as it is possible to procure money for them. Let them know in the most prevailing manner how absolutely necessary it is for His Majesty's service and the good of their country. All possible care will be taken of them and go on cheerfully with their duty during the general scarcity of money throughout the kingdom."

Mitchell received another order urging him to get the *Lenox* and *Hampton Court* into condition to proceed into the Soundings with all despatch.[204] Meanwhile Rear Admiral Benbow and his squadron returned from their cruise to join *Lenox* at Portsmouth. They had met the West Indies convoy that had been attacked by the French, and had tried to intercept the attackers but were forced to return to port for want of provisions. Benbow received new orders from the Admiralty to set sail again and cruise in the Soundings between 10 and 100 leagues west of Scilly to

meet a rich homebound Virginia fleet of merchantmen, as it was known that a French squadron was at sea. He was to reinforce his squadron with the fourth rates *Anglesey* and *Plymouth* as they passed Plymouth. *Lenox* was to be the flagship of Benbow's small squadron, which consisted of the *Hampton Court* and *Kent*, two of *Lenox*'s sisters; the fourth rate *Dreadnought*; and two fireships, the *Firebrand* and the *Shambles*. At 9 o'clock in the morning of the 22nd Admiral Benbow came aboard and as Rear Admiral of the Blue hoisted his blue flag at the mizzen masthead. *Lenox*'s complement of men was made up to strength before sailing when 150 men were brought aboard from the *Queen*. The shortage of money restricted the squadron to only a month's provisions at short allowance.[205]

The squadron sailed into St Helens where they anchored as the tide turned. The next day they set sail again but off Dunnose an admiral's boat from Portsmouth came up to *Lenox* with new orders for Benbow. He was to take into his care three merchant ships: the *Tavistock*, *Bedford* and *Dorothy*.[206] The squadron returned to St Helens where they found the three East Indiamen bound for the East Indies. The delay cost a day but by the 26th they were off Plymouth Sound where they lay by and sent in boats, with orders for the fourth-rate ships that were ready to sail out and join them. The calm weather obliged the squadron to anchor off Rame Head in eighteen fathoms of water. The following morning the wind sprang up and they were able to make sail and continue westwards. They were joined by two fourth rates – the *Medway* off the Lizard, and the *Weymouth* off the Scillies. Both ships were short of beer and *Lenox* issued some of hers to them. The wind continued to build and *Lenox* reefed her topsails but other ships suffered damage; the *Hampton Court* split her fore topmast and an East Indiaman broke her main topsail yard.

With the first day of June the wind increased to a storm. On the 2nd, a lieutenant in a boat from the *Kent* came aboard and gave an account of a serious leak in his ship and reported that the foremast had sprung. The *Weymouth* then flew a distress signal as she had also sprung a leak. Most of the ships in the navy had spent years on active service at sea and were becoming structurally very weak. The next day *Lenox* lay under a main sail and mizzen; by the 4th the wind had moderated to a gale when a ship from Cork bound for the West Indies joined the convoy.

That evening they were able to set their foresails and main topsails. On the morning of the 6th the leaking *Kent* was sent home in company with the *Shambles* fireship. The *Kent* was later taken to Rotherhithe where she was taken apart and rebuilt as a new ship, eventually launched again in 1699. As soon as *Kent* left "at ten that morning we saw four sail SSW of us three leagues standing to the westward, they bore down upon us" and at 11 am after seeing the strength of the English opposition "they tacked to the eastward showing them to be enemy".

On the 8th, the convoy lay 120 leagues west of the Scilly Isles, a position considered safe from French privateers. In accordance with his orders, Benbow made a signal for the commanders of the merchantmen to come aboard the *Lenox*, where they were informed that he would now leave them. The squadron then turned eastwards. Two days later during a gale from the NW the *Dreadnought* carried her main and mizzen topmasts by the lee. Later the same morning a ship was seen to leeward and *Lenox* gave chase for an hour, but then she too suffered damage when her sprit sail topmast and fore topgallant mast were carried away.

Benbow was now in position to meet the incoming Virginia convoy. On the 11th he met the cruising *Hastings*, which informed him of nine sail seen bearing NE. Two days later, at 5 o'clock in the morning, the Scilly lighthouse was seen and a boat was sent in to gain information. The Virginia fleet had not yet been observed so Benbow stood off all night and at daybreak tacked and stood to the eastward. The following day a French privateer appeared and the *Medway* unsuccessfully give chase. The squadron was now very short of provisions and Benbow decided to make for Plymouth to replenish. It was a fortuitous decision, for as they neared the Lizard a large body of ships was seen, which turned out to be the long sought-after Virginia merchant fleet combined with a West India convoy. *Lenox*'s master, John Burrington, counted 120 ships.

The wind died away, slowing the fleet, but it was reinforced by a cruising squadron of six men-of-war under Vice Admiral Mitchell near the Start. He gave orders to Benbow to enter Plymouth in company with the merchantmen, which they did on the 22nd.

Lenox entered the Hamoaze and was taken alongside a hulk so that her stores could be removed into it before she was taken into dock for cleaning. The upper tier of guns was removed but the following morning orders were received from the Admiralty instructing Benbow to accompany the Virginia and Barbados merchant fleet into the Downs.[207] Work on *Lenox* was immediately suspended and the guns were quickly put back on board. At the same time Benbow received orders from the Admiral of the Fleet, George Rooke, to leave *Lenox* for the 80-gun *Cornwall* and join a squadron off Dunkirk. *Lenox* spent the remaining few days of June making ready for sea. She received provisions, gammoned the bowsprit, and replaced her best bower anchor. On 1 July *Lenox* was got out into the Sound and anchored in seven fathoms. The following day she set sail for the Downs in company with the *Hampton Court*, whose captain now acted as commodore, the *Dreadnought*, and the Virginia and West Indian merchant fleet. Slowly they made their way up-Channel and as they passed the Isle of Wight, *Lenox* took the slow-sailing sternmost ship in tow. She did not cast it off until they arrived in the Downs during a gale on the 6th. The following day the merchant fleets continued their voyage up the Thames to deposit their wealth in London, while the warships remained at anchor in the Downs for a further six days.

During this period *Lenox* received provisions from Dover, including beer, eighty bags of bread, thirty barrels of beef, flour, oatmeal, and butter. *Lenox*, *Hampton Court* and the rest of the small squadron of warships weighed anchor on 12 July, bound for Spithead to join the cruising squadrons in the west. Sailing with contrary winds they anchored off Dover on the 15th and Dunnose the 16th and finally Spithead on the 19th, where they remained for the rest of the month.

Lenox was heeled and scrubbed and received twenty-seven men from the *Sheerness* before the Commodore, in the *Hampton Court*, made the signal to sail into St Helens in company with the *Firebrand* fireship.

The Admiralty soon received intelligence that a number of French warships were at sea. Considering the matter at their board meeting of 5 August, they decided to strengthen the cruising warships with the addition of *Hampton Court* and *Lenox*.[208] Orders were accordingly sent to the two ships to join the main fleet under Sir George Rooke in Torbay. They weighed on 8 August and arrived the next day to join a powerful squadron consisting of the first rates *Britannia*, *Royal William* and fifteen other English and Dutch warships. George Rooke aboard the *Britannia* ordered *Lenox* on to Plymouth for her long overdue docking and cleaning. At 6 o'clock the next morning *Lenox* weighed anchor and sailed alone to the westward. She anchored in Plymouth Sound on the 10th and warped into the Hamoaze the next morning to be moored against the dock. During the next ten days the ship was made ready for docking. The sails were sent ashore, yards lowered, guns got out, powder and shot taken out, provisions and cables put aboard a hulk, anchors sent ashore, casks taken out and finally the ballast removed.

Ships queued up to enter the dry dock to be cleaned, an operation which normally took one day. On the 21st it was the *Milford*'s turn, followed by the *Rochester* and then the *Constant*. *Lenox* followed; she was hauled in on the 24th and taken out at night the same day. The next morning began the long process of taking back on board all the provisions, stores and ballast, a task that was completed on the 30th.

Lenox was now ready to sail, but still desperately short of men. Captain Greenway wrote to the Admiralty concerning his plight. The Admiralty sent a copy of Greenway's letter to Rooke and asked him to supply *Lenox* with marines and men.[209] Rooke was unable to help and a few days later on 6 September new orders arrived for Captain Greenway to sail to Spithead to join the fleet.[210] She was delayed by contrary winds until 10 September. While waiting, she received twenty-seven men from the *Pendennis* and thirty from the *Exeter*. When at last the wind changed she set sail to the eastward. She anchored off Torbay where she found Vice Admiral Mitchell, who had arrived the day before in command of a squadron of 8 third rates, six fourth rates and three fifth rates, six Dutch warships and two fireships under the Count of Nassau.

Mitchell received orders at Torbay to proceed with his squadron to Cape St Vincent as it was heard that a French squadron had been observed there. Mitchell had only two months' provisions on board and the Dutch even less. He immediately wrote letters outlining these concerns, which were carried by *Lenox* to Spithead where she anchored at 11 o'clock the next morning. Captain Greenway delivered them to Admiral George Rooke who then gave Greenway orders from the Admiralty to sail with the *Rye*, a new 32-gun fifth rate, to cruise in the Soundings for the protection of trade.[211] It was now late in the year for the three-deckers to be at sea and many of their crews were now transferred to other ships, fifty men being sent to *Lenox* from the *Britannia* and fifteen sent from the *Sandwich*.

Then everything changed; on the evening of 16 September, while it rained and blew a gale, "the whole fleet English and Dutch fired their guns on news of a peace". News had arrived of the peace signed at Ryswick. The terms were advantageous to the English and Dutch. Louis XIV agreed to recognise William III rather than James II as King of England and conceded commercial benefits to the two seagoing allies.

In the war at sea, France had started on the offensive and achieved early success, but after its defeat at Barfleur and La Hogue in 1692 it had become defensive and was forced to rely on commerce raiding. At the start of the war the English fleet was not far from equal strength with either the Dutch or French fleets. Now the French were defeated and the Dutch exhausted. The English, although suffering financial problems, were in a much better state than their rivals. Peace also gave the Admiralty time to rebuild almost all of the warships built before the war started. This was just as well, for it is doubtful many of them could have continued at sea much longer without being docked for many months for repairs, as *Lenox* had been in 1696. The terms of the treaty would not take effect until 18 October; until then it was necessary to provide escort for merchant ships for another month.

Lenox and the *Rye* weighed anchor on 20 September to sail out from Spithead into St Helens. As *Lenox* hove up her best bower anchor she found that half the stock had been broken off. Gales continued blowing for two more days and *Lenox* was obliged to lower her yards. After the winds moderated on the evening of the 22nd, the yards were squared and she was hove short on her anchor to make ready to sail. The next morning she got under way but that night the wind suddenly increased in force, breaking the mizzen topmast in three pieces, part of which fell on the gunner's head, badly injuring him.

Lenox took two more days tacking in gales trying to sail past the Isle of Wight, during which the foresail was split. They sighted Vice Admiral Mitchell's squadron to the west as *Lenox* put into Torbay to put her injured gunner and eight sick men ashore. After sending the sick into Dartmouth, *Lenox* set reefed topsails and endeavoured to carry clear of the land and head westwards, but the wind once more made progress difficult. *Lenox* hauled in her topsails but a few hours later the head of the mainmast sprung, forcing them to sail with the wind which took them east into Portland Road. The following morning *Lenox*, *Rye*, and the prize *Rainbow*, a 32-gun fifth rate recently taken, attempted to clear the land, but the wind blowing from the SSE forced them back again. The following day, and no doubt to the relief of Captain Greenway, the wind changed to blow from the NE, allowing the ships to clear land and safely arrive in Plymouth Sound on the 28th.

A pilot came aboard to take *Lenox* into the Hamoaze for repairs, where she was taken alongside a masthulk at 2 o'clock in the afternoon of the 30th. Within two hours the main topmast and main yard were removed, followed in the morning by the mainmast. When the main yard was lowered and examined it was also sprung. As *Lenox* lay at Plymouth she received her last orders from the Admiralty: "Notwithstanding any orders to the contrary you are required and directed with the first opportunity of wind and weather to sail with the ship under your command to Blackstakes and there to make all possible dispatch in putting out her guns, stores and provisions and then to proceed to Chatham where she is to be laid up and paid off. In your passage to the river you are to convoy any ship bound your way."[212]

As repairs progressed, news reached Plymouth on 2 October of the destruction by Prince Eugene of the mighty 100,000-strong Turkish army of Mustapha II. The battle took place in early September at Zenta, a town on the right bank of the river Tisa, in what was then Hungary. It ended for good the Muslim invasion of Western Europe; in reality, a victory more important than that recently celebrated over the French. The significance was not lost on the crew of *Lenox*, for that night they went ashore and lit bonfires fore and aft of their ship to express their joy.

The repaired mainmast was taken in on 9 October and work started to rig it. Two days later *Lenox* parted from the hulk and the main topmast was got up. Then the repaired main yard was brought on board and also rigged. On the 14th *Lenox* weighed and got out into the Sound where she met a convoy of ten merchant ships waiting to be escorted to the east; by 8 o'clock that evening they had made sail and headed out to sea. On the 17th *Lenox* and her convoy arrived and anchored at St Helens where four of the merchant ships ended their voyage and *Lenox* received a new longboat and fresh water from the dockyard. The same day the remaining ships weighed and sailed for the Downs. Contrary winds obliged them to anchor with each ebb tide and it was not until 21 October that they arrived. Upon arrival they found Sir Cloudesley Shovell in his ship *Defiance*, along with several other men-of-war of the English fleet.

The war was now officially over and all that remained for *Lenox* was for the ship to be taken to the anchorage at Chatham. A pilot came aboard and she immediately weighed to begin the difficult voyage past the sand banks. The first day she anchored off North Foreland but it took another three days before she reached the Nore, where gales sprang up and *Lenox* was obliged to lower her yards and topmasts. The weather eased three days later and she raised them again but a sudden storm caught her crew unawares. The mizzen mast was split, requiring the mizzen yard and topmast to be taken down. While at the Nore, Captain Greenway wrote to the Navy Board on behalf of the men who had sailed with him in both the *Edgar* and *Lenox*:

"I am like wise to acquaint your Honours that I am importuned very much by my men which were turned over with me last year out of the *Edgar* into the *Lenox* to request the *Edgar*'s books to be sent down that they have their pay for the time they served in her, she was but eleven months in pay and they say it will be a great hardship for them to seek their money when they are dispersed, some of them living in the north, and some in the west of England. They being, many of them that has followed me, all, or most part of the war, if it may not be thought an ill precedent, do beg this favour for them".[213]

Greenway's next ship was the *Northumberland*, wrecked in the great storm of 1703 in which he and his crew were all lost.[214] A pilot came aboard and *Lenox* was taken to Kitts Hole on 1 November where her guns and some of her stores were taken out. She was then moved to Upnor Castle where the sails were unbent and the ship unrigged. By 11 November her crew had left for the last time and she was run out of commission. A report of 4 March 1698 estimated that large repairs, or a rebuild, would cost £10,344.[215]

It is worth mentioning that for her seven years of service, which relates to about 3,000 crew years, about 115 men died from all causes. On these facts, the average life expectancy aboard her was about twenty-six years. Of all the men who died aboard her, very few were killed in action. Another interesting statistic concerns the number of men court-martialled and punished. A search through all the court-martial papers for the period reveals nothing concerning *Lenox*, however, although most of her sister ships feature at least once or twice. Although *Lenox* was fortunate, stories of brutality aboard ship did not apply in the late seventeenth century. Men were sometimes hanged for mutiny but it was not until 1703, when two seamen were executed for desertion, that it was recorded as being the first instance in a very long time that maritime law was used against such offenders.[216] The whip was used as a court-martial punishment, the maximum penalty being 120 strokes. During the war one man in the navy died during the punishment and an inquiry was held to determine the reason. It was concluded that the injuries inflicted could not have caused his death as no-one else had died and that he must have had other medical problems. Humanity pervades through many of the court-martials, young men "not knowing any better" were often leniently treated. Clear proof was needed before conviction and many men were acquitted although they were probably guilty.

Lenox's final voyage started on 13 November 1699, when 142 men took her back to Deptford.[217] It was the first time she had been there since she was built and during the intervening period she had deteriorated into a worn-out wreck. She had survived many of the men who built her, although some remained, including Frank Hosier, the storekeeper. Nevertheless, she had done her duty, and would be rebuilt in 1701 at Deptford dockyard, the place of her birth, to be made ready for service again during the next century. Indeed, she went on to serve in a different guise until 1756, but that is another story.

APPENDICES

APPENDIX 1 – PRINCIPAL DIMENSIONS

The actual dimensions of *Lenox* are inserted for comparison, the main source being her building list of 1678.

NA, ADM 106/36
Principal Dimensions for His Majesty's Ships of War of the Three Superior Ranks

	1st Rate	2nd Rate	3rd Rate
	ft.ins	ft.ins	ft.ins
Length of the keel	———	———	———
Lenox touch actual			131 0
Length of the gundeck from the rabbet of the main stem to the rabbet of the stern post	165 0	158 0	150 0
Lenox actual			151 6
Greatest breadth from the outside to the outside of the plank	46 0	44 0	39 8
Lenox actual			39 8
Breadth at the main transom within the plank	29 4	28 4	25 6
Breadth at the top of the stern at the gunwale	19 6	18 9	17 0
Breadth of the ship at the top of the side in the waist from outside to outside of the timber	32 0	31 3	30 4
Breadth at the beakhead at the top of the gunwale being the foremost toptimber	28 0	27 4	26 0
Depth in the hold from plank to plank	19 2	18 2	17 0
Lenox actual			17 0
Height between the gundeck and second deck from plank to plank at the side	7 0	6 10	7 3
Height from the second to the upper deck between plank and plank at the side	7 1	6 11	———
Number of ports on the lower deck on each side	14	13	13
Lenox actual			13
Bigness of the ports on the lower deck fore and aft	3 9	3 6	3 6
Depth of the same	2 11	2 9	2 9
Number of ports on the second deck on each side	14	13	———
Bigness of the ports on the second deck fore and aft	3 3	3 2	———
Depth up and down	2 9	2 8	———
Number of ports on the upper deck on each side	14	13	13
Lenox actual			12
Bigness of the ports on the upper deck fore and aft	2 9	2 8	2 8
Depth up and down	2 6	2 5	2 5
Number of ports on the quarter deck in all	12	10	12
Lenox actual			10
Bigness of the same fore and aft	2 5	2 4	2 4
Depth up and down where the work is not cut	2 1	2 0	2 0
Ports on the forecastle cut through the beakhead or over the gunwale	4	4	4
Lenox actual			0
Memorandum :- not to exceed in draught of water when all provisions are on board for the allowance of men according to the establishment	21 0	20 0	18 0
Lenox actual			18 6

APPENDIX 2 – SCANTLING LIST

The official *Scantlings of the 30 Sail of Ships* (NA, ADM 7/827) relating to the new third-rate ships of 1677 are reproduced in the left-hand column. The proposed lists, which the Master Shipwrights Isaac Betts and Daniel Furzer both called *The Principal Dimensions and Scantlings of Timbers for the 30 New Ships to be built in prosecution of the late Act of Parliament*, are listed in the right-hand columns (NA, ADM 106/329, f70 and NA, ADM 106/327, f215 respectively). The Betts and Furzer lists are identical in format, indicating that they were requested by the Navy Board in the form of a standard questionnaire to which the shipwrights only had to fill in the dimensions.

The following scantling lists have been compiled using the clearest or fullest description provided by the originals. For instance the Furzer list that mentions "drumhead capstans" was preferred to the *Scantling of the 30 Sail*, which simply says "capstans". This is perhaps the earliest mention of the drumhead capstan that had very recently replaced the crabb capstan. Where numbers of items are given, for instance two teir of carlines, they relate to one side of the ship only.

The *Scantlings of the 30 Sail of Ships* was the official list issued to the king's and merchant shipbuilders. The surviving copy is very poor and appears to have been made shortly after 1685. Unfortunately it contains a number of muddled inconsistencies that are not easy to interpret; there are numerous mistakes; sometimes a line is repeated and sometimes the description is incomplete, especially concerning the knees of the gun deck. It was, however, similar in layout to the earlier lists sent to the Navy Board by Betts and Furzer. Sir John Tippetts and Sir Anthony Deane made the official list in consultation with the master shipwrights' lists. Although the scantling lists were too late to be applied to *Lenox*, they are all very similar and reflect current shipbuilding practices that would have been employed at Deptford dockyard.

Fortunately the problems with the official scantling list were resolved with the aid of the Betts and Furzer lists and the contracts for the *Yarmouth* and *Cumberland* of 1694. The *Yarmouth* was another ship of 70 guns built to almost identical dimension as those of 1677 (NA, ADM 106/3071). Most of the dimensions in the various sources agree with each other or are very close. Fragments of 1677 contracts made with merchant builders also exist in the form of letters written to the Navy Board. For example, old Jonas Shish describes "the dimensional changes that Mr Castle would like to make to the strakes near the wales" (NA, ADM 106/333, f340).

Among the more glaring inconsistencies between the Betts and Furzer lists is the difference in the length of the scarph of the false stem. The 1ft scarph given by Betts is probably correct, while the 8ft scarph mentioned by Furzer must be a mistake. Betts alone, from his own initiative, gives the dimensions of the wales that are similar to those of the *Yarmouth* contract.

	Scantling list of the thirty ships	Furzer's suggested scantling list for thirty ships	Betts' suggested scantling list for thirty ships
NA, ADM ref.	7/827	106/327 f215	106/329 f70
FRAME	ft.ins	ft.ins	ft.ins
Length of the gun deck.		150 0	150 0
Breadth at the transom.		25 6	25 6
Keel. Length of the keel.	125 0	125 0	124 0
Keel. Square in the midships.	1 6½	1 4¾	1 5
Scarph tabled in ye keel.	4 6	4 6	4 6
Breadth from outside of the board plank.	39 8	39 8	39 8
Depth in hold from ceiling to underside of the gun deck.	17 0	16 8	16 9
Rake of the stem.	22 0	22 0	23 0
Stem. Breadth of the same at the head.	1 4	1 4	1 4
Fore and aft of the same.	1 6	1 6	1 6
False Stem. Thickness of ye false stem.	0 10½	0 10½	0 9
Breadth of the same.	2 4	2 4	2 6
Scarph long.		8 0	1 0
Sternpost. Rake of the Sternpost.	5 0	5 0	5 9
Breadth at the head.	1 10	2 1	1 8
Fore and aft at the head.		1 10	1 10
Fore and aft at the foot.	2 4	2 4	2 6
The post within it fastened to the mainpost fore and aft and as broad as the mainpost.		1 8	
Each arm of the knee at the post long.	8 0	8 0	7 0
1st Main.	1 2	1 2	1 2
2nd.	1 0	1 0	1 0
Transoms 3rd.	0 11½	0 11½	1 0
Sided. 4th.	0 11½	0 11½	1 0
5th.	0 10½	0 10	1 0
Fashion pieces sided.	0 10½	0 11½	1 1
Space of the floor timbers or Timber and Room.	2 3	2 3	2 3
Floor Timbers. Floor timbers fore and aft.	1 1¼	1 1¼	
Depth of the same on ye keel.	1 1¼	1 4½	1 4
In and out at ye wronghead.	1 0¼	1 0¼	1 0
Naval Timbers. Naval timbers fore and Aft.	1 1¼	1 1¼	1 1½
Scarphs of the same.	6 8	6 8	6 6
Timbers in and out at ye gundeck.	0 10¼	0 10¼	0 10
Upper Futtocks. Fore and aft of ye same.	1 0¼	1 0¼	1 1
Length of the scarph.	6 6	6 6	6 0
Top Timbers. Top timbers sided at ye head.	0 8	0 8	0 8
Sided at ye foot.	0 11½	0 11½	1 0
In and out at ye gunwale.	0 4¼	0 4¼	0 4
Keelson. Not more than 5 pieces. Depth in the middle.	1 4	1 4	1 5
Breadth in the Midships.	1 6	1 6	1 6
Hause pieces broad at least.		2 4	2 6
WALES			
Depth of ye Lower Wale.			1 2
Thickness of ye same.			0 10
Depth of ye Upper Wale.			1 2
Thickness of ye same.			0 10
Six Strakes of thick stuff without board, two above, two between & two below ye Wales.			0 6
Chain Wales. Depth.			0 10
Thickness.			0 6
HOLD			
One strake of plank next to ye limber board.	0 4	0 4	0 4
7 strakes of sleepers in ye hold whereof 5 are thick.	0 7½	0 7½	0 8
Thickness of ye other two.	0 5	0 5	0 5
To be broad.		1 4	1 3
Thickness of ye middle bands.	0 8½	0 7½	0 7
Broad, tabled under the beams.	1 6	1 4	1 3
The rest of the footwaling in hold broad.	1 6		
The rest of the footwaling thickness.		0 4	0 4
Orlop Beams. Orlop beams Fore and Aft.	1 4	1 4	1 4
Up and down of ye same.	1 3	1 3	1 2½
Floor Riders. Floor riders Fore and Aft.	1 6	1 6	1 5
Depth on the keelson.	1 0	1 0	1 1
Depth at the wrongheads.	1 1	1 1	1 0
Scarph of the same.	7 0	7 0	7 0
Futtock Riders. Length of the futtock riders.	14 0	14 0	12 0
Depth at the beam.	2 0	2 0	2 0
And sided.	1 3	1 3	1 0
(Furzer and Betts say the rider is at the end of the beam)			
Knees. Knees Sided to the Beam.	0 10	0 10	0 10
(Orlop) Fore & Aft.	5 0	5 0	3 0
Square of the pillars in hold.	0 8¼	0 8¼	0 8
Steps. Steps for the Main mast deep.	1 6	1 6	2 10
Thickness of ye same Fore Aft.	2 5	2 5	2 8

	Step for the Fore mast deep.	1 6	1 6	
	Breadth of the same.	2 2	2 2	2 6
Brest Hooks.	Four brest hooks in hold.			
	Deep.	1 2	1 2	1 1
	Long.	14 6	14 6	14 0

GUNDECK

Brest Hooks.	Two brest hooks between decks			
	Deep.		1 2	1 0
	Long.		14 0	14 0
Knees under Ports.	2 Knees to each Beam under the Ports Sided.	0 10	0 10	0 10
	Shortest length of each Arm.		3 6	3 0
	Depth of the same in the throat.	1 8		
Riders between Ports.	1 rider and 1 Knee between the Ports at end of every other Beam, Length of Rider.	13 6	13 6	12 0
	Breadth of the same by the beam.	2 0	1 8½	2 0
	Broad	1 1½	1 1½	1 0
Carrick Bitts.	Square of the Fore bitts.	1 6	1 4	1 4
	Square of the After bitts. (With cross pieces of equal bigness as the bitts and knees of equal goodness. Furzer and Betts)	1 6	1 6	1 6
Clamps.	Two strakes of clamps thick.	0 8½	0 8½	0 9
	Broad.		1 6	1 5
Partners.	Thickness of partners for ye main mast	0 10	0 10	1 0
	Thickness of partners for ye fore mast.	0 9	0 10	1 0
Beams.	Gundeck beams sided.	1 4½	1 4½	1 5
	Depth of the same.	1 3¼	1 3¼	1 3
	Hatchway beams asunder.	8 0	8 0	8 2
Carlines.	Two tier of oak carlines			
	Broad.	0 10	0 10	0 10
	Deep.	0 10	0 9	0 8
	Long carlines sided.(Hatchway)	0 7½		
	Depth of the same.	1 2		
	Short carlines sided.(Hatchway)	0 7½		
	Depth of the same.	0 6		
Spirket	Wales 2 strakes.		0 6	0 6
Ledges.	The ledges to lie asunder.	0 9	0 9	
	Ledges sided.	0 5	0 6	0 5½
	Depth of the same.	0 4	0 5	0 4½
Water Ways.	Thickness in the chine.	0 6½	0 6½	0 6
	Breadth of the same.	1 2	1 2	1 3
Plank.	The rest of the gundeck plank to the hatchway.	0 4	0 4	0 4
	Upon the gundeck between bitts and main partners.		0 3	0 3
	Hatchway plank.		0 2	0 2
	Memorandum: every plank and waterway hath two spikes in each beam. (And in each ledge 2 treenails Furzer and Betts).			

UPPER GUNDECK

Beams.	Depth of the upper deck beams.	0 9	0 9¼	0 10½
	Breadth of the same.	1 0	1 0	1 0
	Distance between the Beams (except near the hatchways). (Not to exceed 5⅕ foot asunder. Betts and Furzer)	5 0		
	Rounding of the deck.	0 10	0 10	0 10

Knees.	Double kneed arms length.		4 0	4 0
	Sided.			0 7
Clamps.	Thickness of the clamps.	0 5	0 5	0 6
	Breadth of the same.	1 5	1 5	
Carlines.	Two tire of carlines			
	Squaring of the long carlines.	1 0	1 0	1 0
	Depth of the short carlines.	0 6	0 6	0 6½
	Breadth of the same.	0 8	0 8	0 8
Ledges.	Sided.	0 4	0 4	0 4½
	Depth	0 4	0 4	0 3½
Spirket Plank.	Thickness of the spirketting.	0 4	0 4	0 4
	Plank on the deck thick. (Spruce according to Furzer)	0 2¾	0 3	
Water ways.	Thickness of the waterways.	0 5	0 5	0 4
	Breadth of the same.	1 2	1 2	
Standard.	Three pair of standards sided.	0 8½	0 8½	0 8
	Arms long at least.		5 6	5 0
String	Depth of the string prickt home to the side.	0 8	0 8	0 10
	Thick of the string.		0 10	0 10
Partner.	Main Partners.			0 9
	Fore Partners.			0 7
	Mizon Partners.			0 5
	Plank between ports.			0 3
	Turned Pillars square.			0 7
Capstans.	Main Drum fashioned diameter in the barrel.	2 9	2 9	2 2
	Jeer Drum fashioned.	1 10	1 10	1 8

QUARTERDECK

Beams.	Quarter deck beames sided.	0 8	0 8	0 8
	Depth of the same.	0 6	0 6	0 6
	Rising under the beams in the wake of the cabins.			
	Thick.	0 8	0 8	0 9
	Depth.	0 11	0 11	
Knees.	Knees in the wake of the forecastle and steerage sided.	0 5½	0 5½	0 6

FORECASTLE

Beams.	Beams sided.	0 8		
	Depth of the same.	0 6		
Clamps.	Thickness of the clamps.	0 8		

BOLTS

Bigness of the iron Floor bolts.		0 1¼	0 1³/₈
Bigness of bolts for the Orlop.		0 1⅛	0 1¼
Ditto Gun Deck ports.		0 1¹/₁₆	0 1¼
Ditto Upper deck ports.		0 1	0 1
Ditto Quarter Deck Forecastle ports		0 0¾	0 0¾

APPENDIX 3 – BUILDING CONTRACT OF THE *YARMOUTH*

No copy of a building contract for any of the third-rate ships of 1677 survives. The nearest available source is that of the *Yarmouth* (NA, ADM 106/3071). She was launched sixteen years after *Lenox*, but was nevertheless built to the same dimensions except for a minor reduction of 3in in the depth of the hold. *Yarmouth* was one of three ships built to replace losses suffered by the 1677 third rates. During 1689 *Pendennis* was wrecked while *Lenox* lay in ordinary at Chatham. In 1690 the *Anne* was run ashore in Rye Bay and burned by her crew after being badly damaged during the Battle of Beachy Head. In the same year the *Breda* blew up as she was moored alongside *Lenox* in Cork harbour.

The *Yarmouth* contract is dated 23 January 1691 and is made out to Nicholas Barrett, a merchant shipbuilder. The *Yarmouth* was built at Harwich and launched four years after construction began. A contract for *Lenox* would not have been made out as she was built by warrant in the King's own yard. The *Yarmouth* contract is reproduced in its entirety for the first time, headings have been added where missing, for easier reference.

23rd January 1691

Yarmouth [in a later hand]

Contract with Mr Nicholas Barrett for the building another 3rd rate ship at Harwich

To be launched last (of) December 1691
11:0:0 pounds per ton

This indenture made the three and twentieth day of January in the year of our lord one thousand six hundred and ninety between the Principal Officers and Commissioners of their Majesties' Navy (for and on the behalf of their Majesties) of the one part and Mr Nicholas Barrett of Wapping in the county of Middlesex of the other part. Witnesseth that the said Nicholas Barrett for the considerations hereafter expressed doth covenant, promise and grant to and with the said Principal Officers and Commissioners (for and on the behalf of their Majesties) that he, the said Nicholas Barrett, his executors, administrators, servants or assignees shall and will at their own proper cost and charges well and workmanlike erect and build off the stocks, for the use of their Majesties at their yard at Harwich one good and substantial new ship or frigate of good and well seasoned timber and plank of English oak and elm.

Overall Dimensions
And that the said ship or frigate shall contain in length upon the gun deck from the rabbet of the post to the rabbet of the stem, one hundred and fifty feet. Breadth from outside to outside of the plank thirty nine feet eight inches. Depth in hold from the top of the ceiling to the upper edge of the gun deck beam sixteen feet nine inches. Breadth at transom twenty five feet six inches. The rake forward at the harpin to be reckoned at three fifths part of the main breadth. The rake aft five feet nine inches to the main transom.

Keel
The keel not to be made of more than four pieces, to be sixteen inches and a half broad in the midships and fifteen inches and a half up and down. To be sheathed with a four inch plank well fastened for a false keel. To have four feet six inches scarph tabled in the keel and to be well bolted with eight bolts of inch and half quarter inch auger.

Stem
To have a firm substantial stem of sixteen inches thwartships and seventeen inches fore and aft with a sufficient false stem of nine inches thick and two feet six inches broad with scarphs one feet long to the false stem and not less than four feet to the main stem.

Sternpost
To have a substantial sternpost of two feet two inches broad at the head and one feet ten inches fore and aft at the head and two feet six inches below on the keel fore and aft and another post within it to be fastened to the main post of sixteen inches fore and aft and as broad as the main post.

Rising Wood
Unto which shall be joined the rising wood sufficient for the run of the said ship and also a long armed knee of six feet long at the least. Each arm to be well bolted with an inch quarter and half quarter auger fastening the same to the keel and to the sternpost at every twenty two inches length at furthest.

Timber and Room
The space of timber and room to be no more than two feet three inches.

Floor Timbers
The floor timbers of the said ship to be thirteen inches and one quarter of an inch fore and aft and sixteen inches and one half inch up and down upon the keel and twelve inches and one quarter of an inch in and out at the wrongheads or twelve inches full when wrought and to be twenty three feet long in the midships.

Naval Timbers [Sometimes known as the first or lower tier of futtocks]
The naval timbers to fill the rooms being at least thirteen inches and one half inch fore and aft and to have at least six feet nine inches scarph.

Middle Futtocks [Sometimes known as the second tier of futtocks]
The middle futtocks to have six feet six inches scarph.

Gun Deck Timbers [Sometimes known as the third or upper tier of futtocks]
The timbers at the gun deck to be ten inches in and out and twelve inches fore and aft and to have at least six feet scarph.

Keelson
To have a substantial keelson of not more than five pieces to be sixteen inches up and down and eighteen inches broad in the midships and to end at the stem and sternpost in proportion to run fore and aft. Each scarph to be three feet long at the least and to be well bolted with inch, quarter and half quarter auger through every other timber and to bolt every other floor timber through the keel and one bolt through the stem.

Toptimbers
The toptimbers to be sided allow twelve inches, at the head eight inches, and to be in and out at the gunwale or top of the ship's side four inches and a half inch.

Ceiling
To put in one strake of four inch plank next the limber board and seven strakes of sleepers in hold on each side the wrongheads, three of them eight inches, two of seven inches and a half inch and the other two of five inches thick and fifteen inches broad and to run fore and aft. To have two strakes of middlebands on each side of seven inches thick and fifteen inches broad and to run fore and aft. To have two strakes of clamps on each side fore and aft under the beams of the gun deck of eight inches and one quarter of an inch thick and seventeen inches broad each and to be hooked one into the other to prevent reaching. To have an opening of six inches under the clamps for air. All the rest of the footwaling or ceiling in the hold to be good four inch English oak plank.

Orlop Beams
To put in ten beams for the orlop, to be sixteen inches fore and aft and fifteen inches up and down, five of them to be placed before the mast and the other abaft.

Riders
To have five bends of floor and futtock riders. The futtock riders to be of fifteen inches the floor riders to be eighteen inches fore and aft and twelve inches deep upon the keelson, fifteen inches deep at the wrongheads. The floor riders to be bolted with nine bolts of inch, quarter and half quarter of an inch auger. The futtock riders to have seven bolts of inch quarter and half quarter of an inch auger. Each rider and to have seven feet scarph upwards and downwards.

Orlop Platform
And to make platforms upon the orlop beams for stowing cables and other stores and to lie five feet nine inches from the gun deck between plank and plank. To have one knee and one rider at each end of the beams or double kneed, but if riders then the riders to be each thirteen feet long, twelve inches sided and sixteen inches deep at the beam, fayed into the beam and to be well bolted with eight bolts of inch quarter and half quarter of an inch auger in and out, and two into the beams with one knee fore and aft at each end of every beam. The shortest arm to be three feet in length and to be ten inches sided bolted with six bolts to each knee of inch and three eights auger.

Mainmast Step
To have a saddle for the step of the mainmast of two feet seven inches fore and aft and depth sufficient for the same.

Gun Deck Beams
To have a pillar in hold under every beam of the gun deck and orlop eight inches square. The gun deck beams to be sixteen inches and one quarter of an inch broad and fifteen inches up and down and to be placed one beam under each port of the gun deck and one beam between each port of the gun deck, excepting in the main hatchway, which must be eight feet asunder and to be kneed with two knees at each end of the beams, one lodging the other hanging, where the beams fall under the ports of not less than three feet long each arm and ten inches sided to be bolted with six bolts of inch and one quarter of an inch auger. The beams that fall between the ports to be in like manner double kneed at each end, if knees can be procured, but if so many knees cannot be had then to have one knee at each end of the beams and of the bigness of the other knees before mentioned and one rider at each end of the said beams, to be thirteen feet long, thirteen inches broad, and twenty inches depth at the beams, to be bolted with eight bolts of inch and three eights auger.

Cross Pillars
To have five pairs of cross pillars in hold of ten inches square and to be well bolted to the beams and riders and to be kneed at the upper ends well bolted.

Gun deck
And to have a double tire of carlines on each side fore and aft of oak to be nine inches thick and ten inches broad and the ledges to lie within nine inches one of another and six inches broad five inches deep. The waterways to be six inches in the chine in thickness and fourteen inches broad. All the rest of the gun deck as far as the hatchways from the side to be good four inch English oak plank well seasoned and of good lengths the said plank and waterways to be treenailed, spiked with two good spikes in each beam and two treenails in each ledge. To put out nine leaden scuppers on each side of the gun deck.

Storerooms

To make as many hatches in the hatchways as shall be convenient of two inch plank with the hatchway abaft the mast for the stowing of provisions and the hatchway to the steward room and for boatswain's and gunner's store rooms and powder rooms which said store rooms and powder rooms are to be built of ordinary deals of such bigness and continuance as equals any of their Majesties' ships of the like burthen.

Hawse Pieces

To make a manger on the lower deck to have four scuppers in it of lead, two whereof to be four inches diameter and to put in four hawse pieces not less than two feet three inches broad each and to cut out four hawse holes in them.

Bitts

To place two pair of conic bitts eighteen inches square the aftermost, and sixteen inches square the foremost pair with cross pieces to the same of equal bigness and two pair of knees suitable to the said bitts and to bolt them with five bolts in each knee of inch and half quarter auger.

Breasthooks in Hold

To have four breasthooks in hold fourteen inches deep and fourteen feet long each and seven bolts in each breasthook of inch and three eights auger.

Foremast Step

To have a step for the foremast two feet four inches broad and of sufficient depth and length bolted with eight bolts of inch and quarter auger.

Gun Deck Spirket

To have two strakes of spirket wales on the lower deck of six inches thick from the waterway unto the lower edge of the ports fore and aft.

Gun Deck Gun Ports

To cut out thirteen ports on each side the same deck three feet six inches broad and two feet nine inches deep with four ports abaft between the transoms and to make and hang portlids with hooks and hinges and to fit and drive two ringbolts and two eyebolts to each port of inch and quarter auger for the guns.

Gun Deck Fittings

To place partners for the main and fore mast of ten inches thick and a pillar for the main capstan to be iron bound for the end of the spindle to stand on in the hold. A step for the mizzen mast on the keelson. To have three inch plank upon the gun deck between the bitts and the main partners in the wake of the hatchways and to raise the hatches above the deck and to have turned pillars under the beams on the upper deck as shall be found convenient and placed upon a four inch plank for the pillar to rest upon the gun deck. To make a staircase up into the quarterdeck and the stairs and ladders to all the conveniences. To bring on two breasthooks between decks fourteen feet long fourteen inches deep and to fasten them with seven bolts in each hook of inch and three eights auger.

Transoms

To have as many transoms abaft below the ports as may lie within eighteen inches one of another. The wing transom to be fourteen inches thick and the rest not less than twelve inches thick and one transom at the upper edge of the ports under the helm port to take hold of the sternpost. All the said transoms to be well kneed with long armed knees as is usual, fastened with six bolts in each knee by inch and three eights auger.

Upper Deck Gun Ports

To make twenty four ports on the upper deck (that is to say) twelve on each side, two chase ports forward and four right aft to be two feet eight inches broad and two feet five inches deep each port and to garnish them with carved work fore and aft.

Upper Deck Clamps

To bring on clamps fore and aft of six inches thick and fifteen inches broad under the beams of the upper deck and to shut up between decks with four inch oak plank fore and aft.

Upper Deck Beams

The beams of the said deck to be eleven inches up and down and thirteen inches broad fore and aft to lie between and under each port and not to exceed five feet asunder excepting in the wake of the hatchways and the beams over the main capstan and under the bulkheads to lie as near as conveniently they may. To be in height between the said decks between plank and plank seven feet three inches in the midships. The beams to round ten inches to go flush fore and aft all the said beams to be double kneed with four knees to each beam of seven inches and half an inch sided, the shortest arm three feet long, the hanging arm to come down to the spirketting under the port and to be well bolted with three bolts in each arm with bolts of inch and half quarter auger.

Upper Deck

To have two tier of carlines on each side fore and aft, to be twelve inches square, the long carlines and the other short carlines eight inches broad and six inches up and down with sufficient ledges of four inches square to lie not more than nine inches asunder. To lay the said deck with good three inch oak plank in the wake of the guns, the rest with the like plank or with good dry Prussia deals, to answer the said plank in thickness. To have a waterway five inches thick fourteen inches broad.

Upper Deck Fittings

The Spirketting to be of four inch plank fore and aft. To have a string of English oak of six inches deep and ten inches thick, to be pricked home to the outside plank and to make the lower sill of the upper ports to be well spiked and treenailed through and between the timbers. To have coamings, head ledges with grating hatches before and abaft the mast to vent the smoke of the ordnance. To fit topsail sheet bitts, jeer bitts or knightheads cats and supporters, a davit and clasp of iron. To fit partners for the main and jeer capstans and partners for all the masts, and to put out nine scuppers on each side of the upper deck.

Capstans

To make a main capstan (drum fashion) thirty inches diameter in the barrel and a jeer capstan (drum fashion) of twenty two inches diameter in the barrel with capstan bars and iron pauls sufficient for the said capstan.

Quarter Deck and Forecastle Beams

To make a large quarterdeck and a large forecastle. The beams of the same to be eight inches fore and aft and six inches and a half up and down and to lie within two feet one of another and each other beam to be kneed with one up and down knee at each end of good length and six inches sided and bolted with five bolts in each knee by a three quarter of an inch auger.

Quarter Deck and Forecastle Fittings

To have round bulkheads in the bulkheads of the said forecastle and quarterdeck for four cabins next the side and in the midships of the bulkhead of the forecastle. To place the cookroom for roasting and boiling and to set all the bulkheads upon oaken plank fayed on the deck for the foot of the stanchions of ten inches broad and four inches thick, laid with tar and hair and the seams leaded. To have two ports in the bulkhead of the forecastle and two in the bulkhead of the steerage and twelve ports on the quarterdeck, six ports on each side of two feet four inches wide and two feet deep. The beams in the wake of the bulkheads to be double kneed at each end. To have three pair of standards on the upper deck in each bulkhead one pair, to be nine inches sided and not less than three feet and a half feet long each arm and to have one pair in the bulkhead of the coach to be bolted with six bolts with inch and half quarter auger. The quarter deck to be laid with two inch oak plank well seasoned next the side, and the rest with Prussia deal of like thickness. To have a rising of elm under the beam of the great cabin of eight inches thick and ten inches deep the beams to be dovetailed and bolted into the same. To make and hang with port lids about the whole ship with substantial hooks and hinges. To have a transom abaft under the windows in the cabin and one under the ports and the same to be kneed and bolted with six bolts in each knee and to have an open balcony abaft out of the great cabin with rails and bannisters, and to have as large a roundhouse and coach as the works with conveniency will bear. To be completely fitted with bulkheads joiner's work and doors to the same.

Rooms in Hold

To make all platforms in hold with bulkheads and partitions viz for the powder room and gunner's store rooms, sailroom, boatswain's carpenter's and steward's store room and steward room, a fish room and a store room for the captain's provisions and as many cabins for lodgings as shall be convenient. To make a large bread room and sheath the same with lead or tin plate, the lead or plate thereof to be at their Majesties' charge.

Without Board Planking

Without board the ship is to be planked up from the keel ten feet in height with elm, oak or beech plank of four inches thick and from thence up to the chainwales with four inch oak plank excepting six strakes which is to be six inches thick and fourteen inches broad in the midships and to lessen in thickness in proportion toward the stem and stern as is usual, ending at the stem and stern in four inch plank viz two strakes below the wale, two between the wales and two above the wales. To have two formed wales of fourteen inches up and down and nine inches and a half of an inch thick and to have two chainwales for the conveniency of the chain plates and bolts and to go fore and aft both to be six inches thick and ten inches broad to be chocked between the timbers with oak in the wake of the chain bolt. To have one strake of three inch oak plank between the chainwales ten inches broad and the wale upwards so high as the waist from the upper chainwale to be wrought up with three inch English oak plank and the quarter with well seasoned Prussia deals of two inches thick.

Head

To have a fair head with a firm and substantial knee and cheeks treble rails, trail board, beast, brackets, keelson, cross pieces and standard. To have catheads and supporters under them.

Stern

To have a fair lower counter with rails and brackets and open galleries garnished with carved works. To have a house of office in the gallery windows and casements into the cabin. To have a fair upright and to put in it a complete pair of King's arms or other ornament of like value maskheads, pilasters and forms.

Rigging Attachments

To have a pair of chesstrees fore, main and mizzen chainwales well bolted, chain bolts and chain plates sufficient for the shrouds and backstays of all the masts.

Gripe

To have a sufficient gripe well bolted, stirruped and dovetailed and a stirrup on the skeg well bolted.

Rudder

To make and hang on a complete rudder with six pair of braces, gudgeons and pintles a muzzle for the head and a tiller thereto.

Decoration

To gunwale and planksheer the said ship fore and aft and to put on brackets, hancing pieces and to garnish them complete. They are likewise to do and perform all the carved work, painting and guilding answerable to their Majestie's ships of the like bigness in the Navy. The gilding work being intended to be only the lion in the head and the King's arms in the stern. The head stern and galleries in carved work not inferior to those of the thirty ships formerly built and to find and provide all materials for the same.

Details

Likewise to do and perform all the joiner's works finding deals, locks, iron bars, hinges for store rooms, steward room's, doors, settlebeds and cabins. The cabins to be as many and as well adorned in all respects as any of their Majestie's ships of the like burthen and to equal them in all respects both withinboard and without. They are to find all plumber's work, lead and leaden scuppers and all glazier's work of stone ground glass with sash lights, scuttles for the cabin windows and all painter's work for painting and guilding as aforesaid within and without board and to do and perform whatsoever belongs to the carpenters to do for the finishing and completing the hull (without masts and yards) in like manner as is usually done and performed to the like ships built in their Majestie's own yards and to set the masts at his or their own charge with the help of the boatswain (that is to say) heel the masts wedge them and shut them in.

Materials and Delivery

And the said Nicholas Barrett, for himself his executors and assignees doth covenant and grant to and with the Principal Officers and Commissioners of the Navy that he will at his or their own cost and charges find and provide all manner of iron work of the best Spanish iron or what shall be equal to the same in goodness and all spikes, nails, brads, likewise all timber planks, boards and treenails of Sussex well seasoned which are to be all mooted from Prutia deals above the chainwale down to the keel. To find white and black oakum, pitch, tar, rozin, hair, oil, brimstone and all other materials that shall be needful to be used or spent in or about the work and premises aforesaid for the complete finishing the said ship and in like manner to discharge and pay all manner of workmanship touching all and every part of the work herein expressed and hereafter expressed or to be done and performed and to finish complete and launch the said ship or frigate, and to deliver her safe on float in the river of Harwich unto such person or persons as shall be duly and sufficiently authorised by the said Principal Officers and Commissioners to receive her for the use of their Majesties by the last day of December next coming after the date hereof.

Quality

And it is further agreed that the said Principal Officers and Commissioners shall have liberty to appoint such person or persons as they shall see fit to inspect and oversee the building of the said ship which person or persons shall have free liberty at all times to discharge his or their duty therein without any lett or molestation and if at any time during the building the said ship or frigate herein contracted for, according to the dimensions proportions and scantlings herein expressed and set forth or intended to be expressed and setforth there shall be found and discoursed of the said person or persons any unsound, insufficient timber, plank or other materials used in the building of the said ship or which shall be of different scantling from what the same ought to be by this present contract, or that any insufficient workmanship or such as is not answerable to this contract shall be performed on the same that then after due notice thereof given in writing by the said surveyor or surveyors unto the said Nicholas Barrett or to his chief master workman under him on the said ship there shall be an effectual and speedy amendment reforming of all and every such default in stuff and workmanship whereby the same may be made agreeable to this contract in dimensions goodness and workmanship and the said amending or reformation shall be certified in writing by the said surveyor or surveyors to the said Principal Officers and Commissioners of the Navy for the service of their Majesties in this behalf. And the said Nicholas Barrett do further oblige himself, his heirs, executors and administrators and every one of them to comply with the said Principal Officers and Commissioners.

Supply of Stores

To transport from the river of Thames to Harwich free of charge to their Majesties' the masts, yards, rigging, sails and sea stores for Boatswain and carpenter necessary to be provided by the office of the navy for the said ship the same to be got ready to be embarked in the River of Thames by the said Principal Officers or their substitutes by the first of October next or thereabout after the date of this contract.

Draught

And also the said Nicholas Barrett for the well building and finishing the said ship in all respects shall produce unto the Navy Board his master workman together with a fair draught or design of a ship according to the dimensions herein set forth for the approval of the one and for the correction and approval of the other which afterwards is and shall be the design and draught which he shall cause his master workman to follow in the framing and building the said ship intended and agreed as aforesaid to be built.

Payment

And the said Principal Officers and Commissioners of the Navy for and behalf of their Majesties shall according to the custom of the office of the Navy sign and mark out bills to the Treasurer of the Navy to be paid to the said Nicholas Barrett his executors, administrators or assignees after the rate of eleven pounds per ton for each ton the said ship shall measure with the dimensions and limitations before

expressed and the main breadth to be reckoned as is hereby mutually condesended and agreed unto. To be four inches on each side without the timber and tonnage to be cast according to the accostomed rule of Shipwrights Hall. The said money to be paid in manner form and payments following (that is to say) sixteen hundred pounds at and upon the sealing this contract. The like sum of seventeen hundred pounds more when the floor of the said ship shall be laid across and bolted. The sum of two thousand five hundred pounds more when the second futtocks to the said ship shall be all fast. The sum of eleven hundred pounds more when the orlop beams shall be fast and footwaling in. The sum of eleven hundred pounds more when the gun deck beams are all kneed and fastened. The sum of one thousand pounds more when the upper deck beams are kneed and fastened and both her upper and lower decks shall be laid. The sum of one thousand pounds more when her quarter deck and forecastle beams shall be all kneed and fastened and the remainder when the said ship shall be launched and delivered on float as aforesaid. It is further agreed that if the said ship shall be exceeded in dimensions and scantlings contrary to what is herein before agreed that no satisfaction or allowance shall be made for such overwork or increase in scantling unless the said increase of works or scantling have been made by order first given therein in writing by the said Principal Officers and Commissioners. In witness whereof to and part of these put to indentures the said Nicholas Barrett hath set his hands and seal and to the other part thereof the said Principal Officers and Commissioners of the navy have accordingly to the custom of their Majesties' Navy set their hands and caused the common seal of the office of the Navy to be affixed the day and year first above written.

Witness

Edmund Dummer   Nicholas Barrett
N .Dalkes

APPENDIX 4 – JOHN SHISH'S BUILDING REPORTS TO THE NAVY BOARD, 1677

NA, ADM 106/323, f260: 13th July 1677
New Ship at the head of the Dock. The Keele laid and two floor timbers crossed. The lower piece of Stem is up and are now getting the Stern post up. The upper piece of Stem & the frame of the Stern I hope will up tomorrow in the afternoon.

NA, ADM 106/329, f23: 20th July 1677
New Ship at the head of the dry Dock. The frame of the Stern is up & fashion pieces. The Knee of the main post is fayd & fast & one piece of rising hood. Her Stem is up and 28 floor timbers are in their places upon the Keele.

NA, ADM 106/329, f25: 27th July 1677
New Ship at the head of the Dock. There is about 2/3rds of her floor in with the rising hood (excepting one piece). The Ribbons at the Floor Sirmarks are about fore & aft. Two pieces of Kelson are upon the floor timbers in the midships but not faid and are now in hand with the frame Bends.

NA, ADM 106/323, f276: 3rd August 1677
New Ship at the head of the dry Dock. All the rising timbers both afore and abaft are in & bolted & the rest of the floor is completed. The rising hood is in and fast and are in readiness with the half timbers.

NA, ADM 106/329, f27: 10th August 1677
New Ship at the head of the dry Dock. All our Frame Bends are in and most part of the Keelson is faid & fast and are in hand faying the half timbers.

NA, ADM 106/329, f29: 17th August 1677
New Ship at the head of the dry Dock. All her half timbers are in and 8 of her Harping timbers. We have dubbed down her Floor and are this day beginning to plank.

NA, ADM 106/329, f31: 24th August 1677
New Ship at the head of the dry Dock. All the lower Futtocks and Harping timbers are in. The Keelson is perfected and two strakes of plank on each side are about and fast. We are in hand with another strake of plank on each side and are now making Stage to get the middle tier of Futtocks in.

NA, ADM 106/323, f296: 31st August 1677
The ship in the dry Dock hath all her middle tier of Futtocks in and with what expedition we can the upper tier are getting in also and we have completed six strakes of planking on each side.

NA, ADM 106/329, f33: 7th September 1677
New Ship at the head of the dry Dock. Her Bilge is plank out and all her upper Futtocks are in and are now bringing her lower wales about.

NA, ADM 106/329, f35: 14th September 1677
New Ship at the head of the dry Dock. The lower wales is about & fast and are planked up to the lower Futtock heads on both sides. The next work we go about is to raise the toptimbers and shall go in hand with the same within two or three days.

NA, ADM 106/329, f37: 21st September 1677
New Ship in the dry Dock. The ship is planked on both sides, two strakes above the lower Futtock heads and have brought on and fastened one strake of thick stuff on each side under the lower wale and part of the next strake beneath. We are now dubbing down within board and going in hand to bring on her footwaling.

NA, ADM 106/329, f39: 28st September 1677
New 3rd Rate Ship at the Head of the dry Dock. We are part shut in with plank under the lower wale on one side and shall shut in on both sides by tomorrow night if weather permits. There is 3 strakes of Footwaling brought on and fast on each side and are in hand with another strake. We are now making stage to bring up her toptimbers.

NA, ADM 106/329, f41: 5th October 1677
New Ship in the dry Dock. We have shut in with plank under the lower wale on both sides and almost dubbed square down. There is likewise eight strakes of Footwaling brought on, on each side & square. Are in hand with another and the Limber boards are faid. There is also fifty top timbers up in their places & fast. There is wanting 3 foot & 2 foot & a half treenails to drive of the work under water.

NA, ADM 106/329, f43: 12th October 1677
New Ship in the dry Dock. The major part of the toptimbers are up & in their places and 10 strakes of Footwaling on each side brought on and fast. We are now dubbing down in order to bring on her Middlebands and getting the rest of the toptimbers up.

NA, ADM 106/329, f45: 19th October 1677
New Ship in the dry Dock. All the toptimbers are up. We have faid & fastened one strake of thickstuff on each side above the lower wale and have brought on two strakes of Middlebands on each side and all the footwaling perfected underneath it.

NA, ADM 106/329, f47: 26th October 1677
New 3rd Rate Ship in the Dock. The upper wales is brought on and are shut in between the wales. Two hawse pieces is up and have faid and perfected 4 strakes of Middlebands.

NA, ADM 106/329, f49: 2nd November 1677
New Ship in the Dry Dock. The Hawse pieces are all up and two strakes brought on above the upper wale. There is 7 Orlop Beams in their places & 8 or 9 Knees faid to them & all the Footwaling underneath the Orlop perfected.

NA, ADM 106/329, f51: 9nd November 1677
New Ship in the Dock. All the Orlop Beams are in and Kneed fast there is 4 strakes brought on above the upper wale on each side. Three or four of the Counter timbers are up. The Knee of the head is made ready and faid a piece for the Lyon and another for the trail board.

NA, ADM 106/3538, part 1: 16th November 1677
New Ship in the Dry Dock. One strake of Gun deck Clamps on each side is faid & fast and are now faying of our Floor riders & some breasthooks. The Counter timbers are all up and are now planking of the lower Counter. One Chain wale on each side is brought on and fast and are in hand with the strake of plank above it.

NA, ADM 106/3538, part 1: 23rd November 1677
New Ship in the Dock. The Gun deck Clamps are in & perfected with all the footwaling. There is 10 Gun deck Beams up in their places, five floor riders & 2 Transom Knees faid. The lower Counter is planked. We have shut in under the lower Channel wale, brought on one strake above it and are now in hand with the upper channel wale. We are now beginning to Knee some of the Gun deck Beams.

NA, ADM 106/3538, part 1: 30th November 1677
New Ship in the Dock. There is one strake of plank on each side brought on above the upper Channel wale. All the Gun deck Beams are in and are in hand with the futtock riders.

NA, ADM 106/3538, part 1: 4th December 1677
New Ship in the dry Dock. There is 3 strakes of plank brought on without board above the channel wale on each side. The upperdeck clamps are faid and fast. The upright of the stern is birthing up. The Gun deck ledges are faying and are now getting the Knee of the Head and Gripe up. Both bilges are caulked square out.

NA, ADM 106/323, f337: 7th December 1677
New Ship in the Dock. There is 4 pair of Transom Knees faid and about half the Gun deck Knees with two long Carlings fore and aft on each side the hatches and are now in hand in faying of her Futtock riders.

NA, ADM 106/3538, part 1: 14th December 1677
New Ship in the Dock. The Knees of the Gun deck Beams are almost completed. The major part of the Gun deck Port cills & Carlings are faid. Two of the Main Bitts are in and the Futtock riders near finished.

NA, ADM 106/3538, part 1: 21st December 1677
New Ship in the dry Dock. The Knees of the Gun deck Beams are all faid. The main Bitts are in and 10 upperdeck Beams are up in their places. The timbers of the upright in the Stern are up and are in hand with the Knee of the Head.

NA, ADM 106/3538, part 1: 28st December 1677
New Ship in the dry Dock. There is two strakes of plank brought on without board above the upper channel wale on each side. the Gun deck ports are cut out on both sides & all her upperdeck Beams are in.

NA, ADM 106/3538, part 1: 11th January 1677/8
New Ship in the dry Dock. The Ship is planked up to the Gunwale in the waist. The upperdeck ports cills are faying. The major part of the Gun deck waterways & Spirketting is faid & fast. The Knee of the Head, Gripe and false Sternpost are fast & 2 lower Cheeks of the Head faid.

NA, ADM 106/3538, part 1: 18th January 1677/8
New Ship in the Dry Dock. All the upperdeck port cills are faid with the two long Carlings on each side the Hatchway on the upperdeck. Half the Ceiling between the Gun deck ports is completed. Four Transoms (aft) above the Gun deck is faid. The two lower Cheeks of the Head are fast. The trailboard & lacing of the Head is up and are in hand with the two upper cheeks of the Head. Faying the upperdeck Knees and fitting the rails for the sides fore and aft.

NA, ADM 106/3538, part 1: 25th January 1677/8
New 3rd rate Ship in the dry Dock. The upperdeck Carlings are all faid. The major part of the upperdeck Knees & Ledges are faid. All the riders are completed. One great rail on each side without board fore & aft is brought on and fast. The lyon of the Head is up. There is four rails in the stern perfected and some carved work faid there.

NA, ADM 106/3538, part 1: 1st February 1677/8
New 3rd rate Ship in the dry Dock. The upperdeck Knees are almost completed. The upperdeck waterways are very near faid and part of the said deck is laid. The Gun deck waterways & spirketting is all finished. The Step for the foremast is in. The main Capstan is in hand. The upper rails fore & aft her sides are up. The upper cheeks of the Head is faid and are now going forward with the Galleries.

APPENDIX 5 – THE BUILDING LIST OF THE *LENOX*

The building list of the *Lenox* is preserved in the Pepys Library, Magdalene College, Cambridge, reference PL1339. It is a small manuscript book of 50 folio pages, 8in by 3in with dark blue Morocco binding and gilt edges. This is not a proposal nor an estimate, but the listing of all the materials actually used in the construction of the ship. The end page is signed by W Fownes, Clerk of the Cheque at Deptford dockyard, who prepared the account that remained with Samuel Pepys. In two codicils to Samuel Pepys' will, provision was made for his library, of which the book became a part, to be preserved "intire in one body at either Trinity or Magdalene College Cambridge". The book has never been published before; it is reproduced here in its entirety. The original seventeenth-century spelling is retained for authenticity. Modern spelling has been inserted in brackets where the meaning is obscure, and some of the abbreviations and truncations have been extended where necessary.

The building list would have been written at Deptford for the Navy Board, from where the list, or a copy of it, found its way into the hands of the Secretary of the Admiralty, Samuel Pepys. It does not include details of guns or gunner's stores; these were dealt with by a separate body, the Board of Ordnance. The building list includes the hull, masts, rigging, wages, and six months' stores for boatswain and carpenter.

> The Avoirdupois weight system is used: hundredweight (cwt), quarter (qr), and pounds (lb).
> 28 pounds = 1 quarter
> 4 quarters = 1 hundredweight
>
> Rope lengths are measured in fathoms and cables.
> Circumference is measured in inches.
> 6 feet = 1 fathom
> 100 fathoms = 1 cable's length
>
> Costs are in pounds (£), shillings (s), and pence (d).
> 12 pence = 1 shilling
> 20 shillings = 1 pound
>
> Timber is measured in loads, nominally a cart load, and cubic feet.
> 50 cubic feet = 1 load
>
> Volume of liquids measured in barrels (barr), firkins (fir), and gallons (gall)
> 9 gallons = 1 firkin
> 4 firkins = 1 barrel

His Ma:ts 3rd Rate Shipp ye *Lenox*

	ft : inches
Length by ye keel	131 : 0
Bredth by ye beam	39 : 8
Depth in hold	17 : 0
Draught of water	18 : 6

30973

Tons 1096

79524

Built in Deptford Dock by Mr: John Shish master Shipwrt: there since ye 25th June 1677 being ye first Shipp finished of the 30 ordered by Act of Parliamt: Launched ye 12 of April 1678 & sailed to Chatham ye 12th May follow:

folio 2

Lenox: Stores expended in building her.

		Quantity Cwt qr lb	Rate £ s d	Sume £ s d
Blacking.		10 barr	6d ye:bar	0 : 5 : 0
Brimstone.		4 : 1 : 0	:18: 8	3 :19 : 4
Scrubbing Brushes.		: : 1	: :	0 : 4 : 9
Brass Rowles.	2	0 : 0 :24	: :14	1 : 8 : 0
Cop(Copper)				
Sheets.	9 1/3	2 : 0 :24	: :17½	18 : 1 : 8
" Funnells.	4	2 : 1 :13	: :	19 : 6 5½
" Nails.		0 : 0 : 6	: :20	0 :10 : 0

		Quantity Cwt qr lb	Rate £ s d	Sume £ s d
Charcole.		3 bushe	: : 8	0 : 2 : 0
Canvas				
Holland Duck.		21 yards	: :20	1 :15 : 0
" Lincolnshire.		20	: :16	1 : 6 : 8
" Ipswich.		129	: :11¼	6 : 0 :11¼
" Old.		88	: : 4	1 : 9 : 4
Candles.		1 : 1 :19	: : 5½	3 :12 :10½
Decaid(decayed)				
Ensignes.	16	brr: 2		0 :11 : 8
Lashing line.		26 coyles	:16: 6	21 : 9 : 0
Deep sea line.		1	: :	0 : 6 : 6
Junke.		2 : 1 :22	: 8: 0	0 :19 : 6¾
Tard marlin.		0 : 0 : 8	: : 3 6/7	0 : 2 : 6 6/7
Twine.		9¼	: :10	0 : 7 : 8½
	inch	: :		: :
New Rope of	6½	8 fath:	1:10: 0 cwt	1 : 4 : 0
"	4	129		7 :16 : 7
"	3	99		3 : 7 : 2
Old Rope of	4	196	: 6: 0 cwt	2 : 7 : 7
"	3½	60½		0 :11 : 2
"	3	160		0 :11 : 8
"	2½	60		0 : 5 : 1
"	2	110		0 : 8 : 3
	inch			folio 3
Old Rope of	2	10 fath:	: 6: 0 cwt	0 : 0 : 8
"	1½	8	: :	0 : 0 : 4½
"	1	10	: :	0 : 0 : 3¾
½ houre glass.		1	: :	0 : 0 : 9
Glew(Glue).		1 : 0 :18	3: 0: 0	3 : 9 :7 7/20
Spun haire.		5 : 0 :14	2: 0: 0	10 : 5 : 0
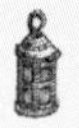 Tin hand Lanthorns.		5	2: 5	0 :12 : 1
Lead Milled No	2	46 : 0 : 0	1: 5: 0	57 :10 : 0
"	4	4 : 2 :21	1: 3: 0	} :
"	5	3 : 0 : 0		}8 16 9¾
"	6	5 : 3 : 4	} : :	: :
"	8	3 : 2 : 0	}1: 2: 0	10 : 9 : 9 ¼
"	-	0 : 1 : 0	} : :	: :
" Old.	-	0 : 1 : 0	:13: 0	0 : 3 : 3
" Sheet.		10 : 0 :18	:18: 0	9 : 2 :10¾
" Scuppers.	55	23 : 0 :13	}1: 2: 0	28 : 5 : 3¾
" Pisdale pipe.	-	2 : 2 : 9	} : :	
Leather Liquord.		0 : 1 :24	: :10	2 : 3 : 4
" Bucketts.		6	: 3: 8	1 : 2 : 0
" Scupper.		1	: :	0 : 2 : 4
Ocam(Oakum)Black.		39 : 2 : 0	:12: 0	23 :14 : 0
" White.		57 : 2 : 0	:18: 0	51 :15 : 0
" Rent.		9 : 1 :11	1: 0: 0	9 : 6 :11½
Oyle(Oil).		45 gallons	: :16	3 : 0 : 0
Pitch.		10½ barr	1: 7: 6¾	14 : 9 : 4 7/8
Rozin(Resin).		7 : 2 : 0	: 9: 6	3 :11 : 3
Sodder.		0 : 1 : 8	: : 9	1 : 7 : 0
Sope(Soap).	12furk	6 : 3 :16	: : 3½	14 : 9 : 6
Tallow.		10 : 0 :19	2: 8: 0	24 : 8 : 1 ½
Tarr.		11 barr	:16: 8	9 : 3 : 4
				folio 4
Thrumes.		0 : 2 :10	: : 8	2 : 4 : 0
Grindstones.	5		: 5: 6	1 : 7 : 6
Rubstones.	7		: 2: 6	0 :17 : 6
Bricks.		2000	:18: 0	1 :16 : 0
Lime.		2: :	:11: 0	1 : 2 : 0
Sand.		2 load	: 2: 0	0 : 4 : 0
Salt.		2 bushl	: 3: 0	0 : 6 : 0
Tyles(Tiles) Paving.		12	: :	0 : 1 : 6
" Plaine.		500	: 2: 0	0 :10 : 0
Auger bitts extra of			: :	
inch & upwards.		4	: :15	0 : 5 : 0
Ord:ry		120	: :12	6 : 0 : 0
Bolts Plate.		4	: :15	0 : 5 : 0
" Spring Plate.		40	: : 6	1 : 0 : 0

		Quantity Cwt qr lb	Rate £ s d	Sume £ s d
" Rain.	48	8 : 1 :25	} : :	
Hanging Clamps.	6	0 : 1 :20	}1: 8:10	14 : 7 : 4
Iron Crows.	7	1 : 1 :18	} : :	
		Quantity		
Hinges Cross}		17 pair	: 3: 2	2 :13 :10
" Dozen} Garnetts.		41 pair	: :12	2 : 1 : 0
" Dove Tayles.		66 pair	: : 4	1 : 2 : 0
" Side.		34 pair	: :13	1 :16 :10
" Scuttle.		10 pair	: :11½	0 : 9 : 7
"Side with rising joints		6 pair	: 4: 0	1 : 4 : 0
" Esses.		36 pair	: :10	1 :10 : 0
" Butt.		12 pair	: : 5	0 : 5 : 0
" Lamb heads.		21 pair	: : 4½	0 : 7 :10½
Rivetts to ye Butt hinges.		100	4½ per doz	0 : 3 : 1½
White Handles.		12	: :12	0 :12 : 0
Boat hooks.		4	: :	0 : 4 : 0
Handles for two				
hand saws.		12	: : 3	0 : 3 : 0
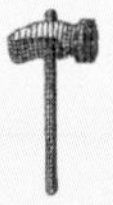 Clench hammers.		16	: :12	0 :16 : 0
				folio 5
Hasps &		20	: : 4	0 : 6 : 8
Staples.		24	: : 2	0 : 4 : 0
Spring Latches.		10	: :14	0 :11 : 8
Iron Ladle.		1	: :	0 : 3 : 0
Locks Cupboard		18	}: :22	3 :17 : 0
" Settle.		24	} : :	
" Hanging.		17	: :14	0 :19 :10
" Spring double.		28	: 2: 4	3 : 5 : 4
" " Single.		20	: 1: 4	1 : 6 : 8
" Stock Extra.		8	: 2: 4	0 :18 : 8
" " Ordinary.		14	: 1: 6	1 : 1 : 0
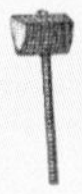 Double headed Mawls.		10	: 3: 3	1 :12 : 6
Saile Needles.		4	: :	0 : 0 : 7 1/3
Double white tin plates.		2 dozen	: 7: 4	0 :14 : 8
Single Topp Plates.		12	: : 6	0 : 6 : 0
Reeming Irons.		72	: 2: 8	9 :12 : 0
Dutch Rings.		50	: : 3	0 :12 : 6
Hatch Rings & Staples.		18 pair	: : 6	0 : 9 : 0
Port Rings				
with eyes rag'd		52	: : 8	1 :14 : 8
Scrapers.		4	: :15	0 : 5 : 0
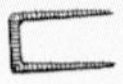 Staples.		12	: :	0 : 2 : 0
Port Staples.		18	: : 4	0 : 6 : 0
Two hand sawes.		2	: 4: 0	0 : 8 : 0
		Cwt qr lb		
Spikes.		21:1: 4	}1: 8: 0	145:18 :10½
Nailes Weight		82:3:23½	}	
" 40		8:2:11		
" 30		4:2:17		
" 24		3:1:17		
" 20		4:1:16½		
" 10		5:1:24		
		26:2:11½	1:10: 0	39:18 : 0 6/7
				folio 6
Nails 6		4 : 2 :16½	1:12: 5¼	7: 10: 8¼
" 4		3 : 2 : 6	2: 3: 6¾	7: 14: 9½
" 3		0 : 1 :23	2:16:10¾	1: 5:10 2/3
" 2		0 : 1 :21¼	6	1: 4: 7½
" Lead.		1 : 2 :18	} 5¾	4: 12:11½
" Scupper.		0 : 0 : 8	}	
" Tacks.		0 : 0 : 1	12	0: 1: 0
" Clamps.		3 : 2 : 2		
" Ribbon.		0 : 3 :19		
" Deck double.		9 : 0 : 8		

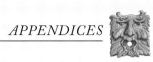

	Quantity Cwt qr lb	Rate £ s d	Sume £ s d
" " Single.	1:0:0		
" Port double.	1:2:13		
" " Single.	0:3:12		
" Sheath : 20	2:1:10½		
" 10	1:0:26		
	20:2:6½	4½	40: 15: $5\frac{5}{8}$
Bradds of 1½ inch.	1:2:2	} 9½	9: 17:11
" 1 inch.	0:2:24	}	
Bolts for Chain Pumps.	0:0:3	}1: 8: 10	0: 2: 3¾
Small Wedges for Pump wheels. 12	0:0:6	}	
Hand Screws.	18	3: 0: 0	54: 0: 0
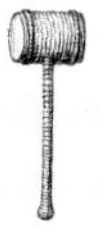 Reeming Beetles.	24	18	1: 16: 0
Broomes Double.	69	20 dozn	0: 9: 7
" Single.	91	10 dozn	0: 5:11¾
Basketts Ballast.	39	4	0: 13: 0
" Naile.	6	3¼	0: 1: 7½
" Rozin.	1		0: 4: 6
" Small Prickles.	7	8	0: 4: 8

folio 7

	Quantity	Rate s d	Sume £ s d
½ Capstan Barrs.	24	5: 0	6: 0: 0
Bucketts Double.	21	1: 3	1: 6: 3
" Single.	10	:12	0: 10: 0
Baulks.	26	:18	1: 19: 0
inch	26	:18	1: 19: 0
Dead Eyes. 11	8	11 ft	0: 6: 8
" 10	4		0: 3: 0⅓
" 9	2		0: 1: 4½
" 7	6	9½	0: 2: 9¼
Deales			
New England.	1720 feet	2½p foot	17: 18: 4
" Spruce.	352	12: 0	211: 4: 0
" Ordinary.	1923	} :12	100: 17: 6
" Slitt.	94½	}	
" Battens whole.	155	} :4	6: 12: 0
" " Slitt.	241	}	
Wainscotts.	2		2: 10: 0
Hand Spikes.	102	: 5½	2: 6: 9

New England	inches feet No		
Masts } Mainmast	33 : 95 : 1		100: 8: 6
dec:d (decayed) for a Paunch.	23 : 70 :½		9: 0: 0
dec:d at ye head but made a bowsprit.	30 : 88 : 1		49: 0: 0

Bought of & made by Mr Gray.	Masts Length Diameter feet inches	Yards Length Diameter feet inches	
Main top mast.	60 : 17½	49½ : 11¾	
" top gall:mast.	24 : 7½	25 : 6	
Fore mast.	85½: 27⅓	78 : 18	330: 0: 0
" Top mast.	52½: 15	42¾ : 10½	
" Top gall:mast.	21 : 6	21 : 5	
Sprit top mast.	18⅓: 6½	27¾ : 6	
Mizon mast.	84¾ : 18½	79½ : 12½	
" Top mast.	31½: 9	25½ : 6	
Crossjack yard.		54 : 9¼	

folio 8

	Hands		Rate £ s d	Sume £ s d
Gothenburg masts for shores etc.	11	5	3 : 9 : 0	17: 5: 0

	Hands	No	Rate £ s d	Sume £ s d
"	10	16	2 :11: 0	40: 16: 0
"	8	15	1 : 8: 0	21: 0: 0
"	7	11	:19: 0	10: 9: 0
"	6	2	:13: 0	1: 6: 0
Decaid (Decayed)	10	1		0: 13: 0
"	8½	3		1: 1: 0
Dec:d (Decayed) Bowspritt.		1		0: 6: 0

Mast of ye	inches feet No	Rate	Sume
Cleavland ya:(yacht).	18 : 54 : 1		1: 6: 0
Top masts of	13 : 42 : 2		1: 10: 0
"	10 : 40 : 6	: 9: 0	2: 14: 0
"	10 : 36 : 1		0: 8: 6
"	9 : 36 : 2		0: 13: 0
"	9 : 29 : 1		0: 6: 0
"	8 : 30 : 1		0: 5: 0
"	6 : 25 : 3		0: 7: 6
"	5 : 18 : 2		0: 4: 0
Yards of	12 : 48 : 4	:12: 0	2: 8: 0
"	11 : 53 : 1		0: 11: 0
"	11 : 44 : 1		0: 10: 6
"	11 : 40 : 1		0: 10: 6
"	10 : 46 : 1		0: 9: 6
Boat Oares.	6	: 3: 4	1: 0: 0
Turned Pins.	20	: 2: 6 doz	0: 4: 2

	inches			
Planke Oake of	1½	620 feet	4:10: 0 ye load	6: 19: 6
"	2	2652 feet		39: 15: 7
"	3	10028 feet		225: 12: 7
"	4	20823 feet		624: 13: $9\frac{3}{5}$

folio 9

		No	Rate £ s d	Sume £ s d
Shovells.		6	: :15	0: 7: 6
Scoops.		4	: :10	0: 3: 4
Sparrs Midling.		16	: :7	0: 9: 4
" Small.		10		0: 1:11
Bearing Tubb.		1		0: 9: 6
Wedges Elme.		3912	:18: 0	35: 4: $1\frac{22}{25}$
Kneed.		6 bushells	: 3: 9	1: 2: 6
	Feet			
Treenails Ruffe of	1	4150	: 2: 3¾	4: 15:11½
" Mooted of	1	169	: 3: 1¾	0: 5: 3
"	1½	6582	: 5: $7\frac{3}{5}$	18: 10: 9¼
"	2	9884	: 7: 9	38: 5:11¾
"	2½	10002	:14: 0	70: 0: 3¼
"	3	2014	:16: 6	16: 18: 3½
		Loads Feet		
Timber Elme.		74 : 15	2:15: 0	204: 6: 6
" Firr.		99 : 4	2: 2: 6	210: 10:10½
Oak stuff of 8 inch.		42 : 45	2:16: 3	120: 13: 1½
" Streight.		733 : 28		2063: 2: 9
" Compass.		657 : 24	2:18: 4	1917: 13: 0
" Knees Raking.		41 : 30	3:10: 0	145: 12: 0
" " Square.		32 : 12½	4:10: 0	145: 2: 5½
				7613: 1: 8½

Timber returned and expended on ye 2nd Rate			
Elme	2 : 3	2:15: 0	5: 10: 0
Oak Compass	3 : 0	2:18: 4	8: 15: 0
		Remaines	7599: 6: 8½

folio 10

	Quantity	Cwt qr lb
Iron worke from Mr Loader.		
Bolts Drive.	7	0: 2: 7
" Rain.	35	7: 0:21
" Sett.	89	5: 2:16
" Sett4 & Drive.	4	0: 1:24
Axletree & Winch for a grindstone.	1	0: 0:20
Axletrees 2 Winches 4 for Chain Pumps.		2: 0:21

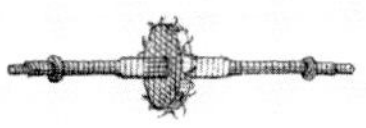

	Quantity	Cwt qr lb
Clamps.	9	0: 1:19
Hanging Clamps.	28	1: 2:15½
Dogge to haul timber etc.	3	0: 0:11
Fetts.	4	0: 0:19
Hammers Clench.	6	0: 0:24
Sett.	1	0: 0: 5½
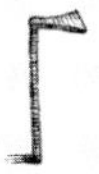 Limber.	5	0: 0:16
Reeming Irons.	4	0: 0:26
Double headed Mauls.	19	2: 0:20
Nagshead for a Tiller.	1	0: 2: 2
Plates for Round Topps.	4	0: 3:25
Belaying Pins.	2	0: 0:27
Shank painter.	1	1: 0: 1
Pump Chains.	2	4: 0:27
8 Hoops for Pump		0: 2: 3
16 Sprockets "		
Other Iron worke.		424: 1:20

		Rate £ s d	£ s d
Totall	453: 1: 6	1 : 8: 10	653: 10: 3
Shutts	170	1½	1: 1: 3
			8253: 18: 2½

folio 11

	No	Rate s d	Sume £ s d
Iron worke from Mrs Beckford.			
Spring bolts large.	11	}: 2: 0	1: 18: 0
" " extra.	8	}	
Doore bolts.	25	: :12	1: 5: 0
Windoore barrs. 72	215½ feet	: : 4	3: 11:10
Stay barrs for Windoore shutters.	24	: : 1	0: 2: 0
Eyes for ditto.	24	: : 1	0: 2: 0
New Casements. 10	80 ft 8 inchs	: 2: 0	8: 1: 4
Casement stay barrs	10	: : 4	0: 3: 4
" Eyes	18	: : 1	0: 1: 6
" Hooks	26 Pr	: : 2	0: 4: 4
" Hinges	4 Pr	: : 7	0: 2: 8
" Hinges with squares	6 Pr	: :12	0: 6: 0
" Large Hinges	10 Pr	: : 7	0: 5:10
" Locks	1	: :	0: 1: 6
" " Large	13	: :	0: 19: 6
Curtaine Hooks	3 pr	} : : 2	0: 1: 2
Hooks for Shutters.	4 pr	}	
Hooks & Hinges "	6 pr	: : 7	0: 3: 6
Girdles for Pooplanthorns.	3	: 6: 8	1: 0: 0
Stays for Pooplanthorns.	6	: 1: 8	0: 10: 0
Hasps for Pooplanthorns.	3	: : 4	0: 1: 0
Large side hinges with rising joynts.	9 pr	: 4: 0	2: 4: 0
Ditto extra.	2 pr	"	
Hold Fasts for Joyners benches.	12	: 3: 0	1: 16: 0
Nutts for a bed.	10	: : 3	0: 2: 6
Pins for a bed.	4	: : 1	0: 0: 4
			8277: 7: 6½

Column 1 — folio 12

	No	Quantity	Rate s d	Sume £ s d
From Mrs Beckford.			: :	
Pins with chains			: :	
for 2 Capstans.		22	: :12	1: 2: 0
Plates for a bed.		10	: : 4	0: 3: 4
" to line stayrs		9 40½ feet	: :	0: 13: 6
Port Rings with eye. Raggd	4		: : 8	0: 2: 8
"		Large 48	: :	1: 12: 0
Rivetts.		178	: : 2	1: 9: 8
Roves.		8	: : 1	0: 0: 8
Curtain Rodds.	3	17 feet	: : 3	0: 4: 3
Large Squares.		12 pr	: :	0: 3: 0
Staples Round.		83	: : 2	0: 13:10
Staples Square.		32	: : 4	0: 10: 8
Screws.		12	: : 6	0: 6: 0
		Cwt qr lb		
Bolts for Capps.	38	1: 0: 0		
Cleat to belay Rope.	36	0: 3: 2		
Cross Garnetts.	4 pr	0: 1: 5		
Door hooks&hinges.	19 pr	1: 2:24		
Hooks for ye				
Gundeck.	26	0: 2: 7		
Hooks for ye				
Port ropes.	6	0: 0: 18		
		4: 2: 0	4d	8: 8: 0
Mr Foley.				
Ribbon Nails.		9: 2: 6 }		
Clench Na: & roves.		0: 0: 18 }	4½p	21: 6: 7¹/₈
3 inch Bradds.		5200 1: 0: 4½ }		
Mr Hardwin				
New Muscovia glass				
Diamond Quatries.		80 pcs 252 feet		
		140 inch	1 : 8	21: 1: 3
Mend: Muscovia glass				
with Diamond				
Quatries		12	1½	0: 1: 6
Single tin plates				
expended in ye Cook				
Room & Gallarys.		506	5½	11: 11: 1
Mr Edgell		inches	No	
Blocks Double.		26	1	0: 9: 4²/₃
		20	2	0: 12: 2²/₃
		16	1	0: 4: 5¹/₃
Blocks Single.		26	1	0: 4: 8¹/₃
		20	2	0: 6: 1¹/₃
		16	1	0: 2: 2¹/₃
				8348: 16: 7⁵/₈

folio 13

	No	Rate £ s d	Sume £ s d
Mr Edgell.			
Globe Balls.	36	: : 3	: 9: 0
Collums & Pillars of 7 feet	20	: 2: 4	2: 6: 8
" " 5 feet	10	: :12	0: 10: 0
" " 3 feet	59	: : 4	0: 19: 8
" Flower Potts. 2 feet	46	: : 3	0: 11: 6
New Chain Pumps.	2	3:15: 0	7: 10: 0
Cases.	2	: 4: 0	0: 8: 0
Wheels.	3	: 4: 0	0: 12: 0
Pump Dales.	2	25 foot : :18	1: 17: 6
Chains leatherd.	3	: 4: 0	0: 12: 0
Lign: vitre sheeves	74 }	: :	: :
" Pins.	58 } 4: 0: 19	: : 6	11: 13: 6
" Rowle.	1	: :	0: 3: 8

Column 2

	No	Rate £ s d	Sume £ s d
Whipstaffe of 12 ft	1	: :	0: 2: 6
Shivers Cottin.	54	: : 8	1: 16: 0
Supporters for		: :	: :
Lanthorns.	3	: 1: 8	0: 5: 0
Trucks for stumps.	2	: :	0: 2: 0
			8378: 15: 7⁵/₈
And for Carvers		: :	
worke.		160: 0: 0	
Painters worke.		98: 6: 7	
ye wages of			:
Shipwrights.		2517: 1: 7	
Caulkers.		125:10: 8	
Ocam boys.		19: 7: 2	
Joyners.		229: 3: 5	
Houscarpenters.		5. 6. 0	
Wheelwrights.		10: 0: 0	in all
Plummers.		1:10: 0	3888:17: 1
Bricklayers.		6: 9: 0	
Scavellmen.		57: 7: 1	
Labourers.		316:15: 2	
Teams.		50:18: 0	
Sawyers.		547: 9: 0	

folio 14

Expended in fitting of the Bilge ways for Launching which will serve for other Shipps		Rate £ s d	Sume £ s d
	Load feet		
Timber Elme.	1: 10	2:15: 0	3: 6: 0
" Firr.	10: 20	2: 2: 6	22: 2: 0
Oake Streight.	59: 46	2:16: 3	168: 10: 6
" Compass.	8: 5	2:18: 4	23: 12: 6
" 7 inch stuffe.	21: 0	2:16: 3	59: 1: 3
Oake Planke of	:	: :	: :
4 inch	: 582 feet	4:10: 0	17: 9: 2²/₅
3 inch	: 200 "	: :	4: 10: 0
2 inch	: 98 "	: :	1: 9: 4³/₄
Iron worke from	:	: :	: :
Mr Loader.	Cwt qr lb	: :	: :
Bolts.	80 7: 2:12½	} : :	: :
Splitting Wedges.	28 2: 1:13	} 1: 8:10	14: 9: 8¹/₄
			314: 8: 6²/₅
More used for straping of Blocks & for launching her which may serve for other Shipps			
	inches Fathoms	1:10: 0	: :
Hawsers of	8 136		31: 1: 8
	5 60		5: 7: 1
	2 30		0: 9: 7
	1½ 155		1: 14:10
White Hawsers 4			: :
strand 6 No of	6 630		74: 5: 0
Lashing.	2 Coyles		1: 13: 0
Spun yarne.	0: 3: 0	:18:	0: 13: 6
Old Canvas.	52 yards	: : 4	0: 17: 4
Shovells.	12	: :15	0: 15: 0
Tallow.	0: 0: 4	: 5¹/₇	0: 1: 8⁴/₇
			116: 18: 8⁴/₇

Lenox ye Boatswane Rigging & present use.   folio 15

folio 15

	No	Rate £ s d	Sume £ s d
Blacking.	12 barrels	: : 6	0: 6: 0
Broomes.	30	10d doz	0: 2: 1
Bucketts double.	1	: :	0: 1: 3
" Single.	4	: :	0: 4: 0
" Old leather.	4	: : 8	0: 2: 8
Ballast basketts.	4	: : 4	0: 1: 4
Old canvas.	278 yards	: : 4	4: 12: 8

Column 3

	No	Rate £ s d	Sume £ s d
Candles.		0: 0: 12 : 5½	0: 5: 6
Junke of 15 inches.	6 fathoms	: 8: 0	1: 3:10
Lashing line.	2 Coyles	: :	1: 13: 0
Spunyarne.	42: 0: 10	18: 0	37: 17: 7¹/₄
Tard lines.	49	1: 4	3: 5: 4
" marlin.	0: 2: 22	1:16: 0	1: 5: 0½
Twine.	0: 0: 8½	10	0: 7: 1
Oyle.	6 galls	: 1:14	0: 8: 0
Rozin.	2: 0: 0	: :	0: 19: 0
Tallow.	2: 3: 19	2: 8: 0	7: 0: 1½
Tarr.	3⁵/₁₂ barr	:16: 8	2: 16:11¹/₃
Thrums.	0: 0: 2	: :	0: 1: 4
Hatchetts.	1	: :	0: 1: 2
Woolding nails.	0: 1: 9½	: : 6	0: 18: 9
Rag'd staples		: :	: :
for yards.	24	: : 4	0: 8: 0
Scrapers.	12		0: 15: 0
Thimbles.	90		1: 10: 0
" Large.	20	: 1: 4	1: 6: 8
" for top ropes }			: :
"Fish pend But sling }	18		1: 4: 0
Brass shivers.	6	3: 1: 3 : 1: 2	21: 8: 2
Rundlett for Oyle			: :
of 6 galls.	1		0: 2: 0
Scoope.	1		0: 0:10
Small Sparr.	1		0: 0: 2¹/₄
			90: 7: 6

folio 16

Trucks for	No	Rate	Sume : £ s d
Vaine Spindles.	2		0: 1: 0
" Stumps.	2		0: 2: 0
" Shrouds.	48	3¾	0: 15: 0
" Grt: for studding sail boomes.	4	12	0: 4: 0
			91: 9: 6⁵/₆
	Cwt qr lb		: :
Tackle hooks.	30 1: 1: 20		
" " for ye mizon Burtons	2 0: 0: 6		
" " main & Fore top mast Burtons.	4 0: 0: 12		
			: :
From Mr Loader.			: :
Double thimbles.	14 0: 2: 27		
Thimbles for 5 tops & slings.	21 0: 1: 15		
Large thimbles.	18 0: 0: 20		
Tackle hooks with Thimbles.	28 0: 1: 0		
Luff tackle hooks & thimbles.	4 0: 0: 21		
Puttock hooks & thimbles.	22 0: 3: 7		
Puttock hooks Puttock Plates.	6 1: 0: 5		
Puttock plates for ye			: :
Fore }			: :
Main } Shrouds	28 3: 0: 7		
Mizon }			: :
Blocks Iron bound topp.	6 3: 1: 26		
" " " Single topp.	4 1: 3: 3		
" " " Catt.	2 1: 2: 10		
Iron Togle or Fidd for ye slings to ye Sprit saile yard.	1 0: 0: 11		
			: :
Racks	1 pair		0: 1: 16
Hooks to sling cables.	24 0: 0: 26		
Clapper for ye bell.	1 0: 0: 9		
	17: 3: 17	1:8:10	25: 16: 2
Bell bound.	1		0: 10: 0

folio 17

Rope New of-	inches fathoms	
	5 : 20	
	4 : 100	
	3½ : 70	

Column 1

	inches	fathoms
	3	230
	2½	30
	2	50
	1½	390
	1	200
	¾	250
Old-	3½	40
	3	60
For ye Bowspritt	:	
	:	
Sheets Cablet	3½	60
Pendants	4½	4
Cluelines	2½	32
Buntlines	2	28
Lifts	3	48
Standing lifts	4	8
Lanyards	2	6
Slings	6	5
Halyards	3	28
Braces	2½	68
Pendants	3	3½
Horses in ye head best worn	4	8
for ye Bowsprit.	4	9½
Woold:(Wolding)ye bowsprit	3	55
best worn	6	90
Sprit topmast	:	*folio 18*
	:	
Shrouds	2½	18
Lanyards	1½	9
Lifts	1½	17
Braces	1½	44
Pendants	2	3
Tye	2½	3½
Halyards	2	9
Cluelines	1½	32
Parrell Ropes	2	3
Pendants	2	3
Falls of back stays	1	18
Foremast		
Pendants }	7	18
Runners } of Tackles	5½	26
Falls }	3½	152
Shrouds	7	150
Lanyards	4	56
Stay	14	12½
Coller	10	3
Lanyard	4	12
Lifts	3	66
Sheets Cab:(Cablet)	5	70
Tacks tap & Cab	7	36
Braces	3	52
Pendants	3½	5
Bowlines	4	42
Bridles	3½	6
Buntlines	2	76
Leechlines	2½	40
Clue garnetts	3	60
Parrell Ropes	4	12 *folio 19*
Jeeres	5½	84
Puttock Shrouds	4	30½
Catharping leggs	2½	14
" Falls	2	16
Knaveline for ye pells	1½	26
Luffe hook rope	5	9
Horses for ye yard bestworn	5	11
Fore topmast		
Pendants of ye top rope	7	13
Falls of ye top rope	4	42
Pendants of burton tackle	4	6
Falls of burton tackle	2	32
Shrouds	4	66
Lanyards	2½	24
Standing backstays	4	76

Column 2

	inches	fathoms
Lanyards	2½	10
Stay	5	16½
Lanyard	3	9
Lifts	2	60
Sheets	6	42
Braces	2	58
Pendants	2½	4
Bowlines	2½	50
Bridles	2	10
Cluelines	3	70
Runners or jeere	4	20
Halyards	3	50 *folio 20*
Buntlines	2	36
Leechlines	2	14
Parrell Ropes	3	8
	:	
Fore topgall mast	:	
Shrouds	2	14
Lanyards	1	8
Stay	2	19
Lifts	1½	20
Braces	1½	54
Pendants	2	2
Bowlines	1	38
Bridles	¾	8
Cluelines	1½	56
Tye	2½	3½
Halyards	1½	28
Parrell Ropes	2	3
Main Mast:		
Pendants}	7½	20
Runners }of Tackles	5½	8
Falls }	3½	152
Pendants	7	10
Guy of ye Garnett	5	10
Fall of ye Garnett	3½	36
Shrouds	7½	183½
Lanyards	4	64
Stay	16	19
Coller	12	8
Lanyards	5	12
Lifts	2½	70 *folio 21*
Sheets Cab:(Cablet)	6	76
Tacks tap & Cab:(Cablet)	8	38
Braces	2½	72
Pendants	3½	10
Bowlines	4	42
Bridles	3½	12
Buntlines	2	90
Leechlines & Pendants	2½	20
Clue garnetts	3	64
Parrell Ropes	5	14
Jeeres	6	100
Puttock Shrouds	4	36
Catharping leggs	2½	16
Falls	2	18
Bowline	2	10
Luffe Tackle	2½	12
Knaveline for ye pells	2½	28
Horses for ye yard bestworn	5	15
Woolding ye main mast	3	252
False stay for ye main stay	:	
stay sail ½ worn	5	14
Main topmast		
Pendants of ye top rope	7	30
Falls of ye top rope	3½	90
Pendants of burton tackle	4	7
Falls of burton tackle	2½	36
Shrouds	4½	90
Lanyards	2½	35
Standing backstays	5	120
Lanyards	3	21

Column 3

	inches	fathoms	
			folio 22
Stay	5½	24	
Lanyard	3	9	
Lifts	2½	66	
Sheets	6½	58	
Braces	2½	66	
Pendants	3	5	
Bowlines	3½	56	
Bridles	3	16	
Cluelines	3	80	
Tye or jeere	4½	23	
Halyards	3	58	
Buntlines	2	42	
Leechlines	2	16	
Parrell Ropes	3½	7	
Main topgall mast			
Shrouds	2½	18	
Lanyards	1½	8	
Stay	3	22	
Lifts	1½	74	
Pendants	2	3	
Bowlines	1½	60	
Bridles	1	6	
Cluelines	2	56	
Tye	3	4	
Halyards	2	30	
Parrell Ropes	2	4	
Standing backstays	2½	46	
Lanyards	1½	4	
Flagg staffe stay	2	17	*folio 23*
Mizon Mast			
Shrouds	5	105	
Lanyards	3	35	
Stay	5½	13½	
Puttock Shrouds	2½	12	
Pendants of burton tackle	5	6	
Falls of burton tackle	2½	34	
Trusses	2½	24	
Sheet	3½	24	
Tack	2	4	
Bowlines	3	10	
Brailes}	2½	55	
}	2	55	
Jeere	5	44	
Parrell Ropes	4	4	
Crossjack yard			
Standing lifts	3	8	
Lanyards	2	5	
Braces	2	40	
Slings	4	5	
Mizon Topmast			
Shrouds	2½	30	
Lanyards	1½	15	
Stay	3	9	
Lifts	1½	30	
Braces	1½	40	
Pendants	2	3	
Bowlines	1½	34	
Bridles	1	6	
Cluelines	2	40	*folio 24*
Tye or Jeere	3	10	
Halyards	1½	28	
Parrell Ropes	2	4	
Sheets	3	34	
	:		
Other necessary Ropes	:		
Catt Ropes	4½	52	
Pendants of fish tackle	7	7½	
Fall of fish tackle	3½	35	

Column 1

	inches	fathoms
Stoppers at ye Bow	6½	24
Shank painters	5½	20
Stoppers at ye Bitts	9	10
Lanyards	3½	14
Vyall Cablett	10	36
Pendant of ye Winding tackle	10	12
Fall of ye Winding tackle	5	48
Buoy }	7½	60
Boat }	6	30
Guesses } Rope Cablet	3½	32
Pinnaces }	4½	25
Guesses }	3	30
Butslings	5	9
Gun slings	7	6
Hogs head slings	4	14
Ratline of shrouds	1½	200
" "	1	200
" "	¾	200
Robans & Earing of sails	1½	200
" " "	¾	200
Entering Ropes white	3½	18

folio 25

Necessary Ropes

	inches	fathoms
Stream Anchor buoy rope	4½	18
Kedge Anchor buoy rope	4	18
Poop ladders	6	24
Puddrings for ye yards &	:	
stops of ye tops sheets.	6	24

	inches	fathoms
for seazing of Blocks }		
Racking of Parrells }		
Seazing of stops between }	1½	200
Deck Robans & earings for }		
ye main saile. }		

	inches	fathoms
Lashing jeere blocks to ye }		
lashing ye main top mast }		
stay block to ye head of }		
ye Foremast. }	:	
Lashing topsail jeere block }	2½	150
ye Fore bowline block }		
Main tops:l bowline }		
block—— }		
Main jeere block }		

	inches	fathoms
Strapping Winding tackle blocks	6½	12
Lashers for ye Fore jeere blocks at ye mast head	4	30
for ye Fore jeere block on ye yard.	3	12
Main jeere block at ye mast head.	4½	30
Main jeere block on ye yard.	3	12

folio 26

Abstract of ye Cordage

*Rate: £1: 10s: 0d per cwt

	Inches	Fathoms	*	Sume £ s d
New Cable laid	16	19 :		15: 17: 9
	14	12½:		8: 4: 0¾
	12	8 :		3: 18: 7
	10	51 :		17: 11:10
	9	10 :		2: 17: 0
	8	38 :		8: 11: 0
	7½	60 :		11: 18: 6
	7	36 :		6: 4: 2
	6	106 :		13: 10: 3
	5	70 :		6: 6: 0
	4½	25 :		1: 17: 0
	3½	92 :		4: 2: 9
	3	30 :		0: 19:10
Hawser laid		:		: :

Column 2

inches	fathoms:	sume £ s d
7½	203½:	40: 12: 0
7	234½:	23: 2: 2
6½	94 :	14: 2: 0
6	147 :	17: 6: 3
5½	195½:	20: 18:10
5	413½:	36: 18: 4
4½	217 :	16: 5: 6
4	690½:	51: 15: 9
3½	739 :	34: 6: 1
3	1344½:	45: 12: 2
2½	1017 :	21: 15:10
2	879 :	14: 2: 5
1½	1591 :	17: 17:11
1	476 :	3: 11: 4
¾	658 :	2: 9: 4

Abstract of Rigging

folio 27

*Rate: £1: 10s: 0d per cwt

	Inches	Fathoms	*	Sume £ s d
Hawser laid				
White	3½	18 :		0: 16: 8½
best worne	6	90 : : :10:0		3: 10: 8
	5	26 :		0: 15: 5
	4	17½:		0: 4: 3
1/2 worne	6	48 :		1: 17: 8
	5	14 :		0: 8: 4
Old	3½	40 :		0: 12: 0
	3	60 :		0: 13: 6

471: 13: 8¼

An Account of ye Blocks
with their particular uses
given by Captain John
Kirk Master Attendant.

T stands for Treble Block
D stands for Double Block
L stands for Longtackle Block
S stands for Single Block
LV stands for Lignum Vitae shieves
Ash stands for Ash shieves
BS stands for Brass shieves
IR stands for Iron bound

		inches	No	
Spritsaile				
Halyard	L	27	1	LV
	S	13	1	LV
Sheet	S	11	2	
Cluelines	S	10	2	
Lifts	S	10	4	
Longhead	S	62	2	with LV shieves and 6 pins

folio 28

		inches	No	
Spritsaile Brace	S	10	6	LV
Sprittops : Halyard	"	10	2	Ash
Cluelines	"	6	4	"
Lifts	"	6	4	"
Brace	"	6	6	"
Cranelines	"	5	10	"
Foremast : Jeere	D	22	2	LV
Jeere	"	21	2	"
Pendant	S	18	2	"
Tackle	L	28	4	"
Tackle	S	19	2	"
Tackle	"	18	2	"
Quarter	"	19	2	"
Topsail sheets	"	19	2	"
Fore sheets	"	17	2	"
Cluegarnetts	"	11	6	"
Lifts	"	11	4	"
Cluelines	"	10	4	"
Brace	"	10	4	"
Bowlines	"	14	2	"
Studd saile	"	10	2	"
Buntlines	L	18	3	"
"	S	1	3	"

Column 3

		inches	No	
Leechlines	"	9	9	Ash
Seazing blocks	"	10	6	"
Crowfeet	"	6	2	"
Knaveline	"	7	2	"
Catharping	"	7	12	"

folio 29

		inches	No	
Foretopmast				
Jeere	"	17	2	LV
Cluelines	"	10	4	"
Brace	"	9	6	"
Bowline bowspt	T	10	1	LV
Burton	L	17	2	"
"	S	10	2	"
Stay	L	18	1	"
"	S	10	1	"
Topsl halyard	L	28	1	"
"	S	20	1	"
Topp	"	19	2	Bs
Topp tackle	D	17	3	LV Ir
Lift	S	8	4	LV
Leechlines	"	8	2	"
Buntlines	"	8	3	"
Reeftackle	"	7	4	Ash
Foretopgallt mast	"			"
Halyards	"	6	2	"
Cluelines	"	6	4	"
Brace	"	6	8	"
Bowline	"	6	4	"
Stay	"	6	1	"
Lift	"	6	4	"
Main Mast	"			"
Jeere	D	24	2	LV
"	"	22	2	"
Pendant	S	18	2	"

folio 30

		inches	No	
Tackle	L	30	4	"
"	S	18	4	"
Garnett	L	30	1	"
"	S	19	1	"
Quarter	"	19	2	"
Topsail sheet	S	20	2	LV
Main sheet	"	18	2	"
Clugarnetts	"	11	6	"
Lifts	"	11	4	"
Braces	"	10	4	"
Studding sail	"	10	2	"
Buntline	L	19	3	"
"	S	11	4	"
Leechline	"	10	9	Ash
Seazing	"	11	6	"
Lufftackle	L	20	2	LV
"	"	19	1	"
"	S	10	3	"
Knave line	"	8	2	Ash
Carthharping	"	8	14	"
Crow foot	"	7	2	"
Main top mast	"			"
Jeere	S	18	2	LV
Cluelines	"	11	4	"
Brace	"	10	2	"
Bowline	"	11	2	"
Burton	L	18	2	"
"	S	10	2	"

		inches	No	
Main Topmast				folio 31
Stay	L	19	1	"
"	S	11	1	"
Stay block } fore top }	"	17	1	"
Reef tackle	"	7	4	Ash
Lifts	"	8	4	LV
Topp block	"	21	2	Bs
Topp Tackle	D	17	3	LV Ir
Topsl halyard	L	28	1	LV
"	S	21	1	"
Studdsl main } & Fore }	S	8	4	LV
Buntline	"	9	3	"
Staysl halyard		8	2	Ash

		inches	No	
Main topgall mast				
Halyards	S	8	2	"
Cluelines	"	6	4	"
Bowlines	"	6	6	"
Brace	"	6	8	"
Lift	"	6	4	"
Stay	"	8	1	"
Mizon mast	"			"
Jeere	D	18	1	LV
"	"	17	1	"
Sheet	L	18	1	"
"	S	11	1	"
Trusses	L	18	1	"
"	S	10	1	"
Brayle	"	9	12	Ash
Bowline	"	8	2	"
				folio 32
Crossjack				
Halyards	S	11	1	LV
Braces	"	8	6	"
Quarter	"	11	2	"
Mizon topmast	"			"
Cluelines	S	6	2	Ash
Brace	"	6	4	"
Bowline	"	6	4	"
Lift	"	6	4	"
Stay	S	7	1	Ash
"	"	6	1	"
Burton	L	19	2	LV
"	S	10	2	"
Quarter	"	10	2	"
Sheet	"	13	2	"
Single block	"	8	1	"
Halyards	L	16	1	"
Other necessary				
Blocks	"			"
Fish Tackle	L	30	2	"
Winding tackle	T	31	1	"
" "	D	24	1	"
" "	S	20	1	"
For ye Davitt	T	20	1	"
Catt blocks	D	19	2	LV Ir
The Abstract follows	"			"
in folio 33	"			"
				folio 33

Lenox. Abstract of ye Blocks With Lig:vitie shives & pin

		inches	No	£ s d
Winding Tackle	T	31	1	1: 11: 0
" "	D	24	1	0: 16: 0
" "	S	20	1	0: 6: 8
Blocks Treble		10	1	0: 2: 3½
Iron bound : Catt	D	19	2	0: 11: 7⅓
" " Double		18	3	0: 15: 9
" "		17	3	0: 14: 10½
Double blocks		24	2	0: 16: 0
" "		22	4	2: 8: 1⅓
" "		21	2	0: 13: 5
" "		18	1	0: 5: 3
" "		17	1	0: 4: 11½
Long tackle		30	5	1: 8: 1½
" "		28	6	1: 10: 4
" "		27	1	0: 4: 10½
" "		20	2	0: 6: 1⅓
" "		19	7	1: 0: 3⅚
" "		18	8	1: 1: 0
" "		17	2	0: 4: 11½
" "		16	1	0: 2: 2⅔
Longhead with 6 shieves				: :
and 6 pins.		62	2	0: 12: 11
Single blocks		21	1	0: 3: 4¼
" "		20	4	0: 12: 2⅔
" "		19	9	1: 6: 1½
" "		18	14	1: 16: 9
" "		17	5	0: 12: 4¾
" "		14	2	0: 3: 6
				folio 34

	inches	No	£ s d
Single Blocks	13	3	0: 4: 7¼
" "	11	40	1: 13: 7¼
" "	10	47	1: 15: 10⅚
" "	9	9	0: 6: 2¼
" "	8	24	0: 14: 8
Single with			
Brass shivers	21	2	0: 6: 8½
" " "	19	2	0: 5: 9⅔
Ash shivers	11	6	0: 5: 0½
" "	10	17	0: 12: 11⅚
" "	9	21	0: 14: 5¼
" "	8	23	0: 14: 0⅔
" "	7	25	0: 13: $4\frac{5}{12}$
" "	6	76	1: 11: 8
" "	5	10	0: 3: 5⅔
Totall			30: 13: 7¾
			folio 35

	inches	No	p ft	£ s d
Deadman Eyes for ye			(price per foot)	
Spritsail standing lifts.	9	4	}11	: :
" ye horses for ye Bowspritt.	9	4	}	0: 5: 6
" in ye head.	10	4		0: 3: 0
Sprittops:ll for ye Shrouds.	7	6	}	
" with long Puttock plates			}	
Iron bound.	7	6	} 9½	0: 5: 6½
Foremast for channel plates				
Iron bound.	13	14	}	
" ye Shrouds.	13	14	}12	1: 10: 4
" Stay & Coller.	20	2	5:6	0: 11: 0
" Horses for ye yard.	8	2	}	
" Spritsll sheet to run in.	8	2	}	
			}	
Foretopmast for ye Shrouds.	8	10	}10	0: 7: 9⅓
" Puttock Plates				
Iron bound.	9	10	11	0: 6: 10½
" Back stay Plates " "	8	4	10	0: 2: 2⅔
				: :
Foretogallt mast Shrouds.	6	6	9½	0: 2: 4½
				: :
Main mast Stay & Coller.	23	2	5:6	0: 11: 0
" " Staysail stay.	8	2		0: 11: 0
" " Horses on ye yard.	9	2	11	0: 1: 4½
" " Shrouds.	14	16	}	
" " Chain Plates				
Iron bound.	14	16	}12	1: 17: 4
Maintopmast Shrouds.	9	12	}	
Puttock Plates Iron bound.	9	10	}	
Back stay Plates Iron bound			}	
fixt to ye ship sides.	9	6	}11	0: 19: 3
Maintopgallt mast Shrouds.	6	6	9½	0: 2: 4½
Mizonmast Stay.	9	2	5:6	0: 11: 0
Trusses.	12	1		0: 5: 6
				folio 36
	inches	No	p ft	£ s d
Deadman Eyes for ye			(price per foot)	
Mizonmast Shrouds.	10	10	}	
" Chain plates Iron bound.	10	10	}11	0: 15: 3½
				: :
Crossjack- Standinglifts.	8	4	10	0: 2: 2⅔
				: :
Mizontopsll Shrouds.	7	6	}	
" Puttock Iron bound.	7	6	} 9½	0: 5: 6½
				9: 16: 6⅔
Parrells for ye Foremast of	29	1		0: 15: 0
" Mainmast.	31	1		0: 15: 0
" " Topmast.	21	1		0: 6: 8
" Mizonmast.	21	1		0: 8: 6
" Foretopmast.	19	1		0: 6: 8

	inches	No	p ft	£ s d
" Topgalltmast	11	1		0: 2: 9
" Maintopgalltmast	11	1		0: 2: 9
" Sprittopmast.		1		0: 2: 9
" Mizontopmast.		1		0: 2: 9
Totall				3: 2: 10
Deadeyes in folio 36				9: 16: 6⅔
Blocks in folio 34				30: 13: 7¾
Lign:vitie shivers* & pins est				42: 0: 0
Cordage in folio 27				471: 13: 8¼
Iron worke in folio 16				26: 6: 2
Other provisions in folio 16				91: 9: 6⅚
In all				675: 2: 5½
ye wages of ye Riggers.				83: 17: 7
				759: 0: 0
				folio 37

Lenox. A proportion of Stores for ye Boatswain for 6 months.

	Quantity		Rate	£ s d
Bells Cabbin.	1		s: d	0: 2: 0
" Watch.	1	1:0:11	:16	8: 4: 0
Blacking.	12 barr		:6	0: 6: 0
Large Brushes.	8		4: 9	1: 8: 0
Awning of best ½				
worn Canvas.	205 yards		:4	3: 8: 4

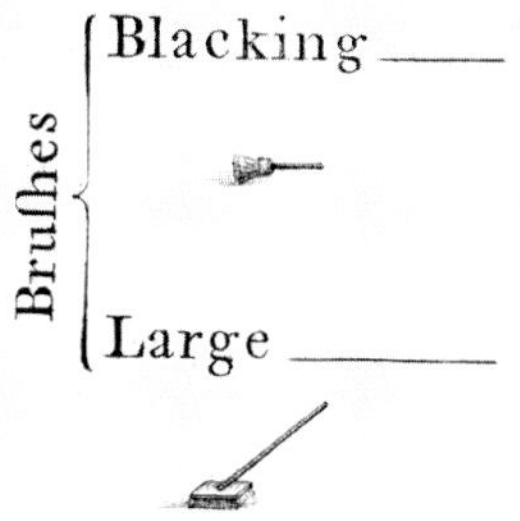

	Quantity		Rate	£ s d
Canvas French.	100 yards		:12	5: 0: 0
" Ipswch 4 bolts ye est.	128 "		:11¼	5:12: 6
" Vittry	40 "		: 9¾	1:12: 6
" Old	20 "		:4	0: 6: 8
Colours Ensignes of Bewps of	20	breadths 3	:5½yd	15: 2: 6
" Flaggs	18	" 2		8: 5: 0
" "	12	" 3		4: 19: 0
" Pendants	31	yards 4		4: 10: 9
" Vanes		" 4		1: 9: 4
Hammaccoes.	600		:18	45: 0: 0
Kerzie for Accommodation				: :
Wast Cloath & toparms.	240 yards		2:10	34: 0: 0
Compasses.	12		8: 0	4: 16: 0
		Cwt qr lb		
Copper covers for furnaces.	2	0:1:14	1: 5½	3: 1: 3
" Funnell	1	0:2:10		2:15: 5
" Furnaces	2	5:2:17	1: 7	50: 2: 3
" Hood	1	0:2: 0		4: 8: 8
" Kettles Fish	1	0:3:14	1: 6	7: 7: 0
" " Small	1	0:1:14		3: 3: 0

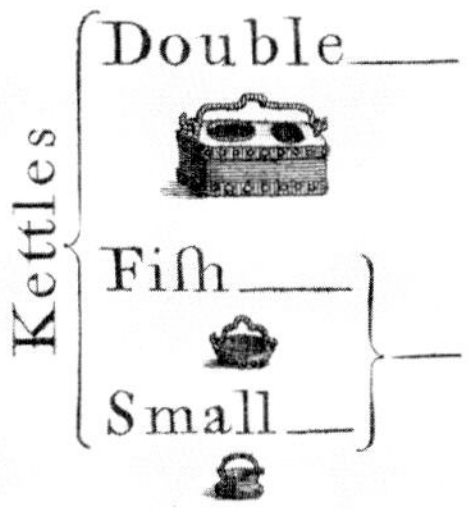

folio 38

	inches	Nr	1:10:10	£ s d
Cordage Cable laid.	18	7		721:17: 6
	12	1		49: 2: 6

	(Cables length) inches	Nr	at	£ s d
	7½	1		19:17:6
	7	1		17:5:0
	6	1		12:15:0
	4	1		6:0:0
	3½	1		4:10:0
Hawser laid.	7½	1		21:0:0
	6½	1		15:15:0
	6	1		12:7:6
	5	1		9:7:6
	4	1		6:7:6
	3½	2		9:15:0
	3	2		7:2:6
	2½	3		6:15:0
	2	3		5:1:3
	1½	4		4:10:0
	1	4		3:0:0
	¾	2		0:15:0
Junke of	15	50 fath	8:0	9:19:0
Lines Cabbin.		24 "		0:12:0
" Deepsea.		3 "		0:19:6
" Lashing.		2 Coyles		1:13:0
" Logg.		14	8	0:9:4
" Tard.		24		0:12:0
" White.		16		0:18:8

folio 39

		Cwt qr lb		£ s d
Marlin Tard.		0:0:24	1:16:0	0:7:8½
" White.		0:0:14	1:17:0	0:4:7½
Rope Netting.		2 Coyles		2:14:0
" Old for Rounding etc.		500 fath		5:0:0
Spun yarne.		1:0:0		0:18:0
White Twine.		0:0:24	: :10	1:0:0
Glasses Watch.		1		0:4:0
" ½ Watch.		1		0:2:0
" Houre.		18		0:12:0
" Minute.		6		0:4:0
Anchors of	1	48:0:0	2:8:0	115:4:0
	1	46:0:0	2:8:0	110:8:0
	1	44:0:0	2:4:0	96:16:0
	1	40:0:0	2:2:0	84:0:0
	1	11:0:0	1:13:0	18:3:0
	1	4:0:0	1:10:0	6:0:0
	1	2:0:0	1:10:0	3:0:0
Handirons.		1 pr		0:2:6
Billows with 7 Shackles.		1 pr		0:9:6
Boom Irons.		4	3:7¼	0:14:5
Topp Chains.	2	2:2:0	1:8:10	3:12:1
Creeper.	1	0:1:0		0:7:2½
Crows.	3	0:3:6		1:3:2
Esses for Kettles.	2	0:0:5		0:1:31¼
Fender.	1	0:0:24		0:6:2
Fidds Splicing.	6	2:2:8		3:14:1½
" Toppmast.	2	2:2:0		3:12:1

folio 40

				£ s d
Fire Forks.		1		0:1:6
" Shovel.		1		0:1:6
" Tongs.		1 pr		0:1:6
Grapn: Boats of	1	0:3:0	1:8:10	1:1:7½
" "	1	0:2:0		0:14:5
Fire & Chain.	2	2:1:4		3:5:10¾
Furnace doore.	1	0:0:18		0:4:7½
Grometts & Staples		80 pr	22p yard	0:12:2⅔
Hatchetts.		6	14	0:7:0
Hooks Boat.		6	2:0	0:12:0
" Cann.	2 pr	0:0:14	1:8:10	0:3:7¼
" Fish.	2	0:1:20		0:12:4¼
" Flesh.		1		0:0:8
" Gaming.	2	0:0:4		0:1:0¼
" Kettle.	1	0:0:8		0:2:0½
" Tackle.	6	0:1:4		0:8:2¾
Hanging Locks.		5	14	0:5:10
Marlin Spikes.	16	0:1:15	1:8:10	0:11:0¾
Mawls for Topmasts.	2	0:1:26		0:13:10¾

				£ s d
Nails for Hammaccoes.	2400	1:1:4	4¼d	2:11:0
Needles Bolt rope.		6		0:2:0
" Saile "		54	1:10doz	0:8:3
Oven lidds.	1	0:0:24	1:8:10	0:6:2
Pitch Potts.	1	0:1:2	2½	0:6:3
Puttock Plates.	6	0:2:17½	1:8:10	0:18:11
Scrapers.		40	15	2:10:0
Vane Spindles	4	16 ft	10	0:13:4

folio 41

				£ s d
Spitts.	4	0:1:8	4	0:12:0
Thimbles.		30	4	0:10:0
Lanthorns Hand.		2	2:5	0:4:10
High Diam				
" Poop of 5½ft 3⅓ft		2	}	
" 6½ft 4ft		1	}	30:0:0
Leads Deepsea.	3	0:2:10	1:4:0	0:14:1¾
Hand.	6	0:1:24		0:11:1¾
Leather Backs.	¼	0:0:14	10	0:11:8
Bucketts.		30	3:8	5:10:0
Hose.	1	0:1:17¾	2:2	4:19:1½
Oyle.		8 gls	16	0:10:8
Rozin.		1:0:0		0:9:6

Sailes.	No	Clo	Dep	Gores	Yards	Canvas	Rate	
Spritsail						(d yd)	: :	
Courses	2	24	7		168	Noyalls	15½	10:17:0
					168			10:17:0
Topsaile	1	17½	9½		166¼	Ipswich	14½	10:0:10⅝
Fore							: :	
Courses	2	31	12½	3	390½	Duck	23½	38:4:81¾
					390½			38:4:8¾
Topsaile	2	23	16	8	376			36:16:4
					376			36:16:4
Topgallt	1	11	6		66	Vittry	13	3:11:6
Main								: :
Courses	2	36	14	4½	508½	Duck	23½	49:15:9¾
					508½			49:15:9¾
Topsaile	2	27	18	10	496			48:11:4
					496			48:11:4
Topgallt	1	13	7		91	Vittry	13	4:18:7
Mizon							: :	
Courses	2	11½	19		218½	H:Duck	23½	21:7:10¾
					218½			21:7:10¾
Topsaile	1	17	9½		161½	Ipswich	14½	9:15:1¾

Lenox. Boatswain Stores. **folio 42**

	No	Clo	Dep	Gores	Yards	Canvas	Rate	£:s:d
Staysail								
Main	1	11	13		143	Noyalls	15½	9:4:8½
Mizon	1	9½	11		104½	Ipswich	14½	6:6:3¼
Topstaysail							: :	
Main	1	7	12		84	Vittry	13	4:11:0
Fore	1	6	11		66			3:11:6
Studdingsail								: :
Main	2	7	15½		108½	Ipswich	14½	6:11:1½
					108½			6:11:1½
Topsails	2	6	19		114	Vittry	13	6:3:6
					114			6:3:6
Boatsail							: :	
Main & Fore					32			1:14:8
					70	Old		1:15:0

		Cwt qr lbs		
Brass Shivers.	14est	7:2:16	14	49:18:8
		No		
Grindstone.		1		0:5:6
Tallow.		1:0:0		2:8:0
Tarr.		2 barr		1:13:4
Thrumes.		0:0:12	8	0:8:0
Tin Trucks.		4	:8:0	1:12:0
Ballast Basketts.		28	4p each	0:9:4
Bedsteads.		1		0:16:0

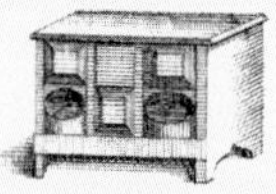

Bittacles		2		1:4:0

		No		£ s d
Blocks Catt Iron bound.		3	1:10:0	4:10:0
Deadman eyes spare.		20		0:16:8
Spare blocks.		60		3:0:0
Winding tackle Treble.			1	0:18:1
" " Double.		1		0:8:0
" " Single.		1		0:3:0

folio 43

Boats Long boat.	1	32 ft	:18:0	28:16:0
" Pinnace.	1	33 ft	:15:0	24:15:0
Boat hook Sparrs.		6		0:1:2
Bowls.		6	8½	0:4:3
Brooms.		72		0:5:0
Bucketts.		30		1:10:0
Buoys Cann.		4		4:0:0
" Wood.		3		0:15:0
Comaunders.		8	:1:6	0:12:0
Fidds for Splicing.		6	:1:8	0:10:0
Formes.		2		0:6:0
Hand Spikes.		30	: :7	0:17:6
Oares Barge.		18	:5:0	4:10:0
" Boat.		42	:4:0	8:8:0
Rundlet of 8 galls.		1		0:3:6
Scoops		4	: :10	0:3:4
Serving Malletts.		4		0:4:0
Shovells steelshod.		30	: :15	1:17:6
Tables Spanish Wainscot.		3	:14:0	2:2:0
Thoalwood.		6 Cutts	: :3	0:1:6
	Totall			2325:18:4⅝

folio 44

Lenox. Carpenters Stores
for 6 Months.

				£ s d
Canvas Ipswich.		25 yds	11¼	1:3:5¼
" Old.		40 yds	4	0:13:4
Junke.		1:2:0	:8:0	0:12:0
Lashing for Ports.		1 Coyle		0:16:6
Netting Rope.		30 fath		0:8:0
Twine.		0:0:0½		0:0:5
		Cwt qr lbs		
Bolts Chain.	5	0:1:22		
" Plates.	5	1:2:17		
" Drawn.	2	0:1:20		
" Drive.	2	0:0:15		
" Eye.	3	0:0:19		
" Hatch barr.	4	0:0:10		
" Ring.	2	0:0:24		
" Sett.	2	0:0:5		
		3:0:20	1:8:10	4:11:7¾
Caulking Irons.		2	:12	0:2:0
Cold Chizels.		2		0:2:0
Hanging Clamps.	2	0:0:14 }		
Esses for Shrouds.	3	0:1:14 }		0:14:5
Clench hammers.		2		0:2:0
Hasps & Staples.		10 pr		0:1:6⅓
Hinges Cross garnetts.		3 pr	:3:2	0:9:6
" Dove Tails.		4 pr		0:1:4
" Esse hinges.		4 pr		0:3:4
" Port hinges.	4	0:2:0		
" Port hooks.	4	0:2:0		0:14:5
" Scuttle.		2 pr	11½	0:1:11

folio 45

				£ s d
Locks Hanging.		3	:14	0:3:6
" Spring Double.		1		0:2:4
" " Single.		2		0:2:8
		Cwt qr lbs		
Logger heats.	2	0:1:16		0:11:3¾
Maul Double headed.	1	0:0:11		0:2:10
Small Spikes.		1:3:0	1:8:10	2:9:0
Nails 40d	300	0:0:27		
30	500	0:1:6		
24	1000	0:1:22		
20	1000	0:1:7		
10	1500	0:1:5		
		1:2:11	1:10:0	2:7:11¼

	Cwt qr lbs	£ s d
6	2000 0: 1: 0	0: 8: 1¼
4	2000 0: 0:11	0: 4: 3¼
3	2000 0: 0: 7	0: 3: 6½
Pump.	1200 0: 0: 2²/₅ : : 7	0: 1: 4⁴/₅
Lead.	2000 0: 0:24 5¾	0:11: 6
Scupp.	2000 0: 0:11	0: 5: 3¼
Port.	95 0: 0:11²/₅ 4½	0: 4: 0½
Rove & Clench.	350pr 0: 0:12½ 5³/₈	0: 5: 4½
Pitch Ladle.	1	0: 3: 0
" Pott.	1 0: 1:10 2½	0: 7:11
Double white tin plates.	18 7¹/₃	0:11: 0
Rings & Forelocks.	0: 0: 8 1: 8:10	0: 2: 0½
Rother Irons.	8 pr : 2: 6	1: 0: 0
Sawes Two hand.	1	0: 4: 0
" Whipp.	1	0: 5: 6

folio 46

	Cwt qr lbs	£ s d
Port Shackles.	5 0: 0:16 1: 8:10	0: 4: 1
" Staples.	15 4p each	0: 5: 0
Tin hand lanthorn for ye Storeroom.	1	0: 2: 5
Lead Scuppers.	2 0: 3:10 1: 2: 0	0:18: 5½
" Sheet.	1: 2: 0 :18: 0	1: 7: 0
Leather liquord 1¼ buck.	0: 2:14 : :10	2:18: 4
" Scuppers.	20 : 2: 8	2:13: 4
Ocam Black.	3: 2: 0 :12: 0	2: 2: 0
" White.	0: 1:17 :18: 0	0: 7: 2¾
Pump Chain Spare.	1 2: 0:13 1: 8:10	3: 1: 0
" Bolster.	1 0: 0: 5	0: 1: 3¼
" Esses.	18 0: 0:22	0: 5: 7¾
" Fidds.	1 0: 0:16	0: 4: 1¼
" Hooks.	18 0: 0:18	0: 4: 7½
" Sprocketts.	10 0: 0:22	0: 5: 7¾
" Swivells.	10 0: 0:16	0: 4: 1¼
" Winches.	1 0: 1: 0	0: 7: 2½
 " Wheele. 1		0: 4: 0
Grindstone.	1	0: 5: 6
Tallow.	2: 0: 0	4:16: 0
Tarr.	3 barr	2:10: 0
Temperd stuffe.	3 barr	4: 2: 6
Thrumes.	0: 0:12 : : 8	0: 8: 0
Anchor stocks of	18ft 2	5: 8: 0
Ballast Basketts.	4	0: 1: 4

folio 47

	Quantity	£ s d
Elme board.	200 ft	1: 0: 0
Bucketts.	2	0: 2: 0
Deadman Eyes Iron bound.	3	0: 7: 6

		£ s d
Deales.	25	1: 5: 0
Fishes for masts.	2	0:16: 0
	inch	
Oaken Planke of	4 30 ft 4:10: 0	0:18: 0
	3 40	0:18: 0
	2 50	0:15: 0
	1½ 40	0: 9: 0
Sparrs for Brushes.	6	0: 1: 2
Cant.	4	1: 8: 0
" Small.	12	0: 2: 4
Studding Saile Booms.	4	2: 8: 0
Spare Topmast of	18 hands 1	24: 0: 0
Treenails.	2ft 200	0:15: 6
Totall		90: 6: 7¼

folio 48

Del to ye Gunner for Port Rope.	inch	Quantity	Rate	£ s d
Rope of	3	1 Coyle	1:10: 0	3:11: 3
	2½	1 Coyle		2: 5: 0
White	2	20 fath		0: 6: 5
	1½	20 fath		0: 4:10
Tard Marlin		0: 0:10	1:16: 0	0: 3: 2½
Twine to whipp Ropes		0: 0: 1		0: 0:10
Totall				6:11: 6½

[FOLIO]   folio 49

Abstract of the whole Charge of ye *Lenox*.

Lenox Charge of	£ : s : d	£ s d
[13] Stores expended in building her		8378: 15: 7
Carvers worke.	160: 0 : 0	
Painters worke.	98: 6 : 7	258: 6: 7
Wages of Shipwrights.	2517: 1 : 7	
Caulkers.	125: 10: 8	
Ocam boyes.	19: 7: 2	
Joyners.	229: 3: 5	
Housecarpenters.	5: 6: 0	
Wheelwrights.	10: 0: 0	
Plummers.	1: 10: 0	
Bricklayers.	6: 9: 0	
Scavellmen.	59: 7: 1	
Labourers.	316: 15: 2	
Teams.	50: 18: 0	
Sawyers.	547: 9: 0	
	: :	

A more correct Acct: of ye Wages under ye hand of ye Clerk of ye Checque remaining with Mr Pepys   3888: 17: 1

[FOLIO]	£ s d	£ s d
[14] Expended on her Launching which will serve other Shipps. On Timber & plank for ye Launch	314: 8: 6	
Hawsers to launch her.	116: 18: 8	431: 7: 2
	: :	
Rigging for severall provicions	91: 9: 6	
Iron worke	26: 6: 2	
[27] Cordage.	471: 13: 8	
[34] Blocks.	30: 13: 8	
[36] Lign vita shivers est	42: 0: 0	759: 0: 0
Dead eyes.	9: 16: 7	
Parrells.	3: 2:10	
Wages.	83: 17: 7	
	: :	
[43] Seastore for 6 months Boatswain.	2325: 18: 5	
[47] Carpenter.	90: 6: 7	
[48] Gunners Port rope.	6: 11: 6	2422: 16: 6
		16139: 2:11

The following folio appears to be the "more correct account of wages" referred to in folio 49:

An Account of ye Wages Earned upon his Ma:ts Ship *Lenox* built at Deptford betwixt ye 1st July 77 & 12th April 1678.

	£ s d
Shipwrights.	2517: 1: 7
Caulkers.	125: 10: 8
Joyners.	229: 3: 5
Housecarpenters.	5: 6: 0
Wheelwrights.	10: 0: 0
Plummers.	1: 10: 0
Bricklayers.	6: 9: 0
Scavelmen.	59: 7: 1
Labourers.	316: 15: 2
Ocam boys.	19: 7: 2
Teams.	50: 18: 0
Riggers.	83: 17: 7
Sawyers.	547: 9: 0
	3972: 14: 8

W Fownes

APPENDIX 6 – CONTRACT FOR THE CARVING OF THE THIRD RATE SHIPS OF 1677 BUILT AT DEPTFORD AND WOOLWICH

NA, ADM 49/24, f57

21st January 1677/8
Contracted the day and & year above said W:th Ye Principal Officers & Commissioners of His Ma:ts Navy by me Jos: Helby Carver & I do hereby oblige myself at my own proper cost & charges His ma:ts finding timber well and workmanlike to perform upon all His Ma:ts new Ships of the Third rates now in building in His Ma:ts yards at Deptford and Woolw:ch in part of the 30 Ships appointed to be built by Act of Parliam:t in such manner & with such time as the Ma:r Shipwrights of ye said yards shall direct The severall Carved works hereafter particularly expressed Viz:

	Lower Counter Brackerts	6
	Side Brackets	2
	Mask Heads	8
	Second counter Brackets	8
	Badges	7
	Gallery Badges	2
The Stern	Lower Cabin Lights Brackets	8
	Gallery Badges	2
	Second Counter Brackets	8
	Badges	7
	Round House lights	8
	Vazines (Terms)	2
	Large Taffrai	1
	Rother Face	1
	Gallery bottom pts	2
	Lower Brackets	18
	Badges	14
Gallery	Light Brackerts	18
	Brackerts above the Lights	14
	Turritt Brackets	14
	Pieces a top the Stool	2
	Upper Tier of Ring Ports	24
	Quarter Deck Ring Ports	14
	Poop Halfe Rings	4
Side	ForeCastle ½ Rings	4
	Lights	2
	Hancing pieces	8
	Main Chestrees	
	Supporters to the Catts	2
	Catt faces	2
	Catt sides	None
	End of the upper Rails	2½
Head	Head Brackets	12
	Half Brackets	2
	Lyon of ye Head	1
	Stem top	1
	Trayle board	1
	Tack faces	2
ForeCastle bulkhead forward	Brackets	6
	Half Brackets	2
	Ports Round	4
ForeCastle bulkhead abaft	Belfry Cheeks	2
	Cap	1
	Brackets	10
Steerige of ye bulkhead	Brackets	10
Roundhouse bulkhead	Brackets	6
Gangways in the waist	Before	2
	Abaft	2
	Top sail sheet Bitts	4
	Jeer Bitts	4

Halliard Blocks	3
Winding stair String	2
Great Cabin large cornice the whole depth round	1
Roundhouse ye same	

I do further oblige myself that all the aforesaid carved works shall be equal in goodness to the carved works performed on His ma:ts ship the Defiance. For and in consideration of £160. To be paid one third part in hand by way of imprest at the beginning of each ship, One third part thereof when halfe of the afore mentioned works are performed & the remainder when Certificate shall be produced from the Master Shipwright that the whole work is in every respect duly and completely finished according to the true intent & meaning of this Contract.

Jo: Helby

APPENDIX 7 – TIMBER CONTRACTS

NA, ADM 49/24, f3

Extracts from a contract made early in the programme with the timber merchant Sir John Shorter who had a yard next to Sir William Warren's. The contract shows that not only plank but also timbers themselves were imported from the east. Much of this timber would have been used on *Lenox*.

Before 23rd May 1677
Contracted ye day and year abovesaid with the Principal Officers and Commissioners of His Ma:ty's Navy by me Sir John Shorter Knight of London, Merchant and I do hereby oblige myself to deliver into His Ma:ty's Stores at Deptford and Woolwich as ye Principal Officers etc shall direct free from all charge to His Ma:ty ye timber, plank and deals under mentioned, viz.:

Three of four hundred loads of straight East country Oak or out Landish timber. White wood free from Red rotten dead knots or shakes well hewn to shape so as that both ye waines upon ye sides shall not exceed one ½ of ye square of ye said side and every way fit for ye service of his Ma:ty's Navy to be in length from 30 to 45 foot (long) and to meet at 60 (cubic) feet at ye rate of £3:5s per Load. Compass timber with Knees Raking and Square 150 Loads whereof ⅓ part Compass timber or thereabouts none less than 15 feet in each Compass piece and none less than 3 feet each Knee at ye rate of £3:5s per Load for ye Compass £5 per Load for ye square Knees and £4 per Load for ye Raking Knees———. To be paid for the same out of ye money appointed by Act of Parliament for building of 30 ships for his Ma:ty.
John Shorter

NA, ADM 49/24, f5

This huge contract between the Navy and Sir William Warren of Wapping is for enough timber to build two third-rate ships. Much of it went into *Lenox*.

6th June 1677
Contracted the day and year above said with the Principal Officers and Commissioners of his Ma:ty's Navy by me Sir Wm Warren Knight and do hereby oblige myself to deliver into his Ma:ty's Stores at Deptford and Woolwich by equal proportions to each of ye said yards, free of all charge to his Ma:ty the Streight, Compass and Knee timber undermentioned.

Streight & Compass Oake spire timber to meet at 2 Loads in a piece one with another 1000 Loads at L3:2s:6p per Load. More Streight Oake spire timber to meet one Load in a piece one with another 1000 Loads at 53s per Load. More Streight Oake spire timber to meet at 40 ft in a piece one with another 1000 Loads at 51s per Load. Compass Oake timber to meet at 25 ft in a piece one with another no piece less than 15ft at ye rate of £3 per Load 1000 Load. Knees Raking and Square 40 or 50 Loads none less than 3 feet at ye rate of £4:10s per Load ye square and £3:10s per Load ye Raking.

I do further Oblige myself that all ye said Streight Compass and Knee timber shall be good sound and Merchantable well hewen every way fit for his Ma:ty's Navy and that I will deliver 160 Load Weekly every week for ye First 2 months and 80 Loads weekly every week afterwards untill ye whole be delivered ye 1st week to end ye 1st June 80. It is further agreed that an allowance be made of 12p a Load for which of his Ma:ts stores shall be delivered into his Ma:ty's stores at Woolwich over and above ye prices beforementioned as ye also that what of ye said Timber shall be cut or split for convenience of carrying be reckoned into one piece provided ye same be done by direction of one of his Ma:ty's officers appointed hereunto by this board.

It is also further agreed that if amongst ye aforesaid 4000 Load of Timber there shall be Pollard Timber not exceeding 80 pieces the same shall be received into his Ma:ty's stores provided it be Compass timber and be also sound and fit for ye building of ye 30 shipps at ye prices mentioned in this Contract for Streight and Compass spire Timber of ye like scantlings. To be paid for ye same in Course out of money appointed by ye act of Parliament for building 30 shipps for his Ma:ty
William Warren

NA, ADM 49/24, f7

A typical contract for a large parcel of timber for Deptford supplied by Thomas Holland. Thomas Shish did much of the negotiations and marked out the timber as described in chapter 2.

13th June 1677

Contracted the day and year above said by me Thomas Holland of Stifford in Essex, and do hereby oblige myself to deliver into his Majesty's stores at Deptford, one hundred & twenty loads or thereabouts of oak spire timber wherein are several knees, some stem pieces and a considerable parcel of compass all the said timber well hewn not butted and of good scantlings at the top end every way fit for the service of His Majesty's Navy to meet at 70 [cubic] feet in a piece one with another at 55 shillings per load. I do further oblige myself to deliver all the said timber into His majesty's stores aforesaid within two months time from the date hereof whereof 30 loads thereof within 14 days or three weeks. To be paid for the whole out of ye money arising upon the Act of Parliament for building 30 ships, viz. one hundred pounds imprest so soon as the said timber shall be marked by one of His majesty's officers for His Majesty and the remainder in course.

    Thomas Holland

NA, ADM 49/24, f42

A contract for a large amount foreign plank. The defect referred to as "red" is a shortened version of "Red rotten dead knots".

7th September 1677

Contracted the day and year above said with Peter Causton of London merchant to deliver into his Ma:ty's stores at Deptford and Woolwich as he shall be directed ye East Country Oake plank and Prutia Deals undermentioned at ye rate and prices there also expressed Viz:

Good sound Dantzick Crown Plank white wood of 4 inches & 3 inches thick whereof ²/₃ to be 4 inches and ye rest 3 inch plank and to meet at 35 feet in length 8 or 900 Load at ye rate of £4:10 s per Load. Good sound Prutia Deals 600 (Load) whereof ½ to be 36 ft long and ye rest to be 33 feet long all of them to be full 3 inches thick and none to be less than 14 inches at ye top end at ye rate of 15s each deale. I do further Oblige myself to deliver about 120 Load of ye said plank presently with so much more as will make up that ½ ye abovesaid quantity by Xmas next and ye remainder of ye said plank together with ye Prutia deals by Michmas following and that ye said plank & deals shall be square edged in all respects fit for the service of ye 30 ships and that in case that any of ye said Plank or Deals shall be rotten, shaken or red the same to be refused, to have a bill made out at ye delivery of every parcel of ye said plank and deals and to be paid out of ye money appointed by Act of Parliament for ye said building of ye new ships.

    John Causton

APPENDIX 8 – CONTRACTS FOR BOATS OF THE SIZE USED ABOARD *LENOX* AT THE TIME OF HER OPERATIONAL SERVICE DURING WAR WITH FRANCE IN 1690.

NA, ADM 106/3069.

Contracted this 21st Jan: 1689/90 with the Honourable Sr Richard Beach Kn:t one of the Principal Officers & Commissioners of their Majesty's Navy for and on behalf of their Majesty's by me John Smith of Fareham in the County of Southampton Shipwright And I do hereby Oblige myself to build & deliver into their Majesty's Stores at Portsmouth free of all charge by the tenth of March next ensuring one Longboat of dimensions and Scantlings and fitted with the several particulars following

Viz:-

	Length Feet	Breadth feet	Depth feet : inches
Longboat of One	30	9	3 : 8

Breadth of the keele 5 inches, depth 9 inches & 8 inches, distance of the timbers room and space 12 inches, breadth of the timbers fore & aft 2½ inches, Depth of the Gunwales in and out 3½ inches, up and down 3 inches, breadth of the Stem thwart Ships 5 inches, fore & aft 12 inches, Scarph of the timbers 22 inches. Rising 8½ inches & 1¼ inches in and out, footwales 8½ & 1½ inches, keelson 3 inches & 10 inches Windlass 11 inches, Cheeks 4½ inches, thwarts of 3 inches four bound with 4 inch knees and number 3 bolts in each knee of ¾ inch, Lignum Vita Rowles two of 4½ inches & 2½ inches, benches number 3, Rother irons Number 2 pairs: Rother one, Sheets two, Ringbolts four of full inch, and two of ¾ inch, Eares of 4 inches & Chocks, futtock Riders 3 pair well fastened, the said Boats to be wrought with a wale of 4 inches deep & 3 inches in & out with a Rail of 2 inches up & down & one inch thick at the Rate of Eighteen shillings per foot.

I doe oblige myself that the said boat shall be wrought in good substantial & workmanlike manner with good dry well seasoned oak board of inch thick for which I am to be paid ready money at Portsmouth when a Bill is made out and signed by the Officers of ye yard as is usual.

Copia Vera   Jn: Smith

Rich Beach
    Warrant 25th January 89/90

Contracted ye 20th Jan: 1689/90 with the Honourable Sr Richard Beach Kn:t one of the Principal Officers & Commissioners of their Majesty's Navy for and on behalf of their Majesty's by me Thomas Oxford of Gosport Shipwright: And I do hereby Oblige myself to build & deliver into their Majesty's Stores at Portsmouth by the first of March next the Boats under mentioned of the dimensions and Scantlings and each fitted with the several particulars following

Viz:-

	Length Feet	Breadth feet inches	Depth feet inches
Pinnaces of Two	30	6 : 0	2 : 7

Depth of the keele 5 inches Breadth 4 inches, Scantling of the timbers inch & ½, room and space 13 inches, Depth of the Gunwales up and down 4½ inches, in & out 2 inches and 1¼ inches, Scarph of the timbers 18 inches, Breadth of the Stem thwartships 3½ inches fore and aft 7 inches, Sterne post 3 inches, Rising 4½ inches & one inch in and out, keelson 8 inches & inch & ½ thick to be fitted each with twelve iron knees number 5 bound thwarts with Iron knees and number 2 transom knees and number 3 thwarts of 1½ inch and two of 2 inches with Gangboard, Benches three Linings, Grounds and Mouldings of O,G (ogee) Plansiers turned off, Back board one, bottom board and Scarr board, keele band 26 feet Ring Bolts two, Rother Irons two pair Rother One Once primed at the Rate of fifteen shillings per foot.

I doe oblige myself that the said Boats shall be well and & Workmanlike performed with good dry well seasoned oak board of inch thick for which I am to be paid ready money at Portsmouth when Bills are made out and signed by the Officers of the yard as usual.

Copia Vera   Tho Oxford

Rich Beach
    Warrant 25th January 89/90

Contracted the 2nd Xber [December]: 1689 with the Honourable Sr Richard Beach Kn:t one of the Principal officers & Commissioners of their Majesty's Navy for and on behalf of their Majesty's by me Tho Oxford of Gosport Shipwright, and I do hereby Oblige myself to build & deliver into their Majesty's Stores at Portsmouth free of all charge by the 20th January ye next ensuring the two Yawles under mentioned of the Dimensions and Scantlings there expressed & each fitted with the particulars following

Viz:-

	Length Feet	Breadth feet inches	Depth feet inches
Yawles of Two	23	5 : 9	2 : 5

The Rails of ye upper strake to be made out of ye whole wood, up & down Gumwales Stuck, 3 thwarts bound with Iron knees, The Transom with two Iron knees, the Stateroome stuck with an O:G: (ogee) and plansiers for the Gunwales with two panels on each side, a back board & a locker under ye after bench, linings under the benches Keel thwart ships 4 inches, up & down 4½ inches & 4 inches: Keelson 6 inches broad of 1½ inch plank, timbers of 1½ inches with 13 inches room & space & 18 inches Scarph for the floor timber heads to nail to the lower edge of the binding strake, with bottom boards & Scarr boards, Keele band & Iron bolts & Rings for the Stem & Stern, each to row with Six Oars to be graved & primed to ye waterline & paid with stuff in the inside to the Rising at the Rate of 13 shillings per foot.

I doe oblige myself that the said boats shall be wrought in good Substantial & workmanlike manner with good dry well seasoned oak board of inch thick for which I am to be paid ready money at Portsmouth when a Bill is made out and Signed by the Officers of the yard as usual.

Copia Vera   Tho Oxford

Rich Beach
    Warrant 4th December 89

APPENDIX 9 – SAILMAKERS' CONTRACTS, 1690

NA, ADM 49/25

Saile-Makers Contracts

Duck Great Noyalls, Suffolk & such other canvas Sowed with a Double flat seam	} } yard
Pertrees, Vittery small Noyalls, Ipswich & such other Canvas sowed with a round seam & one in ye Middle	} } yard

To be worked with the best English Twine

Double Ipswich or Small Noyalls Courses	108}
Duck	100} Stitches in a yard
Noyall or Suffolk Courses Topsails Spritsail or Mizens	96} & not under
Vittery, Pertrees, or Small Noyalls into Small sails	90}

All Boltropes to be cross stitched at every 12 inches in length with 3 cross stitches & to be of ye size following

	I Rate incs	II Rate incs	III Rate incs	IV Rate incs	V Rate incs	VI Rate incs
For Main Course	6	5½	5	4½	4	3½
Fore Course	5½	5	4½	3½	3½	3
Main Topsails	5¼	5	4¾	4	3¼	2½
Fore Topsails	5	4¾	4	3½	2¾	2½
Mizons & Spritsails	5	4¾	4	3½	2¾	2½
Main Staysails	3¼	3	3	2¾	2½	2¼

The Boltropes for all other sails to be of such sizes as shall be as sufficient & approved by ye Officers Two reefs to be made in every Topsail from ye 1st Rate downwards & one reef in every course for ships of ye 4th & 5th Rate & 2 course for 6 rates with one Bonnet for each if required.

To be allowed for measuring Canvas into ye Stores. Per Thousand yards

Months certain from 21st Oct 90	No	
Months warning	No	
Interest after 6 Months		

APPENDIX 10 – TABLES CONTRACT

NA, ADM 106/3069

Contract with Henry Ward, 1690

Contracted and agreed this 1st February 89/90 with the Honourable Edward Gregory Esq. one of the Principal Officers and Commissioners of their Majesties Navy for and on behalf of their Majesties by me Henry Ward of Chatham joiner and I do hereby oblige my self at my own proper cost and charge, and with my own materials well and seasonably to make finish and deliver into their Majesties stores at Chatham in good and workmanlike manner within six weeks from the date hereof (that is to say ¹/₃rd part thereof in 14 days, ²/₃rd part thereof in one month and the residue by the time exprest) the tables and hencoops under mentioned for the ships now ordered to be got ready viz:

	Feet	No	
	3 ½	20	With legs of oak, deal tops, deal rails ¹/₃rd part
Tables	3	20	of the top to hang with hinges with a swing leg
			to retain it and a drawer to each at 8s 6d each
	2 ½	40	After the same manner at 8s each
Hencoops		20	Each to be 6 feet 2 inches long 2 feet 7 inches
			High 1 foot 9 inches broad at bottom 1 foot 8
			Inches broad at top-The tops and ends to be
			Whole deal and the back slit deal with a
			Trough at 16s each

To be paid for the same at Chatham upon passing a regular bill as is usual-

Henry Ward

A warrant 3rd February 1689

A copy
Edward Gregory

APPENDIX 11 – DEANE'S PROPOSITION

NA, ADM 106/371 f28

Sir Anthony Deane - His Proposition for Preserving the New Shipps from Decay. London 23rd March 1683

Gentlemen

The duty I owe to His most gracious Ma:ty, the special regard I have for the honour of your Board together w:th the zeal I retain for the welfare of the Royal Navy, gives me confidence to believe you will not only take it in good part, but excuse the trouble if I entertain you w:th a short discourse relating to the Thirty new ships Built by Act of Parliament, which shall be done with all imaginable Candour and faithfulness in hope it will meet with your concurrence, a service to his Ma:ty, and a friendship to your selves for so I intend it.

The first thing I shall observe is this, that a general rumour is spread about that more than ordinary decay is discovered upon part of the new ships, w:ch I believe to be true from what I have been informed by the Officers of ye Navy,

The Second is the different reasons offered by many, and of those some of considerable note of the Cause of the decay, w:ch if allowed may prove of fatal consequence, especially if it should lead into a mistake by w:ch either proper remedy are neglected, or wrong physic applied for cure. For prevention of both, I shall offer my knowledge what it is that will hasten ye decay of a new ship riding in Harbour, and what will infallibly preserve them if timely and duly applied.

To let you properly into the whole matter it will be necessary to state the two Grand reasons alleged for the decay, The first and that which seems most ready in Vogue is that they are built w:th East Country Plank, but the Second is most dangerous coming from those that should be better informed, that they are built in too short time from w:ch there seems to be no remedy, About these I shall confine my discourse and demonstrate that neither of these reasons affect the sudden decay if proper and early remedies be applied for their preservation before I come to the causes of sudden decay and the remedy against it I shall speak of the nature of Oake Timber, by which you will perceive whither what I say has any weight, Oake Timber is a Body made up of a multitude of hollow vessels through w:ch the Sap is conveyed from the Root to the Top of the Tree, besides the Vessels that are perpendicular there is a great number that lye horizontal and although these Vessels are not discernible to our Sight yet by the help of Glasses they appear so plain they may be numbered, So great is the Sap or moisture contained in one Cube foot of green Timber that the very weather will extract ten or twelve pound weight of Liquor in less than a year out of the Vessels of one Cube foot, now Observe what the weather will do [to] the whole of a ships hold that is new will do the same or more, by drawing from the Outside of the ship first all its own Liquor contained in the Vessels in the Timber, Plank wrought upon the upper work from the waters edge which is more than Seventy two tuns weight for a Third rate, and above Eighty five Tuns weight for a Second rate that is discharged of natural Liquor in the Timber of the whole upper works, When the heat of the hold has drawn out of the vessel all its own natural Liquor and moisture, then these empty Vessels are so many Pipes by w:ch new moisture and Air is drawn through continually from the outside of the ship by the heat of the hold till the Vessels themselves decay and putrefy the Timbers and inside which spreads itself like a Leprosy for want of a due supply of Pitch and Tar to stop those Vessels upon the Outside of the ship and to fortify it as tallow or other matter does on Leather or Stared Boots, It is to be noted that Chatham River being narrow lying between two great hills exposes the ships to extreme hazard by the Sun, wind and weather which forces everything the faster if not timely prevented by a more than ordinary care to preserve the ships from sudden decay:

Having taken notice of the nature of Oake Timber I come to explain what it is that will suddenly decay new ships riding in harbour let them be built with English or any other sort of Plank (which in short is this) If you let a new green ship lye in Harbour without heeling them at least twice in a year for the first three or four years after they are Built and not burn off the old Stuff from the waters edge a Strake or two above the upper Wales, they will certainly be decayed at the Breadth, in the Bows, quarters and Buttocks for the reason aforementioned, On the contrary if you do heel new ships for two or three years after they are Built and burn off the old stuff from the waters edge a Strake or two above the Upper wales not less than twice in a year and lay new stuff in its room, it fills up all the hollow Vessels or porousness of the Plank and so fortifies that it stops the moisture from passing through the Plank, The often Burning and melting the old Stuff upon the outside of the Plank so hardens and fills the Vessels that neither the Weather, Worm, or Insects can live or penetrate it, If it shall be objected that to heel a ship twice or often in a year to Bream them as propounded is a great charge, I make answer the twenty one Pounds five shillings will Burn and Bream one of these New third rates twice, and Twenty six pounds twelve shillings will do the like for a Second rate taking fit opportunity and help of the Ordinary to heel them. Then the whole sum for the Twenty nine ships already built amounts but to Six hundred seventy five pounds four shillings to be completely done twice in a year, Therefore let me beg For Gods sake, for His Ma:ts service, England's Safety, and your own Justification (if it not already done) to begin presently and lose no time to heel the said ships three or four Strakes to burn, Bream, Liquor and fill the Starved Vessels or Pores of the Plank w:th Pitch and Tar so often that they will hold no more, wherever it is required without Board, and also within board, w:th such matter as you know is proper for it by Oyle, Colour, or Stuff, It will put a stop to their further sudden decay, and preserve such as are yet untouched, It is my passion for the welfare of the Navy that exhorts from me this in private to you before it be too late,

It is also a false Doctrine to believe that either East Country Plank, or the sudden building the Thirty new ships is any cause of their defects, the following Comparison may clear that point.

First the Dutch, the Dane, the Swede, the French, nay our East India and biggest Merchant ships are and have been for several years last past built w:th East Country Plank, and if any man living can show me an instance of such a sudden decay as some of these ships are reported to be at ye Breadths, Bows, Buttocks and Quarter in so short a time, then I will allow it to be in the Plank, Again to show you that English or any other Country Timber or Plank without timely and due preservation will decay, I can name you a ship no less than a First Rate built all w:th English Timber and Plank perished and decayed in little more than Seven years that she was forced to be Rebuilt at near the same Charge as her first cost, but being built in the Kings own yards it made no noise, I do not say this to reflect upon anything past god knows my heart but for Comparison and Co—ition to seek into Ye two causes of defects and not to mistake one thing for another, If we do it may prove fatal,

Another instance to confirm what I say, Note That when new ships go to Sea they are often heeled to clean, At the same time they Burn and Bream the old Stuff and lay on new, w:ch so hardens and fills the Pores of the Plank that it prevents the sudden decay of ships in the places aforementioned, And also prevents it from Worms or other Insects either to bite or Putrefy it without or within board, I appeal to you Gent of the Board whither I speak truth or not.

I should advise (with your Permissions) that every new ship might have a Sail spread over the hatchways to throw down the wind into the hold to cool them, it will contribute very much to the preservation of those ships.

I am told that several of the new ships has their Graving yet remaining that was put upon them at their launching and that a good deal of it lies above the water, If it be so (w:ch I hope is a mistake) do not wonder at the defect of the new ships Bows and Quarters, for that Brimstone burns the Plank and is very Porous and has no healing or virtuous substance in it, nor would it stick to ye ships side but by the force of Oyle first mixed w:th Rozin, neither of w:ch is good for preservation of Timber or Plank but is used more for its cleanness (being hard) keeps clean ye longer at Sea w:ch in harbour should be avoided, It is also said (how true I know not) that some of the new ships are scraped above the Brimstone in the Buttocks between and below the wales afore and abaft and that they have never been Heeled or Breamed since their being brought into Chatham River, If it should be so I have the more reason to reiterate my former requests, for most certainly the ships will be destroyed and perish for ye reasons I have before mentioned if it be suddenly prevented by proper remedies for their preservation, I am confident if the Lords of the Treasury or the worthy Lords and Gent of the Admiralty were assured that Six hundred seventy five pounds a year would Bream the Twenty nine ships twice in the manner I mentioned, they would desire it to be done were there nothing more than a general satisfaction that all is done that could tend to their preservation, and you could never receive blame to put it into execution as the Essential part of the standing charge of ye Ordinary, Gent, you are the Christians and may apply anything for the preservation of the ships that is within your reach, And therefore as a most Cordial friend I send this my Judgement being afraid that most of this Complaint arises from the overlooking that part of heeling the ships which difficult to justify and perhaps fatal to be any longer delayed, for if inquiry should be made by either his Ma:ty or the Rt Ho:bles Com:s of the Admiralty and that the causes of the defects should be found and allowed from ye reasons I have mentioned, it might be of Injury at least a severe Censure that all was not done that might be expected on your parts, I know that you may w:th great Justice ask me what I have to do w:th the Navy being no more than a private man, and that it might be time enough to give my Opinion if I were called to it, but when I do assure you w:th all faithfulness and friendship unto you I do hope you will receive it with the same kindness as I send it having no other end in this Contemplation than his Ma:ts service, if it meet w:th your concurrence it is well, if not, commit it to the fire.

> I am Gent
> Your very humble serv:t
> signed A. Deane

dated 23 March 1683
read ye 4th April

APPENDIX 12 – LETTER FROM MR LEE, MR FURZER AND MR CHOTHIER

NA ADM 106/370 f51

Letter from Mr Lee, Furzer and Mr Chothier at Chatham concerning the Preservation of the New ships from Decay

Chatham Dock 10th April 1684

Rt Hono:ble In Answer to Yo:r Hon Order of the 7th instant, Wee humbly make our Return followeth, That the sudden and more than ordinary Decay w:ch from Year to year have appeared in many of the XXX Shipps hath very much Employed our thoughts; And put us upon Strictly Observing all Occurrence w:ch might render a reason for the Cause Whereof, in Order to a full and due Discharge of that Duty incumbent on us for their Preservation.

Notwithstanding we can discern no other cause of their so sudden Decay (w:th all humble Submission to better Judgement) than This - That the Timber w:th w:ch they are wrought in those parts where they most Complain, being a very large sort of East country Trees were we conceive past their Growth and declining before their fall, or had lain long after, where by their nature all substance was abated and became Spongy, not unlike to the Grain of a Cane, w:ch after having past the fire in Working, that little State w:ch it had, was we judge thereby Drawn forth, and so Decay thereof soon followed.

But in Order to Yo:r Hon fuller satisfaction therein, we conceive it necessary, not only to give our opinion but likewise those reasons w:ch induce us hereunto, by a particular State of what we have Observed. Viz.

That the Plank w:ch first Appeared Decayed is that of their Bows and Buttocks, w:ch for the Roundness in working is always most Heated and Dried by the fire; And that it was some time after before those large Plank in their Midshipps Complained.

That the outward Surface of the Plank as well in their Bows and Buttocks as Midshipps does yet appear very Sound hard and fair to the Eye, for about an Inch or an Inch and half In, but in searching Deeper do find it Dirt rotten generally throughout the Plank to the very Timber.

That most of these shipps have been Heeled and payed once a year in the manner yo:r Hon:bles prescribe, Yet we find those Equally Decayed w:th them w:ch have not, And we find not any Decays Under Water, the State and Substance of the Plank there, being we conceive preserved by the Moisture thereof; Nor more decays at the Water Edge or in the Wash of it then to 5 or 6 f:t above water.

As to the Inward part of the Shipps, and to placing a Sail over the Grating for Casting the Wind into Hold, We humbly conceive no Occasion requiring it, for that upon Boring and fully Trying their Footwaling, Beams, Clamps; etc We find them no way Defective but in all respects very Sound and good, And we judge the Holds sufficiently Aired by what Wind now comes down the Gratings, the Hatches on their Gun decks being fitted w:th Gratings for that purpose.

From all w:ch for the future preservation of his Ma:ts said shipps, we humbly offer it as our Judgement, and think it of Absolute necessity, that as this East country Stuff is or as shall be found Decayed, it should be forthw:th shifted, and good English Timber or Plank brought on in the room thereof, which will preserve their frames from Infection by it, We finding by Experience that many of their Timbers are already much Stained from the Rottenness of the above said Plank bearing against them and remain

Chatham Dock 10th April 84
      Yo:r Hono:bles most Obedient Serv:s
read ye 11th   Robert Lee
              Dan Furzer
              Robert Chothier

APPENDIX 13 – PETT'S PROPOSALS

NA ADM 106/378 f11

12th May 1685
Sir Phineas Pett's proposals offered for preservation of His Majestys Ships in Harbour.

To my Honoured friends Sir John Tippetts and Sir John Godwin Knights— Principal Officers of His Majestys Navy now at Chatham

Gentlemen Whereas I long since writt to ye Navy Board, for the using of some fitting & proper means for the Preservation of His Ma:ts Shipps at Chatham of the 30 saile built by Act of Parliament, And finding upon the Survey now taken of them here & at Gillingham by our Selves & some others of His Ma:ts Mast:r Shipwr:ts & their Assis:ts that nothing of what I formerly proposed for their Preservation, hath been yet done, I cannot without the betraying of my trust to his Ma:ty, Omit the reminding you of so great a Concern as this is, the Welfare of his Ma:ts Navy- Therefore Desire before wee leave this Place, there being three of us present a Sufficient Number to make a Board, that we may give Warr:t to the Respective Officers for the speedy putting of the seven Particular Undermentioned in Execution for the better Preservation of his Ma:ts said Shipps (many of them having rec:d Prejudice (by riding too light) & found very much decay:d for want of due care taken)-.-

1stly
That the said Shipps be immediately Heel'd Breem'd, 3 or 4 Strakes under water & Pay:d with Blackstuff from thence upwards one Strake above the Upper Wale, & so to continue once or twice every year.

2ndly
That all the new 3rd Rate Ships that wants Graving w:ch was alledged could not be done for want of Docks, may be forthw:th Hauled on Shore on the Graving Places, & the old Graving well burnt off & Payed w:th black stuff & to be graved once in 2 years till their Planks are more seasoned.

3rdly
That their Seams from ye Wale upward be pay'd with Pitch, & their sides to be payd' w:th a good Coat of Stuff, to preserve them from the Weather, and all their Upper Decks, Quart:r Decks, Fore Castles & Roundhouses to be pay'd w:th Tarr & some Pitch mixt w:th it, to give it a Body when they have their Ordinary Repairs.

4thly
To take up a Strake of Plank on ye Gunn Decks of ye Said Ships & that their sides & upper Decks may be well Watered, every Morning & Evening, in the Summer Season, for to Cool ye Ships & preserve them from Rotting.

5thly
That Constant Care be taken in keeping open their Ports & Tarpawlings in fair Weather as also their Hatches & Scuttles, to be always kept open, and to place a Saile to strike down the Wind, into their Holds, to Cool their Holds & preserve them from a Further Decay.

6thly
That their Holds be forthwith Cleared of all their [Gun] Carriages, Standing & Running Rigging w:ch both Heats ye Hold & Damnifies the Carriages and Rigging.

7thly
That the New First & Second Rate Ships be speedily removed to Gillingham into deeper Water, & more Ballast put on board them, as also on board ye 3rd Rate Ships, to keep them from Cambering, most of them having Retched & fetched way, & their Decks Straightened for want of Ballast. And that a 1st Rate may not ride with less than 600 Tons, a 2nd 500 & a 3rd 400. The Quantities on Board each Ship being much short of these Proportions.

8thly
That their Masts be pay'd with Rozin well mixt with Tallow to soake into the Mast once every yeare & that the wedges in the Partners be often shifted & ye Coats kept Tight, to prevent any Dripps going down to rott ye Masts.

9thly
To Rave all ye Ocham out of ye Seames of the Spirketrising to give Aire to ye Spirketrising & Timbers & to lye open till such time as the ships goe to Sea.

I am - Gent:m

Chat:m Dock   Yo:r very Humble
12th May   Serv:t
  1685
          Ph Pett
Read at the Board 20 May 1685

APPENDIX 14 – THE SHIPWRIGHTS' RESPONSE

NA ADM 106/378 f 424

12th May 1685

His Ma:ts Master Shipwrights etc report in answer to Sir Phineas Pett's touching sent requisite for the preservation of the Kings ships at their moorings.

Honorable Sirs In compliance w:th your Warrent of this day, We have considered the Letter from Sir Phineas Pett to yourselves Dated also this day (A copie of which came w:th your said Warrent) and Return our opinions on each Head thereof as they lye in Order Viz.-

1stly
As to Heeling the Shipps (of the XXX) once or twice a year etc. Though we doe not think their Decays to proceed from want thereof, yet it may be convenient, once a year, to Heel Bream & Pay such of them as are not Decayed betwixt Wind & Water to preserve their Plank from being Water Soaked

& keep the seams Tight; But for such of them as are already Defective in those parts, the fire cannot be brought to them w:thout great Danger: Some of them have been soe Heel'd Caulked & Payd, & more might have been, could we have procured the Heeling of them.

2ndly
To bringing the New Third Rate Shipps, on upon the Graving Place & Grave them once in Two years etc We reply, That if the Navy Board be please to give Warrent for bringing thereof such of them as have received their Repair, We deny not but it may be done, though they being very sharp, Large, & Weighty Shipps, much Danger will attend the doeing of it; but for those of them w:ch are not Repaired, It is in our Opinions a work not w:ch safely to be done through Rotteness of their Bows & Buttocks and hazard of fire thereby, therefore must of necessity be brought into Dry Docks for shifting their Works before they can be Breamed.

3rdly
To the third for Paying their seams w:th Pitch etc The Bowes Buttocks & Midships under between, & one Strake above the Waales hath allwaies been Payd w:th Blackstuff, & the seams from thence Upward on Caulking payd w:th pitch till scraping, & their Sides paid w:th Rozin & Tallow as Usual: Their Upperdecks Quar'decks & forecastles, have yearly after caulking been paid w:th 1/3 Pitch & 2/3 Tarr.

4thly
To taking up one Strake on their Gun decks, We say That our present Survey having strictly Searched & Observed the Condition of the XXX Shipps in their Holds doe find the Gun deck Clamps Beams & Ceiling to be very sound & dry as when first Wrought (Except in their Breadrooms which were Lined w:th Lead, much Decayed) w:th which considering the Damage that may Attend them, by water w:ch through opening will be apt to rundown upon their Beams & Knees, doe think the prejudice may be greater than the benefit. To that necessary work of Watering the Decks, We heartily wish that the Orders already given by Sir John Godwin for the same may be duly Observed, as in some of the Shipps it hath been.

5thly
To the Fifth, The Ports & Hatches are generally kept open in convenient weather: And for a sail to cast the wind into the Hold it may not be Inconvenient.

6thly
To the Sixth That their Holds be forthw:th cleared of all the cordage Standing & Running Rigging, w:th the Carriages, It is in our Opinions very necessary to be speedily done.

7thly
To the 7th We yesterday observing the Cambering of his Ma:ts Shipps through their want of Ballast, or Riding in too Shoal water, did on board the St Michaell render our Opinions to Sir Phineas Pett That the said Shipps should be speedily removed into such Depth of Water as may capacitate the First Rates to receive on board. 600 tuns of Ballast The 2nd Rate 500. and the 3rd 400 tuns; And we repeat the same again to be Absolutely necessary.

8thly
To the 8th Some years since the Masts & Yards of his Ma:ts Shipps were paid w:th Rozin & Tallow by the carpenters of the Shipps; but of late years that work having been committed to the care of the Ma:tr Attendent the practice hath been to pay them w:th Tarr, tho we think the first tends most to their preservation.

9thly
To the 9th We observing in our Survey, the upperdeck clamps above the Spirketing & the Gun deck Clamps under the same to be very sound & Good though uncaulked & yet the Spirketing (being caulked & well payed) to be extremly decayed, Did impute the preservation of the said clamps to the Air passing round them through the Open seams, & on the contrary the Decay of the Spirketing to be hastened thro the close caulking thereof, did acquaint Sir Phineas Pett three days since with our notions thereof; And doe think that Raveing the Ocham out of the middle Seam may preserve the Spirketing.

We conclude in Answer to the later part of your Warrent, That in our humble opinions there is noe better way to preserve his Ma:ts Shipps of the XXX from future Decay (of this nature) Than w:th all possible speed to shift all such Timber or Plank as is now decayed on them; Judging the Defective stuff Infects the Sound, and (by continuence together) will dayly spread its effects, to the great Damage of their frames if not preserved. Wee remain

Chatham Dock
12th May 85

Read at the Board the 20 May
              Yo:r Hono:s most Obedient Serv:ts
     John Shish   Tho Shish   Rob:t Lee
     Jos Lawrence   Dan Furzer   Robert Castle

     Thomas Kirkes   William Bagwell
     Aug:st Robinson   John Aework
     Joseph Goodnm   Elias Waff
     John Williams

NOTES TO THE TEXT

INTRODUCTION

1 Anthony Deane, *Deane's Doctrine of Naval Architecture*, 1670, Pepys Library, Magdelene College, Cambridge, PL2910, (Ed. Brian Lavery, Conway Maritime Press, 1981), p14 (Hereafter, *Deane's Doctrine*).
2 Cat. No. 1655-1 National Maritime Museum, Catalogue of Ship Models, Part 1.
3 Brian Lavery, *Ship of the Line, Vol I*, p159.
4 NRS, Samuel Pepys, *Naval Minutes*, p241. (Hereafter, *Pepys' Naval Minutes*).

CHAPTER I – PARLIAMENTARY APPROVAL

1 *Pepys' Naval Minutes*, p243.
2 A Caruana, *The History of Sea Ordnance, Vol 1*, p77, also NA, WO51/15 f219-f220, WO51/16 f123-f124.
3 NRS, Catalogue of Pepysian MSS, Vol I, p43 (Hereafter, Catalogue of Pepysian MSS).
4 E. S. de Beer, ed., *The Diary of John Evelyn Vol 3*, (Oxford, 1955), p296–7, 1 October 1661. (Hereafter, *Diary of John Evelyn*).
5 J. Sheffield, *Miscellanea*, (London, 1933), p56.
6 *Diary of John Evelyn*, 28 Jan 1685.
7 NMM SPB/50.
8 *Pepys's Naval Minutes*, p128.
9 NMM SPB/50.
10 *Pepys's Naval Minutes*, p394.
11 NRS, Catalogue of Pepysian MSS, Vol IV, p115.
12 Ibid, p126.
13 Commons Journal, ix. 321.
14 Pepys Library, PL2265, No. 29.
15 *An Act for Building Thirty Ships of War*, 1677, p145.
16 NRS, Catalogue of Pepysian MSS, Vol III, p380.
17 Ibid, Vol IV, p406.
18 NA, PRO ADM 106/36 24th April 1677.
19 NA, ADM 106/324 f250.
20 NRS, Catalogue of Pepysian MSS, Vol IV, p411.
21 NA, ADM 106/36, 1st May 1677 and NA, ADM 1/5138, 2nd May 1677.
22 NRS, Catalogue of Pepysian MSS, Vol IV, p413-7.
23 Ibid, p662.
24 NA, ADM 106/36, 6th May 1677.
25 NRS, Catalogue of Pepysian MSS, Vol IV, p418-9.
26 NA, ADM 106/36, 10th May 1677.
27 BM ADD MSS 9316 f216 and BM ADD MSS9322 f26.
28 NRS, Catalogue of Pepysian MSS, Vol IV, p422-3.
29 NA, ADM 106/36, 12th May1677.
30 *Pepys's Naval Minutes*, p239 and the Wilton House Model that has a plate mounted in its case.
31 NA, ADM 106/323 f20, 17th May 1677.
32 NRS, Catalogue of Pepysian MSS, Vol IV, p425.
33 Ibid, p427-8.
34 Ibid, p430.
35 NA, ADM 91/1 f280v, Surveyors Report.
36 NA, ADM 106/36, 19th May 1677.
37 BM ADD MS9322 f27.
38 NA, ADM 106/36, 30th May 1677.
39 NA, ADM 106/324 f176-180.
40 NA, ADM 106/323 f222.
41 NA, ADM 106/322 f223.
42 NRS, Catalogue of Pepysian MSS, Vol IV, p431-4.
43 NA, ADM 106/36.
44 NRS, Catalogue of Pepysian MSS, Vol IV, p466.
45 NA, ADM 106/36, 31st July 1677 and NRS, Catalogue of Pepysian MSS, Vol IV, p481.
46 NA, ADM 106/329 f68.
47 NA, ADM 106/327 f217.
48 BM ADD MS 9322 f30v.
49 NA, ADM 2/1748, p99.
50 NA, ADM 106/37.
51 *Pepys's Naval Minutes*, p374.

52 BM ADD MSS 9316, 14th May 1677.
53 Ibid, f222v.
54 NRS, Catalogue of Pepysian MSS, Vol IV, p459.
55 NA, ADM 106/36 27th June 1677.
56 NA, ADM 1/5138 f714 and Bodleian Library, Oxford, Rawlinson A185 f163.
57 Pepys Library PL1339.
58 *An Act for Building Thirty Ships of War*, 1677, p4.
59 Ibid, p7, p76.
60 Latham and Matthews, ed., *Diary of Samuel Pepys, Vol 11*, p220.
61 Document in the private collection of Chris Mackie.

CHAPTER II – DEPTFORD DOCKYARD AND TIMBER SUPPLIES

1 *Diary of John Evelyn*, 17th Jan 1652.
2 Blaise Ollivier, *Remarks on the English Navy*, 1737.
3 NA, ADM 106/323 f174.
4 NA, ADM 106/3069.
5 NA, ADM 42/485-486.
6 Henry Teonge, *His Diary*, (Routledge, 1927), p189 and p192.
7 NA, ADM 106/3118.
8 NA, ADM 106/323 f174.
9 Ibid, f232.
10 NA, ADM 106/323 f85.
11 Pepys Library PL2266 f28.
12 *Diary of John Evelyn*, 3rd March 1668.
13 www.familysearch.com
14 The Shish Memorial, St Nicholas Church Deptford.
15 Latham and Matthews, ed., *Diary of Samuel Pepys, Vol. 10* f397.
16 Pepys Diary, 22 July 1664.
17 Stephen Johnson, 'The Carpenter's Rule', *Eleventh International Scientific Instrument Symposium*, 1991, p39-45.
18 *Pepys's Naval Minutes*, p200-201.
19 NA, ADM 2/1725.
20 *Pepys's Navy White Book*, PL2581. March 26–April 4 1664.
21 *Pepys's Diary*, 24th March 1668.
22 NA, ADM 1/3548 p263.
23 NA, ADM 106/333 f322.
24 Ibid, f308.
25 *Diary of John Evelyn*, 3rd March 1668.
26 Ibid, 13 May 1680.
27 Bodleian Library, Oxford, Rawlinson A186 f151-152
28 C Knight, *Mariner's Mirror, Vol. 81*, p412.
29 www.familysearch.com
30 NA, PRO PROB 11/385.
31 www.familysearch.com
32 C Knight, *Mariner's Mirror, Vol. 81*, p411.
33 www.familysearch.com
34 NA, ADM 1/3548 p263.
35 Ibid, p264.
36 NA, ADM 106/37.
37 NA, ADM 106/38.
38 NA, ADM 1/3548, p629.
39 *Pepys's Navy White Book*. PL2581, March 31 1664.
40 *Diary of John Evelyn*, 17th April 1683.
41 Evelyn to Pepys, 3rd August 1685, published in *Particular Friends*, (Guy de la Bedoyere, 1997), p154.
42 NA, ADM 106/370 f348.
43 Nathan Dews, *The History of Deptford*, 1884, p169.
44 NA, ADM106/3541, Part II.
45 Bodleian Library, Oxford, Rawlinson A178 f264.
46 Ibid, A185 f325.
47 NA, ADM 7/633.
48 NA, ADM 106/333 f312.
49 *Mariner's Mirror*, Vol 27 No. 4, p284.
50 *Diary of John Evelyn*, 13th May 1680.
51 NA, ADM 106/333 f344.

52 Nathan Dews, *The History of Deptford*, 1884, p316.
53 NA, ADM 106/3540.
54 NA, ADM 106/37.
55 NA, ADM 106/38, 9th March 1678.
56 NA, ADM 106/321 f21.
57 NA, ADM 106/341 f132.
58 Kate Loveman, *Historical Journal, Vol. 49* (2006), p893–899.
59 The Building List of *Lenox*, Pepys Library, PL1339, (for transcript refer to Appendix 1 of this book).
60 NA Prob/11/350.
61 NA, ADM 106/323 f323.
62 NA, ADM 106/324 f368.
63 NA, ADM 106/321 f492.
64 NA, ADM 91/1, 23 July 1677.
65 Evelyn to Pepys, 3rd August 1685, published in *Particular Friends* (Guy de la Bedoyere, 1997), p154.
66 NA, ADM 20/24.
67 NA, ADM 18/60, p267.
68 NA, ADM 42/486.
69 NA, ADM 106/37.
70 NA, ADM 20/25.
71 NA, ADM 106/323 f82.
72 NA, ADM 106/3540, Part II.
73 NA, ADM 106/3539, Part II.
74 NA, ADM 49/132 No. 57.
75 NA, ADM 106/37.
76 NA, ADM 7/633.
77 NA, ADM 106/330 f431.
78 NA, ADM 106/331 f217v.
79 NA, ADM 106/330 f431.
80 NA, ADM 42/486.
81 M.A Faraday, Camden Society.
82 NA, ADM 42/486.
83 NA, ADM 18/60 f269 f265.
84 NA, ADM 18/60 f269.
85 NA, ADM 1/3549, p765.
86 NA, ADM 106/3538.
87 Ibid.
88 Catalogue of Pepysian MSS, Vol. 1, p462.
89 NA, ADM 1/3548 f496 f499.
90 NA, ADM 2/1748, p159.
91 *Instructions for Pressing Seamen*, NA, ADM 2/1748, p198-201.
92 NA, ADM 106/326 f154.
93 NA, ADM 106/324 f407-409.
94 NRS, *Samuel Pepys and the Second Dutch War*, p69.
95 NA, ADM 106/3538.
96 NA, ADM 106/325 f148.
97 NA, ADM 49/132, 7th Oct 1698.
98 NA, ADM 18/61, 21st Jan 1678.
99 NA, ADM 106/323 f160.
100 NA, ADM 91/1, 22 June 1677.
101 Pepys's Library, PL1339.
102 Pepys's Library, PL1338.
103 Cumbria Records Office, C/LONS/L Admiralty.
104 NMM POR/B/2.
105 *Pepys's Naval Minutes*, p192.
106 NA, ADM 106/323 f210.
107 The Building List of *Lenox*, PL1339.
108 William Sutherland, *Shipbuilding Unveiled*, 1717, Introduction, pX. (Hereafter, Sutherland, 1717).
109 Ibid, Part II, p38.
110 K. Cartwright and W. Findlay, *Decay of Timber and its Preservation*, (HMSO, 1958), p225.
111 The Building List of *Lenox*, PL1339 (Appendix 1).
112 NA, ADM 106/322.
113 Wreck of the *Anne*.
114 William Sutherland, *The Shipbuilders Assistant*, 1711, p30. (Hereafter, Sutherland, 1711).
115 Samuel Pepys, *State of the Navy*, 1690, p68.
116 NA, ADM 106/3538, Part II.
117 NA, ADM 106/331 f299.

118 Sutherland, 1711, p29.
119 NRS, *Samuel Pepys and the Second Dutch War*, p105.
120 NA, ADM 106/326 f40.
121 Ibid, f50.
122 NA, ADM 106/323 f12.
123 NA, ADM 106/326 f52.
124 Ibid, f61.
125 Ibid, f72.
126 Ibid, f74.
127 NA, ADM 106/323 f204.
128 NRS, Catalogue of Pepysian MSS, Vol. 1V. p472.
129 NA, ADM 106/323 f210.
130 Ibid, f212.
131 Ibid, f214.
132 Ibid, f216.
133 Ibid, f218.
134 Ibid, f230.
135 Ibid, f244.
136 Ibid, f258.
137 NA, ADM 20/23, No. 129 and 138.
138 NA, ADM 106/323 f246.
139 Ibid, f339 & f162.
140 NA, ADM 106/3538, Part II, 23 June 1677.
141 NA, ADM 106/323 f327.
142 NA, ADM 106/331 f311.
143 NA, ADM 106/3538, 18th May 1677.
144 NA, ADM 106/323 f162.
145 Ibid, f256.
146 NA, ADM 106/329 f29.
147 NA, ADM 106/340 f10.
148 Ibid, f21.
149 NA, ADM 20/24.

CHAPTER III – BUILDING *LENOX*

1 *Deane's Doctrine*, p65.
2 NRS, Catalogue of Pepysian MSS, Vol. IV, p454.
3 Ibid, p469 and p471.
4 Catalogue of Pepysian MSS, Vol. I, p266.
5 PRO ADM 106/36.
6 NA, ADM 7/827 and NA, ADM 106/327.
7 NA, ADM 106/3071.
8 Pepys Library, PL1339.
9 *Deane's Doctrine*, p31.
10 *Mariner's Mirror*, vol 52, p180.
11 Frank Fox, private correspondence, 1998.
12 NA, ADM 106/349 f349.
13 Pepys Library, PL2934.
14 NA ADM 106/349 f349.
15 Bodleian Library, Oxford, Rawlinson A172 f26.
16 Pepys Library, PL1074.
17 NA, ADM 18/63.
18 NA, ADM 18/61.
19 NA, WO50/13.
20 NA, ADM 106, f359.
21 NA, ADM 18/61.
22 NA, ADM18/63.
23 NA, ADM 106/349 f359.
24 NA, ADM 106/349 f357.
25 NA, ADM3/277.
26 Edmund Dummer, *Tables of Proportions of Ships*, British Museum, Department of Printed Books, 534 k5. (Hereafter, Dummer).
27 NA, ADM 106/322 f223.
28 NA, ADM 106/323 f222.
29 NA, ADM 106/332 f165.
30 NA, ADM 106/331 f42.
31 BM, POR/B/2.
32 *Deane's Doctrine*, p37.
33 ibid, p45.
34 Thanks to Father Jose Marie of Houston, TX for this. Also Wilton House draught, *Deane's Doctrine*, p70 and draught of *Resolution*, 1708, NMM, box 65.

35 Sutherland, 1711, p82.
36 NA, ADM 106/325 f108, May 16 1677.
37 *Deane's Doctrine*, p70.
38 NA, ADM 7/827, the scantling list for the 30 ships.
39 Sutherland, 1711, p78 (n.b. source says "level'd under", but this is a printer's error for "bevelled").
40 Ibid, p78.
41 True framed model of a 50-gun ship of c. 1715, NMM, draught of sixth rate c. 1670, NMM box 51, and *Deane's Doctrine,* p70.
42 Model of *Boyne*, 1692, NMM.
43 *Deane's Doctrine*, p67–68, and Wilton House draught.
44 *Deane's Doctrine*, p70.
45 Edward Battine, *Method of Building Ships of War*, 1684, NMM, SPB/28, p1.6 (Hereafter, Battine).
46 Mentioned in *Deane's Doctrine*, p71 and fully drawn on the Wilton House draught.
47 Bodleian Library, Oxford, Rawlinson, A 175 f328.
48 *Deane's Doctrine*, p52.
49 There are many references by Blaise Ollivier as to the necessity of keeping parts light, printed in David Roberts, ed., *18th Century Shipbuilding*, 1992. (Hereafter, Ollivier).
50 NA, ADM 49/137.
51 NA, ADM 106/323 f240.
52 NA, ADM 106/323 f240, John Shish to Navy Board.
53 NA, ADM 2/1748.
54 Sutherland, 1711, p25.
55 NA, ADM 106/323 f236, John Shish to Navy Board.
56 NA, ADM 106/322 f237, Phineas Pett II to Navy Board.
57 NA, ADM106/3538 Part 2, fragment of contract for the thirty ships, 21 Jan 1678, Captain Castle's and Mr Johnson's demands.
58 NA, ADM 106/321 f272 & f309, P Pett requesting timber for third rates, 20th July 1677.
59 NA, ADM 106/3071, The contract for *Yarmouth* also specifies four pieces.
60 NA, ADM 106/3538, 9th Aug 1677.
61 NA, ADM 106/321 f213, Phineas Pett II, 22nd June 1677.
62 NA, ADM 106/321 f303.
63 NA, ADM 106/323 f236, John Shish to Navy Board.
64 NA, ADM 106/323 f246, John Shish to Navy Board.
65 NA, ADM 106/321 f303, Pett mentions they were sawn.
66 NA, ADM 7/827, *Scantlings of the 30 Saile.*
67 NA, ADM 106/321 f34, Francis Bayley to Navy Board with progress report of *Northumberland.*
68 Colin Martin, 'The *Dartmouth* wreck', IJNA, 1978. A new keel was fitted to *Dartmouth* in 1678.
69 NA, ADM 106/3071, *Yarmouth* contract.
70 Colin Martin. *Dartmouth* wreck IJNA 1978
71 R. Pering, An Enquiry into Premature Decay in our Wooden Bulwarks, 1812 p30-34
72 Colin Martin. 'The *Dartmouth* wreck', IJNA, 1978.
73 NA, ADM 106/3071, *Yarmouth* contract.
74 NA, ADM 106/321, Pett mentions using elm.
75 BL ADD MSS 9370 f41-45v, contract of the *Warspite.*
76 Ollivier, p84.
77 NA, ADM 106/329 f23 f25, John Shish building report for *Hampton Court.*
78 Deane's Doctrine, p53.
79 Dummer, stem of *Burford.*
80 NA, ADM 106/323 f240, John Shish to Navy Board.
81 NA, ADM 106/323 f260, John Shish to Navy Board.
82 NA, ADM 106/321 f309, Phineas Pett II, 18 Aug 1677.
83 NA, ADM106/323 f260, John Shish to Navy Board, 13 July 1677.
84 NA, ADM 106/3071, *Yarmouth* contract.
85 'Archaeological survey of 17th century ship's timbers', *Mariner's Mirror, Vol. 84*, No. 2, p179. By 1737, Ollivier states that a tabled scarph was in use.
86 NA, ADM 7/827, *Scantling list of 30 ships.*
87 John Smith, *A Sea Grammer*, 1627. Kermit Goell, ed., 1970.
88 Sutherland, 1711, p76.
89 Ibid, p26-27.
90 Pepys Library, Building list of *Lenox*, PL1339.
91 NA, ADM 106/323 f260, John Shish to Navy Board.
92 Wilton House draught and NA, ADM 106/3071, *Yarmouth* contract.
93 William Sutherland, 1717, Part I, p80.
94 Edmund Bushnell, *The Complete Shipwright*, 1669, p10 from Brian Lavery, ed., *Marine Architecture*, 1993. (Hereafter, Bushnell).
95 NA, ADM 106/332 f167, 29 Aug 1678, letter from Phineas Pett II. The *Anne* built at Chatham had chocks between the transoms and some surviving models show these chocks including a fourth rate, c. 1680 Annapolis; Henry Huddleston, Collection of ship models, No. 33; NA, ADM 106/3070, the contract of *Romney*, 1694, also called for half tramsomes.
96 Terminology described in Sutherland, 1717, Part I, p76. Shish called both the false post within and the false post without false posts.
97 NA, ADM106/3538, Part 2, 11 Jan 1678, *Lenox* building report.
98 Thomas Milton, *A Geometrical Plan of His Majesty's Dockyard at Deptford*, 1753. (Hereafter, Milton, 1753).
99 NA, ADM 106/323 f296, 31 August 1677.
100 NA, ADM 106/323 f99, progress letter from William Fownes, the Clerk of the Cheque at Deptford, to the Navy Board, 13 July 1677.
101 NA, ADM 106/322 f246, Phineas Pett II to Navy Board, 25 Aug 1677.
102 Sutherland, 1711, p77-84.
103 John Franklin, *Navy Board Models*, 1989, p17.
104 Bushnell, p15 and Sutherland, 1711, p83.
105 Anon., *The Shipbuilder's Repository*, 1788, p376.
106 Sutherland 1711, p82.
107 Ibid, p84.
108 NA, ADM 106/329 f25 f29, John Shish to Navy Board.
109 NMM, KLT 5, Keltridge, section of a ship.
110 Sutherland, 1717, Part I, p76, and Thomas Blanckley, *A Naval Expositor*, 1750, p 97. (Hereafter, Blanckley).
111 NA, ADM106/323 f260, *Lenox* building report.
112 Sutherland, 1711, p26 and NA ADM 106/329 f29, John Shish to Navy Board, 17 Aug 1677.
113 NA, ADM 7/827, *Scantling list of 30 ships.*
114 Sutherland, 1711, p26.
115 NA, ADM 106/323 f276, John Shish to Navy Board, 3 Aug 1677.
116 NA, ADM 106/325 f108, Timber Account, Thachery Medbury to Navy Board, May 16 1677. Similar frame letters and numbers feature in the Wilton House draught.
117 NA, ADM 7/827, *Scantling list of 30 ships.*
118 NA, ADM 106/329 f23, John Shish building report.
119 NA ADM 106/3071, contract of *Yarmouth,* 1691.
120 Sutherland, 1717, Part I, p80.
121 Ollivier, p 47.
122 Sutherland 1717, Part I, p119.
123 NA, ADM 106/3070, contract of the third rate *Newark,* 1693.
124 Sutherland, 1717, Part I, p76.
125 Ibid, p120.
126 NA, ADM 106/324 f218, Isaac Betts to Navy Board, 26 July 1677.
127 NA, ADM 106/329 f25, John Shish building report.
128 Sutherland 1711, p26
129 NA, ADM 106/323 f276, John Shish to Navy Board, progress of *Lenox,* 3 Aug 1677.
130 NA, ADM 106/329 f25 NA, ADM 106/323 f276, John Shish building reports.
131 NA, ADM 106/514.
132 NA, ADM 106/323 f276, John Shish to Navy Board, progress of *Lenox,* 3 Aug 1677.
133 NA, ADM 106/329 f5, Phineas Pett II to Navy Board.
134 Sutherland, 1711, p26.
135 NA, ADM106/3538, Part 2, Fragment of contract for the 30 Ships, 21 Jan 1678, Captain Castle's and Mr Johnson's demands.
136 J Nutting, *Dissection of the body of a first rate man-of-war*, c. 1712.
137 Sutherland, 1711, p26 and draught of the *Resolution,* 1708, NMM.
138 NA, ADM 106/329 f25, John Shish building report.
139 NA, ADM 106/323 f276, John Shish to Navy Board, progress of *Lenox,* 3 Aug 1677.
140 NA, ADM 106/329 f27, John Shish building report.
141 NA, ADM 106/323 f242, Thomas Shish to Navy Board, 15 June 1677.
142 Sutherland, 1711, p78.
143 Ibid, p26.
144 The archaeological remains of the *Anne* of 1678.
145 NA, ADM 106/329 f37, John Shish called them crosse spawls, later known as cross spales.
146 Sutherland 1711, p26.
147 Ibid, p39.
148 NA, ADM 106/322 f167.
149 NA, ADM8/3, Building progress of ship at Harwich, 17th June 1692.
150 Sutherland, 1717 Part II, p51.
151 NA, ADM 106/329 f31 John Shish building report.
152 Wilton House draught of *Essex.*
153 NA, ADM 7/827, *Scantling list for the 30 ships.*
154 NA, ADM 106/3538 Part II, Phineas Pett II progress report, 13 Dec 1677.
155 NA, ADM 106/329 f43, Shish building report.
156 NA, ADM106/325 f108v.
157 True framed model of a 50-gun ship of c. 1715, NMM.
158 *Dimensions of Old Ships*, 1755.
159 NA, ADM 106/3538 Part II, 13 Dec 1677; NA, ADM 106/332 f167, 29 Aug 1678, Phineas Pett II to Navy Board; NA, ADM8/3, Progress of *Sussex,* 10 June 1692; and NA, ADM 106/331, Part 1 f20, William Bagwell to Navy Board.
160 Sutherland, 1711, p84.
161 NA, ADM 106/3071, contract for fourth rate, 1673.
162 Thomas Fagge, *Bends of a third rate*, c. 1677-90; Keltridge, draught of midship flat, NMM KLT 5; and NA, ADM 106/3070, contract of *Romney,* 1694.
163 ADM 106/332 f167, Phineas Pett II building report.
164 W.G Perrin, ed., *Autobiography of Phineas Pett I*, (NRS, 1918), pLXXVI.
165 True framed model of a 50-gun ship of c. 1715 in NMM; draught of the *Resolution,* 1708, NMM.
166 Colin Martin, 'The *Dartmouth* wreck', IJNA, 1978, p45. The heads and heels of the timbers are all chocked.
167 Bodleian Library Oxford, Rawlinson C429.
168 Sutherland, 1711, p46.
169 NA, ADM 106/329 f29, John Shish building report.
170 NA, ADM 106/3071, *Yarmouth* contract and the archaeological remains of the *Anne* of 1678.
171 NA, ADM8/3, progress report of *Norfolk* makes it clear that the first plank on was the garboard strake.
172 Sutherland, 1711, p46.
173 NA, ADM 106/329 f31 and NA, ADM 106/323 f296, John Shish building reports.
174 Sutherland, 1711, p7-8.
175 Ibid, p46-49.
176 NA, ADM 106/356 f 222.
177 NA, ADM106/370 f51, Robert Lee and Daniel Furzer to Navy Board, 10 April 1684.
178 NA, ADM 106/326.
179 Building list of *Lenox*, Pepys Library, Magdelene College, Cambridge, PL1339.
180 Mungo Murry, *The Sixty Gun Ship*, 1768.
181 NA, ADM106/370 f65, Robert Lee, report on timber for *Lenox*'s buttocks, 12 July 1684.
182 NA, ADM 106/323 f296.
183 Building list of *Lenox*, Pepys Library, Magdelene College, Cambridge, PL1339.
184 Sutherland, 1717 Part II, p54.
185 The archaeological remains of the *Anne* of 1678.
186 Colin Martin, 'The *Dartmouth* wreck', IJNA, p48.
187 NA, ADM106/3542 Part II, Instructional letter from Navy Board to Daniel Furzer.
188 Building list of *Lenox*, Pepys Library, Magdelene College, Cambridge, PL1339.
189 NA, ADM106/323 f319.
190 NA, ADM106/3538 Part 2, 23 Nov 1677.
191 NA, ADM106/3538 Part 2, 30 Nov 1677.
192 NA, ADM106/3538 Part 2, 4 Dec 1677.
193 NA, ADM 106/323 f296, John Shish to Navy Board, progress of *Lenox,* 31 Aug 1677.
194 NA, ADM 106/323 f306, John Shish to Navy Board, 26 Sept 1677.
195 NA, ADM106/329 f33.
196 NA, ADM 106/3071 *Yarmouth* contract.
197 NA, ADM106/336 f322.
198 Sutherland, 1711, p51-53.
199 NA, ADM 106/33538 Part 2, Phineas Pett II to Navy Board, 21 Feb 1678.
200 NA, ADM106/333 f340, fragment of the contract for *Elizabeth,* mentioned by Jonas Shish in a letter to the Navy Board, 22 July 1678.
201 NA, ADM106/329 f37, *Lenox* building report.
202 NA, ADM106/329 f41, *Lenox* building report.
203 NA, ADM106/329 f37, *Lenox* building report.
204 NA, ADM 106/3070, contract of the *Romney,* 1694.
205 The archaeological remains of the *Anne* of 1678.
206 NA, ADM 7/827, *Scantlings of the 30 Saile.*
207 BL ADD MSS 9370 f41-45v, contract of the *Warspite* and NA, SP29/169 f77-87, contract of the *Edgar.*
208 NA, ADM 106/329 f49, John Shish building report *Lenox.*
209 NA, ADM 106/329 f35 f39, John Shish building report *Lenox.*
210 Thomas Milton, *A Geometrical Plan of His Majesty's Dock Yard at Deptford*, 1753.
211 NA, ADM 7/827, The scantling list for the 30 ships.
212 True framed model of a 50-gun ship of c. 1715, in NMM.
213 NA, ADM 106/327 f217, Furzer's proposed scantling list for the 30 ships.
214 True framed model of a 50-gun ship of c. 1715, in NMM.
215 Ollivier, p360.
216 NA, ADM106/329 f49, *Lenox* building report.
217 NA, ADM 106/323 f321, John Shish to Navy Board, 7 Nov 1677.
218 Blanckley, pp6, 130.
219 Building list of *Lenox*, Pepys Library, Magdelene College, Cambridge, PL1339.
220 Ollivier, p163.
221 BM ADD MSS 9303.
222 NA, ADM 106/3538 Part 2, John Shish to Navy Board, 16 Nov 1677.
223 NA, ADM 106/332 f167, Phineas Pett II used only two whole beams on the 3rd rate *Berwick,* 29 Aug 1678.
224 Sutherland, 1717 Part II, p88.
225 Sutherland, 1711, p42.
226 NA, ADM106/3538 Part II, John Shish building report for *Hampton Court,* 18 Jan 1678 .
227 NA, ADM106/330 f328, ledges were fitted to the orlop of *Restoration*, and NA, SP29/169 f77-87, contract of the *Edgar* of 1668.
228 NA, ADM 106/3538 Part 2; NA, ADM106/329 F51; *Lenox* building report, 9 Nov 1677.
229 Sutherland, 1711, p62.
230 Sutherland, 1717 Part 1, p 128.
231 Sutherland, 1711, p62.
232 Sutherland, 1717 part 1, p 85.
233 NA, ADM 106/3538 Part 2, John Shish to Navy Board, 11 Jan 1678.
234 Sutherland, 1711, p63.
235 Bushnell, p10.
236 William Keltridge, *His Book*, 1675, NMM AND/31, p145.
237 Battine, p14.
238 Sutherland, 1717, Part 1, p85.
239 NA, ADM 106/3538 Part 2, John Shish to Navy Board, 25 Jan 1678.
240 NA, ADM 106/3538 Part 2, John Shish to Navy Board, 1 Feb 1678.
241 Sutherland, 1717 Part 1, p86.
242 Ibid, p85-86.
243 Sutherland, 1711, p64.
244 NA, ADM 106/329 f51 and NA, ADM 106/3538 Part 2, building report *Lenox,* 9 and 16 Nov 1677.
245 NA, ADM 106/3538 Part 2, building report *Lenox,* 23 Nov 1677.
246 NA, ADM 106/3538 Part 2, building report *Lenox,* 21 Dec 1677 and 18–25 Jan 1678.
247 NA, ADM 7/827, Scantling list for 30 ships; NA, ADM 106/327 f217, Furzer's proposed scantling list for the 30 ships and NA, ADM 106/3071 contract of the *Yarmouth.*
248 ADM 106/3071 contract of the *Yarmouth.*
249 NA, ADM 106/3538, Part II.
250 NA ADM 7/694, p192.
251 Wilton House draught of the *Essex* of 1679.
252 NA, ADM 7/827, Scantling list of the 30 ships.
253 NA, ADM 106/3538 Part 2, John Shish to Navy Board, 23 Nov 1677.
254 NA, ADM106/3071, contract of *Yarmouth*; NA, SP29/169 f 77-87, contract of *Edgar*, and BL ADD MSS 9307 f41 -45v contract of *Warspite.*
255 Dummer's draught of an English Man-O'-War, Pepys Library, PL2934.
256 Sutherland, 1717 Part 2, p84.
257 Bodleian Library, Oxford, Rawlinson, C429 and Sutherland, 1717 Part 1, p77.
258 Sutherland, 1717 Part 1, p77.
259 NA, ADM 106/3538 Part 2, John Shish to Navy Board.
260 True framed model of a 50-gun ship of c. 1715.
261 NA, ADM 106/327 f217, Furzer's proposed scantling list for the 30 ships.
262 NA, ADM106/3071, contract of *Cumberland*, 1694.
263 Crutches are shown in Dummer's draught of an English Man-O'-War, Pepys Library, PL2934 and Blanckley, p47.
264 NA, ADM 106/3071, contract of *Cumberland*, 1694.
265 Bushnell, p10.
266 NA, ADM 7/827, The scantling list for the 30 ships.
267 NA, ADM 106/3538 Part 2, John Shish to Navy Board, 16, 23 and 30 November 1677.
268 NA, ADM 106/3538, Phineas Pett II to Navy Board, progress of the third rate *Anne,* 21 Feb 1678.
269 William Keltridge, *His Book*, 1675, NMM, AND/31, p123-4.
270 NA, ADM 106/3071, contract of the *Yarmouth.*
271 NA, ADM 106/3538 Part 2, John Shish to Navy Board, 30 November 1677.
272 NA, ADM106/323 f337, *Lenox* building report.
273 NA, ADM 106/323 f337, NA, ADM 106/3538 Part 2, 14th Dec 1677 and 18th Jan 1678, *Lenox* building reports.
274 NA, ADM 106/327 f217, Furzer's proposed scantling list for the 30 ships.
275 NA, ADM 7/827 Scantling list of the 30 ships.
276 Ibid.
277 NA, ADM 106/3538 Part 2, John Shish to Navy Board, 4 December 1677.
278 NA, ADM 106/3538 Part 2, John Shish to Navy Board, 14 and 28 December 1677.
279 True framed model of a 50-gun ship of c. 1715 in NMM. The arrangement of the timbers around the ports is consistent with the progress reports of John Shish.
280 NA, ADM 106/3538 Part 2, John Shish to Navy Board, 11–18 Jan and 1 Feb 1678.
281 NA, ADM 106/3538 Part 2, 30 Nov and 4 Dec 1677 *Lenox* building report.
282 NA, ADM 106/3538 Part 2, John Shish to Navy Board, 21 and 28 December 1677.
283 ADM 106/3070, contract *Romney,* 1694.
284 NA, ADM106/333 f306, survey of *Yarmouth.*
285 NA, ADM 106/3538 Part 2, John Shish to Navy Board, 18 Jan 1678.
286 NA, ADM 106/3538 Part 2, John Shish to Navy Board, 25 Jan 1678.
287 NA, ADM 106/3538 Part 2, John Shish to Navy Board,. 1 Feb 1678
288 NA, ADM 106/3071, contract of the *Yarmouth.*
289 NA, ADM 106/3538 Part 2, John Shish to Navy Board, 11, 25 Jan and 1 Feb 1678.
290 NA, ADM 7/827, Scantling list of the 30 ships.
291 William Keltridge, *His Book*, 1675, NMM, AND/31, and *Dimensions of Old Ships,* 1755.
292 NA, ADM 106/3538 Part 2, John Shish to Navy Board, 1 Feb 1678.
293 Blanckley, p10.
294 NA, ADM 106/3538 Part 2, John Shish to Navy Board, 14–21 Dec 1677.
295 NA, ADM 106/3538 Part 2, John Shish to Navy Board, 1 Feb 1678.
296 NA, ADM 106/327 f217, Furzer's proposed scantling list for the 30 ships and "half capstan bars" mentioned in *Lenox* building list, PL1339.

297 The Wilton House model of *Essex* of 1679.
298 Blanckley, p140.
299 NA, ADM 49/24 f24.
300 NA, ADM 18/11, 5 April 1652.
301 NA, ADM 7/827, *Scantlings of the 30 Sail.*
302 Sutherland, 1717 Part 1, p63.
303 NA, ADM 106/321 f29.
304 Building list of *Lenox*, Pepys Library, Magdelene College, Cambridge, PL1339. Both the *Vienna* model and Wilton House model have 12 capstan bars for the jeer capstan.
305 Building list of *Lenox*, Pepys Library, Magdelene College, Cambridge, PL1339.
306 NA, ADM 106/3538 Part 2, John Shish to Navy Board, 1 Feb 1678.
307 Bodleian Library, Oxford, Rawlinson A464 and BM, Stowe, 144 f12.
308 Bodleian Library, Oxford, Rawlinson A189 f152.
309 ADM 106/3566, Survey of 1684 and Battine, p18.
310 Building list of *Lenox*, Pepys Library, Magdelene College, Cambridge, PL1339.
311 NA, ADM 106/3566, Survey of *Essex,* 1684.
312 Robert Peacock, archaeological remains of the *Northumberland.*
313 Bodleian Library, Oxford, Rawlinson C429.
314 Sutherland, 1717, Part II, p239 and Robert Peacock, archaeological remains of the *Northumberland,* seem to agree with this dimension.
315 Blanckley, p126.
316 NMM, draught of *Resolution* of 1708 .
317 Phillips print, *Section of a First rate* and NMM draught of *Resolution* of 1708.
318 NA, ADM 106/337 f52.
319 Building list of *Lenox*, Pepys Library, Magdelene College, Cambridge, PL1339,
320 Robert Peacock, archaeological remains of the *Northumberland.*
321 Ibid,
322 Bodleian Library, Oxford, Rawlinson C429,
323 True framed model of a 50-gun ship of c. 1715 in NMM.
324 Henry Mainwaring, *Dictionary*, c. 1617, G.E Mainwaring and W.G Perrin, ed., (NRS, 1921), p237.
325 Blanckley, p97.
326 Pepys Library, Magdelene College, Cambridge, PL1338.
327 Wilton House model of *Essex*; John Franklin, *Navy Board Models*, 1989, p182.
328 Building list of *Lenox*, Pepys Library, Magdelene College, Cambridge, PL1339.
329 The *Vasa*, although dating some fifty years before *Lenox*, was the same size and the dynamics of her steering arrangement are similar to *Lenox*'s. *Vasa*'s steering gear is described by Olaf Pipping, *Mariner's Mirror, Vol. 86*, p19.
330 Building list of *Lenox*, Pepys Library, Magdelene College, Cambridge, PL1339 and the Wilton House model of *Essex* 1679 also has a 12-foot whipstaff. Described by John Franklin, *Navy Board Models*,1989, p182.
331 Blanckley, p186 .
332 Building list of *Lenox*, Pepys Library, Magdelene College, Cambridge, PL1339.
333 Wilton House model of *Essex*, 1679; John Franklin, *Navy Board Models*, 1989, p182.
334 Bodleian Library, Oxford, Rawlinson C429.
335 Building list of *Lenox*, Pepys Library, Magdelene College, Cambridge, PL1339.
336 Blanckley, p9.
337 Colin Martin, *Scotland's Historic Shipwrecks*; wreck of the *Swan,* 1653, p63.
338 NA, ADM 7/827, *Scantlings of the 30 Sail.*
339 NA, ADM 106/3071, contract of the *Yarmouth.*
340 NA, SP29/169 f77-87, contract of the *Edgar*, and BL ADD MSS 9370 f41-45v, contract of the *Warspite.*
341 NA, ADM 106/3071, contract of the *Yarmouth.*
342 Van de Velde the Elder, Museum Boijmans van Beuningen, Rotterdam, No. 364.
343 Scheepvaart, Amsterdam, No. 6.
344 Battine, p12.
345 NA, SP29/169 f77-87, contract of the *Edgar.*
346 Bodleian Library, Oxford, Rawlinson C429.
347 Wreck of the *Stirling Castle*, rebuilt 1700 and wrecked 1703.
348 Blanckley, p37 and p60.
349 NA, ADM49/24, ADM20/23 and building list of *Lenox*, Pepys Library, Magdelene College, Cambridge, PL1339.
350 Building list of *Lenox*, Pepys Library, Magdelene College, Cambridge, PL1339.
351 Colin Martin, 'The *Dartmouth* wreck', IJNA, p35.
352 Bodleian Library, Oxford, Rawlinson C429.
353 NA, ADM106/3566.
354 Building list of *Lenox*, Pepys Library, Magdelene College, Cambridge, PL1339.
355 NA, ADM106/371 f197.
356 Building list of *Lenox*, Pepys Library, Magdelene College, Cambridge, PL1339.
357 NA, ADM106/3541 Part 1.
358 NA, ADM106/337 f52.
359 Museum Boijmans van Beuningen, Rotterdam, No. 362.
360 Building list of *Lenox*, Pepys Library, Magdelene College, Cambridge, PL1339.
361 NA, ADM49/24 f115.
362 NA, ADM106/3546.
363 Thanks to Dr Douglas McElvogue of the *Mary Rose* trust for pointing this out to me.
364 NA, ADM 106/3566, survey of ships, 1684.
365 NMM POR/B/2.
366 Dummer's draught of an English Man-O'-War appears to show a lining under the cabin beams, Pepys Library, Magdelene College, Cambridge, PL2934.
367 Sutherland, 1717 Part I, p82.
368 NA, ADM 106/330 f328.
369 NA, ADM106/330 f328.
370 NA, ADM 106/3566.
371 Catalogue of Pepysian MSS Vol. 1, NRS, p191.
372 NA, ADM 106/330 f328.
373 Bodleian Library, Oxford, Rawlinson C429.
374 NA, ADM49/123.
375 Pepys Diary July 23 to 24 August 1666.
376 Dummer's draught of an English Man-O'-War, Pepys Library, Magdelene College, Cambridge, PL2934 and the Phillips print show cornices, mouldings and wainscot work in high status cabins.
377 NA, ADM49/123, cabin establishment 1686.
378 Bodleian Library, Oxford, Rawlinson C429.
379 NA, ADM106/331 f272.
380 BM ADD MSS 9320 f14.
381 Colin Martin, 'The *Dartmouth* wreck', IJNA, p34.
382 Building list of *Lenox*, Pepys Library, Magdelene College, Cambridge, PL1339.
383 NA, ADM 18/11 p112.
384 Robert Peacock, wreck of the *Stirling Castle*, rebuilt 1700 and wrecked 1703.
385 NA, ADM49/123.
386 Many thanks to Robert Peacock and Seadive Organisation for sharing their information with me.
387 NA ADM106/3069, contract with Henry Ward, 1690..

Chapter IV – Carvings and Finishing

1 Admiralty Library Portsmouth MSS 207.
2 NA, ADM 20/22.
3 NA, ADM106/38.
4 M.S Robinson, *Catalogue of the Elder and the Younger Willem Van de Velde*, 2 Vols., 1990, No. 304, No. 306 and No. 305.
5 Vincenzo Coronelli, Cat. of prints translated by Mario M. Witt, 1970, No 68a and 68.
6 NA, ADM 18/60 f218.
7 Catalogue of Pepysian MSS, Vol IV, NRS, p519-521.
8 NA, ADM 49/23 f139.
9 NA, ADM106/325 f302.
10 NA, ADM106/325 f311.
11 Van de Velde drawing, NMM 1080.
12 *Pepys' Naval Minutes*, NRS, p243.
13 Van de Velde drawing, NMM 1102, inscribed 1675.
14 The Peace of Breda medal 1667 by John Rottier.
15 Manning and Walker, *British Warship Names*, 1959.
16 NA, ADM 106/336 f42.
17 NA, ADM 106/331 part1 f42.
18 NA, ADM 106/349 f359.
19 Catalogue of Pepysian MSS, Vol IV, NRS, p565.
20 NA, ADM 49/24 f57-58.
21 Pepys Library, Magdelene College, Cambridge, PL1339.
22 NA, ADM 106/336 f35 and f42, Isaac Betts at Harwich; NA, ADM 106/355 f129, Phineas Pett at Chatham; NA, ADM 106/332 f165, Daniel Furzer at Portsmouth; all sent letters requesting directions for carved work.
23 Battine, 1684, NMM, SPB/28, p14.
24 NA, ADM 106/332 f167.
25 Battine, 1684, NMM, SPB/28, p13.
26 NA, ADM 106/354 f146.
27 NA, ADM 49/24 f57.
28 NA, ADM 106/332 f167.
29 NA, ADM 18/60 f278.
30 Sutherland, 1717, Part II, p109.
31 NMM AND/31 p266 and NA, ADM 20/57 p69.
32 NA, ADM 106/3060 and NA, ADM 20/57 p69.
33 NA, ADM 20/57 p69.
34 NMM AND/31 p266 and contract of *Cumberland* NA, ADM 106/3071.
35 Sutherland, 1717, Part II, p196.
36 NA, ADM 106/378 f92.
37 NA, ADM 106/3538 Part 1, Contract with Mary Harrison to paint ships at Portsmouth, 15 March 1683.
38 Pepys Library, Magdelene College, Cambridge, PL1339.
39 Sutherland, 1717, Part II, p191.
40 The contracts for the *Yarmouth* and *Cumberland* NA, ADM 106/3071 state that the lion of the head and the arms in the stern must be gilt while the contract for the *Warspite,* BL ADD MSS 9307 f41-45v and that of *Edgar*, NA, SP29/169 f77-87, state that the gilding must be the equal of earlier ships.
41 NA, ADM 95/14 p35, Reference to contract of *Dartmouth.*
42 NA, ADM 106/382 f302.
43 Pepys Library PL1339 Building list of *Lenox.*
44 Colin Martin, 'The *Dartmouth* wreck', IJNA, 1978.
45 Bodleian Library, Oxford, Rawlinson MSS C429.
46 Thomas Blanckley, *Naval Expositor*, c. 1730, p170.
47 Ibid, p146.
48 Pepys Library, Magdelene College, Cambridge, PL1339.
49 *Mariner's Mirror*, Vol. 91, No. 1, p80.
50 Sutherland, 1717, Part II, p186.
51 Catalogue of Pepysian MSS, Vol. IV, NRS, p185, 575, 584.
52 Brian Lavery, *Arming and Fitting of English Ships of War*, (Conway, 1987), p57.
53 NA, ADM 106/378 f424.
54 NA, ADM 106/378 f424, NA and ADM 106/3541 part 2.
55 NA, ADM 106/335 f45.
56 NA, ADM 106/335 f43-45.

Chapter V – Ship's Boats

1 Building list of *Lenox*, Pepys Library, Magdelene College, Cambridge, PL1339.
2 NA ADM 106/323 f79.
3 NA ADM 106/339 f265.
4 NA ADM 106/384 f195.
5 Building list of *Lenox*, Pepys Library, Magdelene College, Cambridge, PL1339.
6 NA ADM 95/13.
7 NA ADM 106/384 f195 and NA ADM 95/13.
8 NA ADM 106/389 f375.
9 It could be reasoned that two small boats were added to the complement. However, the numerous contracts for boats of the period mentions longboats, pinnaces and yawls but there is never a mention of jolly boats or anything else, NA ADM 106/3069.
10 NA ADM 106/384 f195.
11 Contracts of boats, NA ADM 106/3069.
12 Ibid.
13 Building list of *Lenox*, Pepys Library, Magdelene College, Cambridge, PL1339.
14 Ibid.

Chapter VI – Anchors

1 Building list of *Lenox*, Pepys Library, Magdelene College, Cambridge, PL1339.
2 Wilton House model, probably the *Essex* has this arrangement. The nomenclature is given in Thomas Blanckley, *Naval Expositor*, c. 1730, p2.
3 Mainwaring, *The Seaman's Dictionary*, NRS , p225.
4 Stephen Martin-Leake, *The Life of Sir John Leake*, 1750, p69.
5 Captain John Smith, (Kermit Goell, ed., 1970), *A Sea Grammar,* Chap. 7.
6 Boteler's *Dialogues*, NRS, p189.
7 NA ADM 49/24.
8 Wilton House model, probably the *Essex.*
9 NA, ADM 106/3250.
10 Ibid.
11 Building list of *Lenox*, Pepys Library, Magdalene College, Cambridge, PL1339.
12 Building list of *Lenox*, Pepys Library, Magdalene College, Cambridge, PL1339.
13 Calculated from Edward Battine, *Method of Building Ships of War*, 1684, p80; NMM SPB/28 and Building list of *Lenox*, Pepys Library, Magdalene College, Cambridge, PL1339, p38.
14 Edward Battine, *Method of Building Ships of War*, 1684, p80; NMM SPB/28 and William Keltridge, *His Book*, 1675, p218, NMM AND/31.
15 Thomas Blanckley, *Naval Expositor*, c. 1730, p26.

Chapter VII – Launching

1 NA, ADM 106/3538 Part 1, 5 March 1678.
2 NA, ADM 106/332 f180. The *Anne*, built at the head of the double dock at Chatham also required ways for launching,
3 Pepys Library, Magdalene College, Cambridge, PL1339 f49.
4 NA, ADM 106/331 f295.
5 Pepys Library, Magdalene College, Cambridge, PL1339 f14.
6 Launching of the *Royal Prince* in 1610, Rev S Denne, *The Life of Phineas Pett*, p259.
7 NA, ADM 1/3548 f 753.
8 Catalogue of Pepysian MSS Vol IV, NRS, p577.
9 Drawing by Van de Velde, NMM, No. 1109, Launching of possibly *Lenox* or *Defiance.*
10 Launching of the *Royal Prince* in 1610, Rev. S Denne, *The Life of Phineas Pett*, p261.
11 NA, ADM 1/3552 f739, Account of launching of the first rate *Britannia* at Chatham 1682. It is not certain that *Lenox* was held at the stern of the dock but it is probable.
12 NA, ADM 51/4170, log book of the *Dreadnought* and NA, ADM 51/4215, log book of *Harwich* at Plymouth.
13 *Diary of John Evelyn*, 24 Jan 1682.
14 Rosemary Baird, *Mistress of the House*, 2003, p65-73.
15 Bryan Bevan, *Charles the Second's French Mistress,* 1972, p63.
16 Rosemary Baird, *Mistress of the House, 2003,* p65-73.
17 NA, ADM20/23 f496.
18 NA, ADM106/3538 Part 1.
19 James Vernon, *Letters Illustrating the Reign of William III.*
20 Drawing by Van de Velde, NMM 1109. Launching of *Lenox* or possibly *Defiance.*
21 NA, ADM 20/23 p743.
22 Private correspondence, Dr David Davies and Mrs Rosemary Baird.
23 Launching of the *Royal Prince* in 1610, Rev. S Denne, *The Life of Phineas Pett*, p262-3.
24 NA, ADM 106/3523.
25 *Further Correspondence of Samuel Pepys*, NRS,1929, Pepys to Deane, 14 December 1667.
26 Flagon given to Jonas Shish for building the *London* 1670, Museum of London.
27 NA, ADM 20/41 p111 and ADM 18/65 p289.
28 NA, ADM 106/3523, The Duke of York's arms is probably a mistake as the Duke was no longer Lord High Admiral. The arms were probably those of the King.
29 *Further Correspondence of Samuel Pepys*, NRS, p186 and *Pepys Diary,* 26 October 1664.
30 BM ADD MS9322 f56v.
31 *Pepys' Naval Minutes*, NRS, p394.
32 London Gazette, April 13 1678.
33 NA, ADM 106/38.
34 NA, ADM 106/339 f165.
35 NA, ADM 1/3551 f471.
36 BM ADD MS9322 f56v.
37 NA, ADM 2/1740.
38 *Diary of John Evelyn*, 13 May 1680.
39 NA, PROB 11/385.
40 Journal of Edward Gregory.
41 NA, PROB 11/385.
42 The Shish Memorial, St Nicholas Church, Deptford.

Chapter VIII – Shipwright Productivity

1 Pepys Library, Magdalene College, Cambridge, PL2265
2 NA, ADM 42/485 and 486.
3 Pepys Library, Magdalene College, Cambridge, PL1339
4 NA, ADM 42/485 and 486.
5 NA, ADM 106/323 f79.
6 NA, ADM 106/323 f337.

Chapter IX – Sails and Rigging

1 NA, ADM 106/3548 p845.
2 NA, ADM 106/3538 Part 1.
3 Catalogue of Pepysian MSS, NRS, Vol IV, p415.
4 Battine, 1684, p37; Sutherland, 1711, p108 and Henry Bond, *The Art of Apparelling and Fitting of any Ship*, 1655.
5 NA, ADM 7/827.
6 NA, ADM 49/24 f79.
7 NA, ADM 49/24 f157.
8 Samuel Manning, *New England Masts*, NMM monograph, No. 42, 1979, p9.
9 Sutherland, 1711, p109.
10 Building list of the *Duchess*, Cumbria Records Office D/Lons/L.
11 Pepys Library, Magdalene College, Cambridge, PL1339.
12 NA, ADM 49/24 p40.
13 NA, ADM 106/340 f24.
14 NA, ADM 49/24 p40.
15 NA, ADM 106/358 f525.
16 Battine, 1684, p45.
17 Pepys Library, Magdalene College, Cambridge, PL1339.
18 Building list of the *Duchess*. Cumbria Records Office D/Lons/L.
19 NA, ADM 49/24, p40.
20 NA, ADM 106/3071 contract for *Cumberland* of 1694.
21 NA, ADM 18/60 p253.
22 Edward Hayward, *The Sizes and Lengths of Rigging*, 1655.
23 NA, ADM 42/1978.
24 NA, ADM 106/3538 Part 1.
25 Pepys Library, Magdalene College, Cambridge, PL1339.
26 NA, ADM 49/24.
27 Cumbria Records Office C/LONS/L Admiralty.
28 Sailmakers Contract, NA, ADM 49/25.
29 Pepys Library, Magdalene College, Cambridge, PL1339.
30 Cumbria Records Office C/LONS/L Admiralty.
31 NA, ADM49/25 Sailmakers contract.
32 Sir Henry Mainwaring, *The Seaman's Dictionary,* NRS , p104.
33 Bodleian Library, Oxford, Rawlinson A189 f102.
34 NA, ADM106/381 f186 for Portsmouth and NA, ADM 106/380 f19 for Chatham.
35 Michael Robinson, *The Paintings of the Willem Van de Veldes*, No. 467.

Chapter X – Laid Up in Ordinary

1 NA, ADM 106/3539 Part 2.
2 NA, ADM 42/1-4.
3 NA, ADM 106/3540 Part 2.

4 NA, ADM 106/3548 p534.
5 NA, ADM 1/3548 p633.
6 NA, ADM 106/3538 Part II.
7 NA, ADM 42/486.
8 NA, ADM 106/323 f82.
9 NA, ADM 106/39.
10 NA, ADM 1/3548 p913.
11 NA, ADM 106/3538 Part 1.
12 NA, ADM 106/3548 p845.
13 NA, ADM 106/39, 30 April 1678.
14 NA, ADM 106/336 f289.
15 NA, ADM1/3548 P893.
16 Building list of *Lenox*, Pepys Library, Magdelene College, Cambridge, PL1339.
17 NA, ADM 33/106.
18 NA, ADM 1/3548, p913.
19 NA, ADM 33/106.
20 F. B. Cockett, *Early Sea Painters*, 1995, p51.
21 NA, ADM 8/1.
22 NA, ADM 106/342 f421; ADM 106/350 f198; ADM 1/3553.
23 NA, ADM 106/331 f272.
24 NA, ADM 33/106.
25 NA, ADM 2/1748.
26 London Guildhall Library 30004/5, Trinity House minute book 1677-85 p88.
27 NA, ADM 106/331 f283.
28 NA, ADM 106/331 f291.
29 NA, ADM 106/330 f267.
30 NA, ADM 106/331 f295.
31 NA, ADM 106/331 f305.
32 NA, ADM 106/3119.
33 NA, ADM 106/3540 Part 2.
34 NA, ADM 106/331 f307.
35 NA, ADM 106/3537 Part 2.
36 NA, ADM 106/331 f307.
37 *Diary of Samuel Pepys*, 12 July 1663.
38 NA, ADM 18/81.
39 NA, ADM 106/3537 Part 2.
40 NA, ADM 1/3549 p371.
41 NA, ADM 1/3550 p437.
42 NA, WO 50/13 p25.
43 NA, ADM 106/330 f433.
44 NA, ADM 106/330 f461.
45 NA, ADM 42/1.
46 NA, ADM 106/3540 Part 2.
47 NA, ADM 106/341 f 241.
48 NA, ADM 106/341 f 248.
49 NA, ADM 106/341 f 303.
50 NA, ADM 1/3549 p940.
51 NA, ADM 1/3550 p143.
52 NA, ADM 106/383 f114.
53 Pepysian MSS Adm letters VIII 296.
54 Catalogue of Pepysian MSS, Vol IV, NRS, p620.
55 BM, ADD MS9322 f48v.
56 NA, ADM 106/371 f310.
57 *Pepys' Naval Minutes*, p219.
58 NA, ADM 106/341 f350.
59 NA, ADM 106/342 f487.
60 Boteler's *Dialogues*, NRS, p227.
61 NA, ADM 106/341 f354.
62 NA, ADM 106/342 f421.
63 NA, ADM 3/277, 18th June 1679.
64 NA, ADM 1/3550 p23.
65 BM ADD MS9322 f6v f50.
66 NA, ADM 106/330 f496.
67 NA, ADM 3/277, 12th July 1679.
68 Private correspondence, Dr Peter Lefevre, April 2003.
69 NA, ADM 106/350 f198.
70 NA, ADM 106/350 f196.
71 NA, ADM 1/3551 p605.
72 NA, ADM 1/3551 p757.
73 NA, ADM 106/3538 Part 1.
74 NA, ADM 106/371 f265.
75 Bodleian Library, Oxford, Rawlinson A171 f145.
76 NA, ADM 106/3539.
77 NA, ADM 106/349 f153.
78 NA, ADM 106/356 f 222.
79 NA, ADM 106/356 f271.
80 NA, ADM 106/356 f293.
81 *Diary of John Evelyn* 29 April, 25 May and 12 June 1681.
82 NA, ADM 106/356 f295.
83 NA, ADM1/3552 p175.
84 NA, ADM 1/3552 p603.
85 NA, ADM 1/3556 p783.
86 NA, ADM 106/346.
87 NA, ADM 106/350 f207.
88 NA, ADM 1/3551 p934.
89 NA, ADM 1/3552 p29.
90 NA, ADM 106/3541 Part 1.
91 NA, ADM 1/3552 p25.
92 NA, ADM 106/3539 Part 2.
93 NA, ADM3/278 p67.
94 NA, ADM 106/51.
95 NA, ADM 42/2.
96 NA, ADM 106/361.
97 NA, ADM 106/359 f544.
98 NA, ADM 106/3541 Part 2.
99 NA, ADM 106/359 f544.
100 NA, ADM 106/361 f173.
101 NA, ADM 106/361 f175.
102 NA, ADM 106/361 f178.
103 *Pepys' Naval Minutes*, p160.
104 NA, ADM 106/361 f189.
105 NA, ADM 106/361 f215.
106 NA, ADM 106/361 f219.
107 NA, ADM 1/3553 p361.
108 NA, ADM 106/364 f 544.
109 NA, ADM 106/3538 Part 1.
110 *Diary of John Evelyn*, 23 Dec 1683.
111 *Diary of John Evelyn*, 9 and 24 Jan; 5,8 Feb; 28 March; 4 April 1684.
112 NA, ADM 106/371 f184.
113 NA, ADM 106/3541 Part 1.
114 NA, ADM 106/57.
115 NA, ADM 106/3566 p16.
116 BM ADD MSS 9307 and Bodleian Library, Oxford, Rawlinson, A295 f76.
117 NA, ADM 106/58, 6 June 1684.
118 NA, ADM 106/370 f65.
119 NA, ADM 106/375 f25.
120 Samuel Pepys, *State of the Navy*, 1690, p25.
121 NA, ADM 106/371 f310.
122 NA, ABM 106/364 f552.
123 NA, ADM 106/370 f65.
124 NA, ADM 106/371 f355.
125 NA, ADM 106/364 f349.
126 *Diary of John Evelyn* 2 and 13 July; 10 and 24 Aug 1684.
127 NA, ADM 106/370 f75.
128 NA, ADM 106/375 f25.
129 NA, ADM 95/14 p104.
130 Bodelian Library, Oxford, Rawlinson, A464 f11 and f124.
131 M.S Robinson, *Catalogue Van de Velde Drawings*, NMM, 1958, p101 and p145.
132 NA, ADM 1/3554 p312.
133 NA, ADM 1/3554 p247.
134 NA, ADM 1/3554 p311.
135 BM ADD MS 9322 f103.
136 Pepys Library, Magdalene College, Cambridge, PL1534.
137 NA, ADM 106/371 f28.
138 Private correspondence, David Woodbridge, Director of the Institute of Wood Science.
139 NA, ADM 106/370 f51.
140 Arthur Bulger, *HMS Victory, Building, Restoration and Repair* (HMSO, 1966) and private correspondence, David Woodbridge, Director of the Institute of Wood Science.
141 NA, ADM 106/370 f51.
142 NA, ADM 106/323 f296.
143 NA, ADM 106/326 f81.
144 NA, ADM 106/326 f88.
145 NA, ADM 106/3541 Part 2.
146 NA, PC 2/17 p320 and Samuel Pepys, *State of the Navy*, 1690, p59.
147 BM ADD MS 9303, 17 April 1686.
148 Samuel Pepys, *State of the Navy*, 1690, p87.
149 NA, ADM 106/359 f544.
150 *Prevention of Decay in Wood Boats, Forest Products Research Bulletin*, No. 31, 1954, p1.
151 K.Cartwright and W.Findlay, *Decay of Timber and its Preservation*, (HMSO, 1958), p9.
152 Ibid, p41.
153 Ibid, p10.
154 Ibid, p26.
155 Ibid, p253.
156 Ambrose Bowden, *A Treatise of Dry Rot*, 1815, p62.
157 K.Cartwright and W.Findlay, *Decay of Timber and its Preservation*, (HMSO, 1958), p219.
158 Ibid, p143.
159 Samuel Pepys, *State of the Navy*, 1690, p72.
160 K.Cartwright and W.Findlay, *Decay of Timber and its Preservation*, (HMSO, 1958), p166.
161 Ibid, p231.
162 NA, ADM 106/387 f187.
163 Roger Fisher, *Heart of Oak*, 1771, p114.
164 NA, ADM 106/381 f267.
165 Journal of Edward Gregory, p15.
166 Private correspondence, David Woodbridge, Director of the Institute of Wood Science.
167 Roger Fisher, *Heart of Oak*, 1771, p113-114.
168 NA, ADM 106/371 f312.
169 NA, ADM 1/3553 p643. The position of the mooring and direction of the cables is given.
170 NA, ADM 106/371 f312-318.
171 NA, ADM 106/342 f439.
172 NA, ADM 106/371 f313.
173 Thomas Blanckley, *Naval Expositor*, c. 1730, p107 shows this arrangement although the ground tackle employs more chain than that used in the 1680s.
174 NA, ADM106/342 f459.
175 NA, ADM 106/342 f467.
176 NA, ADM 3/278 p32.
177 NA, ADM 1/3551 p933.
178 NA, ADM 106/371 f311; ADM 1/3553 p645.
179 NA, ADM 1/3553 p641.
180 NA, ADM 49/132, 22nd Oct 1683.
181 NA, ADM 1/3553 p657.
182 NA, ADM 1/3553 p681.
183 Tangier Papers of Samuel Pepys, NRS, 1935, p323.
184 NA, ADM 106/371 f310.
185 Tangier Papers of Samuel Pepys, NRS, 1935, p323.
186 Bodleian Library, Oxford, Rawlinson A179 f87.
187 Bodleian Library, Oxford, Rawlinson A186 f259.
188 NA, ADM 91/1 p18.
189 NA, ADM 106/380 f52.
190 Bodleian Library, Oxford, Rawlinson, A177 ff1-76
191 NA, ADM 106/389 f506.
192 *Diary of John Evelyn*, 1 January 1685.
193 NA, ADM 106/375 f21.
194 NA, ADM 1/3664 p669.
195 Bodleian Library, Oxford, Rawlinson A464 f59 and NA, ADM 106/378 f11.
196 Bodleian Library, Oxford, Rawlinson A464 f61 and NA, ADM 106/378 f424.
197 NA, ADM 106/378 f 92.
198 Bodleian Library, Oxford, Rawlinson A464 f49.
199 NA, ADM 106/378 f428.
200 Bodleian Library, Oxford, Rawlinson A464 f29.
201 Samuel Pepys, *State of the Navy*, 1690, p33.
202 Ibid, p48-55.
203 BM, ADD MSS 9307 f203.
204 C.S Knighton, *Pepys's Later Diaries*, p198.
205 NA, ADM 106/389 f506.
206 Admiralty Library Portsmouth, MSS207, 12 April 1676.
207 NA, ADM 106/132 f82.
208 NA, ADM 49/132.
209 BM, Harley, MS 7476, f66v.
210 BM, ADD MS 9322 f131.
211 *Diary of John Evelyn*, 2–20 June; 13 July 1686.
212 NA, ADM 106/3541 Part 2, 12th July 1686.
213 NA, ADM 18/65 p430.
214 NA, ADM 18/65 p207.
215 NA, ADM 106/63, 2nd Feb 1686.
216 NA, ADM 106/69B.
217 NA, ADM 42/4.
218 NA, ADM 1/3556 p47 and BM ADD MS 9307 10 Nov 1686.
219 Journal of Edward Gregory,.
220 NA, ADM 106/380 f128.
221 NA, ADM 106/389 f375.
222 NA, ADM 106/380 Part 2 f19.
223 NA, ADM 1/3555 p46.
224 BM, ADD MS9303 f20.
225 NA, ADM 106/350 f207.
226 NA, ADM 1/3553 p661.
227 NA, ADM 106/375 f127.
228 NA, ADM 106/3540 Part 2.
229 NA, ADM 106/3539.
230 NA, ADM 8/1.
231 NA, ADM 106/380 Part 2 f31.
232 NA, ADM 106/380 Part 2 f33.
233 NA, ADM 106/382 f92.
234 NA, ADM 95/14 p237..
235 NA, ADM 18/65 p391
236 NA, ADM 18/65 p442.
237 Bodleian Library, Oxford, Rawlinson C429.
238 NA, ADM 106/399 f337.
239 NA, ADM 95/14 p35; NMM, SPB/28 and Battine, 1684, p131.
240 NA, ADM 106/399 f337.
241 *Pepys' Naval Minutes*, p52.
242 Bodleian Library, Oxford, Rawlinson C429.
243 Bodleian Library, Oxford, Rawlinson A215.
244 Bodleian Library, Oxford, Rawlinson C429.
245 John Charnock, *Biographica Navalis*, Vol. II, p57.
246 NA, ADM 106/384 f154.
247 BM Harley MS 7476 f67v.
248 NA, ADM 106/384 f168 & f171.
249 NA, ADM 42/4.
250 NA, ADM 106/69C p36.
251 NA, ADM 106/384 f177 & f179.
252 Samuel Pepys, *State of the Navy*, 1690, p185 and NA T38/657.
253 NA, T38/657.
254 Ibid, p185.
255 NA, ADM 106/387 f 307 and f309.

Chapter XI – Ordnance

1 A. Caruana, *The History of Sea Ordnance*, Vol. I, 1994, p72 and p91.
2 *Journals and Narratives of the Third Dutch War*, NRS, Vol. 86, p188.
3 Catalogue of Pepysian Manuscripts, Vol. 1, NRS, 1903, p233.
4 Catalogue of Pepysian Manuscripts, Vol. IV, NRS , 1923, p361.
5 Ibid, p407.
6 NA, WO 49/111.
7 Catalogue of Pepysian Manuscripts, Vol. IV, NRS, 1923, p417.
8 Bodleian Library, Oxford, Rawlinson, A185 f161.
9 *Dimensions of Old Ships*, p16.
10 Catalogue of Pepysian Manuscripts, Vol. IV, NRS, 1923, p425.
11 Ibid, p475.
12 NA, ADM 49/123.
13 Catalogue of Pepysian Manuscripts, Vol. IV, NRS, 1923, p518.
14 NA, ADM49/123.
15 BM ADD MSS 9316 f226 and Bodleian Library, Oxford, Rawlinson D919.
16 BM ADD MSS 9316 f226.
17 NA, WO 49/111.
18 A. Caruana, *The History of Sea Ordnance*, Vol. I, 1994, p70.
19 NA, WO 49/111.
20 Sarah Barter Bailey, *Prince Rupert's Patent Guns*, 2000, p26.
21 Ibid, p52.
22 Frank Fox, *Great Ships*, 1980, p195.
23 NA, ADM 49/132 3rd April 1694.
24 NA, ADM 106/330 f333.
25 NA, ADM 52/58, Part 2, *Lenox* log books.
26 Catalogue of Pepysian Manuscripts, Vol. IV, NRS, 1923, p549.
27 Frank Fox, *Great Ships*, 1980, p164.
28 NA, WO 50/13 p29.
29 NA, WO 50/13 p25.
30 NA, WO51/21 f56.
31 NA, WO55/39.
32 Bodleian Library, Oxford, Rawlinson A189 f409 and Pepys Library PL2879 pp111-122 reproduced in full in Frank Fox, *Great Ships*, p191.
33 NA, WO55/1762.
34 NA, WO55/1763.
35 NA, WO55/1763 and Staffordshire Record Office, D (W)1778/V/44.
36 Rev Richard Allyn, *Narrative of Victory at La Hogue*, 1692, p10.
37 NA, WO55/1763, 26 July 1689.
38 NA, WO51/41 f116.
39 NA, WO51/40 f99 and WO51/43 f137 and f82.
40 NA, ADM 51/ 3881, Part 1, *Lenox* log books.
41 NA, ADM2/170.
42 A. Caruana, *The History of Sea Ordnance*, Vol. 1, 1994, p37.
43 Ibid, p66 and p77.
44 Bodleian Library, Oxford, Rawlinson A 178 f59.
45 *Pepys Naval Minutes*, NRS, 1925, p225.
46 For full history see Sarah Barter Bailey, *Prince Rupert's Patent Guns*, Royal Armouries.
47 NA, ADM 51/ 3881, Part 1, *Lenox* log books.
48 Ibid.
49 NA, WO55/1763.
50 BL ADD MSS 9289.
51 NA, ADM 51/ 4238, Part 5, *Lenox* log books.
52 Priddy's Hard PHA/3.
53 NA, WO55/1736.
54 David Hepper, *British Warship Losses*, 1994, p19.
55 A. Caruana, *The History of Sea Ordnance* Vol. I, 1994, p148.
56 John Seller, *The Sea Gunner*, 1691, p144.
57 Private correspondence with Mr Charles Trollope.
58 NA, WO55/1763.
59 Frank Fox, *Great Ships*, 1980, p184-185.
60 NA, WO55/1763, 26 July 1689.
61 Charles Trollope, 'Design and Evolution of English cast iron guns', *Journal of the Ordnance Society*, Vol. 17, 2005.
62 NA, ADM 51/3881, Part 1, *Lenox* log books.
63 Private correspondence with Mr Charles Trollope.
64 NA, ADM 51/3881, Part 6.
65 Private correspondence with Mr Charles Trollope.
66 NA, PC2/96 p168.
67 NA, PC2/96 p169.
68 A. Caruana, *The History of Sea Ordnance*, Vol. II, p11.
69 Sarah Barter Bailey, *Prince Rupert's Patent Guns*, 2000, p124.
70 Ibid, p128.
71 Museum Boijmans van Beuningen, MB1866/T362.
72 Museum Boijmans van Beuningen, MB1866/T359.
73 Musee des Beaux-Arts, Besançon, No. 27.
74 Paper by Peter Pershall 1998.
75 Pepys Library, Magdalene, College, Cambridge, PL2943.
76 NA, WO51/46 and 47.
77 NA, WO51/44 f167.
78 NA, WO51/44 f163.
79 NA, WO51/50 f93.
80 NA, WO 51/52 f110.
81 NA, WO51/52.
82 Private correspondence with Mr Charles Trollope
83 NA, WO55/1717.
84 Battine, 1684, p104.
85 NA, WO 53/39.
86 Thanks to Dr Doug McElvogue for identifying the woods.
87 Battine, 1684, p104.
88 NA, WO55/13.
89 Examined by the author in autumn 2003.
90 Thanks to Dr Doug McElvogue for this information.
91 Battine, 1684, p104.
92 NA, ADM7/827.
93 Cumbria Records Office D/Lons/L.
94 Sarah Barter Bailey, *Prince Rupert's Patent Guns*, 2000, p70.
95 Pepys Library, Magdalene, College, Cambridge, PL1339.
96 Ibid.

97 D/Lons/L; Cumbria Archives Carlisle, reproduced in A. Caruana, *The History of Sea Ordnance*, Vol. I, p202.
98 NA, PRO ADM 7/827.
99 Priddy's Hard PHA/3.
100 A. Caruana, *The History of Sea Ordnance* Vol. I, 1994, p165.
101 Priddy's Hard PHA/3.
102 Bodleian Library, Oxford, Rawlinson C429.
103 Sir Henry Mainwaring, NRS, 1921, p181 and the Wilton House draught.
104 Sutherland, 1711, p36.
105 NA, ADM 106/406, 12th Jan 1691.
106 NA, WO53/39.
107 Boteler's *Dialogues*, NRS, 1929, p202.
108 Article submitted to the *R.A Journal* by A Caruana, 1998.
109 John Ehrman, *The Navy in the War of William III*, 1953, p471.
110 NA, WO51/43.
111 BM ADD MSS 9289.
112 Archaelogical remains from the *London* blown up in 1665.
113 John Smith, *A Sea Grammer*, 1627, (Kermit Goell, ed., 1970), p87.
114 Battine, 1684, p99; NMM SPB/28.

CHAPTER XII – SERVICE HISTORY

1 Dr Peter le Fevre, *Mariner's Mirror*, Vol. 78, No. 1.
2 NA, ADM 8/2-7.
3 NA, ADM 33/106-185.
4 NA, ADM51/4238, Part 5.
5 NA, ADM51/3881, Part 7.
6 NA, ADM 106/389 f375.
7 NA, ADM 106/389 f449.
8 NA, ADM 106/5812.
9 BL ADD MSS 9324 f29.
10 PRO ADM 8/2.
11 NMM CHA/l/1.
12 NA, ADM 1/3560.
13 NA, ADM 2/5.
14 John Charnock, *Biographica Navalis*, Vol 2, 1795, p155.
15 NA, ADM 2/5.
16 NA, ADM 106/397 f136.
17 NA, ADM 2/1748, p198-201, *Instructions for Pressing Seamen*.
18 NA, ADM 2/5.
19 NA, ADM 106/397 f138.
20 W.G Day, ed., *The Pepys Ballads, Vol. V*, Catalogue of the Pepys Library at Magdalene College Cambridge, p363.
21 Captain John Smith, *The Seaman's Grammar*, 1627-1690, (Kermit Hoell, ed.) p48-9.
22 NA, ADM 1/3560 p805.
23 NA, ADM 2/5.
24 NA, ADM 106/397.
25 Catalogue Pepysian Manuscripts, NRS, Vol. 1, p240.
26 NA, ADM 106/397 f227.
27 NA, ADM 2/5.
28 NA, ADM 106/384, f154.
29 NA, ADM 2/379.
30 NMM CHA/L/2.
31 NA, ADM 106/397 f140.
32 NA, ADM 51/3810 Journal 2.
33 NA, ADM 106/397 f143.
34 NA, ADM 106/397 f142.
35 NA, ADM 106/397 f145.
36 David Hepper, *British Warship Losses*, p13.
37 Cumbria Records Office D\Lons\L Admiralty.
38 NA, ADM82/13 p9.
39 NA, ADM 106/397 f265-7.
40 NA, ADM 2/6.
41 NA, ADM 106/398 f67-9 Part 1.
42 NA, ADM 106/396 f178.
43 NA, ADM 106/398 f70 Part 1.
44 NA, ADM 106/397 f434.
45 NA, ADM 2/6.

46 NA, PRO 2/7.
47 NA, ADM 1/5253.
48 Peter Le Fevre, 'The Earl of Torrington's Court-Martial', *Mariner's Mirror*, Vol. 76, No. 3, 1990.
49 B.M Stowe, 143.
50 Torrington's speech to the House of Commons, 1710.
51 Stephen Martin Leake, *The Life of Sir John Leake*, 1750, p22.
52 *Pepys' Naval Minutes*, p293.
53 John Charnock, *Biographica Navalis*, Vol 2, 1795, p158.
54 NA, ADM 2/17.
55 NA, ADM 106/406.
56 Ibid.
57 John Charnock, *Biographica Navalis*, Vol 2, 1795, p179.
58 NA, ADM 3/10.
59 NA, ADM 106/408.
60 NA, ADM 106/408.
61 NA, ADM 2/7.
62 Ibid .
63 NA, ADM 8/2.
64 NA, ADM 106/408.
65 Rev. Richard Allyn, *Narrative of Victory at La Hogue*, 1692.
66 NA, ADM 2/8.
67 Ibid.
68 Ibid.
69 NA, ADM 106/3542 Part 1.
70 NA, ADM 106/3543.
71 NA, ADM 2/8.
72 NA, ADM 106/421.
73 Boteler's *Dialogues*, NRS, p208.
74 NA, ADM 106/421.
75 Ibid .
76 Ibid .
77 Ibid .
78 R.C. Anderson, *Lists of men-of-war*, SNR, Occasional Publication, No. 5.
79 Private correspondence, A. Caruanna.
80 NA, ADM 106/421.
81 NA, ADM 2/9.
82 NA, ADM 7/692.
83 NA, ADM 2/9.
84 NA, ADM 33/137.
85 NA, ADM 7/692.
86 NA, ADM 106/421.
87 NA, ADM 2/9.
88 NA, ADM 7/692.
89 Ibid.
90 Philip Aubrey, *Defeat of James Stewart's Armada*, 1989, p95-111.
91 NA, ADM82/16 f56.
92 NA, ADM 51/4384.
93 NA, ADM 7/692.
94 Ibid.
95 NA, ADM82/16 f56.
96 NA, ADM 51/932.
97 NA, ADM 7/692.
98 Ibid.
99 Ibid.
100 Ibid.
101 NA, ADM 106/421.
102 NA, ADM 7/692.
103 NA, ADM 1/5253.
104 NA, ADM 7/692.
105 NA, ADM 3/8.
106 NA, ADM 2/10.
107 Ibid.
108 Ibid.
109 NA, ADM 3/8.
110 Ibid.
111 Ibid.
112 NA, ADM 2/9.
113 NA, ADM 106/436.
114 NA, ADM 3/8.
115 NA, ADM 2/11.
116 John Charnock, *Biographica Navalis*, Vol. 2, 1795, p179.

117 NA, ADM 3/8.
118 NA, ADM 2/11.
119 John Charnock, *Biographica Navalis*, Vol. 2, 1795, p321.
120 NA, ADM 3/8.
121 NA, ADM 106/435.
122 NA, ADM 2/11.
123 NA, ADM 106/435.
124 NA, ADM 3/8.
125 NA, ADM 2/9.
126 NA, ADM 51/3970.
127 NA, ADM 2/14.
128 John Charnock, *Biographica Navalis*, Vol. 3, 1795, p96
129 NA, ADM 2/11.
130 NA, ADM 7/694.
131 Ibid.
132 NA, PRO/11/424.
133 NA, ADM 3/9.
134 NA, ADM 7/694.
135 NA, ADM 7/694, p192.
136 NA, ADM 7/694.
137 NA, ADM 2/12.
138 NA, ADM 3/9.
139 NA, ADM 2/12.
140 Author's collection.
141 NA, ADM 7/694.
142 John Charnock, *Biographica Navalis*, Vol. 2, 1795, p188.
143 NA, ADM 106/436.
144 NA, ADM 2/13.
145 NA, ADM 106/436.
146 Ibid.
147 NA, ADM 106/452.
148 NA, ADM 2/15.
149 NA, ADM 1/5254.
150 NA, ADM 2/15.
151 W.G Day, ed., *The Pepys Ballads, Vol. V*, Catalogue of the Pepys Library at Magdalene College Cambridge, p365.
152 NA, ADM 2/17.
153 NA, ADM 106/473.
154 Ibid.
155 NA, ADM 2/17.
156 NA, ADM 106/473.
157 NA, ADM 2/17.
158 NA, ADM 106/468.
159 Ibid.
160 NA, ADM 106/473.
161 W.G Day, ed., *The Pepys Ballads, Vol. V*, Catalogue of the Pepys Library at Magdalene College Cambridge, p375.
162 NA, ADM 106/468.
163 NA, ADM 2/18.
164 NA, ADM 106/473.
165 Ibid.
166 NA, ADM 2/18.
167 NA, ADM 106/473.
168 NA, ADM 2/18.
169 Ibid.
170 Ibid.
171 Ibid.
172 Ibid.
173 NA, ADM 3/13.
174 NA, ADM 106/473.
175 NA, ADM 2/18.
176 NA, ADM 106/467.
177 John Charnock, *Biographica Navalis*, Vol. 2, 1795, p188.
178 NA, ADM 2/19.
179 Ibid.
180 NA, ADM 106/467.
181 NA, ADM 2/19.
182 NA, ADM 106/467.
183 NA, ADM 106/567.
184 *Mariner's Mirror*, Vol. 57, 1971, No. 4, p451.
185 NA, ADM 106/414.
186 NA, ADM 106/486.

187 NA, ADM 2/20.
188 NA, ADM 106/486.
189 NA, ADM 2/20.
190 Ibid.
191 NA, ADM 106/487.
192 NA, ADM 106/462.
193 NA, ADM 3/13.
194 John Charnock, *Biographica Navalis*, Vol. 2, 1795, p392.
195 NA, ADM 106/488.
196 W.G Day, ed., *The Pepys Ballads, Vol. V*, Catalogue of the Pepys Library at Magdalene College Cambridge, p366.
197 NA, ADM 2/22.
198 NA, ADM 106/488.
199 Ibid.
200 NA, ADM 106/504.
201 Ibid.
202 Ibid.
203 BL ADD MSS 9320.
204 NA, ADM 2/22.
205 NA, ADM 3/13.
206 NA, ADM 2/22.
207 NA, ADM 2/24.
208 NA, ADM 3/13.
209 Ibid.
210 NA, ADM 2/24.
211 Ibid.
212 Ibid.
213 NA, ADM 106/504.
214 John Charnock, *Biographica Navalis*, Vol. 2, 1795, p392.
215 NA, ADM 7/334, p242.
216 Josiah Burchett, *Transactions at Sea*, 1720, p645.
217 NA ADM 39/1716.

VISUAL GLOSSARY

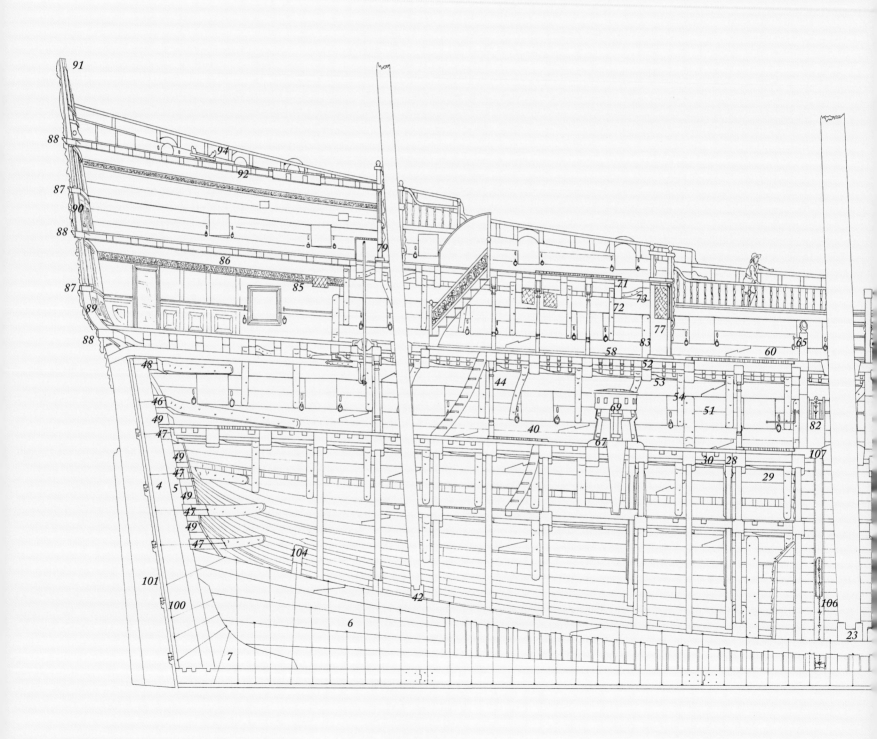

Key

1 Keel
2 Stem
3 False stem
4 Stern post
5 False post within
6 Rising wood
7 Long armed knee
8 Floor timbers
9 Middle futtock
10 Keelson
11 Toptimbers
12 Ceiling, 4in thick
13 Five strakes sleepers 7½in thick
14 Two strakes sleepers 5in thick
15 Middlebands
16 Gundeck clamps
17 Orlop beams
18 Floor riders
19 Futtock riders
20 Orlop platform

21 Orlop lodging knee
22 Orlop hanging knee
23 Saddle for mainmast step
24 Pillar in hold
25 Gundeck beams
26 Gundeck lodging knee
27 Gundeck hanging knee
28 Cross pillars or pointers
29 Gundeck carlines
30 Gundeck ledges
31 Gundeck waterways
32 Gundeck 4in plank
33 Gundeck long carlines
34 Manger
35 Conic bitts
36 Bitt cross pieces
37 Bitt knees
38 Breasthooks in hold
39 Foremast step
40 Gundeck spirket
41 Portlids
42 Mizzenmast step
43 3in plank between bitts and main partner

44 Turned pillars
45 Breasthooks between decks
46 Main or wing transome
47 Transomes
48 Helm port transome
49 Half transomes
50 Upper deck clamps
51 Shut up plank between ports
52 Upper deck beams
53 Upper deck lodging knee
54 Upper deck hanging knee
55 Upper deck shortcarlines
56 Upper deck long carlines
57 Upper deck ledges
58 Upper deck plank
59 Upper deck waterway
60 Upper deck spirket
61 String
62 Head ledges
63 Grating hatches
64 Topsail sheet bitts

65 Jeer bitts
66 Gallows
67 Main capstan partner
68 Jeer capstan partner
69 Main capstan
70 Jeer capstan
71 Quarter-deck beams
72 Quarter-deck hanging knee at every other beam
73 Quarter-deck lodging knee at bulkheads only
74 Forecastle beams
75 Forecastle hanging knee at every other beam
76 Forecastle lodging knee in wake of foremast
77 Steerage bulkhead
78 Forecastle bulkhead
79 Roundhouse bulkhead
80 Beakhead bulkhead
81 Cookroom
82 Pump
83 Upper deck standard at each bulkhead

84 Gundeck standard in wake of masts and bitts
85 Great cabin elm rising
86 Great cabin lining
87 Window transomes
88 Deck transomes
89 Counter timbers
90 Timbers of the upright
91 Taffrail
92 Poop beams
93 Gunwale
94 Kevel
95 Cleat
96 Belfry
97 Spritsail sheet block
98 Cathead
99 False keel
100 False keel post without
101 Rudder
102 Head
103 Pissdale
104 Crutch
105 Orlop standard in wake of main hatch

106 Well
107 Mast partners
108 4in thick elm plank for 10ft in height oak plank above
109 Six strakes 6in thick stuff in midships
110 Lower main wale 9½in thick stuff
111 Upper main wale 9½in thick stuff
112 4in thick oak plank
113 Lower channel wale 6in thick stuff
114 Upper channel wale 6in thick stuff
115 3in thick oak plank
116 Plank sheer
117 Fife rail
118 Great rail
119 Main channel
120 Spur
121 Deadeye
122 Chains

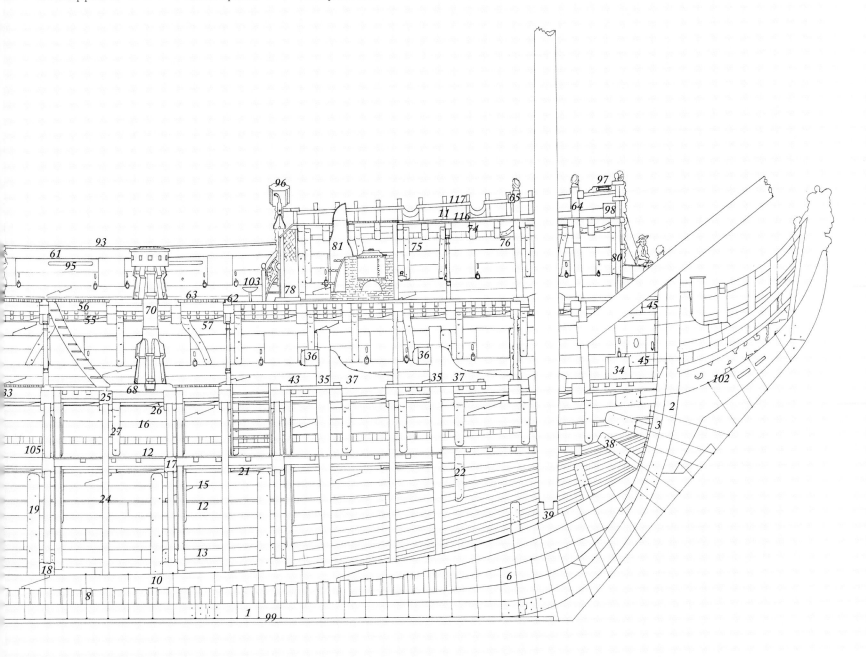

Longboat   *Shallop*   *Ketch*   *Pink*   *Lighter*   *Smack*   *Yacht*

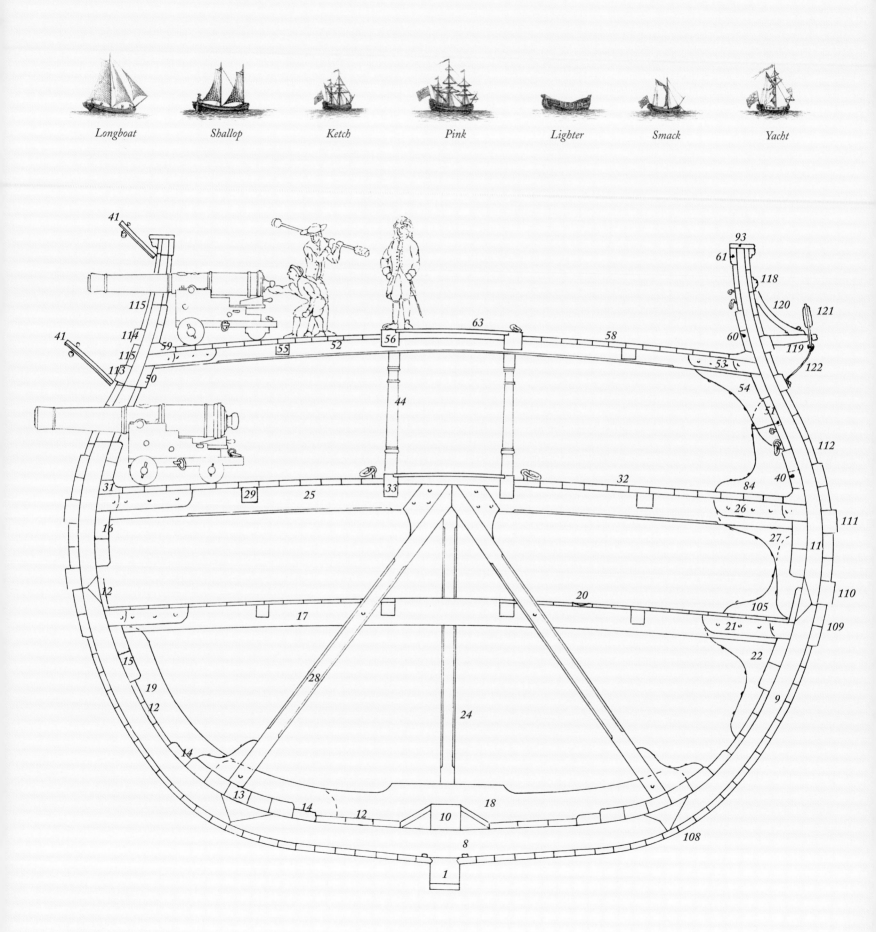

INDEX

Lenox viewed from the starboard beam, by Willem Van de Velde the Elder (Museum Boijmans van Beuningen, Rotterdam)